Wireless Transceiver Circuits

System Perspectives and Design Aspects

Devices, Circuits, and Systems

Series Editor

Krzysztof Iniewski
CMOS Emerging Technologies Research Inc.,
Vancouver, British Columbia, Canada

PUBLISHED TITLES:

FORTHCOMING TITLES:

FORTHCOMING TITLES:

Electrostatic Discharge Protection of Semiconductor Devices
and Integrated Circuits
Juin J. Liou

Gallium Nitride (GaN): Physics, Devices, and Technology
Farid Medjdoub and Krzysztof Iniewski

Laser-Based Optical Detection of Explosives
Paul M. Pellegrino, Ellen L. Holthoff, and Mikella E. Farrell

Mixed-Signal Circuits
Thomas Noulis and Mani Soma

Magnetic Sensors: Technologies and Applications
Simone Gambini and Kirill Poletkin

MRI: Physics, Image Reconstruction, and Analysis
Angshul Majumdar and Rabab Ward

Multisensor Data Fusion: From Algorithm and Architecture Design
to Applications
Hassen Fourati

Nanoelectronics: Devices, Circuits, and Systems
Nikos Konofaos

Nanomaterials: A Guide to Fabrication and Applications
Gordon Harling, Krzysztof Iniewski, and Sivashankar Krishnamoorthy

Optical Imaging Devices: New Technologies and Applications
Dongsoo Kim and Ajit Khosla

Physical Design for 3D Integrated Circuits
Aida Todri-Sanial and Chuan Seng Tan

Power Management Integrated Circuits and Technologies
Mona M. Hella and Patrick Mercier

Radiation Detectors for Medical Imaging
Jan S. Iwanczyk

Radio Frequency Integrated Circuit Design
Sebastian Magierowski

Reconfigurable Logic: Architecture, Tools, and Applications
Pierre-Emmanuel Gaillardon

Soft Errors: From Particles to Circuits
Jean-Luc Autran and Daniela Munteanu

Terahertz Sensing and Imaging: Technology and Devices
Daryoosh Saeedkia and Wojciech Knap

Tunable RF Components and Circuits: Applications in Mobile Handsets
Jeffrey L. Hilbert

Wireless Medical Systems and Algorithms: Design and Applications
Pietro Salvo and Miguel Hernandez-Silveira

Wireless Transceiver Circuits

System Perspectives and Design Aspects

EDITED BY
WOOGEUN RHEE
TSINGHUA UNIVERSITY, INSTITUTE OF MICROELECTRONICS
MICROELECTRONICS & NANOELECTRONICS DEPT.
CHINA, PEOPLE'S REPUBLIC

KRZYSZTOF INIEWSKI MANAGING EDITOR
CMOS EMERGING TECHNOLOGIES RESEARCH INC.
VANCOUVER, BRITISH COLUMBIA, CANADA

CRC Press
Taylor & Francis Group
Boca Raton Llondbon Nsew-Yónkk

CRC Press is an imprint of the
Taylor & Francis Group, an **informa** business

MATLAB® is a trademark of The MathWorks, Inc. and is used with permission. The MathWorks does not warrant the accuracy of the text or exercises in this book. This book's use or discussion of MAT-LAB® software or related products does not constitute endorsement or sponsorship by The MathWorks of a particular pedagogical approach or particular use of the MATLAB® software.

CRC Press
Taylor & Francis Group
6000 Broken Sound Parkway NW, Suite 300
Boca Raton, FL 33487-2742

First issued in paperback 2017

© 2015 by Taylor & Francis Group, LLC
CRC Press is an imprint of Taylor & Francis Group, an Informa business

No claim to original U.S. Government works

ISBN-13: 978-1-4822-3435-0 (hbk)
ISBN-13: 978-1-138-89400-6 (pbk)

**Visit the Taylor & Francis Web site at
http://www.taylorandfrancis.com**

**and the CRC Press Web site at
http://www.crcpress.com**

For my parents and my wife, Soojung

Contents

SECTION I System Design Perspectives

SECTION II   Millimeter-Wave Transceivers

SECTION III   Biomedical and Short-Range Radios

SECTION IV   Modulators and Synthesizers

Preface

Modern transceiver systems require diversified design aspects as various radio and sensor applications have emerged. Choosing the right architecture and understanding interference and linearity issues are important for multistandard cellular transceivers and software-defined radios. A millimeter-wave CMOS transceiver design for multi-Gb/s data transmission is another challenging area. Energy-efficient short-range radios for body area networks and sensor networks have recently received great attention. To meet different design requirements, gaining good system perspectives is important. This book addresses not only comprehensive system design considerations for robust wireless communication but also practical design aspects in state-of-the-art transceivers. In addition to dealing with system architectures and design considerations, a few chapters are devoted to critical building blocks with detailed circuit description and analyses.

The book is divided into four sections dealing with system design perspectives for wireless transceivers, millimeter-wave transceivers, biomedical and short-range radios, and modulators and frequency synthesizers. Section I provides system design considerations in modern transceiver design. Chapter 1 discusses how to design a receiver or a transmitter with passive mixers for the best gain and linearity performance. The passive mixer–based transceiver is another important trend in modern transceiver systems. For multistandard transceiver systems, an interference-robust receiver design with good linearity is a must. Chapter 2 presents an in-depth study of the second-order intermodulation distortion (IM_2) for the design of robust homodyne and low-intermediate frequency receivers, and Chapter 3 shows how to mitigate the performance degradation of the receiver in the presence of out-of-band interference. Chapter 4 presents frequency-translated filters to deal with interferers for the design of surface acoustic wave-less receivers. The following two chapters show different receiver design aspects: Chapter 5 defines the cognitive radio in the narrow sense of an intelligent device that is able to dynamically adapt and negotiate wireless frequencies and communication protocols for efficient communications and describes the different kinds of wideband spectrum sensing architectures. Chapter 6 introduces a direct delta-sigma receiver architecture that transforms a traditional direct conversion front end and a baseband delta-sigma converter into a complete radio frequency-to-digital converter.

For Gb/s data transmission, millimeter-wave transceivers are promising, but achieving low power consumption is critical for mobile applications. Section II covers both systems and circuits for the millimeter-wave transceiver design. Chapter 7 introduces the most up-to-date status of the 60 GHz wireless transceiver development, with an emphasis on realizing low power consumption and small form factor that is applicable for mobile terminals. Chapter 8 describes D-band (110–170 GHz) CMOS circuits to realize low-power, ultrahigh-speed wireless communication systems. As a case study, a 10 Gb/s wireless transceiver with a power consumption of 98 mW is demonstrated using a 135 GHz band. Combined with the contemporary advances in millimeter-wave electronic technology, photonic technology enables significant advances in communications and remote sensing. Chapter 9 presents photonic techniques that have made significant advances in recent years and offers approaches to significantly reduce or eliminate the drawbacks associated with millimeter-wave technologies. For example, generating sufficiently low-noise millimeter-wave signals to enable high modulation format communications signals has proven difficult using electronic techniques. The following two chapters deal with two important building blocks of the millimeter-wave transceiver: In Chapter 10, challenges in the design of and different kinds of millimeter-wave power amplifiers are described. Frequency multipliers are useful in millimeter-wave transceivers to alleviate the difficulty of fundamental oscillator design at high frequencies. Chapter 11 discusses the design issues of frequency multipliers with several circuit examples.

While millimeter-wave transceivers are studied for high-data-rate transmission, low-power transceivers have recently received great attention for body area networks and sensor networks. Section III introduces four energy-efficient short-range radios for biomedical and wireless connectivity applications. Chapter 12 describes ultrawideband (UWB) transceivers for microwave medical imaging. A 65 nm CMOS fully integrated stepped-frequency continuous wave radar is presented as a case study. Chapter 13 discusses design challenges in ultralow-power and ultralow-voltage circuits and presents circuit techniques. Wideband ultralow-power and ultralow-voltage, low-noise amplifiers and a UWB transceiver for wireless sensor networks with a low-complexity synchronization scheme are demonstrated. Chapter 14 presents an energy-efficient sub-GHz transmitter design for biomedical applications. In this work, a 100 Mb/s transmitter design with an energy efficiency of 13 pJ/bit is reported. Chapter 15 describes the design and implementation of a compact, low-power, high-spurious-free dynamic range receiver suitable for ZigBee or wireless personal area network (WPAN) applications. It gives an overview of a low-power receiver based on the *split-LNTA + 50% LO* architecture and also presents a number of ultralow-power and ultralow-voltage circuit design techniques.

Frequency synthesizer is an important block in transceivers, and a digital-intensive phase modulator becomes one of the key building blocks in modern transmitters as emphasized in Section IV. Chapter 16 discusses the all-digital phase-locked loop design as well as the two-point modulation scheme for linear wideband phase modulation. Chapter 17 introduces a hybrid two-point modulator that employs a mixed-signal loop control to achieve analog phase tracking and digital frequency acquisition. Chapter 18 gives a good overview of modern fractional-N frequency synthesizer architectures. In Chapter 19, a 60 GHz frequency synthesizer with a frequency calibration scheme that can support IEEE 802.15.3c, wirelessHD, IEEE 802.11ad, and ECMA-387 TX/RX front end is presented. Chapter 20 gives a good tutorial of the digitally controlled oscillator (DCO), which is the most critical block in the all-digital phase-locked loop.

The book is written by top industrial experts and renowned academic professors. We thank all contributors for their hard work and for carving out some precious time from their busy schedules to write their valuable chapters. Despite some challenges in integrating the material, 20 chapters from nearly 50 contributors have been put together in this book. We sincerely hope that you will find this book useful for your work and research.

Woogeun Rhee
Tsinghua University
Beijing, People's Republic of China
Krzysztof (Kris) Iniewski
Vancouver, British Columbia, Canada

MATLAB® is a registered trademark of The MathWorks, Inc. For product information, please contact:

The MathWorks, Inc.
3 Apple Hill Drive
Natick, MA 01760-2098 USA
Tel: 508-647-7000
Fax: 508-647-7001
E-mail: info@mathworks.com
Web: www.mathworks.com

Editors

Woogeun Rhee received his BS in electronics engineering from Seoul National University, Seoul, South Korea, in 1991; MS in electrical engineering from the University of California, Los Angeles, in 1993; and PhD in electrical and computer engineering from the University of Illinois, Urbana–Champaign, in 2001.

From 1997 to 2001, Dr. Rhee was with Conexant Systems, Newport Beach, California, where he was a principal engineer and developed low-power, low-cost fractional-*N* synthesizers. From 2001 to 2006, he was with the IBM Thomas J. Watson Research Center, Yorktown Heights, New York, and worked on clocking area for high-speed input/output serial links, including low-jitter phase-locked loops, clock-and-data recovery circuits, and on-chip testability circuits. In August 2006, he joined the faculty of the Institute of Microelectronics at Tsinghua University, Beijing, China, and became a professor in 2012. His current research interests include clock/frequency generation systems for wireline and wireless communications and low-power transceiver systems for wireless body area networks. He holds 19 U.S. patents.

Dr. Rhee served as an associate editor for *IEEE Transactions on Circuits and Systems-II* (2008–2009) and a guest editor for *IEEE Journal of Solid-State Circuits*, special issues of November 2012 and November 2013. He has been an associate editor for *IEEE Journal of Solid-State Circuits* and for *IEEK Journal of Semiconductor Technology and Science* from 2009. He is also a member of the Technical Program Committee for the IEEE International Solid-State Circuits Conference (ISSCC) and the IEEE Asian Solid-State Circuits Conference (A-SSCC). He received the IBM Faculty Award in 2007 from IBM Corporation, United States, and is listed in *Marquis Who's Who in the World* (2009–2014).

Krzysztof (Kris) Iniewski is managing R&D at Redlen Technologies Inc., a start-up company in Vancouver, British Columbia, Canada. Redlen's revolutionary production process for advanced semiconductor materials enables a new generation of more accurate, all-digital, radiation-based imaging solutions. Dr. Iniewski is also president of CMOS Emerging Technologies Research Inc. (www.cmosetr.com), an organization of high-tech events covering communications, microsystems, optoelectronics, and sensors.

In his career, Dr. Iniewski held numerous faculty and management positions at the University of Toronto, University of Alberta, Simon Fraser University, and PMC-Sierra Inc. He has published over 100 research papers in international journals and conferences. He holds 18 international patents granted in the United States, Canada, France, Germany, and Japan. He is a frequently invited speaker and has consulted for multiple organizations internationally. He has written and edited several books for CRC Press, Cambridge University Press, IEEE Press, Wiley, McGraw-Hill, Artech House, and Springer. His personal goal is to contribute to healthy living and sustainability through innovative engineering solutions. In his leisurely time, Kris can be found hiking, sailing, skiing, or biking in beautiful British Columbia. He can be reached at kris.iniewski@gmail.com.

Contributors

Karim Allidina
Department of Electrical and Computer
   Engineering
McGill University
Montréal, Québec, Canada

Andrea Bevilacqua
Department of Information
   Engineering
University of Padova
Padova, Italy

Dong Chen
School of Electronic Engineering
University of Electronic Science and
   Technology of China
Chengdu, Sichuan, People's Republic
   of China

SeongHwan Cho
Department of Electrical Engineering
Korea Advanced Institute of Science
   and Technology
Daejeon, South Korea

Thomas R. Clark, Jr.
Johns Hopkins University Applied
   Physics Laboratory
Laurel, Maryland

Hooman Darabi
Broadcom Corporation
Irvine, California

Wei Deng
Department of Physical Electronics
Tokyo Institute of Technology
Tokyo, Japan

Mourad El-Gamal
Department of Electrical and Computer
   Engineering
McGill University
Montréal, Québec, Canada

Mikko Englund
Department of Micro- and
   Nanosciences
Aalto University
Esbo, Finland

Minoru Fujishima
Department of Semiconductor
   Electronics and Integration Science
Graduate School of Advanced Sciences
   of Matter
Hiroshima University
Hiroshima, Japan

Yuan Gao
Division of Integrated Circuits and Systems
Institute of Microelectronics
Singapore, Singapore

Ranjit Gharpurey
Department of Electrical and Computer
   Engineering
The University of Texas at Austin
Austin, Texas

Kaizhe Guo
Department of Electrical Engineering
   (ESAT) Microelectronics
and
Sensors (MICAS) of Katholieke
   Universiteit Leuven
University of Electronic Science and
   Technology of China
Chengdu, Sichuan, People's Republic
   of China

Ramesh Harjani
Department of Electrical and Computer
    Engineering
University of Minnesota
Minneapolis, Minnesota

Chun-Huat Heng
Department of Electrical and Computer
    Engineering
National University of Singapore
Singapore, Singapore

Chih-Ming Hung
MediaTek Inc.
Hsinchu, Taiwan

Youngwoo Jo
Department of Electrical Engineering
Korea Advanced Institute of Science
    and Technology
Daejeon, South Korea

Kai Kang
School of Electronic Engineering
University of Electronic Science and
    Technology of China
Chengdu, Sichuan, People's Republic
    of China

Kimmo Koli
Ericsson
Jorvas, Finland

Chiennan Kuo
Department of Electronics Engineering
National Chiao Tung University
Hsinchu, Taiwan

Salvatore Levantino
Dipartimento di Elettronica,
    Informazione e Bioingegneria
Politecnico di Milano
Milan, Italy

Zhicheng Lin
State-Key Laboratory of Analog and
    Mixed-Signal VLSI and FST-ECE
University of Macau
Taipa, Macao, People's Republic of China

Pui-In Mak
State-Key Laboratory of Analog and
    Mixed-Signal VLSI and FST-ECE
University of Macau
Taipa, Macao, People's Republic
    of China

Rui P. Martins
State-Key Laboratory of Analog and
    Mixed-Signal VLSI and FST-ECE
University of Macau
Taipa, Macao, People's Republic of China

Giovanni Marzin
Blue Danube Labs
Warren, New Jersey

Akira Matsuzawa
Department of Physical Electronics
Tokyo Institute of Technology
Tokyo, Japan

Sven Mattisson
Ericsson
Lund, Sweden

Timothy P. McKenna
Johns Hopkins University Applied
    Physics Laboratory
Laurel, Maryland

Ahmad Mirzaei
Broadcom Corporation
Irvine, California

Ahmed Musa
NTT Microsystem Integration
    Laboratories
Kanagawa, Japan

Jeffrey A. Nanzer
Johns Hopkins University Applied
    Physics Laboratory
Laurel, Maryland

Kenichi Okada
Department of Physical Electronics
Tokyo Institute of Technology
Tokyo, Japan

Kim Östman
Department of Micro- and
    Nanosciences
Aalto University
Esbo, Finland

Dongmin Park
Qualcomm
San Diego, California

Pyoungwon Park
Broadcom Corporation
Irvine, California

Mahdi Parvizi
Department of Electrical and Computer
    Engineering
McGill University
Montréal, Québec, Canada

Woogeun Rhee
Institute of Microelectronics
Tsinghua University
Beijing, People's Republic of China

Jussi Ryynänen
Department of Micro- and
    Nanosciences
Aalto University
Esbo, Finland

Bodhisatwa Sadhu
IBM Thomas J. Watson Research
    Center
Yorktown Heights, New York

Noriaki Saito
Panasonic
Kanagawa, Japan

Carlo Samori
Dipartimento di Elettronica,
    Informazione e Bioingegneria
Politecnico di Milano
Milan, Italy

Naganori Shirakata
Panasonic
Kanagawa, Japan

Teerachot Siriburanon
Department of Physical Electronics
Tokyo Institute of Technology
Tokyo, Japan

Kari Stadius
Department of Micro- and
    Nanosciences
Aalto University
Esbo, Finland

Koji Takinami
Panasonic
Kanagawa, Japan

Takayuki Tsukizawa
Panasonic
Kanagawa, Japan

Olli Viitala
Department of Micro- and
    Nanosciences
Aalto University
Esbo, Finland

Zhihua Wang
Institute of Microelectronics
Tsinghua University
Beijing, People's Republic of China

Ni Xu
Institute of Microelectronics
Tsinghua University
Beijing, People's Republic of China

Section I

System Design Perspectives

1 Analysis, Optimization, and Design of Transceivers with Passive Mixers

Ahmad Mirzaei and Hooman Darabi

CONTENTS

1.1 INTRODUCTION

Most of today's transceivers are designed with passive mixers, which is the motivation behind dedicating this chapter to analyze and understand their operation fundamentals. We study transceivers designed with passive mixers driven by 50% and 25% duty-cycle clocks. It is shown that due to lack of reverse isolation between RF and IF ports, the passive mixer holds a property called impedance transformation. This property of the passive mixers can be utilized to frequency-shift low-Q baseband impedances to synthesize on-chip high-Q filters with center frequencies precisely controlled by the local oscillator (LO) clock, a useful feature to design blocker-resilient receivers. It is also revealed how this lack of reverse isolation can cause problems such as IQ cross talk and different high- and low-side conversion gains. This chapter discusses how to design a receiver or a transmitter with passive mixers for the best gain and linearity performance.

1.2 RECEIVER DESIGN WITH PASSIVE MIXERS

Figure 1.1 shows a typical architecture for a zero- or low-IF receiver [1]. The received signal in the antenna is composed of the desired signal and some unwanted in-band or out-of-band interferers. The desired signal can be very weak, while the out-of-band blockers can be millions of times stronger. To overcome noise of the following

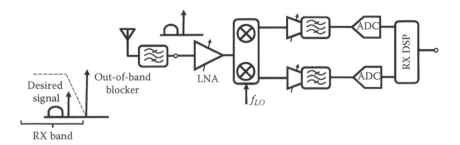

FIGURE 1.1   Role of a down-conversion mixer in receivers.

stages, the weak desired signal requires gain of the low-noise amplifier (LNA) to be set at maximum. Traditionally, an external surface acoustic wave (SAW) filter is placed prior to the on-chip LNA to attenuate the strong out-of-band blockers and prevent them from compressing the LNA. The SAW filter passes the desired signal along with in-band blockers with no substantial attenuation (except with a passband loss called insertion loss). The desired signal then is down-converted into zero or low-IF through an IQ down-conversion mixer. In the IF stage, the in-band blockers, which were almost impossible to be attenuated in the RF, can now be filtered out with low-Q baseband low-pass filters. The composite down-converted signal and blockers pass through this low-pass filter, where in-band blockers as well as residues of out-of-band blockers are attenuated to a level that two IQ analog-to-digital converters (ADCs) with reasonable resolution and power consumption can digitize the baseband signals. The rest of the receiver functionality is performed in the world of digital signal processing (DSP). The down-conversion action relaxes the filtering processing without the need for unrealistically sharp RF filters.

Therefore, inevitably, there is a mixer in the heart of this receiver that performs frequency translation, shifting the desired channel from the RF to the baseband. Since in a linear-time-invariant (LTI) system [2], besides those of the input, no additional frequency components are generated at the output, any mixing system must be either nonlinear or linear; if it is linear, it has to be a time-variant system. The mixers used in modern receivers are linear but time-variant (LTV) systems, which can be in active or passive forms.

Shown in Figure 1.2a is the well-known active mixer. The input is the RF voltage coming from the LNA output. The RF voltage is converted to an RF current that is superposed onto the bias current I_{BB}. The switches of the mixer commutate the RF as well as the bias current, and as a result, the RF signal is down-converted to the baseband. This active mixer has two well-known issues. First, it has been shown in the literature that the output noise of an active mixer due to the flicker noise of the MOS switches is directly proportional to the bias current commutated by the switching pair [3,4]. Therefore, the flicker noise coming from the switches is typically high, magnified by the bias current. This flicker noise is usually problematic in zero-IF receivers, especially in receiving signals with narrow bandwidth such as GSM.*
Another issue of this active mixer is its high LO feedthrough (LOFT), which is again

---

* That is why most of GSM receivers use the low-IF architecture.

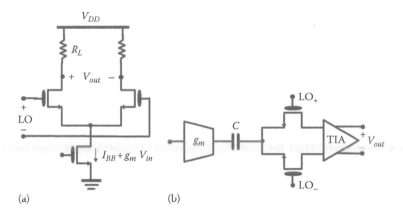

FIGURE 1.2   Active mixer versus current-driven passive mixer. (a) Active mixer. (b) Current-driven passive mixer.

proportional to the bias current. With a double-balanced mixer architecture, the LOFT is reduced, but still, compared to the passive mixer, the LOFT is typically high. Another drawback of this active mixer is the need for additional voltage-to-current conversion besides that of the LNA. This extra V-to-I conversion limits the receiver linearity, especially in SAW-less receivers where the linearity requirements are quite stringent [5]. Finally, the last drawback is the need for high supply voltage due to stack of devices between the supply voltage and ground.

One of the distinguished features of the active mixer is its close-to-infinite isolation from the baseband output to the RF input, which makes it easier to design and understand. If we are willing to give up this isolation, we can come up to mixer topology in Figure 1.2b, the so-called current-driven passive mixer. Setting the commutated dc current to zero leads us to a mixer that only commutates the ac signal current [6–14]. The capacitive coupling of the input transconductance stage to the switching pair guarantees the triode operation of the mixer switches and ensures the mixer switches carry no dc currents. In order to have a small ON resistance, the switches should turn on in the deep triode region and should be driven hard by strong LO voltages. When the LO is high, the triode MOS switch directly connects the input transconductor to the output load through its ON resistance. On the baseband side, the switches are buffered with a transimpedance amplifier (TIA), which is also called current buffer. This current buffer, which ideally has zero input impedance, makes all internal nodes of the receiver chain be low impedance. As a result, voltage swings at all internal nodes are reduced, leading to a great linearity. Furthermore, since there is no bias current commutated by the switches, flicker noise of the switches does not appear at the baseband output. For the same reason, the LOFT is much smaller as well. Also, since there is no stack of devices, this architecture is very friendly for low-voltage applications.

Unlike the active mixers, there is no reverse isolation between the IF and the RF ports of the passive mixers. In this chapter, we will see that due to this lack of reverse isolation the mixer reflects the baseband impedance to the RF and vice versa through a simple frequency shifting.

1.2.1 DESIGN OF RECEIVERS WITH 50% PASSIVE MIXER

Figure 1.3 is an example of a receiver front end with current-driven passive mixers driven by 50% duty-cycle clocks. The LNA is the typical design choice that is inductively degenerated for higher IIP3 and is designed for large transconductance to lower noise contribution of the following stages. Also, as it is observed, the mixers are directly coupled to the LNA and there are no intermediate transconductance stages between the LNA and the IQ mixers. Otherwise, these transconductance stages would hurt the receiver nonlinearity due to current-to-voltage and voltage-to-current conversions. The LNA load, Z_L, is a parallel LC load. Mixers of the two I and Q channels are connected to the LNA through two series capacitors C, and thus the switches of the mixer carry no dc currents. The RF current of the LNA is divided between the two quadrature channels. The current buffers are RC feedback–based, with large shunt capacitors connected between the differential inputs of the op-amps to filter out the received signal components residing at high-frequency offsets from the desired signal. This filtering of strong blockers lowers the voltage swings at all internal nodes and, as a result, improves linearity. The series capacitors C also block the low-frequency IM2 components generated inside LNA. Otherwise, due to mismatches in the mixer switches, these components could leak to the baseband degrading the receiver IIP2.

To understand how the mixing system works, as shown in Figure 1.4, assume that an RF current I_{RF} is commutated by the switches of a mixer clocked at a rate f_{LO} with rail-to-rail square-wave differential LOs. The differential down-converted currents flow into baseband impedances Z_{BB}, low-pass filtered, and are converted to the baseband voltages, V_{BB}, across the baseband impedances. Due to the lack of reverse isolation of the passive mixer, these baseband voltages are up-converted back to the RF, becoming an RF voltage around f_{LO} on the RF side of the switches. If we assume that the RF current is a single tone at $f_{LO} + f_m$, where f_m is a small frequency offset,

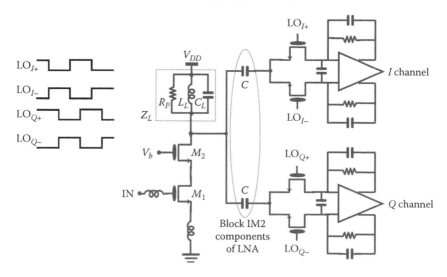

FIGURE 1.3 Example of a receiver front end with 50% current-driven passive mixer.

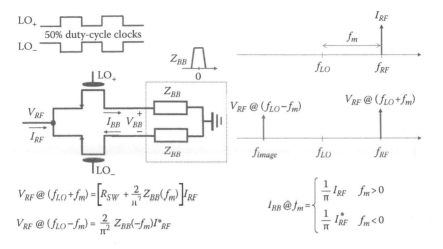

$$V_{RF} @ (f_{LO}+f_m) = \left[R_{SW} + \frac{2}{\pi^2} Z_{BB}(f_m) \right] I_{RF}$$

$$V_{RF} @ (f_{LO}-f_m) = \frac{2}{\pi^2} Z_{BB}(-f_m) I^*_{RF}$$

$$I_{BB} @ f_m = \begin{cases} \dfrac{1}{\pi} I_{RF} & f_m > 0 \\ \dfrac{1}{\pi} I^*_{RF} & f_m < 0 \end{cases}$$

FIGURE 1.4   Governing equations in a 50% passive mixer.

it can be shown that around f_{LO}, the resulting RF voltage has two major frequency components: (1) $f_{LO} + f_m$, called the *main* frequency, at the incident RF frequency, and (2) $f_{LO} - f_m$ called the *image* frequency. Phasors of these components of the RF voltage are given by these expressions:

$$V_{RF} @ (f_{LO} + f_m) = \left[R_{SW} + \frac{2}{\pi^2} Z_{BB}(f_m) \right] I_{RF} \qquad (1.1)$$

$$V_{RF} @ (f_{LO} - f_m) = \frac{2}{\pi^2} Z_{BB}(-f_m) I^*_{RF} \qquad (1.2)$$

in which * is the convolution sign [2]. With a small switch resistance, the magnitude of the image component of the resulting RF voltage can be as big as the main component. Also, the down-converted baseband currents versus the stimulus RF current can readily be shown to be

$$I_{BB} @ f_m = \begin{cases} \dfrac{1}{\pi} I_{RF} & f_m > 0 \\ \dfrac{1}{\pi} I^*_{RF} & f_m < 0 \end{cases} \qquad (1.3)$$

Assuming that the stimulus is an ideal RF current (Figure 1.5), at the incident (main) frequency, the input impedance seen from the RF side is given by the following expression:

$$Z_{in} \cong R_{SW} + \frac{2}{\pi^2} \{ Z_{BB}(s - j\omega_{LO}) + Z_{BB}(s + j\omega_{LO}) \} \qquad (1.4)$$

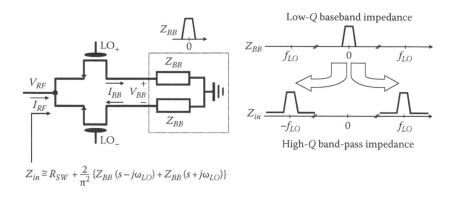

$$Z_{in} \cong R_{SW} + \frac{2}{\pi^2} \{Z_{BB}(s - j\omega_{LO}) + Z_{BB}(s + j\omega_{LO})\}$$

FIGURE 1.5   Impedance transformation of 50% passive mixer.

which is equal to the switch resistance in series with a band-pass filter. This band-pass filter is the same low-pass filter frequency shifted to the RF to become a high-Q band-pass filter. More importantly, the center of this high-Q band-pass filter is precisely controlled by the clock frequency, making it clock tunable.

As shown in Figure 1.6, if the input RF current resides at a frequency offset from f_{LO} larger than the baseband filter's bandwidth, the resulting components of the RF voltage at both main and image frequencies would be low by the same amount dictated by this filtering. On the other hand, if the distance of the RF current from f_{LO} resides in the passband of this filter, the resulting main and image voltages both would be high. In other words, both main and image voltage components experience this built-in high-Q band-pass filter. This property can be utilized to attenuate unwanted blockers. This high-Q filtering would reduce the voltage swing across the switches and, as a result, would increase linearity.

Now, let us consider having another passive mixer circuit in which its LOs are 90° phase-shifted versus the first one (Figure 1.7). If the same RF current is injected into the two mixing systems, the RF voltages at the main frequency will be identical, whereas the image components will be 180° out-of-phase but with the same magnitudes.

Due to this mechanism, as shown in Figure 1.8, in a quadrature down-conversion, a current at the image frequency circulates between the two channels (between

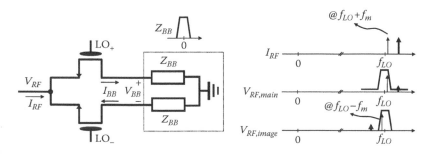

FIGURE 1.6   Image component and high-Q filtering of the 50% passive mixer.

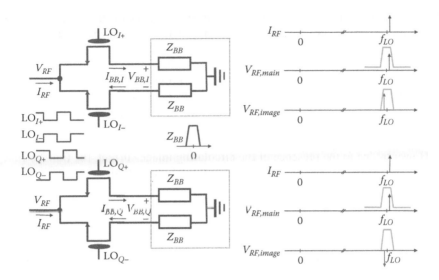

FIGURE 1.7 Current-driven passive mixer driven by 50% IQ clocks.

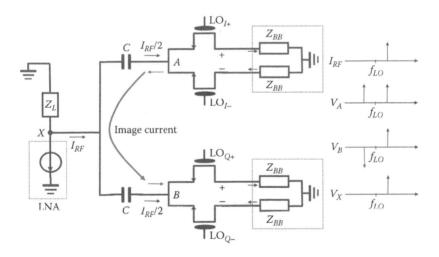

FIGURE 1.8 Circulation of the image current in the IQ receiver.

nodes A and B). No image current passes through Z_L since $V_A + V_B$ is zero at the image frequency. The unwanted image current is down-converted and superposed onto the main baseband currents. The magnitude of the image current is proportional to the baseband impedance Z_{BB} and is inversely proportional to the impedance of the series capacitors C at f_{LO}. If the input impedance of the TIA is not low enough, the large image current would adversely affect the performance of the IQ receiver using this 50% current-driven passive mixer, which will be addressed next.

For now, let us ignore higher-order harmonics. If there was no image current, intuitively, the RF current I_{RF} was supposed to be divided between Z_L and the two I and Q channels inversely proportional to the impedances seen from the three paths. The corresponding baseband currents could be related as in the following expression:

$$I_{BB,I} = -jI_{BB,Q} = \frac{\dfrac{1}{\pi} Z_L(\omega_{LO} + \omega_m) I_{RF}}{2Z_L(\omega_{LO} + \omega_m) + Z_C(\omega_{LO} + \omega_m) + R_{SW} + \dfrac{2}{\pi^2} Z_{BB}(\omega_m)} \quad (1.5)$$

However, due to the presence of the circulating image current, the transfer function for the baseband currents is modified to the following [15]:

$$I_{BB,I} = -jI_{BB,Q} =$$

$$\frac{\dfrac{1}{\pi} Z_L(\omega_{LO} + \omega_m) \left[Z_C^*(\omega_{LO} - \omega_m) + R_{SW} \right] I_{RF}}{\left[2Z_L(\omega_{LO} + \omega_m) + Z_C(\omega_{LO} + \omega_m) + R_{SW} + \dfrac{2}{\pi^2} Z_{BB}(\omega_m) \right] \left[Z_C^*(\omega_{LO} - \omega_m) + R_{SW} + \dfrac{2}{\pi^2} Z_{BB}(\omega_m) \right] - \left[\dfrac{2}{\pi^2} Z_{BB}(\omega_m) \right]^2}$$

$$(1.6)$$

Compared to (1.5), the transfer function (1.6) has these two extra elements: $[Z_C^*(\omega_{LO} - \omega_m) + R_{SW}]$ in the numerator and $[Z_C^*(\omega_{LO} - \omega_m) + R_{SW} + (2/\pi^2)Z_{BB}(\omega_m)]$ in the denominator. These extra elements drastically lead to some interesting and unwanted consequences such as (1) different high- and low-side conversion gains, (2) different high- and low-side IIP2 and IIP3 values, (3) leakage of nonlinearity-generated components from the baseband side of one channel to the other, and (4) reduced conversion gain of the receiver. In the following, each of the aforementioned effects is briefly described:

1. *Different high- and low-side conversion gains*: The first consequence of the image current is having different high- and low-side conversion gains. In other words, as shown in Figure 1.9a, a frequency component of the RF current located at $f_{LO} + f_m$ experiences different conversion gain compared to the one at $f_{LO} - f_m$. Figure 1.9b plots the simulated high- and low-side conversion gains for a conventional design against predictions of (1.6), which are well matched. Note that at negative frequencies, Z_{BB} is a complex conjugate of its values at positive ones, which is why the high- and low-side conversion gains can be totally different in (1.6). Also, from (1.6), it can be proved that high- and low-side conversion gains would not be different if the baseband impedance was symmetric around dc, that is, pure resistive input impedance. This different high- and low-side conversion gains would have also happened even if we did not have the image current. The presence of the image current exacerbates the effect, leading to a lot bigger differences between high- and low-side conversion gains.
2. *Different high- and low-side IIP2 and IIP3 values*: Also, since high- and low-side conversion gains are different, obviously, the high- and low-side IIP3 or IIP2 numbers would be different as well.

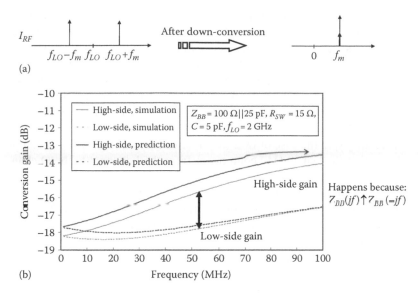

FIGURE 1.9   Impact of IQ cross talk on high- and low-side conversion gains.

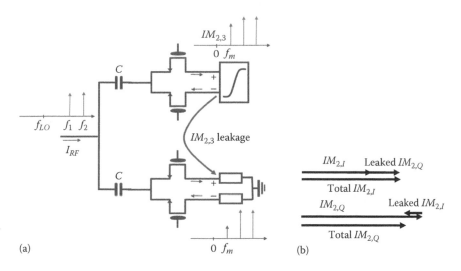

FIGURE 1.10   (a) Impact of IQ cross talk on leakage of nonlinearity-generated baseband components from one channel to the other. (b) Impact of IQ cross talk on IIP2.

3. *Leakage of nonlinearity-generated baseband components from one channel to the other*: Another impact of IQ cross talk is the leakage of nonlinearity-induced low-frequency components on the baseband side of the mixer switches of one quadrature channel to the other. As shown in Figure 1.10a, assume that the baseband loads in the *I* and *Q* paths are identical except the one in the *I* channel (upper channel in Figure 1.10a) that has second- or third-order nonlinear terms. Now assume that the RF input is composed of

two tones at f_1 and f_2. They are down-converted and become two baseband tones at $|f_{LO} - f_1|$ and $|f_{LO} - f_2|$. Due to nonlinearity of the current buffer in the I channel, low-frequency intermodulation components are generated on the corresponding baseband side. However, when the Q-side voltage is monitored, despite its load being perfectly linear, the same intermodulation components exist and could be as strong as the one in the I channel if the receiver is not designed properly. The effect generally causes unexpected IIP2 and IIP3 values.

Interestingly, in the case of IM2 (receiver IIP2), mathematically, it can be shown that the leaked IM2 component from the I channel to the Q channel and vice versa adds up constructively in one channel to the existing IM2 component generated in that channel and destructively in the other one (Figure 1.10b). Therefore, due to the leakage effect, the IIP2 of the receiver measured for one channel is improved and that of the other one is degraded.

4. *Reduced conversion gain of the receiver*: Another impact of the image current is lowering of the receiver conversion gain. In other words, in the presence of the image current, the conversion gain is smaller. If we call the conversion gain without image current (Equation 1.5) $I_{BB}(ideal)$ and the conversion gain with the image current included (Equation 1.6) $I_{BB}(actual)$, it can be shown that

$$\left|\frac{I_{BB}(actual)}{I_{BB}(ideal)}\right| = \begin{cases} 1 & \frac{2}{\pi^2}Z_{BB} \to 0 \\ \left|\frac{Z_C^*(\omega_{LO})+R_{SW}}{2Z_L(\omega_{LO})+2R_{SW}}\right| \le 1 & \frac{2}{\pi^2}Z_{BB} \to \infty \end{cases} \tag{1.7}$$

$$\left|\frac{I_{BB}(actual)}{I_{BB}(ideal)}\right| = \begin{cases} 1 & C \to 0 \\ \left|\frac{R_{SW}}{R_{SW}+\frac{2}{\pi^2}Z_{BB}(\omega_m)}\right| < 1 & C \to \infty \end{cases} \tag{1.8}$$

So, due to the image current, the conversion gain in general is smaller and the noise figure degradation of the receiver can be substantial.

1.2.2 DESIGN OF RECEIVERS WITH 25% PASSIVE MIXER

It is known that receivers designed with 25% duty-cycle passive mixers offer superior performance compared to the ones with 50% passive mixers [16,17]. In this section, it is explained why this is the case. Figure 1.11 is an example of a receiver front end with a current-driven passive mixer driven by 25% duty-cycle clocks. The LNA is inductively degenerated and is designed for high transconductance

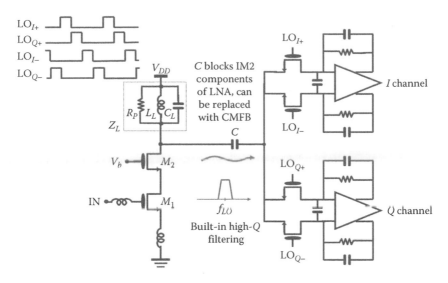

FIGURE 1.11   Receiver front end with 25% current-driven passive mixer.

in the signal band to lower noise contribution of the following stages. The mixer is directly coupled to the LNA with just a single capacitor of size C^*, and there are no intermediate transconductance stages between the LNA and the complex IQ mixer. The LNA load, Z_L, is a parallel LC load, and the complex IQ mixer is connected to the LNA through the series capacitor C. The current buffers are feedback RC. In each quadrature channel, a large shunt capacitor is connected between the differential inputs of the op-amp to filter out the components residing at high-frequency offsets (where out-of-band blockers can be very strong) right at this point. This lowers the voltage swings at all internal nodes and, as a result, improves the receiver linearity.

Again, due to the nature of the passive mixer, assuming the baseband impedances are low-pass, the impedance seen by the LNA from the complex mixer is a high-Q band-pass filter (Figure 1.11). This built-in high-Q on-chip filter would act as an on-chip saw filter that can be utilized to attenuate unwanted blockers.

The series capacitor C blocks any possible IM2 components generated inside the LNA; otherwise, the receiver IIP2 can be degraded due to device mismatches in the mixer switches. Also, depending on the LNA's architecture, let us say if it is simply an inverter, the series capacitor C can be replaced with a common-mode feedback inside the LNA [18].

As shown in Figure 1.12, an RF current I_{RF} is commutated by the switches of the mixer clocked at a rate f_{LO} with rail-to-rail 25% duty-cycle quadrature LO clocks. The differential down-converted currents flow into baseband impedances Z_{BB}, low-pass filtered, and become baseband voltages across the baseband impedances. Due to the lack of reverse isolation of the passive mixers, these baseband voltages are

---

* In a differential implementation, two series capacitors exist.

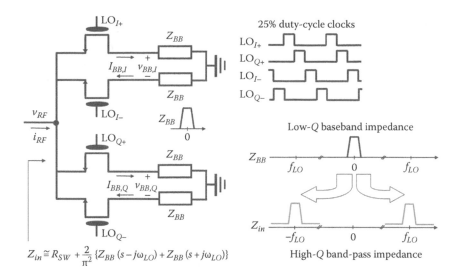

FIGURE 1.12 Impedance transformation of 25% passive mixer.

up-converted back to the RF around the LO on the RF side of the switches. Assuming that the input is an ideal RF current source, it can be shown that input impedance seen from the RF side is given by this expression:

$$Z_{in} \cong R_{SW} + \frac{2}{\pi^2}\{Z_{BB}(s - j\omega_{LO}) + Z_{BB}(s + j\omega_{LO})\} \tag{1.9}$$

which means that the baseband impedance is equal to the switch resistance in series with a band-pass filter. This band-pass filter, in fact, is the same low-pass filter that is frequency-shifted to the RF to become a built-in high-Q band-pass filter. The main advantage of this scheme is that, unlike the 50% duty-cycle mixer, no image component is created. Thus, the receiver does not suffer from a strong IQ cross talk.

Figure 1.13 illustrates how a receiver with 25% passive mixer can be modeled. The LNA is modeled with a current source I_{RF}, with its output impedance given by Z_L. The single-ended input impedance seen from the current buffer is denoted by Z_{BB}. The quadrature 25% rail-to-rail clocks that drive the mixer switches are also shown in that figure. At any moment, only one switch, from either the I or the Q channel, is ON, and the LNA current flows to the corresponding channel, which is why there is no significant IQ cross talk. In other words, no image current can be circulated from the I channel to the Q one.

The LNA current is divided between the impedance seen from the LNA load and the impedance seen from the switching system in series with the capacitor C. Thus, the RF current that is commutated by the switches is given by this expression:

$$I_C \cong \frac{Z_L(\omega)}{Z_L(\omega) + Z_C(\omega) + R_{SW} + \frac{2}{\pi^2}Z_{BB}(\omega - \omega_{LO})} I_{RF} \tag{1.10}$$

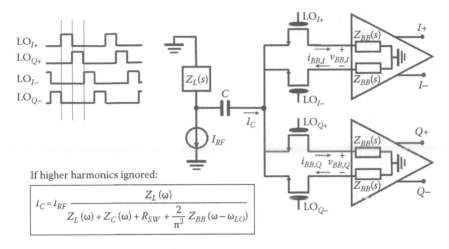

If higher harmonics ignored:

$$I_C = I_{RF} \frac{Z_L(\omega)}{Z_L(\omega) + Z_C(\omega) + R_{SW} + \dfrac{2}{\pi^2} Z_{BB}(\omega - \omega_{LO})}$$

FIGURE 1.13 Receiver model with 25% passive mixer.

This expression is just an approximation, and soon the exact equation will be derived, which would include the effects of higher-order harmonics. As mentioned earlier, with the 25% mixer, the IQ cross talk is weak. Consequently, due to the lack of image current, high- and low-side conversion gains are matched better. Also, since there is no image current, the IIP2 and IIP3 numbers are also better, the conversion gain is higher, and the receiver noise figure is lower.*

In the conventional design, the series capacitor C is sized just large enough to exhibit a low impedance at f_{LO}, and the LNA load is an LC tank tuned at f_{LO} (Figure 1.14a). This way, the RF current (LNA output current) is just delivered to the mixer with no gain and, in fact, with some attenuation depending on the size of R_p (or the inductor Q). However, we can optimize the design utilizing the fact that in any parallel RLC tank, at resonance, the current of an inductor or a capacitor is Q times larger than the total tank current (Figure 1.14b). Thus, for maximum conversion gain, there is an optimum for the size of the series capacitor C, and that is a function of inductor loss R_p and the switch resistance R_{SW}. It can be shown that for maximum conversion gain,

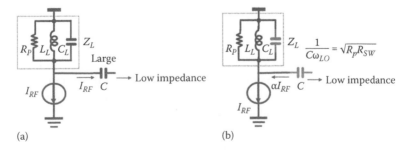

FIGURE 1.14 Sizing of components in a narrowband receiver with 25% passive mixer. (a) Conventional design. (b) Optimum design.

---

* Also, input-referred noise from other stages is less [17].

the series capacitor, C, must have an impedance equal to $\sqrt{R_p R_{SW}}$. Also, the parallel of the series capacitor, C, and the LC load, Z_L, must resonate at f_{LO}, which determines the size of C_L. With this choice of components, the LNA signal current is magnified by $\sqrt{R_p/R_{sw}}$, which is a linear and low-noise gain with appreciable magnitude.

So far, in deriving transfer functions, we have not included higher-order harmonics. If we do not include harmonic up- and down-conversions, the transfer functions from the RF current to the baseband voltages at the input of TIAs are given by the following equation:

$$V_{BB,I} = -jV_{BB,Q} = \frac{Z_L(\omega_{LO}+\omega_m)Z_{BB}(\omega_m)}{Z_L(\omega_{LO}+\omega_m)+Z_C(\omega_{LO}+\omega_m)+R_{SW}+\dfrac{2}{\pi^2}Z_{BB}(\omega_m)} \frac{2\sqrt{2}}{\pi}e^{j\frac{\pi}{4}}I_{RF}$$

$$(1.11)$$

where the input RF current is assumed to be a tone at $\omega_{LO}+\omega_m$. However, if we include higher-order harmonics, the transfer function becomes more complicated, given by

$$V_{BB,I} = -jV_{BB,Q}$$

$$= \frac{Z_L(\omega_{LO}+\omega_m)Z_{BB}(\omega_m)}{Z_L(\omega_{LO}+\omega_m)+Z_C(\omega_{LO}+\omega_m)+R_{SW}}$$

$$\times \frac{\dfrac{2\sqrt{2}}{\pi}e^{j\frac{\pi}{4}}I_{RF}}{1+\dfrac{2}{\pi^2}Z_{BB}(\omega_m)\sum_{k=-\infty}^{+\infty}\dfrac{1}{(4k+1)^2[Z_L((4k+1)\omega_{LO}+\omega_m)+Z_C((4k+1)\omega_{LO}+\omega_m)+R_{SW}]}}$$

$$(1.12)$$

indicating that now we are dealing with an infinite summation in the denominator. Overall, due to the harmonic up- and down-conversions, the conversion gain is expected to be slightly lower.

For both optimum and conventional designs, the transfer functions are simulated in Spectre-RF and plotted in Figure 1.15a and b. In both cases, the simulated transfer functions perfectly match with the accurate transfer function given in (1.12); therefore, legends have been omitted. Also, the simplified transfer function in (1.11), which was derived without including harmonic effects, is also plotted for both cases. As observed, except for an insignificant error around the LO, the simplified transfer function is a good and convenient approximation, without having to deal with harmonics and the associated complicated equations.

1.3 TRANSMITTER DESIGN WITH PASSIVE MIXERS

Figure 1.16 depicts a generic IQ transmitter for wireless applications. The transmitter is composed of two digital-to-analog converters (DACs): one for the I channel and one for the Q one. The two IQ DACs receive their inputs from the DSP unit, and they are

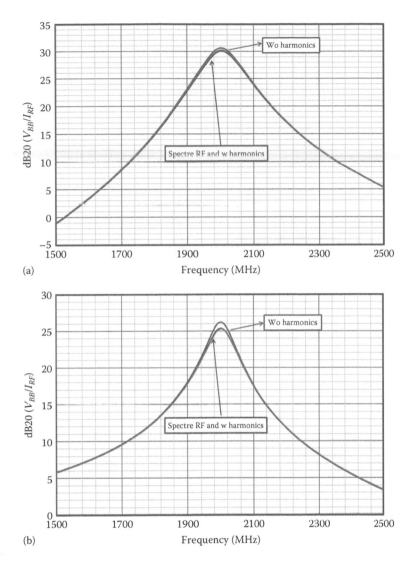

FIGURE 1.15 Spectre-RF simulation results of the transfer function against predictions with and without harmonics (Equations 1.11 and 1.12). (a) Optimum design. (b) Conventional design.

followed by two low-pass filters which are also called reconstruction filters. The low-pass filters are followed by a complex up-conversion mixer, in which the baseband information is up-converted to the LO frequency. The mixer output is followed by an on-chip power-amplifier (PA) driver to drive the off-chip PA. Sometimes, there is an external SAW filter, prior to the PA, to attenuate out-of-band unwanted emissions such as receive-band noise in FDD systems. Finally, the external PA is connected to the antenna typically through an antenna switch. Normally, the transmitter requires some gain control as well. For example, in a wideband code division multiple access (WCDMA) transmitter, at least 70 dB gain control with 1 dB resolution is required.

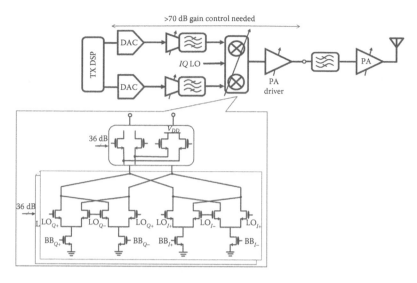

FIGURE 1.16   Transmitter design with an active mixer.

Until recently, most of the transmitters were using active mixers for up-conversion (Figure 1.16). The active mixers are also called Gilbert cells. In order to embed part of the gain controllability in the transmitter, the Gilbert section is usually composed of smaller unit cells connected in parallel. This adds a lot of capacitive loading for the high-frequency LO lines,* increasing power consumption of the LO generation block.

This architecture has a few other serious drawbacks. First, the LOFT does not scale with the transmitter gain. The reason is that if one of the mixer units is disabled to lower the transmitter gain, the switches of that mixer unit are still clocked by the LO clocks. Consequently, the LOFT will not scale down along with the transmitted signal. Another drawback is that the image-rejection ratio (IRR) of the transmitter varies with the gain control too. So, when we disable one mixer unit, we need to recalibrate the IRR. Furthermore, the transmitter layout is complicated, because each mixer unit has four input LO lines, four input baseband lines, and two output RF lines. And considering that the layout must be extremely symmetric and the lines must be shielded for good LOFT, the transmitter layout becomes extremely complicated.

And finally, the transmitters designed with active mixers suffer from typically poor out-of-band noise, including the RX-band noise. Therefore, in this section, we aim to design the transmitters with passive mixers too.

1.3.1 Design of Transmitters with 50% Passive Mixer

Let us first study the designing of transmitters with 50% passive mixers. The first architecture for such a transmitter is shown in Figure 1.17. The baseband signals in the I and Q channels after experiencing the low-pass filtering are individually

---

* Four lines, two differential lines for the I channel clocks, and another pair of differential lines for the Q channel.

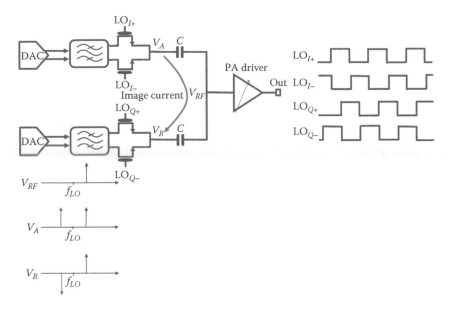

FIGURE 1.17 Transmitter design with 50% passive mixer.

up-converted to the RF through two passive mixers driven by 50% duty-cycle clocks. Switches of the mixers are working in the voltage mode. Therefore, the output impedance seen from the low-pass filters must be very low, in order to have the highest conversion gain. The RF outputs cannot be shorted together, because at any given moment, one switch in the I channel and one switch in the Q channel are simultaneously ON. Thus, direct connection of the RF sides of the I and Q mixers would be catastrophic as the low-pass filters have low output impedance and the mixers are operating in the voltage mode.* To mitigate the problem, the two mixers are capacitively coupled to the following PA driver (capacitors C in Figure 1.17). The size of these capacitors must be small enough to ensure the two low-pass filters do not drastically load each other, yet large enough not to lead to an excessive gain loss.

The entire gain control of the transmitter can now be embedded inside the PA driver. Now, LOFT scales down with the transmitter gain control. If the transmitter gain is lowered by 1 dB, the LOFT scales down by the same factor, which is a very attractive feature. We can calibrate the transmitter for the LOFT[†] once, and the calibration results can be used over the entire gain settings. Also, since by changing the gain control the mixer does not experience any change, the IRR remains unchanged as well. This is an attractive feature too, because we can calibrate the IRR once and use the resulting calibration coefficients over all gain settings.

To implement the gain control, similar to the conventional transmitter design with an active mixer, the PA driver needs to be composed of small unit cells. But unlike the design with an active mixer, the unit cells would have only two differential RF

---

* It resembles as if two different voltage sources are shorted to each other.
† To find proper baseband dc offsets to counterbalance the LOFT.

inputs and two differential RF outputs, making the layout floor planning a lot easier. Additionally, since the mixer is now compact, it would result in less power consumption in the LO generation block.

The transmitter in Figure 1.17, however, suffers from a serious drawback, that is, circulation of the image current between the two I and Q channels. Assume that it is intended to transmit a single tone at $f_{LO} + f_m$ from the PA driver output, where f_m is a small frequency offset. The RF voltage at point A, which is the RF side of the passive mixer of the I channel, is composed of two major frequency components at $f_{LO} \pm f_m$ with equal magnitudes. At point B, which is the RF side of the passive mixer of the Q channel, these two tones with similar magnitudes are present too. The problem is that at the image frequency, $f_{LO} - f_m$, the components at points A and B are 180° out-of-phase. This causes an image current circulating between the two channels through the two series capacitors C, which exacerbates the difference between high- and low-side up-conversion gains.* The image current is proportional to the size of the series capacitors, C. Therefore, a smaller C would lower the image current, but the penalty would be less conversion gain.

There is another way to design the transmitter with 50% passive mixer that does not have the circulating image problem. The architecture is shown in Figure 1.18. The two RF outputs in the I and Q channels are buffered first, for example, by placing two separate transconductors (G_m) units, in which their outputs are shorted together. Compared to the design in Figure 1.17, in this architecture, the PA driver units would be more complicated, that is, four RF inputs as opposed to two.[†]

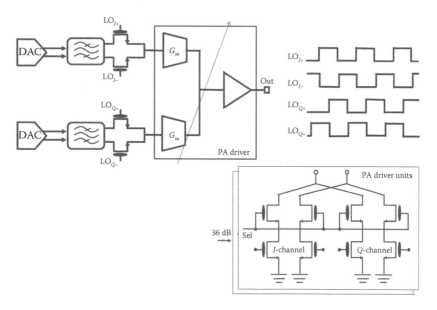

FIGURE 1.18 Alternative way of transmitter design with 50% passive mixer.

---

* We will shortly see that in a transmitter with 25% passive mixer, even though there is no circulating image current, still the transmitter may suffer from having different high- and low-side conversion gains.
† The implementation is assumed to be differential.

Considering that the two transconductors in the I and Q channels in the PA driver units could be randomly mismatched with respect to each other, the IRR would now vary over different gain settings. As a result, each gain setting may require its own calibration procedure. Similarly, the LOFT may need to be calibrated over various RF gain settings. Despite the above mentioned drawbacks, good aspects of the transmitter design with passive mixers such as low noise and applicability to low supply voltages are preserved.

1.3.2 Design of Transmitters with 25% Passive Mixer

Another way of eliminating the image current is to design the transmitters with 25% passive mixer, as shown in Figure 1.19. In fact, the transmitter with 25% duty-cycle passive mixer is the most optimum design in all aspects [19], except its need for 25% LO clocks as opposed to 50% duty-cycle clocks that are typically available. Its architecture is simple, especially in terms of layout floor planning, and leads to the minimum amount of routing parasitics. Also, switches of the passive mixer carry no dc currents, making it superior to the active mixer in terms of power consumption as well as noise, especially flicker noise. Additionally, the passive mixer–based architecture is perfect for low-voltage applications, because there is no device stacking. Furthermore, if the gain control is performed on the RF side, since the passive mixer does not experience any change over different gain settings, LOFT and IRR need to be calibrated only for one given gain setting. The calibration results can be applied to other gain settings in order to get satisfactory LOFT and IRR performance. Such a single-point calibration minimizes the calibration time. Also, the units in the PA driver are very simple, as there are only two RF inputs and two RF outputs (assuming differential implementation). As mentioned earlier, the only drawback is the need for low-noise 25% duty-cycle nonoverlapped clocks, which can be generated from 50% LO clocks by simply using four AND gates [5].

In order to understand the operation of the transmitter with the 25% passive mixer, we can simplify the transmitter as illustrated in Figure 1.20. From the baseband side, we can use the Thevenin theorem to replace the I and Q low-pass filters

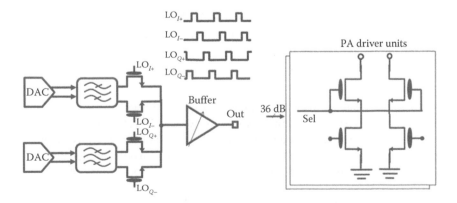

FIGURE 1.19   Transmitter design with 25% passive mixer.

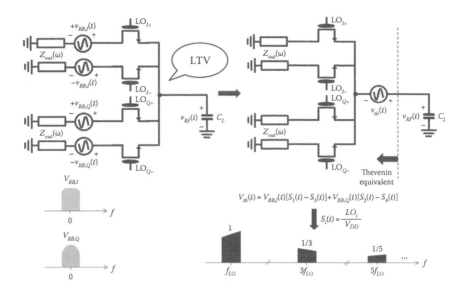

FIGURE 1.20   Simplified model of the transmitter with 25% passive mixer.

with their Thevenin equivalent. Z_{out} is the single-ended output impedance seen from the outputs of the low-pass filters. $\pm v_{BB,I}$ is the open circuit voltage of the I channel, which is the voltage at the low-pass filter output when it is not connected to the mixer switches. Obviously, this voltage is the DAC voltage in the I channel that has experienced the low-pass filter transfer function. Similarly, $\pm v_{BB,Q}$ is the open circuit voltage at the output of the Q channel. On the RF side, the input impedance seen from the PA driver is assumed to be capacitive, which is a practically good assumption. The size of this capacitor is equal to C_L. Now, we assume that switches can be modeled as an open circuit in the OFF mode and as a resistor of size R_{SW} in the ON mode, making the system LTV. The Thevenin theorem remains valid for LTV systems as well. Therefore, the four baseband voltages can be replaced with a single RF voltage in series with the switching system that is composed of four switches and four identical baseband impedances Z_{out} with no series baseband voltage sources anymore (the circuit on the right-hand side of Figure 1.20). On the RF side, the Thevenin voltage is readily found, which is given by the following equation:

$$v_{th}(t) = v_{BB,I}(t)[S_{I+}(t) - S_{I-}(t)] + v_{BB,Q}(t)[S_{Q+}(t) - S_{Q-}(t)] \qquad (1.13)$$

in which $S_{I+}(t)$ is defined to be 1 when the corresponding switch is ON and 0 otherwise. Thus, $S_{I+}(t)$ is a periodic signal with a duty cycle of 25%. Similarly, $S_{I-}(t)$, $S_{Q+}(t)$, and $S_{Q-}(t)$ are defined. The spectrum of the Thevenin voltage, $v_{th}(t)$, is graphically explained in the frequency domain (Figure 1.21a). The baseband signals in the I and Q channels are assumed to have different frequency spectrums. Of course, we know that the baseband signals in the I and Q channels are typically statistically independent signals. The spectrum of the equivalent RF Thevenin voltage is also depicted. The part of the spectrum around the LO is the desired up-converted signal.

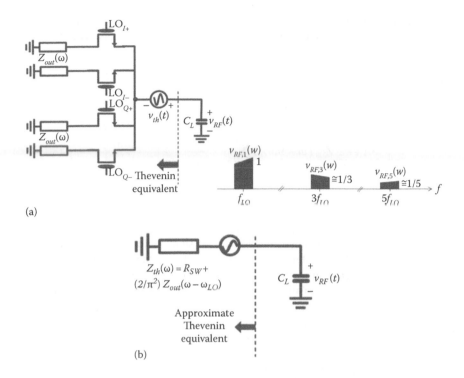

FIGURE 1.21 Transfer function from the baseband to the RF. (a) Thevenin equivalent circuit. (b) Simplified Thevenin equivalent circuit ignoring harmonics.

However, there are some up-converted components around higher-order odd harmonics that are undesirable. The up-converted frequency components around $3f_{LO}$ are those of around f_{LO}, but mirrored around the center along with a scaling factor of 1/3. Around $5f_{LO}$, the scaling factor is 1/5 and no mirroring takes place. And similar trend is repeated for the remaining higher-order odd harmonics of the LO. The fact that the spectrum around $3f_{LO}$ is mirrored adversely impacts the linearity performance of the transmitter, and we will come back to this later on.

Now, as shown in Figure 1.21a, due to the nature of the LTV system, all of the frequency components of the Thevenin voltage around f_{LO}, $3f_{LO}$, $5f_{LO}$, ... can contribute to the final RF signal across Z_L around f_{LO}. It can be shown that the final RF voltage across the RF impedance around f_{LO} is equal to the following expression [19]:

$$V_{RF,1}(\omega) = \frac{\sqrt{2}}{\pi} \frac{Z_L(\omega)}{Z_L(\omega) + R_{SW}}$$

$$\times \frac{e^{j\pi/4} V_{BB,I}\left(\omega - \omega_{LO}\right) + e^{-j\pi/4} V_{BB,Q}\left(\omega - \omega_{LO}\right)}{1 + \dfrac{2}{\pi^2} Z_{out}\left(\omega - \omega_{LO}\right) \displaystyle\sum_{p=-\infty}^{+\infty} \dfrac{1}{\left(4p+1\right)^2 \left[Z_L\left(4p\omega_{LO} + \omega\right) + R_{SW}\right]}}$$

$$(1.14)$$

in which Z_L is the impedance of the RF load (which is the capacitor C_L in Figure 1.21a) and R_{sw} is the switch resistance. Also, the RF voltage around $3f_{LO}$ is given by

$$
V_{RF,3}(\omega) = \frac{\sqrt{2}}{3\pi} \frac{Z_L(\omega)}{Z_L(\omega) + R_{SW}}
$$

$$
\times \frac{e^{-j\pi/4}V_{BB,I}\left(\omega - 3\omega_{LO}\right) - e^{j\pi/4}V_{BB,Q}\left(\omega - 3\omega_{LO}\right)}{1 + \dfrac{2}{\pi^2} Z_{out}^*\left(\omega - 3\omega_{LO}\right)\displaystyle\sum_{p=-\infty}^{+\infty} \dfrac{1}{\left(4p+1\right)^2\left[Z_L\left(4p\omega_{LO} - \omega\right) + R_{SW}\right]}}
$$

$$(1.15)$$

which is the mirror of the frequency components around f_{LO}. Since Z_L is simply a capacitor and there is no resonance (inductor in shunt with C_L), for the maximum conversion gain for the transfer function, the switch resistance R_{sw} and $(2/\pi^2)Z_{out}(0)$ must be much smaller than the impedance of Z_L at the LO frequency. This means that the output impedance of the baseband filter must be low. Typical numbers for R_{sw} and $Z_{out}(0)$ could be 10 Ohm or less. It must be mentioned that in deriving (1.14) and (1.15), Z_{out} of the baseband low-pass filter was assumed to be low pass with the cutoff frequency much smaller than the LO frequency. This approximation is valid for most of practical designs.

Equations 1.14 and 1.15 are not friendly to deal with and also they are not intuitive. A good approximation is shown in Figure 1.21b, in which the switching portion of the Thevenin equivalent in Figure 1.21a is approximated with an LTI impedance. This new Thevenin impedance is approximated by the following equation:

$$
Z_{th}(\omega) = R_{SW} + \frac{2}{\pi^2} Z_{out}\left(\omega - \omega_{LO}\right) \tag{1.16}
$$

Thus, the equivalent Thevenin impedance is equal to R_{sw} in series with $(2/\pi^2)Z_{out}(\omega - \omega_{LO})$, where the baseband output impedance, Z_{out}, is frequency-shifted to the LO along with a scaling factor of $(2/\pi^2)$. This is an LTI-based approximation, and there is no switching system anymore in the model, simplifying the calculations. The transfer function from the baseband to the RF voltage across C_L predicted by this LTI approach is given by

$$
V_{RF,1}(\omega) \cong
$$

$$
\frac{\sqrt{2}}{\pi}\left\{e^{j\pi/4}V_{BB,I}\left(\omega - \omega_{LO}\right) + e^{-j\pi/4}V_{BB,Q}\left(\omega - \omega_{LO}\right)\right\} \times \frac{Z_L(\omega)}{Z_L(\omega) + R_{SW} + \dfrac{2}{\pi^2}Z_{out}\left(\omega - \omega_{LO}\right)}
$$

$$(1.17)$$

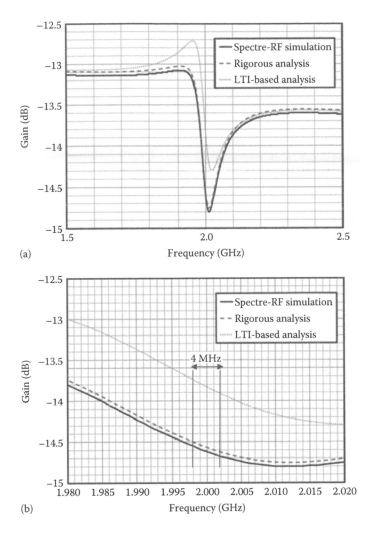

(a)

(b)

FIGURE 1.22 Spectre-RF simulations against predictions for the transfer functions.

Figure 1.22a plots the transfer function obtained from Spectre-RF simulation results and compares it against the LTV prediction of (1.14) and the LTI-based prediction of (1.17), when the input frequency is swept from 1.5 to 2.5 GHz (the LO frequency is 2 GHz). While the LTV-based approach predicts the transfer function with the highest accuracy, with the LTI-based approximation, the error is less than 1 dB. Considering its enormous simplicity, it is worthwhile using it. Figure 1.22b zooms into the frequency range of ±20 MHz over the LO. Evidently, from −20 to +20 MHz, the conversion gain from the baseband to the RF can be different by as large as 1 dB. This ripple would add up to the ripple of the low-pass filter. The requirements on the error-vector magnitude (EVM) of each transmission standard impose some limitations for the allowable ripple. Consequently,

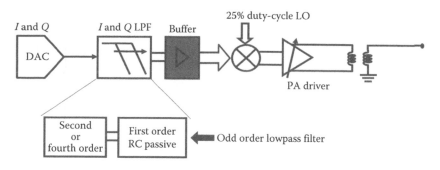

FIGURE 1.23 Interface of the low-pass filter and the passive mixer.

unlike in the design with an active mixer, with the passive mixer design, the high- and low-side conversion gains can be different. We now know that this difference in the high- and low-side conversion gains originates from lack of reverse isolation between the RF and IF ports. The question is, "How can we lower this difference between high-side and low-side conversion gains in the transmitter?" If we look at the conversion gain formula in (1.14), there are two solutions: (1) Output impedance of the baseband filter can be designed to be very low over the desired baseband signal bandwidth, for example, from dc to 2 MHz in the case of WCDMA. This is quite doable as in baseband frequencies as we can always utilize feedback. (2) Over the baseband signal bandwidth, the output impedance, Z_{out}, can be designed to be dominantly real.

As shown in Figure 1.23, the low-pass filter can be followed by a buffer, in order to ensure its output impedance, Z_{out}, is very low (less than 5 Ohm) over the baseband signal's bandwidth, for example, from dc to 2 MHz in the case of WCDMA. This buffer should also be very linear not to adversely impact the linearity performance of the transmitter. The buffer must have very low noise not to degrade the transmitter's out-of-band noise requirements, especially the receive band that usually imposes challenging requirements. Typically, as the last stage of baseband filtering, a first-order RC passive filter attenuates far-out noise originating from the DAC and also noise coming from prior stages of the low-pass filter (complex poles). Therefore, only the buffer stage is the major contributor for the receive-band noise as long as the baseband section is considered. For this purpose, the buffer must be designed to have low noise to begin with. A very good technique would be to design the buffer such that it exhibits very low output impedance across the desired baseband signal bandwidth, while having high output impedance at higher frequencies. This way, the up-conversion gain for far-out noise components would be lower (Z_{out} in [1.17] is large for these components), while the up-conversion gain for the desired baseband signal is maintained high.

It was explained that on the RF side of the passive mixer, which is the PA driver input, besides the desired signal around f_{LO}, there are up-converted signals at $3f_{LO}$, $5f_{LO}$, and so on. Around $3f_{LO}$, the up-converted signal is flipped over its center with respect to the desired signal at f_{LO}, and its magnitude is one-third of the desired signal. Around $5f_{LO}$, the signal is not flipped and the magnitude is one-fifth. Assume that only the desired signal was present at the mixer outputs; the third-order nonlinearity of the PA driver would cause spectral regrowth, described by a metric called the adjacent channel leakage ratio (ACLR). The broadened signal generates frequency components in the adjacent channels, and the ACLR is defined to be the power of the leaked signals to the adjacent channels to the power of the desired signal. Therefore, according to this definition, we can have upper and lower ACLR numbers. The question is, "What happens to ACLR due to the presence of up-converted frequency components around $3f_{LO}$, $5f_{LO}$, ...?" Here, we focus on only the frequency components around $3f_{LO}$. Due to the third-order nonlinearity of the PA driver, the desired signal and the unwanted components at $3f_{LO}$ would mix with each other and create additional unwanted components around f_{LO}. Portions of these components would fall on top of the desired channel, and the rest create frequency components in the adjacent channels, degrading the ACLR performance. Since the up-converted signal around $3f_{LO}$ is mirrored in the frequency domain, mathematically, it can be shown that the generated signal due to the PA driver nonlinearity around f_{LO} is totally a different signal than the desired signal with different statistics. The components falling on top of the desired signal band would impact only EVM, but the effect on EVM is generally negligible. However, the frequency components falling on the adjacent channels can significantly degrade ACLR. For example, in a WCDMA transmitter, this degradation is about 6 dB, meaning that whatever ACLR can be archived for a given PA driver nonlinearity when the input is only the desired signal, with the 3 LO components involved, the ACLR would be degraded by 6 dB (Figure 1.24). The effect makes the already stringent linearity requirements of the PA driver even more challenging. In the design with an active mixer, one can use a tuned LC load to filter out the 3 LO components and beyond before the final signal is applied to the PA driver, but that cannot be easily done in the passive mixer case.*

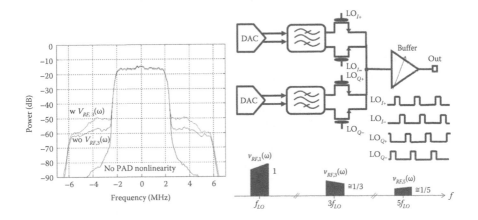

FIGURE 1.24   PA driver linearity requirements.

---

* Having an inductor in parallel with C_L and tuning their center to f_{LO} would not work in the passive mixer, and other techniques must be utilized to filter out up-converted components around higher-order harmonics.

1.4 CONCLUSIONS

The performance of transceivers designed with passive mixers driven by 50% or 25% duty-cycle clocks was studied. How various receiver or transmitter components must be chosen in order to achieve the best linearity and conversion gain performance was demonstrated. It was also shown that since there is no reverse isolation between the RF and IF ports of a passive mixer, the mixer reflects the IF impedance to the RF and vice versa through frequency shifting along with some scaling factors. Due to this lack of reverse isolation, in IQ transceivers, the two I and Q channels can potentially undergo a mutual interaction called the IQ cross talk.

In the receiver, the IQ cross talk causes unequal high- and low-side conversion gains, different high- and low-side IIP2 and IIP3 numbers, lower conversion gain, and increased receiver noise figure. The process as to how to design the passive mixer and its RF and IF loads in order to optimize the receiver for the best performance in terms of linearity, conversion gain, and noise figure and to alleviate the IQ cross talk issue was illustrated. It was also explained that, in general, receivers with 25% duty-cycle current-driven passive mixers outperform the ones designed with 50% passive mixers. With the 25% passive mixer, the IQ cross talk is eliminated, significantly lowering the severity of issues such as unequal high- and low-side conversion gains and different high- and low-side IIP2 and IIP3 values.

Similarly, it was observed that transmitters with 25% duty-cycle passive mixers outperform the ones designed with 50% duty-cycle passive mixers. This chapter clarifies that the circulating image current between the two I and Q channels is the main reason that the issues such as unequal high- and low-side conversion gains are stronger in transmitters designed with 50% duty-cycle passive mixers. With the 25% duty-cycle mixer design, the image problem is eliminated and, as a result, the IQ cross talk is minimized. The main drawback can be the need for generating the 25% duty-cycle quadrature clocks at GHz frequencies, which adds to the transmitter power consumption.

REFERENCES

1. B. Razavi, *RF Microelectronics*, 2nd edn. Upper Saddle River, NJ: Prentice Hall, 1998.
2. A.V. Oppenheim, A.S. Willsky, and S.H. Nawab, *Signals and Systems*, 2nd edn. Upper Saddle River, NJ: Prentice Hall, 1996.
3. H. Darabi and A. Abidi, Noise in RF-CMOS mixers: A simple physical model. *IEEE Journal of Solid-State Circuits* 35(1), 15–25, 2000.
4. S. Zhou and M.F. Chang, A CMOS passive mixer with low flicker noise for low-power direct-conversion receiver. *IEEE Journal of Solid-State Circuits* 40(5), 1084–1093, 2005.
5. A. Mirzaei, H. Darabi, A. Yazdi, Z. Zhou, E. Chang, and P. Suri, A 65 nm CMOS quad-band SAW-less receiver SoC for GSM/GPRS/EDGE. *IEEE Journal of Solid-State Circuits* 46(4), 950–964, 2011.
6. D. Leenaerts and W. Readman-White, 1/f noise in passive CMOS mixers for low and zero IF receivers. In *European Solid-State Circuits Conference*, 2001.
7. E. Sacchi, I. Bietti, S. Erba, L. Tee, P. Vilmercati, and R. Castello, A 15 mW, 70 kHz 1/f corner direct conversion CMOS receiver. In *IEEE Custom Integrated Circuits Conference*, 2003, pp. 459–462.

8. M. Valla, G. Montagna, R. Castello, R. Tonietto, and I. Bietti, A 72-mW CMOS 802.11a direct conversion front-end with 3.5-dB NF and 200-kHz 1/f noise corner. *IEEE Journal of Solid-State Circuits* 40(4), 970–977, 2005.

9. R. Bagheri, A. Mirzaei, S. Chehrazi, E. Heidari, M. Lee, M. Mikhemar, W. Tang, and A. Abidi, An 800-MHz–6-GHz software-defined wireless receiver in 90-nm CMOS. *IEEE Journal of Solid-State Circuits* 41(12), 2860–2876, 2006.

10. T. Nguyen, N. Oh, V. Le, and S. Lee, A low-power CMOS direct conversion receiver with 3-dB NF and 30-kHz flicker-noise corner for 915-MHz band IEEE 802.15.4 ZigBee standard. *IEEE Transactions on Microwave Theory and Techniques* 54(2), 735–741, 2006.

11. N. Poobuapheun, C. Wei-Hung, Z. Boos, and A. Niknejad, A 1.5-V 0.7.2.5-GHz CMOS quadrature demodulator for multiband direct-conversion receivers. *IEEE Journal of Solid-State Circuits* 42(8), 1669–1677, 2007.

12. B. Tenbroekl, J. Strange, D. Nalbantis, C. Jones, P. Fowers, S. Brett, C. Beghein, and F. Beffa. Single-chip tri-band WCDMA/HSDPA transceiver without external SAW filters and with integrated TX power control. In *IEEE International Solid-State Circuits Conference*, 2008, pp. 202–203.

13. B.R. Carlton, J.S. Duster, S.S. Taylor, and J.C. Zhan, A 2.2 dB NF, 4.9–6 GHz direct conversion multi-standard RF receiver front-end in 90 nm CMOS. In *IEEE Radio Frequency Integrated Circuits Symposium*, 2008, pp. 617–620.

14. A. Mirzaei, X. Chen, A. Yazdi, J. Chiu, J. Leete, and H. Darabi, A frequency translation technique for SAW-less 3G receivers. In *IEEE Symposium on VLSI Circuits*, 2009, pp. 280–281.

15. A. Mirzaei, H. Darabi, J. Leete, X. Chen, K. Juan, and A. Yazdi, Analysis and optimization of current-driven passive mixers in narrowband direct-conversion receivers. *IEEE Journal of Solid-State Circuits* 44(10), 2678–2688, 2009.

16. D.L. Kaczman et al., A single-chip tri-band (2100, 1900, 850/800 MHz) WCDMA/HSDPA cellular transceiver. *IEEE Journal of Solid-State Circuits* 41(5), 1122–1132, 2006.

17. A. Mirzaei, H. Darabi, J. Leete, and Y. Chang, Analysis and optimization of direct-conversion receivers with 25% duty-cycle current-driven passive mixers. *IEEE Transactions on Circuits and Systems I: Regular Papers* 57(9), 2353–2366, 2010.

18. M. Mikhemar, A. Mirzaei, A. Hadji-Abdolhamid, and J. Chiu, A 13.5 mA sub-2.5 dB NF multi-band receiver. In *IEEE Symposium on VLSI Circuits*, 2012, pp. 82–83.

19. A.M.D. Murphy and H. Darabi, Analysis of direct-conversion IQ transmitters with 25 duty-cycle passive mixers. *IEEE Transactions on Circuits and Systems I: Regular Papers* 58(10), 2318–2331, 2011.

2 Receiver IP_2 Variability

Sven Mattisson

CONTENTS

2.1 INTRODUCTION

Second-order intermodulation distortion (IM_2) has been much less of a concern historically for the radio receiver designer than third-order intermodulation distortion (IM_3). This has been the case since IM_2 products mostly fell out-of-band for narrowband receivers, while significant levels of IM_3 could easily fall in-band. With the adoption of the homodyne, or zero-IF, and low-IF receivers (see Figure 2.1), this situation has changed. Amplitude-modulated signals will generate IM_2 at baseband frequencies, for example, due to a strong adjacent channel interferer or due to own transmitter leakage. Even if these baseband IM_2 products are generated before the mixer, they may still leak through the mixer due to a nonzero DC offset. When the nonlinearity is in the baseband circuitry, all such IM_2 will directly become co-channel interference.

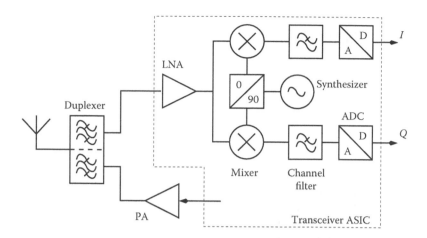

FIGURE 2.1   Simplified block diagram of a homodyne frequency-division duplex receiver.

IM_2 at baseband frequencies is largely due to finite common-mode rejection and as such varies in a random fashion. For high-volume production, it is very important to know the variability of the IM_2 performance of a receiver. To gain understanding of this random process, we will revisit the theory of basic nonlinear effects in a typical radio receiver and apply this insight when developing a statistical second-order intercept point (IP_2) variability model. The model is targeting IP_2 variability rather than IM_2, as the intercept point only requires power-level measurements and not amplitude and phase (sign). Finally, we also show how bounds of this variability can be estimated from a limited number of samples.

2.2   WEAK NONLINEARITY FUNDAMENTALS

The large signal transfer characteristic of an active device can be quite nonlinear, for example, the exponential or square-law behavior of a BJT or long-channel MOS transistor, respectively. In radio receiver circuitry, however, devices in the signal path are operated such that linearity is essentially preserved. We can almost always assume any device transfer function $f(\cdot)$ to be a weak nonlinearity of order 3; see, for example Sansen[5], that is,

$$y = f(x) = a_1 x + a_2 x^2 + a_3 x^3. \qquad (2.1)$$

This cubic polynomial can be used to model the nonlinearity of a transistor transfer conductance (g_m) as well as output saturation, either as isolated effects or in combination. In general there should also be a constant bias term (i.e., a_0) independent of x, but for simplicity and without loss of generality we omit this term.

When a single tone is applied to $f(\cdot)$ we get harmonic responses, but in reality, we operate on modulated signals causing more complex intermodulation products. It is most convenient to study these complex responses via analysis of a two-tone intermodulation test, as will be described in the following.

2.2.1 HARMONIC DISTORTION

To study the effects of the cubic polynomial coefficients, a_2 and a_3, on a narrowband signal, we substitute $x = A\cos(\omega_A t)$ in (2.1) and get

$$f\left(A\cos(\omega_A t)\right) = a_1 A\cos(\omega_A t) + a_2 A^2 \cos^2(\omega_A t) + a_3 A^3 \cos^3(\omega_A t)$$

$$= a_2 \frac{A^2}{2} + \left(a_1 A + a_3 \frac{3}{4} A^3\right)\cos(\omega_A t)$$

$$+ a_2 \frac{A^2}{2}\cos(2\omega_A t) + a_3 \frac{A^3}{4}\cos(3\omega_A t). \qquad (2.2)$$

This equation shows the generation of harmonics due to the nonlinearity $f(\cdot)$. For example, we see that second-order distortion (a_2) causes a DC shift by $a_2 \dfrac{A^2}{2}$ even though our $f(\cdot)$ does not contain any constant term. Furthermore, gain expansion is given by the cubic term in the $\cos(\omega_A t)$ expression, or

$$G_{\omega_A} = \frac{f\left(A\cos(\omega_A t)\right)|_{\omega_A}}{A\cos(\omega_A t)} = \frac{\left(a_1 A + a_3 \frac{3}{4} A^3\right)\cos(\omega_A t)}{A\cos(\omega_A t)} = a_1\left(1 + \frac{a_3}{a_1}\frac{3A^2}{4}\right). \qquad (2.3)$$

For example, when $a_1 a_3 > 0$, the gain will increase with signal level (e.g., in a class C amplifier), or when $a_1 a_3 < 0$, the gain will decrease (e.g., in a saturating amplifier). Finally, second- and third-order harmonic distortions are given by

$$HD_2 = \frac{f\left(A\cos(\omega_A t)\right)|_{2\omega_A}}{f\left(A\cos(\omega_A t)\right)|_{\omega_A}} = \frac{a_2 \dfrac{A^2}{2}}{a_1 A + a_3 \dfrac{3}{4} A^3} \approx \frac{a_2}{2a_1} A,$$

$$\qquad (2.4)$$

$$HD_3 = \frac{f\left(A\cos(\omega_A t)\right)|_{3\omega_A}}{f\left(A\cos(\omega_A t)\right)|_{\omega_A}} = \frac{a_3 \dfrac{A^3}{4}}{a_1 A + a_3 \dfrac{3}{4} A^3} \approx \frac{a_3}{4a_1} A^2.$$

2.2.2 INTERMODULATION DISTORTION

When two or more significant signals are present, harmonic mixing products will appear. By letting x consist of two tones, that is,

$$x = A\cos(\omega_A t) + B\cos(\omega_B t),$$

and substituting this in (2.1), we can study harmonic as well as intermodulation distortion. Then we get

$$f\left(A\cos(\omega_A t) + B\cos(\omega_B t)\right)$$

$$= a_1\left(A\cos(\omega_A t) + B\cos(\omega_B t)\right)$$

$$+ a_2\left(A\cos(\omega_A t) + B\cos(\omega_B t)\right)^2$$

$$+ a_3\left(A\cos(\omega_A t) + B\cos(\omega_B t)\right)^3$$

$$= a_2\frac{A^2}{2} + \left(a_1 A + a_3\left[\frac{3}{4}A^3 + \frac{3}{2}AB^2\right]\right)\cos(\omega_A t)$$

$$+ a_2\frac{A^2}{2}\cos(2\omega_A t) + a_3\frac{A^3}{4}\cos(3\omega_A t)$$

$$+ a_2\frac{B^2}{2} + \left(a_1 B + a_3\left[\frac{3}{4}B^3 + \frac{3}{2}A^2 B\right]\right)\cos(\omega_B t)$$

$$+ a_2\frac{B^2}{2}\cos(2\omega_B t) + a_3\frac{B^3}{4}\cos(3\omega_B t)$$

$$+ a_2 AB\left[\cos(\omega_A t + \omega_B t) + \cos(\omega_A t - \omega_B t)\right]$$

$$+ a_3\frac{3AB}{4}\left[A\cos(2\omega_A t + \omega_B t) + A\cos(2\omega_A t - \omega_B t)\right.$$

$$\left. + B\cos(2\omega_B t + \omega_A t) + B\cos(2\omega_B t - \omega_A t)\right]. \qquad (2.5)$$

We see the same harmonic components as in (2.2) but here repeated for both ω_A and ω_B. In addition to the harmonic components, we also see intermodulation components and cross-modulation components where the amplitude of one carrier influences the level of another, and vice versa. The intermodulation terms are due to a_2 at $\omega_A \pm \omega_B$ and due to a_3 at $2\omega_A \pm \omega_B$ and $2\omega_B \pm \omega_A$, while the cross-modulation terms are due to a_3 at ω_A and ω_B, respectively.

Specifically looking at the intermodulation products in (2.5), we can identify some to be due to a_2 and the others due to a_3, resulting in IM_2 and IM_3, or

$$IM_2 = a_2 AB\left[\cos(\omega_A + \omega_B t) + \cos(\omega_A t - \omega_B t)\right]$$

$$IM_3 = a_3\frac{3AB}{4}\left[A\cos(2\omega_A t + \omega_B t) + A\cos(2\omega_A t - \omega_B t)\right.$$

$$\left. + B\cos(2\omega_B t + \omega_A t) + B\cos(2\omega_B t - \omega_A t)\right]. \qquad (2.6)$$

We also see that the intermodulation products occur at all combinations of sum and difference frequencies and that they depend on the input signal amplitudes. By comparing (2.4) and (2.6), we also recognize that

$$IM_2 = 2HD_2 \text{ and } IM_3 = 3HD_3. \tag{2.7}$$

2.2.3   CASCADED NONLINEARITIES

When input and output nonlinearities are evaluated separately or when we are cascading two nonlinear devices, we can find the overall nonlinearity by cascading two polynomials. Such cascade formulas may be derived by evaluating the nonlinearities recursively; see, for example Macdonald[3]. For example, when two weakly nonlinear amplifiers are cascaded, the second amplifier will have the nonlinear output of the first as its input, or in polynomial form

$$y = b(a(x)) = c(x). \tag{2.8}$$

When the coefficients of $a(\cdot)$ and $b(\cdot)$ are known, we can solve for the coefficients in $c(\cdot)$ by identifying like terms and get

$$c_1 = b_1 a_1$$
$$c_2 = b_1 a_2 + b_2 a_1^2 \tag{2.9}$$
$$c_3 = b_1 a_3 + b_2 2 a_1 a_2 + b_3 a_1^3.$$

For simplicity, we have ignored higher-order terms, and we can do this without loss of generality as we are only interested in weak nonlinearities.

From (2.9), we see that when cascading two strictly square nonlinearities (i.e., $a_3 = b_3 = 0$) we will still get IM_3 as $c_3 \neq 0$, but cascading two strictly odd nonlinearities (i.e., $a_2 = b_2 = 0$) does not cause any even-order distortion as also $c_2 = 0$ in this case.

2.2.4   DIFFERENTIAL NONLINEARITIES

A differential circuit can be considered to consist of two similar single-ended circuits in parallel, sharing a common virtual ground. The differential circuit then sees the full input signal, while the single-ended parts see one-half of the input each but with opposite polarities. The weak polynomial describing each half circuit will in general not be the same as that describing the differential circuit.

Then, by considering the even and odd symmetries of a function $f(\cdot)$, that is, when $f(\cdot) = f_{even}(\cdot) + f_{odd}(\cdot)$, we have

$$f_{even}(x) = f_{even}(-x),$$
$$f_{odd}(x) = -f_{odd}(-x),$$
(2.10)

and due to the fact that a differential amplifier is driven by a differential signal, we can deduct that odd-order distortion products add up while even orders cancel.

Since symmetry cancels the even terms, we would expect $a_2 \to 0$, but mismatch causes some residual second order resulting in a finite IM_{2D}, or

$$IM_{2D} \approx \epsilon IM_{2S},$$
(2.11)

where ϵ is the matching accuracy (often on the order of 0.1%–1%).

2.2.5 Feedback around a Nonlinearity

Feedback is a well-known technique to enhance linearity, but it also occurs as unintentional series and shunt feedback due to parasitic effects.

Assuming a linear feedback β around a reference variable $f(x_e)$ (see Figure 2.2), we get

$$y = g(x) = f(x_e) = f\left(x - \beta g(x)\right)$$
(2.12)

with the feedback factor $F = 1 + \beta \dfrac{\partial f(0)}{\partial x_e}$; see, for example Sansen[5]

To find $g(x)$ as a function of $f(x_e)$ and β, neglecting DC terms, we assume $f(\cdot)$ and $g(\cdot)$ to be cubic polynomials. As before, we have

$$y = f(x - \beta y),$$
(2.13)

but this is an implicit relation in $y(\cdot)$, so it is easier to solve the inverse relation

$$x = \beta y + f^{-1}(y) = g^{-1}(y)$$
(2.14)

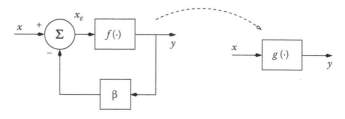

FIGURE 2.2 Mapping of a simple feedback model to a single nonlinearity.

and then use series inversion[2] to express $g(\cdot)$ in the a_i terms of $f(\cdot)$. By identifying terms of like powers, we get

$$g_1 = \frac{a_1}{F}$$

$$g_2 = \frac{a_2}{F^3} \tag{2.15}$$

$$g_3 = \frac{a_3 F - 2(F-1)a_2^2/a_1}{F^5},$$

where $\varGamma$ is the feedback factor (i.e., $1 + \beta\, a_1$). We see that the linear gain drops by F and consequently also the input to $f(\cdot)$, resulting in reduced distortion levels for a fixed input signal level (i.e., constant x).

By normalizing the nonlinear components of $g(\cdot)$ to the linear gain, we get

$$\frac{g_2}{g_1} = \frac{a_2}{a_1 F}\frac{1}{F}$$

$$\frac{g_3}{g_1} = \frac{a_3}{a_1 F}\frac{F - 2(F-1)a_2^2/a_1}{F^3}, \tag{2.16}$$

where we see both the gain reduction and the linearity improvement clearly. These equations refer the linearity improvement to the *input*, but by looking at the last fraction of (2.16), we get the output-referred improvements (e.g., $1/F$ for IM_2) as we then adjust for the lower gain we get when feedback is applied.

2.2.6 Intercept Points

When specifying the linearity performance of a circuit, it is not sufficient to quote the ratio of a distortion component to a fundamental component since this ratio depends on the signal levels. However, by letting the two input signals in (2.5) be equal in amplitude, but different in frequency, and extrapolating one of the IM_2 and IM_3 levels until it intercepts the extrapolated fundamental tone (Figure 2.3), we get the second- and third-order intercept points, respectively, or

$$IP_2 = \frac{a_1}{a_2}$$

$$IP_3 = \sqrt{\frac{4a_1}{3a_3}}. \tag{2.17}$$

These intercept points offer a convenient and unambiguous way to characterizing a week nonlinearity. If $a_1 \cdot a_3 < 0$, then IP_3 will be imaginary according to the definition in (2.17). In most texts, the absolute value of IP_3 is used, but it is convenient to

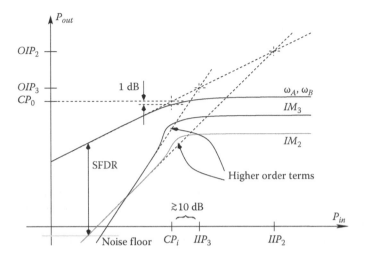

FIGURE 2.3 Intercept point definitions. The spurious-free dynamic range (SFDR) is the maximum ratio of signal power to any noise and distortion products. IIP_x and OIP_x represent the input- and output-referred intercept points, respectively.

use the complex value* as an imaginary IP_3 corresponds to gain compression and a real one to gain expansion.

Using the intercept points in (2.17), we can recast (2.5) to

$$\frac{f\left(A\cos\left(\omega_A t\right)+B\cos\left(\omega_B t\right)\right)}{a_1}=\frac{A^2}{2IP_2}+A\left(1+\frac{A^2+2B^2}{IP_3^2}\right)\cos\left(\omega_A t\right)$$

$$+\frac{A^2}{2IP_2}\cos\left(2\omega_A t\right)+\frac{A^3}{3IP_3^2}\cos\left(3\omega_A t\right)$$

$$+\frac{B^2}{2IP_2}+B\left(1+\frac{B^2+2A^2}{IP_3^2}\right)\cos\left(\omega_B t\right)$$

$$+\frac{B^2}{2IP_2}\cos\left(2\omega_B t\right)+\frac{B^3}{3IP_3^2}\cos\left(3\omega_B t\right)$$

$$+\frac{AB}{IP_2}\left[\cos\left(\omega_A t+\omega_B t\right)+\cos\left(\omega_A t-\omega_B t\right)\right]$$

$$+\frac{AB}{IP_3^2}\left[A\cos\left(2\omega_A t+\omega_B t\right)+A\cos\left(2\omega_A t-\omega_B t\right)\right.$$

$$\left.+B\cos\left(2\omega_B t+\omega_A t\right)+B\cos\left(2\omega_B t-\omega_A t\right)\right],$$

$$(2.18)$$

---

* If in doubt, the coefficients a_1, a_2, and a_3 can always be used.

where we now refer the intermodulation products to the input by dividing $f(\cdot)$ with a_1 (i.e., the relative distortion is the same at input and output, less a scale factor a_1).

Since linearity in practice is characterized by intercept points rather than polynomial coefficients, (2.18) is more convenient to use than (2.5).

By using the cascade coefficients, as derived in (2.9), we can find the intercept points of a cascade as

$$\frac{1}{IP_{2c}} = \frac{1}{IP_{2a}} + \frac{a_1}{IP_{2b}},$$

$$\frac{1}{IP_{3c}^2} = \frac{1}{IP_{3a}^2} + \frac{a_1^2}{IP_{3b}^2}. \tag{2.19}$$

Both IP_2 and IP_3 of the second stage can be referred back the input by simple signal- and power-gain scaling, respectively. It is also interesting to note that combining an expanding and a compressing nonlinearity (i.e., when one of IP_{3a} and IP_{3b} is imaginary) we can increase the cascade IP_3. In the limit, IP_{3c}^2 can be infinite but then we need to consider higher-order terms as well.

2.2.7 CROSS MODULATION

Circuit gain will depend on the levels interfering signals due to circuit nonlinearities. This effect is shown by the $\cos(\omega_A t)$ term in, for example, (2.19).

A strong blocking carrier (ω_B) will reduce, or expand, the gain at $\omega_A t$ when the blocker level B approaches IP_3. In the case where B has a constant envelop, for example, an unmodulated or phase-modulated carrier, then gain will just change, but when the blocker has a time-varying amplitude (i.e., AM), we will get cross modulation of the blocker envelop onto ω_A. Amplitude modulation of ω_B is similar to what is also called triple beat in the literature, when three test tones are used.

The situation in which the interferer is amplitude-modulated can be analyzed by letting $B \rightarrow B(t)$ in (2.18). The output signal at ω_A will then be

$$y_{\omega_A} \approx a_1 A \left(1 + \frac{2B(t)^2}{IP_3^2} \right) \cos(\omega_A t). \tag{2.20}$$

From this equation, we see that the amplitude modulation at ω_B is transferred onto the signal at ω_A, hence, this effect is called cross modulation.

A common cross-modulation scenario is when there are two interferers: one AM-modulated with a bandwidth of BW and another possibly AM-modulated but within a BW away from the desired signal. For example, the transmit signal of a 4G handset will leak to its input and this signal may cross modulate on top of a nearby narrowband signal (e.g., a 2G base station signal or the handset's own receiver LO leakage). Cross modulation on top of the receiver's own LO leakage will look exactly like IM_2 due to $B(t),^2$ but it is really an IM_3 component as it depends on IP_3 (i.e., a_3). Even though it can be hard to distinguish from IM_2, we will disregard it here as its mechanisms and statistics will be very different from those of a genuine IM_2 (i.e., one that is due to a nonzero a_2 polynomial coefficient).

2.3   RECEIVER MIXER IM_2 MECHANISMS

As derived in the previous sections, we know that any IM_2 product is a common-mode signal due to an even-order nonlinearity. That is, it is only due to device nonlinearities causing single-ended IM_2 and layout or cross-talk asymmetries limiting common-mode suppression. To reduce IM_2, we need to reduce mismatches and enhance device linearity (without impairing mismatch in the process). Leakage between mixer ports will also generate IM_2 in this fashion. As mentioned in Section 2.2.7, cross modulation can be mistaken for IM_2, but it is an odd-order nonlinearity phenomenon and will in general require other solutions than what is effective for IM_2, for example, lower LO leakage, and will not be covered here.

Baseband IM_2 at an RX mixer output is primarily a common-mode signal due to mixer device output nonlinearities. With mismatches, a fraction of this signal is converted to a differential signal, and in a well-designed balanced circuit, this differential IM_2 has a Gaussian distribution with a zero mean (i.e., no systematic mismatch). Also, when the stages in front of the mixer generate significant amounts of IM_2 around zero Hz, some of this will also leak through the mixer, for example, due to LO asymmetry, and end up as co-channel interference after the mixer. Normally, these latter contributions are less significant as a balanced RF driver stage is often used in front of the mixer.

We can model IM_2 generation as in Figure 2.4. Here two current sources represent the single-ended IM_2 currents driving the baseband load resistors. The mixer imbalance is introduced by adding small IM_2 current and mixer load resistor

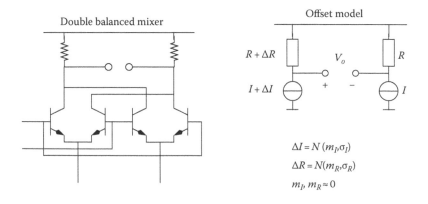

FIGURE 2.4   Conceptual receiver mixer and its IM_2 model.

offsets, ΔI and ΔR, respectively. Although the model is depicted as a Gilbert mixer, the results we will derive in the following are, to a first order, valid for most mixer structures, for example, passive MOS mixers.

With I and $I + \Delta I$ being the single-ended IM_2 currents out of the mixer devices, the differential IM_2 can be found to be

$$V_o = (I + \Delta I)(R + \Delta R) - I R = I\Delta R + R\Delta I + \Delta I\Delta R$$

$$\approx I\Delta R + R\Delta I = I\,N(m_R, \sigma_R) + R\,N(m_I, \sigma_I)$$

$$= I R N\left(\frac{m_I}{I} + \frac{m_R}{R}, \sqrt{\left(\frac{\sigma_I}{I}\right)^2 + \left(\frac{\sigma_R}{R}\right)^2}\right) \approx I R N(0, \sigma) = \widehat{IM}_2, \qquad (2.21)$$

where we have assumed $\Delta I/I$, $\Delta R/R \ll 1$.

Then, by letting the variance of the IM_2 equal $P_i^2/\widehat{IP_2}$ (W), we can find random IM_2 samples as

$$IM_2 = N(0,1)\sqrt{\frac{P_i^2}{\widehat{IP_2}}}\ (V), \qquad (2.22)$$

where $N(\bar{m}, \sigma)$ is a Gaussian random variable of mean $\bar{m}$ (≈ 0) and standard deviation σ. A set of unbiased samples can be obtained by scaling a Gaussian set $M \in N(0, 1)$ with $\sqrt{P_i^2/\widehat{IP_2}}$.

We can plot histograms of simulated IM_2 sample sets, where IM_2 is represented as a linear voltage, linear power, and logarithmic power (dBm), respectively; see Figure 2.5. From these plots, it is clear that only the linear voltage representation has a Gaussian distribution. This also means that we cannot recover the Gaussian distribution of IM_2 from data that lack sign information, for example, due to power-based measurements.

Further, by substituting our simulated set of IM_2 values in

$$IP_2(i) = 10\lg\left(\frac{P_i^2}{IM_2(i)^2}\right), \qquad (2.23)$$

we get a set of IP_2 values. Representing these as linear and logarithmic powers (see Figure 2.6), we get something that resembles typical IP_2 characterization data distributions (in the dBm case at least, see e.g. Figure 2.14).

Fitting a Gaussian distribution to the IP_2 (dBm), as in the right graph of Figure 2.6, will not be very accurate. In particular, the sample size in the lower bin (i.e., the

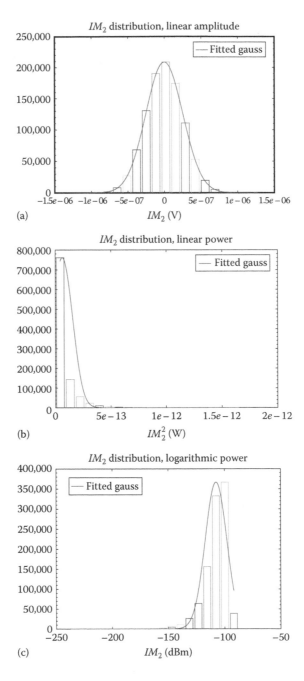

FIGURE 2.5 Histograms for simulated IM_2 variations with a Gaussian distribution fit superimposed: (a) IM_2 as a linear voltage (V), (b) as a linear power (W), and (c) as a logarithmic power (dBm).

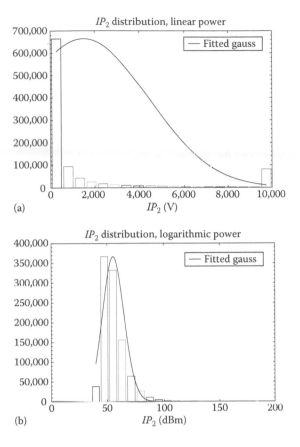

FIGURE 2.6 Histograms of simulated IP_2 variations with a Gaussian distribution fit superimposed: (a) IP_2 as linear power (W) and (b) as a logarithmic power (dBm).

worse IP_2 samples) is greatly exaggerated by the fitted Gaussian, leading to a very pessimistic lower bound of the IP_2 variability range. In the following, we will show how to achieve a better metric and worst-case estimate.

2.4 STATISTICAL *IP₂* MODEL

Let, as before, the mixer outputs I and Q (but not $I+jQ$) be Gaussian with zero mean, that is,

$$IM_2 = N(0,\sigma) = N(0,1)\sigma, \tag{2.24}$$

with the IM_2 amplitude in volts. Then, in dB units, we get

$$IP_2 = 10 \lg\left(\frac{P_i^2}{IM_2^2}\right) = 10 \lg\left(\frac{P_i^2}{N^2(0,1)\sigma^2}\right), \tag{2.25}$$

where P_i is the power of either one of the two equally strong two-tone test signals.

Defining $\widehat{IP_2}$ as an IP_2 factor (in the same spirit as the well-known noise factor) with $\widehat{IP_2} = 10\lg\left(\dfrac{P_i^2}{\sigma^2}\right)$, we get the randomized IP_2 as

$$IP_2 = \widehat{IP_2} - 10\lg\left(N^2(0,1)\right). \tag{2.26}$$

Using the fact that $N^2(0, 1)$ is not Gaussian but a χ_1^2 distribution in one parameter (see e.g. Abramowitz[7]), we can finally estimate the expected $\overline{IP_2}$ as

$$\overline{IP_2} = E\left\{\widehat{IP_2} - 10\lg\left(\chi_1^2\right)\right\} = \widehat{IP_2} - 10\,E\left\{\lg\left(\chi_1^2\right)\right\}. \tag{2.27}$$

The expected mean of $\lg(\chi_1^2)$ is known and is

$$E\left\{10\lg\left(\chi_1^2\right)\right\} = -10\frac{\ln(2)+\gamma}{\ln(10)} \approx -5.52, \tag{2.28}$$

with γ being the Euler constant (0.577). That is, the average IP_2 for several sample circuits (I or Q but not $I + jQ$) is, in dB units,

$$\overline{IP_2} = \widehat{IP_2} + 5.52. \tag{2.29}$$

A very important observation is that, because of the χ^2 distribution properties, the variation of IP_2 is centered around, but independent of, its mean value. The probability of IP_2 being lower than its mean $\overline{IP_2}$ plus Δ is

$$P(IP_2 < \overline{IP_2} + \Delta) = 1 - P\left(\chi_1^2 < 10^{\frac{\Delta-5.52}{10}}\right). \tag{2.30}$$

Figure 2.7 shows simulated IP_2 values with a χ_1^2 distribution fitted. From the graph, one can see that less than 1 ppm of the samples will have an IP_2 more than 19.3 dB below $\overline{IP_2}$ (or some 30 ppm at least 17.9 dB below). For example, if we require $IP_2 > 38$ dBm worst case, then to guarantee at most 1 ppm yield loss, it is sufficient that $\overline{IP_2} \geq 57.3$ dBm (or 55.9 dBm for 30 ppm yield loss).

2.4.1 $\overline{IP_2}$ WITH NONZERO $\overline{IM_2}$

The χ_1^2 model assumes $\overline{IM_2} = 0$, which is not the case when the circuit is not perfectly balanced.

Plotting $\overline{IP_2}$ for a varying $\overline{IM_2}$ (see Figure 2.8) shows a clear reduction in $\overline{IP_2}$ with increasing $\overline{IM_2}$. The reason for this is of course that the random contribution to IM_2 becomes less significant as $\overline{IM_2}$ increases.

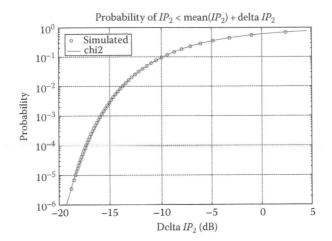

FIGURE 2.7   Plot of χ_1^2 and simulated IP_2 sample set probability vs. offset from IP_2 mean value.

By normalizing the IP_2 mean to its σ, we can check how much IM_2 bias can be tolerated before our model accuracy degrades. In Figure 2.8a, we sweep m as a fraction of σ from 0 to 1 and plot the resulting average IP_2. In Figure 2.8b, we vary m and σ along a constant offset contour (i.e., sweep m while adjusting σ to keep $m^2 + \sigma^2$ constant). For $|m| \lesssim \dfrac{\sigma}{2}$, we get a constant $\overline{IP_2}$ contour (see the lower plot). This indicates that systematic IM_2 contributions below σ/2 will not disturb our model. Also, as a consequence, there is no need to accurately find m and σ to predict any yield impact when $m < \sigma/2$, and the χ_1^2 distribution can be used as if $m \approx 0$ (Figure 2.9).

2.4.2 IP_2 FOR VECTOR SIGNALS

The scalar IP_2 is measured on either of I or Q, but at the system level, it is really the vector $I + jQ$ that matters. Previous sections showed the distribution of the scalar IP_2 and now we will see how this relates to its vector value.

In a two-tone test, we have the input signal

$$v_{in}(t) = v_s \cos\big((\omega_{LO} + \omega_s)t\big) + v_i\big(\cos(\omega_i t) + \cos((\omega_i + \Delta\omega)t)\big), \qquad (2.31)$$

where v_s is the amplitude of the wanted signal and v_i is the amplitude of the two interferers separated by Δω. In a radio receiver, after downconversion and low-pass filtering, we have the baseband signals

$$v_{BB_I}(t) = v_s \cos(\omega_s t) + IM_2 \cos(\Delta\omega t),$$

$$v_{BB_Q}(t) = v_s \sin(\omega_s t) + IM_2 \cos(\Delta\omega t), \qquad (2.32)$$

where the IM_2 amplitude proportional to v_i^2 as well as imbalance and where we assume $\sigma_I, \sigma_Q \approx \sigma$.

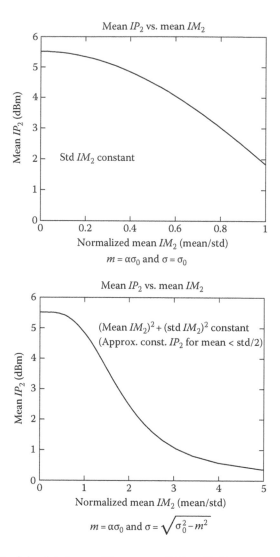

FIGURE 2.8 Plot of simulated mean IP_2 vs. mean IM_2/σ with (a) constant σ and (b) constant IP_2 (i.e., constant $m^2 + \sigma^2$).

For the scalar case, we showed in (2.25) and (2.27) that

$$IP_{2s} = 10\lg\left(\frac{P_i^2}{IM_2^2}\right) = \widehat{IP_2} - 10\lg\left(\chi_i^2\right).$$

In the complex domain, the vector signal amplitude is $1 + j$ times the scalar signal amplitude while the I and Q IM_2 variations (i.e., mismatches) are uncorrelated. The scalar

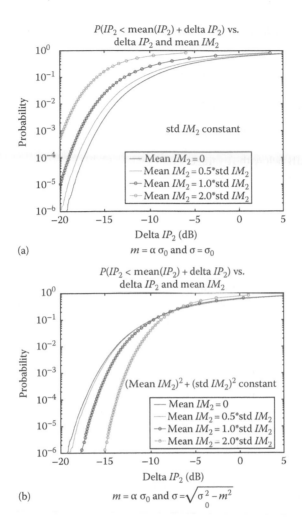

FIGURE 2.9 Plot of simulated IP_2 probability distribution vs. mean IM_2/σ with (a) constant σ and (b) constant IP_2 (i.e., constant $m^2 + \sigma^2$).

ratio P_i/IM_2^2 will, thus, correspond to a vector ratio of $\left(P_i\,|\,1+j\,|^2\right)\big/\left(IM_{2I}^2 + IM_{2Q}^2\right)$ and, with $\sigma_I = \sigma_Q = \sigma$, we get

$$IP_{2v} = 10\lg\left(\frac{P_i^2\,|\,1+j\,|^2}{IM_{2I}^2 + IM_{2Q}^2}\right) = \widehat{IP_2} + 10\lg\left(\frac{2}{N^2(0,1) + N^2(0,1)}\right)$$

$$= \widehat{IP_2} + 10\lg(2) - 10\lg\left(\chi_2^2\right), \tag{2.33}$$

with, as before, $\widehat{IP_2} = 10\lg\left(P_i^2/\sigma^2\right)$. From (2.33), we may conclude that in the complex domain, the IP_2 will be a χ_2^2 rather than χ_1^2 distribution.

The expected mean of $\lg\left(\chi_2^2\right)$ can be evaluated numerically, and we find

$$E\left\{10\lg\left(\chi_2^2\right)\right\} \approx 0.50\,\text{dB}. \tag{2.34}$$

That is, the average measured $IP_{2\nu}$ for $I + jQ$ is

$$\overline{IP_{2\nu}} \approx \widehat{IP_2} + 3.01 - 0.50 = \widehat{IP_2} + 2.51, \tag{2.35}$$

which is 3.01 dB below the average scalar IP_{2s} in (2.29).

As in the scalar case, the $IP_{2\nu}$ variation is also independent of its mean value and the probability of $IP_{2\nu} < \overline{IP_{2\nu}} + \Delta$ is

$$P\left(IP_{2\nu} < \overline{IP_{2\nu}} + \Delta\right) = 1 - P\left(\chi_2^2 < 10^{-\frac{\Delta+0.50}{10}}\right). \tag{2.36}$$

The expected average of IP_{2s} is 3 dB higher than $IP_{2\nu}$, but the required back-off from $\widehat{IP_2}$, to achieve similar yield, is around 6 dB larger. The worst-case IP_2 will roughly be 3 dB worse in the scalar case for the same $\widehat{IP_2}$ (i.e., same σ). This is so because the IM_{2I} and IM_{2Q} amplitudes are uncorrelated and a 3 dB improvement should be expected for the vector case (Figures 2.10 and 2.11).

2.4.3 Worst-Case IP_2 Estimation

When determining statistical properties from a small data set, it is difficult to assess the variability of the data. A possible way to overcome this difficulty is to use statistical bootstrap[6], a form of resampling. The basic principle of statistical bootstrap is to use the available sample set as a discrete representation of the real distribution.

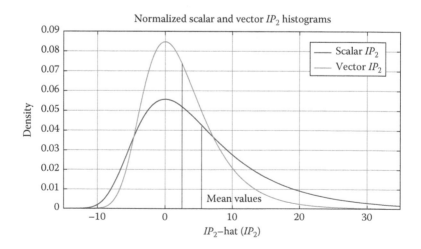

FIGURE 2.10 Comparison of scalar and vector IP_2 distributions. IP_{2s} has a higher mean but also larger spread than $IP_{2\nu}$, assuming $\sigma_I = \sigma_Q = \sigma$.

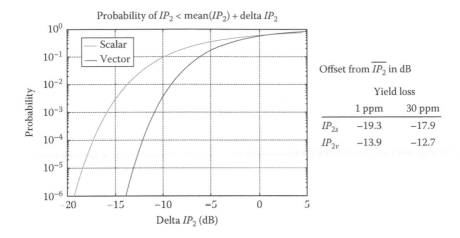

FIGURE 2.11 Plot of scalar and vector IP_2 probability vs. offset from IP_2 mean.

Then, by randomly extracting n samples, with replacement, out of the set of N, we can estimate a new data set (for small sets, we usually use all available samples, i.e., $n = N$). By repeating this extraction m times, we can generate a number of estimated data sets that can be used to better estimate the statistics of the actual samples. As we are resampling with replacement, one sample can be picked more than once while another may be unused. Because of this, each extracted subset will have slightly different populations, mimicking the variability in the real data.

For example, to estimate a lower bound, that is, the worst case, based on a -4σ limit of a Gaussian parameter, we do the following:

1. Measure parameter x on n samples.
2. Calculate a lower bound of x as $lb_x = \bar{x} - 4\sigma_x$.
3. Resample with replacement from x to get m sets of simulated data of length n in vectors z_i (where $i = 1, ..., m$).
4. Calculate the m lower bounds of z_i and assign them to $lb_{z\,i}$.
5. Estimate the lower bound limits of the samples from the distribution of the differences between each $lb_{z\,i}$ and lb_x, where the differences are assumed to be Gaussian.

In the following, we will apply the previous algorithm adopted for a receiver IP_2. As was shown previously, the distribution of IP_2 is not Gaussian when uncorrelated mismatches dominate, and we will apply the more appropriate χ^2 distribution. Since we often measure RF signals as power levels (i.e., in dBm), without phase information, we base the presented method on dB units throughout.

First, we need to define how the lower bound, or the worst case, of an IP_2 data set x is calculated. In the case when the variable is Gaussian, the lower bound corresponding to a yield loss of y is

$$lb_{Gauss}(x) = \bar{x} - \sigma(x)\Phi_{Gauss}^{-1}(1-y). \tag{2.37}$$

That is, we solve $P(X < lb_{Gauss}(x)) = y$ for its lower bound based on a certain yield loss. For example, $\Phi_{Gauss}^{-1}(1-y) \approx 4$ for $y = 31.67$ ppm, which corresponds to the 4σ example used in the algorithm outline.

The χ^2 distribution has a different lower bound than the Gaussian and a nonzero mean value, and we must subtract this mean (i.e., bias) from the sample mean to remove any bias in the bound estimate. The mean of a χ_1^2 distribution is 1.0, but as we measure IP_2 in dB units, the bias will be $\widehat{IP_2} - E\{10\lg(\chi^2)\}$. In (2.29), this bias was shown to be $\widehat{IP_2} + 5.52$ dB for scalar IP_2 data (i.e., χ_1^2). The IP_{2s} lower bound for a scalar data set x can now be estimated as

$$lb_{\chi_1^2}(x) = \bar{x} - 5.52 - 10\lg\left(\Phi_{\chi_1^2}^{-1}(1-y)\right) \qquad (2.38)$$

and for the corresponding vector data set as

$$lb_{\chi_2^2}(x) = \bar{x} + 0.50 - 10\lg\left(\Phi_{\chi_2^2}^{-1}(1-y)\right). \qquad (2.39)$$

The Φ^{-1} terms in these equations represent the *test margin* needed to guarantee a certain maximum yield loss. In Table 2.1, the required margins for some yield loss targets are listed.

2.4.4 EXAMPLE USING SIMULATED IP_2 DATA

To test the proposed lower bound estimate method, we can apply it to a large simulated data set of a known distribution. To this end, we have generated a set of 10^7 samples of IP_{2s}-like data based on (2.25) and the chi2rnd function of the octave program[8].

By randomly picking samples from the full data set, 10 random subsets of 30 samples each are created. These subsets correspond to measurement results, for example, typical characterization lots, whereas the full data set describes the *actual* distribution.

TABLE 2.1

Required Scalar and Vector IP_2 Back-Off in dB from $\widehat{IP_2}$ (First Two Rows) and $\overline{IP_2}$ (Last Two Rows) vs. Yield Target

Loss (y)	1 ppm	10 ppm	100 ppm	0.1%	1%
$10\lg\left(\Phi_{\chi_1^2}^{-1}(1-y)\right)$	13.8	12.9	11.8	10.3	8.2
$10\lg\left(\Phi_{\chi_2^2}^{-1}(1-y)\right)$	14.4	13.6	12.6	11.4	9.6
$\bar{x} - lb_{\chi_1^2}(x)$	19.3	18.4	17.3	15.9	13.7
$\bar{x} - lb_{\chi_2^2}(x)$	13.9	13.1	13.2	10.9	9.1

TABLE 2.2

Results Using Simulated *IP*$_{2s}$ Data Assuming a χ_1^2 Distribution

IP$_{2s}$ Data		Mean	Std	n	Min Sample	Lower Bound χ^2	Lower Bound Normal
All		55.5	9.65	10^7	35.7	37.6	16.9
Subset	1	56.3	10.8	30	44.5	34.4–41.9	−2.75 to 23.0
	2	56.7	9.50	30	44.8	35.5–42.1	11.1 to 24.5
	3	55.6	10.1	30	43.4	34.0–41.2	4.68 to 22.3
	4	55.0	12.2	30	41.8	32.4–40.8	−19.4 to 29.0
	5	54.9	8.94	30	38.4	33.8–40.2	9.95 to 26.6
	6	53.9	11.0	30	42.0	31.9 39.4	6.35 to 24.7
	7	55.7	10.4	30	45.7	33.9–41.1	−1.38 to 23.3
	8	58.2	12.6	30	42.2	35.7–44.4	−7.86 to 21.1
	9	57.4	11.2	30	43.7	35.5–43.4	4.22 to 18.2
	10	54.1	8.71	30	42.6	33.1–39.1	9.99 to 26.9

With the data in each data subset, we then estimate a lower bound testing both χ^2 and Gaussian distributions for the lower band calculation. These bound estimates can then be compared with the actual outcome of the complete data set; see Table 2.2. The yield loss is defined to correspond to 4σ of a Gaussian, that is, 31.67 ppm loss or 316 failing samples in the full data set.

Figures 2.12a and b show the Gaussian and χ^2 fits to our full data set, respectively. The χ^2 fit is quite close when all the samples are used. If we look at just one subset, like in a real measurement situation with a limited number of samples, we may get something like Figure 2.13. Each subset will look differently, but in this case, we still see that the χ^2 fit predicts the histogram peak as well as the samples with very high *IP*$_{2s}$.

The first line in Table 2.2 shows the statistics for the whole data set. In particular, we note that out of the 10^7 samples the lowest *IP*$_{2s}$ of the 316 *failing samples* is 35.7 dBm and the highest failing *IP*$_{2s}$ is 37.6 dBm. The typical Gaussian assumption (see (2.38)) would predict a lower bound around 16.9 dBm, which clearly is wrong as this number is much lower than the known limit.

Looking at the subset data in Table 2.2, it is also evident that the χ^2-based method accurately puts the bound limits around the actual limit for each subset (i.e., 37.6 dBm). Limits vary significantly for the Gaussian approximation. The upper limit of the lower bound range is too low, typically by some 10 dB, and the worst case of the lower limit is 57 dB off. Using vector data yields similar results although the Gaussian approximation is not as bad in this case. This is because the χ_2^2 distribution is closer to a Gaussian than what the χ_1^2 is (cf. the central limit theorem).

The estimated bound interval in Table 2.2 is given with 90% confidence so that on average one subset in 10 will fail to enclose the actual limit. In those cases, the miss

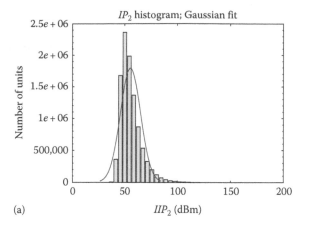

(a)

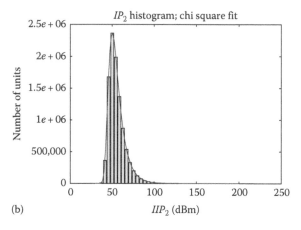

(b)

FIGURE 2.12   Histogram for 10^7 simulated IP_{2s} samples using (a) Gaussian and (b) χ_i^2 fitting models.

will only be by a fraction of the bound interval width. Results will also vary slightly when rerun due to the random nature of the simulated data.

   To conclude, the lower bound will typically be conservative and will not overestimate the actual limit by more than a fraction of the bound range.

2.4.5   MEASURED TEST CIRCUIT IP_2 RESULTS

We have collected data from two different fabricated test chips: two batches based on a BiCMOS architecture[1] and one batch based on a CMOS architecture.[4]

2.4.5.1   BiCMOS Test Circuits

Figure 2.14 shows the BiCMOS results (P1 and R1 samples). For each batch, 6 samples have been measured across extreme conditions (ETC; supply, temperature, and

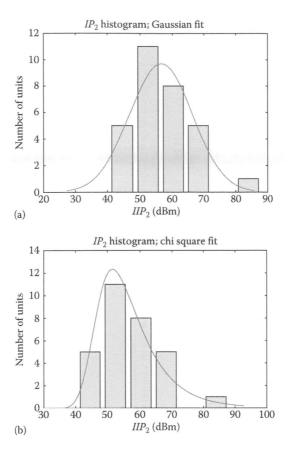

FIGURE 2.13 Exemplary subset (30 samples out of 10^7) using (a) Gaussian and (b) χ_i^2 fitting models.

channel) and 30 samples at nominal conditions (NYC) for 3GPP band 1–6, 8, and 9. Figure 2.15 shows the data for the NTC sample subset. In most cases, the superimposed χ_2^2 fit is quite good.

Applying our variability model and worst-case estimation procedure to the BiCMOS data, we get the results shown in Table 2.3. The worst case is based on a 134 ppm loss limit (corresponding to 4σ for a Gaussian distribution). A 1 ppm loss limit lowers the worst case by another 1.9 dB. The table lists the IP_{2v} specification limits, sample mean, σ, min and max values, and worst-case estimations for all samples as well as for NTC samples only. In no case is the worst-case estimate exceeding any measured worst-case values. At the same time, the worst-case estimate is not far below the observed sample minimum IP_{2v} values.

2.4.5.2 CMOS Test Circuit

The worst-case analysis of the high-band diversity receiver of 15 samples of our CMOS test circuit (M1) is summarized in Table 2.4. The objective of the

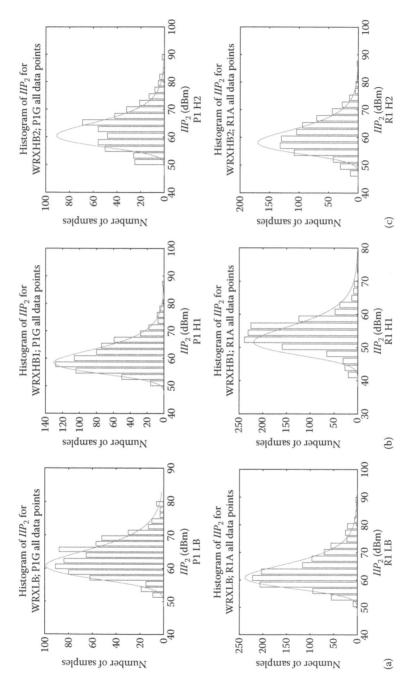

FIGURE 2.14 Plot of vector IP_2 distribution of P1 and R1 BiCMOS test circuits with χ_2^2 fit superimposed: (a) low-band input, (b) high-band 1, and (c) high-band 2, measured under extreme conditions.

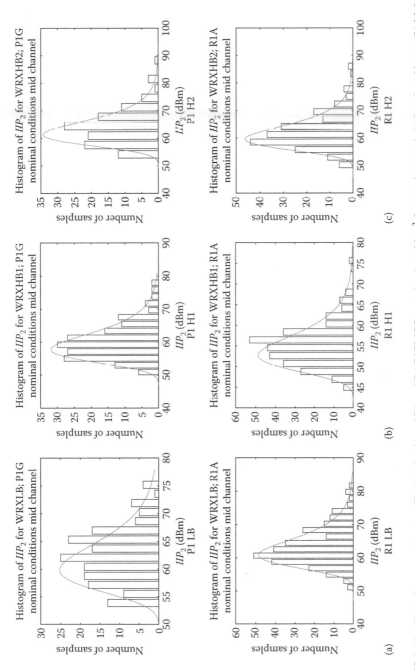

FIGURE 2.15 Plot of measured vector *IP₂* distribution of P1 and R1 BiCMOS test circuits with χ_2^2 fit superimposed: (a) low-band input, (b) high-band 1, and (c) high-band 2, measured under nominal supply voltage and at room temperature.

TABLE 2.3

Worst-Case Estimation of Vector IP_2 from BiCMOS Test Circuit Samples P1 and R1

	R1			P1		
	LB	HB1	HB2	LB	HB1	HB2
Spec	47	38	43	47	38	43
Mean	63.27	54.13	60.53	63.16	60.81	62.82
Std	6.26	4.77	6.39	5.21	5.74	7.07
Min	50.13	40.70	45.87	51.10	49.85	49.77
Max	90.14	76.59	87.95	80.18	88.34	90.11
WC all	48.25	39.11	45.51	48.14	45.80	47.80
WC NTC only	48.23	40.13	46.96	47.30	45.12	48.64

TABLE 2.4

Measured Scalar IP_2 Test Results for CMOS Test Circuit M1 vs. l- and m-Bias Settings

Bias		Data Range					WC Bounds	
l	*m*	Min	Max	Mean	Std	WC	Min	Max
24	8	48.20	81.90	59.69	8.50	42.54	39.48	45.45
25	8	48.00	82.00	58.79	8.55	41.64	38.44	44.46
26	8	48.00	66.10	56.69	5.10	39.53	37.76	41.33
27	8	48.40	63.70	56.07	4.07	38.92	37.32	40.52
28	8	49.10	63.30	55.22	3.26	38.06	36.77	39.37
29	8	50.20	65.70	54.72	3.24	37.56	36.19	38.76
24	9	50.10	82.70	60.95	9.05	43.80	40.09	47.12
25	9	51.20	79.60	61.84	8.10	44.69	41.47	47.74
26	9	50.60	77.20	62.28	7.21	45.12	42.21	47.84
27	9	48.60	76.20	62.84	7.95	45.69	42.83	48.44
28	9	49.10	76.20	62.69	7.83	45.53	42.79	48.24
29	9	49.60	82.20	62.61	8.62	45.45	42.35	48.40

Note: IP_2 values in dBm and worst-case (WC) bounds calculated assuming 134 ppm yield loss using the χ_1^2 distribution.

measurements was to find what mixer LO bias settings resulted in the best IP_2 and what performance could be achieved at this setting. From the table, it is clear that the parameter setting combination m-bias = 9 and l-bias = 27 gives the best worst-case estimate as well as the highest lower bound. The l-bias variable, however, does not appear to be as critical as the m-bias parameter (Figures 2.16 and 2.17).

Comparing Tables 2.2 and 2.4, it is evident that the M1 samples show a smaller standard deviation than do the simulated data. This is probably related to systematic

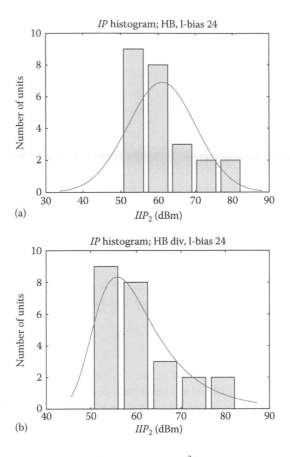

FIGURE 2.16 Histogram using (a) Gaussian and (b) χ_1^2 fitting models for measured IP_{2s} data from CMOS test circuit M1 samples at bias settings m-bias = 9 and l-bias = 24.

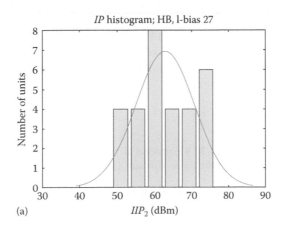

FIGURE 2.17 Histogram using (a) Gaussian. *(Continued)*

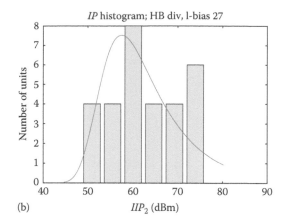

(b) IIP_2 (dBm)

FIGURE 2.17 (CONTINUED) Histogram using (b) χ_1^2 fitting models for measured IP_{2s} data from CMOS test circuit M1 samples at bias settings m-bias = 9 and l-bias = 27.

effects in the IM_2 generation, for example, LO leakage or layout-induced offsets. From the table, we may conclude that M1 has a worst-case IP_2 in the range 42.8–48.4 dBm, assuming a 134 ppm yield loss.

2.5 SUMMARY

We have shown that both scalar and vector IP_2 values should be characterized by their mean values when measured as logarithmic powers (e.g., in dBm). In particular, any measured standard deviations will not depend on circuit imbalance, when the $\overline{IM_2} \approx 0$, but only on the χ^2 distribution properties.

Based on $\overline{IP_2}$, we have shown what test margin, or back-off, is required to achieve a certain minimum yield. We have also developed a lower-bound estimation method based on statistical bootstrap.

Finally, the model and methods have been validated with both simulated data sets and with measured results from two different test circuits in BiCMOS and CMOS technologies, respectively.

We would like to acknowledge Bengt Lindoff for his valuable advice on statistical modeling and for suggesting the use of bootstrap for estimating bound ranges. Numerous Ericsson colleagues in Lund, Sweden, have also contributed with discussions and with measurement results.

REFERENCES

1. O. Gaborieau, S. Mattisson, N. Klemmer et al. A SAW-less multiband WEDGE receiver. In *IEEE International Solid-State Circuits Conference Digest of Technical Papers*, pp. 114–115, 2009.
2. A. Jakobschuk Feedback and nonlinear distortion. In *Proceedings of the IEEE*, p. 1370, October 1981.
3. J.R. Macdonald. Nonlinear distortion reduction by complementary distortion. In *IRE Transactions on Audio*, pp. 128–133, September–October 1959.

4. M. Nilsson, S. Mattisson, N. Klemmer et al. A 9-band WCDMA/EDGE transceiver supporting HSPA evolution. In *IEEE International Solid-State Circuits Conference Digest of Technical Papers*, pp. 366–368, 2011.
5. W. Sansen. Distortion in elementary transistor circuits. *IEEE Transactions on Circuits and Systems II: Analog and Digital Signal Processing* 46:315–325, March 1999.
6. B. Efron and R.J. Tibshirani. *An Introduction to the Bootstrap*. Boca Raton, FL: CRC Press, 1994.
7. M. Abramowitz and I.A. Stegun (eds.). *Handbook of Mathematical Functions with Formulas, Graphs, and Mathematical Tables*, 9th edn., Dover, NY, pp. 940–943, 1972.
8. J.S. Hansen. (June 2011). *GNU Octave. Beginner's Guide*. Packt Publishing. ISBN 978-1-849-51332-6.

3 Mitigation of Performance Degradation in Radio Receiver Front Ends from Out-of-Band Interference*

Ranjit Gharpurey

CONTENTS

---

* This chapter includes portions reprinted with permission from the following publication: R. Gharpurey, Linearity enhancement techniques in radio receiver front-ends, *IEEE Transactions on Circuits and Systems Part 1, Regular Papers*, 59(8), 1667–1679, August 2012, © 2012 IEEE.

3.1 INTRODUCTION

Modern radio systems typically utilize predefined frequency channels for communications. The channelization allows for isolation of signals in the frequency domain. In practice, nonlinearity and time-varying circuit behavior in radio receiver front ends can lead to the appearance of spurious energy within the desired signal channel from out-of-band signals,* which can degrade the received signal. The desired signal may itself be modified by signals located in other parts of the spectrum.

In this chapter, we identify key mechanisms in radio receiver front ends that are responsible for the aforementioned degradation and discuss several circuit and architectural approaches to mitigate these phenomena. To facilitate this discussion, we use the front end of a direct downconversion receiver shown in Figure 3.1. The receiver is assumed to use a front-end band-select filter that allows all desired frequency channels employed by a communication system to pass through while attenuating undesired frequency bands. The front-end filter is assumed to be passive, for example, a surface acoustic wave (SAW) filter, with an ideally linear response. Thus, the filter itself cannot cause any signal degradation due to nonlinear or time-varying behavior.

The first active circuit seen by the input in the receiver of Figure 3.1 is a front-end low-noise amplifier (LNA). The primary role of the LNA is to amplify the input signal while adding minimal noise of its own. The amplified signal level at the output of the LNA reduces the impact of additional noise from circuits that follow the LNA in the receiver. Nonlinearity in the LNA devices can cause degradation of the signal quality at the output of the LNA in the presence of energy sources outside of the signal channel. These out-of-band sources thus act as interferers.

The mixers in the signal chain are assumed to downconvert the input signal to baseband, with quadrature outputs. The mixer circuits apply a periodically time-varying gain to the input, at the frequency of an externally applied local oscillator (LO) signal, which translates the input signal from the signal frequency to baseband. Nonlinearity in mixers can lead to signal quality degradation, similar to LNAs. In addition, another concern in these circuits is the potential for downconversion of spectral content in the vicinity of the harmonics of the LO signal.

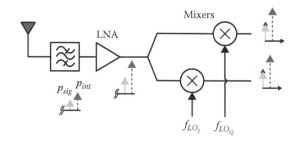

FIGURE 3.1   Direct downconversion receiver front end.

---

* The term *out of band* implies that the energy source is outside the spectrum of the desired signal. This term is also used to refer to signals in other channels that lie within the frequency band specified by a communication standard. Similarly, the term *in-band* is used to signify within the spectrum of the desired signal.

Each radio system presents a specific set of design requirements arising from nonlinearity and harmonic response. Nonlinear response is typically an issue at larger signal and interferer levels. Interferers in the presence of nonlinearity can degrade performance through several mechanisms such as gain compression, intermodulation (IM), cross modulation, AM-to-PM distortion, and desensitization [1]. Depending on specific features of the standard, such as the out-of-band spectrum profile and operating environment, different phenomena can be of significance in the design of the front-end circuits. For example, in the 3G WCDMA system, where the transmitter and receiver operate simultaneously in different frequency bands, several key linearity metrics are impacted by the transmitter leakage into the receiver front end [2–4]. The transmit-to-receive frequency offset can be in the range from tens to hundreds of MHz and varies for different bands. Band I as defined in Ref. [5] employs a transmit-to-receive offset of 190 MHz. For this case, the out-of-band third-order input-referred intercept point (IIP3) due to IM products arising from the transmitter leakage and out-of-band blockers at half and twice the transmit-to-receive frequency separation, namely, at offsets of 95 and 380 MHz, respectively, from the desired signal, is a critical specification. Depending on the allocations for various front-end impairments in specific implementations, the out-of-band IIP3 requirement can be of the order of −3 to 0 dBm. Similarly, the amplitude-varying nature of the transmitter leakage leads to a second-order input-referred intercept point (IIP2) requirement of the order of 45 dBm or more. An in-band IIP3 requirement also arises from interferers at 10 and 20 MHz offsets. Detailed blocker and two-tone interferer information for this standard can be found in Ref. [5].

Several of the degradation sources can also appear simultaneously, for example, in the presence of the aforementioned transmit-to-receive leakage, and a blocker tone at 95 MHz, both second- and third-order nonlinearities of the receiver path contribute to in-band spurious energy, in addition to other phenomena such as reciprocal mixing, which is not discussed here. The amount of degradation that arises from each source has to be apportioned, e.g., [2], which determines minimum requirements for metrics such as IIP2, IIP3, and phase-noise profile of the LO. This distribution of nonideal behavior is a function of several factors, including process technology, supply voltage, power constraints, performance requirements, and the ability to use external components, such as high-performance band-pass filters.

The potential for appearance of spurious in-band energy becomes even more severe in broadband and multimode/multiband radios (e.g., [6]). In multimode systems, the front end has to be designed to satisfy the requirements of multiple standards. In a straightforward implementation, individual front ends can be designed for each standard. This can lead to a large hardware complexity, for instance, due to the requirement for custom external band-pass filters and duplexers for the multiple front ends. A more desirable solution, for reasons of cost and size, would be one that can accommodate requirements of all standards within one signal path. The dynamic range requirements of such a design however can be very challenging, since the front end would need to be designed for the worst-case linearity specification and the most stringent sensitivity specification, among all the standards being considered. Since the front end would necessarily have a wider bandwidth, it would also need to tolerate greater levels of interference. In broadband systems, such as TV tuners, nonlinearity as well as harmonic response of the receiver can cause spurious energy from undesirable channels to appear within the desired channel, e.g., [7].

The choice of process technology is another major factor that can introduce significant design considerations. For instance, in transceivers where the digital section forms a large part of the overall power budget, the use of deep submicron technologies is highly beneficial. The use of these technologies could also be mandated by cost and integration considerations. The supply voltage in such technologies is typically limited, for example, in the range of 1–1.5 V. In integrated system-on-chip-type solutions, this imposes an additional design challenge in satisfying linearity specifications. On the other hand, the use of deep-submicron technologies allows for significant calibration capability, which can help mitigate several nonideal mechanisms, such as mixer harmonic responses and even-order nonlinearity.

Circuit and architectural techniques to mitigate front-end nonlinearity and mixer harmonic response are described below. These include (1) linearization schemes, which seek to correct or compensate nonlinearity within the active devices; (2) active filtering architectures, which seek to mitigate interferers, before these can enter the active signal path; and (3) harmonic rejection techniques in mixers, which minimize the spurious mixer gain from around harmonics of the LO employed. The discussion on front-end linearization and filtering architectures follows the presentation of Ref. [8].

3.2   DEVICE- AND CIRCUIT-LEVEL LINEARIZATION

Front-end nonlinearity is closely tied to the nonlinearity of devices. In MOSFETs, ignoring reactive effects, the drain current i_{ds} shows dependence on the gate-to-source (v_{gs}) and drain-to-source (v_{ds}) voltages [9]. For nonlinearity analysis, the dependence can be expressed as a Taylor series expansion in v_{gs} and v_{ds} [10,11]:

$$i_{ds} = \sum_k g_{m_k} v_{gs}^k + \sum_k g_{ds_k} v_{ds}^k + \sum_p \sum_q g_{mds_{pq}} v_{gs}^p v_{ds}^q \qquad (3.1)$$

where for integers k, p, and q, $g_{m_k} = (1/k!) d^k I_{DS}/dV_{GS}^k$, $g_{ds_k} = (1/k!) d^k I_{DS}/dV_{DS}^k$, and $g_{mds_{pq}} = (1/p!)(1/q!)\left\{ \partial^{p+q} I_{DS}/\partial V_{GS}^p \partial V_{DS}^q \right\}$.

If the impact of v_{ds} can be ignored, for example, in implementations with low-impedance drain and source terminations, the small-signal nonlinearity of the device can be expressed using a Taylor series expansion in v_{gs}. The BJT in the forward active region is modeled by an exponential dependence of the collector current i_c on the base-to-emitter voltage v_{be}, for example, [12]. The collector current can similarly be expressed as a power series in v_{be}, and the Early voltage can be used to model the dependence of i_C on the collector-to-emitter voltage v_{ce}. Reactive nonlinearity, such as that arising from nonlinear device capacitors, can have a significant impact on performance at high frequencies. An accurate analytical model of nonlinearity must include such effects and requires the use of a Volterra series representation [13].

Improving linearity implies that harmonic and IM terms need to be decreased in magnitude. For a given nonlinearity of the form $v_o = a_1 v_{in} + a_2 v_{in}^2 + a_3 v_{in}^3 + \cdots$, this can be achieved by making the ratio a_k/a_1 small or else by canceling the distortion component (e.g., an IM term) at the device output and by subtracting the contributions of two or more nonlinear terms to the same distortion component. Techniques based on these approaches are described in the following sections.

3.2.1 FEEDBACK

Consider a small-signal amplifier with a linear gain a_1 and nonlinear coefficients a_k, which is placed in a negative feedback loop (Figure 3.2), with a linear feedback factor of f. If the closed-loop system is modeled by $v_o = \sum b_k v_{in}^k$, with $k \geq 1$, then the first three coefficients of the closed-loop amplifier can be shown to be given by

$$b_1 = \frac{a_1}{1 + a_1 f}$$

$$b_2 = \frac{a_2}{(1 + a_1 f)^3}$$

$$b_3 = \frac{a_3(1 + a_1 f) - 2a_2^2 f}{(1 + a_1 f)^5}$$

If the input level is held fixed, significant improvement in linearity can be achieved as the loop gain $a_1 f$ is increased. For example, the second-order nonlinear coefficient decreases by a factor of $(1 + a_1 f)^3$ in an absolute sense and by $(1 + a_1 f)^2$ relative to the linear term. Negative feedback also improves linearity for a fixed output level, although the degree of improvement is not as large, for example, the second-order term at the output scales by $1/(1 + a_1 f)$ relative to the linear term. At sufficiently high frequencies, the linear and nonlinear coefficients can show significant variation in magnitude and phase as a function of frequency. Reactive nonlinearity and frequency dependence of linear elements embedded within broadband nonlinear circuits can both lead to such variation. Analysis of such cases requires the use of a Volterra series-based analysis [13,14].

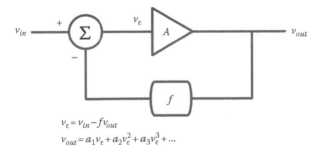

$$v_\varepsilon = v_{in} - f v_{out}$$
$$v_{out} = a_1 v_\varepsilon + a_2 v_\varepsilon^2 + a_3 v_\varepsilon^3 + \dots$$

FIGURE 3.2   Linearity enhancement through the use of feedback.

The expression for b_3 mentioned earlier can be observed to include a contribution from a_2, the second-order nonlinearity of the open-loop amplifier. Consider a sinusoidal signal v_{in} at frequency ω, which is applied to the amplifier with feedback (Figure 3.2). In addition to the signal frequency ω, the input of the amplifier stage (v_e) consists also of a term at 2ω and other harmonics, which are coupled from the output to the input through the feedback path. The second-order and linear terms at frequencies 2ω and ω, respectively, interact through the second-order nonlinear coefficient of the amplifier (a_2) and contribute a term to the harmonic distortion at 3ω. This term is proportional to a_2 and combines with the third-order term arising from a_3, which can lead to an increase or decrease of the overall distortion. Interaction terms [14] can play a similar role in determining the net IM distortion as well. Cancellation of distortion due to interaction can be useful in RF circuits, since linearity improvement can be achieved with potentially small impact on linear gain.

Resistive and reactive feedback have been extensively employed in front-end amplifiers for linearization. An LNA topology employing reactive feedback, which finds widespread use in radio front ends, is an inductively degenerated amplifier with a common-emitter input [15] (e.g., Figure 3.3a)* or a common-source input [1]. This design allows for simultaneous noise and power matching, in addition to enhancing linearity. The degeneration inductor does not incur a dc voltage drop. As an added practical benefit, the inductance of package bond wires can be easily absorbed into the design.

A Volterra series-based analysis of this stage is presented in Ref. [15]. The nonlinear dependence of the collector current i_c on the effective input voltage v_{in} is expressed in terms of Volterra coefficients as $i_c = A_1(j\omega_a) \circ v_{in} + A_2(j\omega_a, j\omega_b) \circ v_{in}^2 + A_3(j\omega_a, j\omega_b, j\omega_c) \circ v_{in}^3 + \cdots$. The Volterra coefficients are evaluated as functions of circuit and device parameters. The model takes into consideration the input impedance connected to the base. The coefficients are expressed

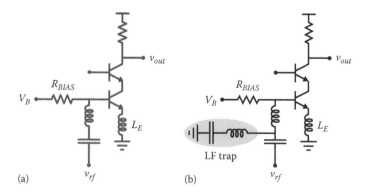

(a) v_{rf} (b) v_{rf}

FIGURE 3.3   (a) Inductively degenerated LNA and (b) one with a low-frequency trap.

---

* The voltage v_{be} across the active device in Figure 3.3a is given by the difference between the base voltage and the voltage across the degeneration element, which itself depends on the output signal, namely, the collector current. As such, inductive degeneration can be seen to provide feedback.

as functions of the frequency variables ω_a, ω_b, etc., which are assigned values corresponding to the input signal frequencies, in order to determine specific IM terms.*

As shown in Ref. [15], one way to enhance IIP3 in a two-tone test with inputs at ω_1 and ω_2 is to ensure that $A_1(j\Delta\omega)/g_m \approx 1$, where $\Delta\omega = \omega_1 - \omega_2$, g_m is the device transconductance, and the Volterra coefficients take into account the emitter path impedance and the effective input impedance placed in series with v_{in}, as mentioned earlier. The use of an inductor for degeneration helps in achieving this. Additional improvement provided by this type of degeneration is also discussed.

Other forms of reactive feedback can also be used for linearization; for example, [16] describes the use of transformers for this purpose.

3.2.2 Out-of-Band Impedance Terminations

A distortion component at a specific frequency can appear at the output through multiple intermediate nonlinear interactions. Some of these nonlinear terms can generate frequencies that are significantly outside the frequency band of interest but ultimately contribute IM components at the desired frequency. The final distortion component can be modified and significantly mitigated by presenting the circuit with specific impedance terminations at these out-of-band frequencies, for example, [17]. A key advantage of out-of-band terminations is that the loading due to these does not impact in-band gain or noise performance.

An example of such a termination is a low-frequency series-resonant LC trap in an inductively degenerated CE stage for improving IIP3 [18,19] (Figure 3.3b). As noted in Ref. [15], the net third-order IM term (IM3) at the output appears from the inherent third-order nonlinearity of the device, as well as a second-order interaction term. By ensuring a low bias impedance at low frequencies, of the order of the difference between the two tones used in the IIP3 test, the IM3 generated by the interaction term can be used to significantly reduce the overall IM3. A small bias resistor, R_{BIAS}, can also be used to reduce the impedance at low frequencies; however, this is not a desirable design choice, since this will also reduce the input impedance seen at RF. A small R_{BIAS} will thus attenuate the input signal, thereby degrading the noise figure. On the other hand, a low-frequency series LC trap, as shown in Figure 3.3b, with a resonance near the difference frequency of the two tones will selectively lower the impedance at this frequency alone while appearing as a high impedance at RF. This avoids loading the signal source and allows for enhancing IIP3 without degrading noise figure. A limitation of this approach is that it is effective for improving IIP3 and not P1dB [18]. For improving P1dB, the trap must present a low impedance at dc. Other techniques for reducing impedance at low frequencies, such as the use of active inductors, are also described in Ref. [19].

Due to the nature of the interaction mechanism with inductive degeneration, it is important that the third-order nonlinear coefficient of the device be positive [19]. This condition is satisfied in BJTs, due to the exponential dependence of the collector

---

* In a two-tone test with physical frequencies ω_1 and ω_2, the third-order intermodulation term at $2\omega_1 - \omega_2$ would be evaluated as $A_3(j\omega_1, j\omega_1, -j\omega_2)$.

current on the base-to-emitter voltage. However, this is not the case for FETs in strong inversion, which exhibit a negative third-order nonlinear coefficient.

The network of Figure 3.3b is also useful for reducing desensitization caused by upconversion of low-frequency noise at the input of the amplifier in the presence of a large close-in interferer, due to second-order nonlinearity of the amplifier [20,21]. A low-frequency trap similar to that employed in Figure 3.3b can be used to attenuate low-frequency noise at the input of the amplifier, thereby reducing the noise that can appear in-band after upconversion.

3.2.3 INJECTION OF IM OR HARMONIC SIGNALS

This approach also exploits interaction terms for enhancing linearity. IM or harmonic signals with frequencies that are distant from the desired frequency of operation are synthesized and applied to the circuit for example, [22,23].

Consider an amplifier with linear coefficient a_1 and second- and third-order nonlinear coefficients a_2 and a_3, respectively. If a two-tone input $\alpha_1\cos(\omega_1 t) + \alpha_2\cos(\omega_2 t)$ is applied to the amplifier, third-order nonlinearity will lead to the appearance of IM3 components of amplitude $(3/4)a_3\alpha_1^2\alpha_2$ at $2\omega_1 - \omega_2$ and $(3/4)a_3\alpha_1\alpha_2^2$ at $2\omega_2 - \omega_1$. If the input is modified to include second harmonics $2\omega_1$ and $2\omega_2$, IM3 terms at frequencies $2\omega_1 - \omega_2$ and $2\omega_2 - \omega_1$ will also result from second-order nonlinearity and will be proportional to a_2. The net IM3 at the output will thus consist of contributions from both second- and third-order nonlinearity. Proper adjustment of the amplitude and phase of the externally applied harmonic terms can be utilized to cancel the net IM3 term. A signal at the difference frequency $\omega_1 - \omega_2$ can also be utilized for this purpose [22].

IM injection has been used in Ref. [23] for linearizing a differential-pair-based LNA and mixer. The tail current is modified to include a component at $\omega_1 - \omega_2$ (Figure 3.4), through the use of a squaring circuit. Interaction of this term with the linear input due to the second-order nonlinearity within the devices of the differential pair is used to reduce the distortion arising from the third-order nonlinearity of the devices. The approach is shown to enhance the IIP3 of an RFE by 10.6 dB at 900 MHz. Since the injection is not in the signal path, there is no degradation of noise figure of the front end.

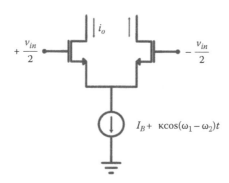

FIGURE 3.4   Second-order IM injection in a differential pair.

The above techniques that inject harmonic or out-of-band signals for improving linearity are similar to out-of-band terminations in that the linearity improvement is achieved through modification of the circuit operation at a frequency band that can be significantly different from the band of interest. In this case, however, customized active circuits are required for the generation of the out-of-band signals.

3.2.4 FEEDFORWARD

The previous approaches for enhancing linearity utilize the interaction between different nonlinear terms. Feedforward techniques utilize a separate auxiliary circuit path to re-create the nonlinearity of the main path. The nonlinear component can then be eliminated through a linear combination of the outputs of the main and auxiliary paths (Figure 3.5). The approach can be implemented at the circuit or device level. A critical challenge in feedforward nonlinearity cancellation is that the gain and phase of the two (or more) paths need to be matched over process and temperature variations, and thus a calibration mechanism is usually essential. The noise of the auxiliary path needs to be minimized at design, since the auxiliary path may not contribute to signal gain, but can generate in-band noise. Examples of specific implementations of the feedforward principle can be found in Refs. [1,24,25].

3.2.5 DERIVATIVE SUPERPOSITION

From Equation 3.1, ignoring the dependence of i_{ds} on v_{ds}, we have $i_{ds} = g_{m_1}v_{gs} + g_{m_2}v_{gs}^2 + g_{m_3}v_{gs}^3 + \cdots$. The dependence of the linear and nonlinear coefficients in this power series on the gate-to-source bias voltage V_{GS}, for an NMOS device, is shown in Figure 3.6. For weak inversion, the third-order coefficient g_{m_3} is positive, while for strong inversion operation, it is seen to be negative. In the transition region from weak to strong inversion, at a specific value of V_{GS}, the third-order term is observed to go to zero [11,26]. By biasing the device at precisely this value of V_{GS}, the third-order distortion of the device can ideally be set to zero. A practical limitation of this approach for IIP3 improvement is that any variation in the threshold voltage of the device or in V_{GS} can significantly degrade performance, compared to this optimal setting.

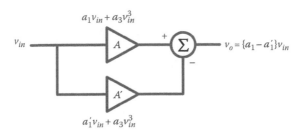

FIGURE 3.5 Feedforward for nonlinearity cancellation.

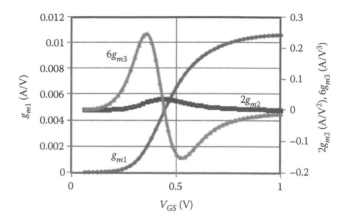

FIGURE 3.6 Simulated NMOS linear and nonlinear coefficients.

Derivative superposition*-based [27–30] approaches employ multiple devices in parallel, which are biased at different values of V_{GS} such that the net g_m (termed $g_{m_3}T$) is nearly zero over a wider range of V_{GS}. The concept is shown in Figure 3.7 using two devices and a linearized g_{m_3} dependence on V_{GS}. This approach makes it possible to achieve a more robust reduction of the third-order nonlinearity of the device and reduces sensitivity of the linearity enhancement on process variation. The approach can be interpreted as feedforward implemented at the device level.

Even with a zero third-order nonlinear coefficient, in practice, reactive feedback, for example, through source degeneration inductance or due to C_{gd}, can limit the achievable improvement at high frequencies due to interaction. For example, the interaction-based contribution of second-order nonlinearity to the third-order distortion can degrade performance. Approaches to counter this degradation can be found in Refs. [29,30], which demonstrate LNAs in 0.35 μm and 0.25 μm, respectively, with corresponding IIP3s of 15.6 and 22 dBm.

Equation 3.1 involves derivatives of nonlinear terms in v_{gs}, v_{ds} and cross terms involving v_{gs} and v_{ds}. Cancellation of the self- and cross terms has also been demonstrated; for example, in Ref. [31], the second-order distortion arising from g_{mds11} is used to cancel that from g_{m_2} in order to enhance IIP2 of a broadband LNA.

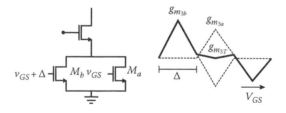

FIGURE 3.7 Derivative superposition with two common-source NMOS devices.

---

* This approach has also been referred to as the multiple gated transistor (MGTR) method in the literature.

The nonlinear coefficients of an MOS device in strong inversion can also be decreased by increasing the overdrive voltage of the device. At sufficiently high overdrive, the transconductance is essentially independent of the device current. This mode of operation can be utilized for enhancing linearity [32]. A consequence of operating the device in this mode, however, is that the g_m/I_D ratio is typically small,* which implies that in order to achieve a given small-signal gain, the current requirement can be relatively high.

In describing the previous techniques for linearization, we have followed the classification utilized in literature. It is useful, however, to recognize that the majority of techniques described earlier fall in one of two classes. The first class utilizes interaction-based IM3 improvement, where the nonlinearity from a second-order term is used to cancel the distortion from the third-order nonlinearity of the device. This approach is used in techniques including feedback, out-of-band terminations, and injection of out-of-band IM and harmonic distortion components. The second broad class of techniques involves feedforward, wherein the nonlinear component itself is replicated in a parallel path and then canceled. This technique was employed in circuit-level feedforward-based designs and derivative superposition-based techniques. It should also be noted that these techniques can provide significantly more improvement in linearity than that available through feedback due to the progressively greater attenuation of nonlinear coefficients with the order of the nonlinearity. This is because the inherent loop gain available at RF is usually fairly limited. On the other hand, since these techniques rely on cancellation of non-linearity through subtraction of two terms, these approaches can exhibit sensitivity to device variations as well as device mismatch.

3.2.6 LINEARITY ENHANCEMENT IN MIXERS

Linearity can be a critical consideration in mixer designs, since the interferer levels at mixer inputs are enhanced by the gain of the front-end amplifier. Even if architectural approaches for improving front-end amplifier selectivity (Section 3.3) are employed, both third-order and second-order nonlinearity metrics can be challenging in practical systems. Minimization of second-order nonlinearity is of particular importance in direct downconversion radios. An AM interferer given by $m(t)\cos(\omega_{int}t)$ after being applied to a second-order nonlinearity $a_2 v_{in}^2$ generates an output of the form $a_2 m(t)^2$ at baseband, which has twice the bandwidth of the original interferer $m(t)$. The interferer can thus degrade the signal quality at baseband. This is not an issue in front-end LNAs that are used to amplify signals with carrier frequencies that are significantly larger than the bandwidth of typical interferers, since baseband components can be filtered using ac coupling between the LNA and the mixer. A brief overview of linearity considerations in active and passive mixers is provided in the following sections.

3.2.6.1 Active Current-Commutating Mixers

A single-balanced and a double-balanced NMOS active current-commutating mixer are shown in Figure 3.8. In each case, the mixer utilizes an input transconductor to convert an RF signal into a small-signal current at RF, which is then frequency-translated

---

* Simulation of an NMOS device in a 0.13 μm process indicated a value in the range of 3–5.

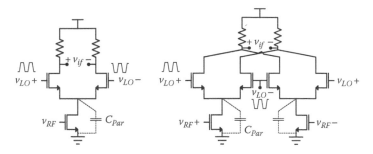

FIGURE 3.8 Active current-commutating mixers.

through commutation in switches driven by a LO. The switching devices in this case are ideally either in saturation or *off* mode, as a function of the LO signal. The small-signal impedance seen by the RF device is set by the source impedance of the switching devices. It is thus proportional to the inverse of the sum of the small-signal transconductance values of the switching devices, which is relatively a low impedance, compared to the output impedance of the RF device and also the reactance of the capacitive parasitics at the output of the RF transconductor (C_{Par} in Figure 3.8). This impedance is also independent of the load at the intermediate frequency (IF), to the first order, since the switch devices are either *off* or in saturation.

An analysis of the nonlinearity of a current-commutating mixer is provided in Ref. [26]. At low frequencies ($f_{LO} \ll f_T$) and low bias currents, the third-order non-linearity is dominated by that of the transconductor. The linearization techniques that were discussed earlier in the context of LNA transconductors can be directly applied to these designs. At high switching frequencies, the switching pair is found to dominate the nonlinearity.

An analysis of IIP2 in current-commutating active mixers is shown in Ref. [33], where several mechanisms for IM2 generation are identified. The IM2 is shown to arise from the second-order nonlinearity of the RF transconductor that appears at the output in the presence of mismatches in the switching devices. The parasitic capacitance at the tail of the switching pair also plays a critical role in the appearance of IM2 products at the output. A low-frequency IM2 term in the current of the RF transconductor modulates the transconductance of the switching devices. This leads to appearance of sidebands around multiples of the LO frequency at the source voltage of the switches, thus leading to sidebands in the current spectrum of the tail capacitor. These current sidebands are then downconverted to provide the baseband IM2 component at the output. For the IM2 of the RF transconductor to appear at the output, a nonzero offset must exist in the switching pair. Spurious leakage of the RF signal into the LO path is also shown to generate a baseband IM2 product through self-mixing and downconversion. This mechanism can be primarily controlled through layout and isolation techniques.

A mechanism that fundamentally limits IIP2 is found to arise within the switching pair and is also related to the parasitic capacitance at the tail of the switching pair and modulation of transconductance of the switching pair. This mechanism too needs the presence of a nonzero offset in the switching pair devices. It is shown that the limit on

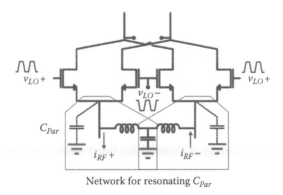

Network for resonating C_{Par}

FIGURE 3.9   Enhancement of IIP2 using a resonant network.

IM2 performance is related to the product of the LO frequency ω_{LO} and the rise time τ at the tail node, which is given by the parasitic capacitance divided by the *on*-state trans-conductance of the switching pair devices. Based on the analysis, an important conclusion is derived that the fundamental limit on IIP2, which depends on $\omega_{LO}\tau$, is expected to improve with process scaling. Enhancement of IIP2 by resonating the parasitic tail capacitance C_{Par} at the LO frequency by employing an LC tank (Figure 3.9) in order to reduce the IM2 contribution of the switching devices is shown in Ref. [34]. Using this approach, a mixer with 78 dBm of IIP2 is demonstrated in a 0.18 μm CMOS process.

Current-commutating active mixers typically employ differential pairs as LO buffers, which results in switching with 50% duty cycle. During the transition region of the LO signal, the switching devices are simultaneously on. This region is critical for noise and linearity performance of the mixer. For instance, the flicker noise in the switching devices, and that of the transconductor in the presence of mismatches of switching devices, appears at the output during the LO transition [35]. In Ref. [36], a technique for reducing flicker noise is demonstrated in a quadrature downconverter that employs double-balanced active mixers. It is shown that this technique also helps to improve the IM2 performance of the mixer. This is the case because IM2 mechanisms also generate components at baseband, and some of these exhibit a similar dependency on differential switch characteristics such as offset, bias current, and transient behavior [33] as flicker noise. Additionally, the conversion gain is also increased by 3 dB.

The approach uses an additional stacked layer of current-steering differential pairs, placed between the RF transconductor and the quadrature downconverting current-steering differential-pair switches. This additional layer is driven at twice the LO frequency ($2\omega_{LO}$) used in the quadrature downconverter switches ω_{LO}. By properly timing the stacked stages, the approach ensures that the I-stage differential pair is switched while the current is steered by the $2\omega_{LO}$ stage toward the Q-stage and vice versa. Consequently, the switching occurs in the quadrature stages, while the current through the devices is zero, thereby lowering the flicker noise at the output. The LOs used in the design lead to switching with effectively a 25% duty cycle instead of a 50% duty cycle, which has also been found to be beneficial in passive mixer designs discussed in the following section.

3.2.6.2 Passive Mixers

Recently, there has been significant interest in the use of passive mixers, for example, [2,3,37–42] in deep-submicron CMOS technologies. The switching devices in these mixers are either in linear mode or in the *off* state, depending on the state of the LO clock applied to their gates. The switching devices do not have a dc flowing through them, and hence they exhibit significantly lower flicker noise compared to basic active current-commutating mixers (Figure 3.8), which makes them well suited for direct downconversion receivers. Additionally, these designs become increasingly attractive at lower supply voltages available in deep submicron technologies, since these do not require vertical voltage headroom that is required in a current-commutating active mixer.

Unlike an active mixer, where the baseband load is isolated from the RF transconductor, the switches in a passive mixer do not isolate the two sides and are bilateral when *on*. Consequently, baseband impedance at the output of the passive mixer is transferred to the output of the RF transconductor. Thus, a large impedance at baseband at the output translates into a large RF impedance at the output of the RF transconductor, and a low impedance at baseband translates into a low RF impedance at the output of the RF transconductor.* In the first case, the passive mixer is said to be operating in the voltage mode, since it translates the RF voltage at the RF transconductor output to baseband (e.g., [41]). The conversion gain of the mixer in this mode is primarily determined by the product of the RF transconductance and the impedance at the transconductor output at RF. In the second case, the passive mixer commutates the current in the RF transconductor, and hence the mixer is said to be operating in the current mode (e.g., [39]). A low impedance can be achieved at baseband by terminating the mixer in a transimpedance amplifier (Figure 3.10) or a common-gate amplifier. The conversion gain here is proportional to the product of the RF transconductance and the transimpedance at baseband. The RF and baseband stages in these designs typically require independent bias currents, unlike an active current-commutating mixer. An example of a passive current-commutating mixer, which allows for bias sharing between the RF and baseband stages, is shown in Ref. [42].

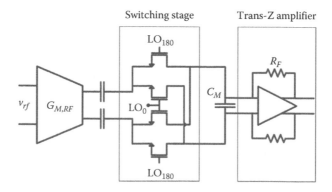

FIGURE 3.10   Passive current-commutating mixer.

---

* The size of the impedance values is relative to the inherent output impedance of the RF transconductor.

A small v_{ds} across the switches helps to improve the linearity of the switches. In current-mode designs, the voltage swing at the mixer nodes is also reduced, which leads to excellent linearity [39]. The key factors impacting the linearity of this mixer include the nonlinear $i_d - v_{ds}$ relationship, nonlinear device capacitance, and finite LO rise and fall times [43]. Furthermore, it has a strong dependence on the load impedance at the output of the switches and the source impedance that is used to drive the switches, which is the output impedance of the RF transconductor. In order to maximize the IIP3 and IIP2 of the mixer [43], the load impedance should be minimized simultaneously near the LO frequency, at baseband, to which the desired signal is downconverted in a direct conversion front end and the IFs to which the interferers are downconverted. It is also shown that for maximizing IIP2, measured with two input signals at ω_1 and ω_2,* the source impedance seen by the switches at $\omega_1 - \omega_2$ should be maximized, for example, through the use of an ac coupling capacitor between the RF transconductor and the switches. For enhancing IIP3, the impedance of the source at RF needs to be maximized, which can be accomplished, for example, by resonating the parasitic capacitance at the output of the LNA with a high-Q inductor. In order for the second-order IM to be observed at the differential output, some degree of mismatch or imbalance must be present between the devices of a switching pair, which can arise from the devices themselves or LO mismatches.

A key aspect to the design of passive mixers is the LO duty cycle that is employed, especially when quadrature outputs are to be derived from a single RF transconductor. The use of a 25% LO duty cycle has been found to improve IIP2 and IIP3 performance in receivers with quadrature outputs [2,44]. This is attributed to the higher impedance seen looking into the RF transconductor from the switches of the passive mixer and the significantly reduced interaction between the I and Q paths of the downconverter in such a topology. Clock overlap with 50% duty-cycle clocks and the lowered impedance looking back into I and Q sides of the mixer, which results from the other side being connected, also has a significant impact on the noise figure, for example, [45] (Figure 3.11).

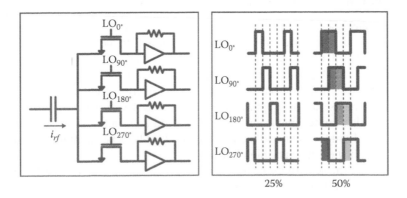

FIGURE 3.11   Passive current-mode mixer with different LO duty cycles.

---

* The IM2 product is measured at baseband at $\omega_1 - \omega_2$.

Similar to the active mixer described earlier, use of 25% duty-cycle LO instead of 50% duty cycle improves the gain by 3 dB.

As mentioned previously, in order for the second-order IM to appear at the output, some imbalance must exist between the devices within the switching pair of the mixer. Reducing this imbalance improves the IIP2. The use of calibration-based approaches for enhancing IIP2 performance in passive mixers is discussed in Ref. [2]. Controllable bias voltages on the gates of the mixer devices are employed for improving IIP2.

3.3   INTERFERENCE REJECTION

Attenuation of interferers in radio receivers is usually implemented at baseband after downconversion, as the selectivity required at RF in typical wireless standards can be significant. For example, in a GSM radio, the carrier frequency is around 900 MHz, the channel bandwidth is 200 kHz, and the separation of large interferers from the carrier frequency is of the order of a MHz. Thus, if a band-pass filter is used to select the desired signal and reject interferers at RF, the required fractional bandwidth of the filter would be of the order of 10^{-3}, which is challenging. An integrated L–C band-pass filter in a standard silicon process, implemented using passive on-chip inductors and capacitors, cannot provide such high selectivity, as the quality factor of the passive components is limited. External passive filters, such as SAW filters, also cannot provide channel selectivity and need to accommodate the entire system bandwidth, since these are not tunable.

In many systems, however, large RF interferers can impose severe performance requirements early in the signal chain. It is thus often beneficial to isolate and select the desired signal within the front end, since this can relax the linearity requirement of the circuits that follow. An overview of approaches that allow for rejection of interferers in the RFE is provided in the following sections.

3.3.1   INTEGRATED LC FILTERS AND Q-ENHANCEMENT

A passive integrated LC filter can provide only limited suppression of interferers in the front end. Typical quality factors of integrated inductors in the GHz band are of the order of 10. For a center frequency of 1 GHz, this translates into a 3 dB bandwidth of 100 MHz, which is not useful for suppression of a majority of interferers that introduce primary linearity requirements in front ends. Passive LC filters can however be useful for suppression of distant interferers, which are sufficiently beyond the filter corner frequency and can be used both in the input matching network to band-limit the input and within integrated loads. A key feature of passive LC filters is that they exhibit very high linearity.

To enhance the selectivity of integrated L–C resonators, so as to allow for filtering of close-in interferers, active negative resistors can be utilized for increasing the effective quality factor. A shunt L–C resonator with a resistance of R_P in parallel, where R_P models the resonator loss, has a quality factor given by $Q = R_P/(\omega_o L)$, where ω_0 is the center frequency of the resonator. If a negative resistance of value $-|R_P'|$ is placed in shunt with this network, the effective Q can be shown to be $Q_{eff} = \left\{ {|R_P'|}/{(|R_P'|-R_P)} \right\} Q$ [46]. In CMOS designs, a negative resistance can be easily implemented using a cross-coupled device pair. A series LC resonator can also be employed with Q-enhancement; for example, [47] demonstrates the use of an effective series L–C notch to reduce the level of transmitter leakage in WCDMA.

In principle, Q-enhancement can provide very high quality factors. The amount of enhancement, however, is limited by the necessity to ensure stability of the resonator in the presence of process and temperature variations. Accurate tuning and frequency tracking techniques for channel selection are also required on-chip. Further, due to the use of negative active resistors, the close-in nonlinearity of the Q-enhanced filter can be degraded compared to a basic L–C tank. The use of active elements for increasing the quality factor can also lead to a higher noise floor, compared to a purely passive filter of similar Q, although as noted earlier achieving such a high Q can pose a technological challenge.

3.3.2 Interference Suppression Using Auxiliary Receivers

The frequency-selective element in Q-enhanced filters mentioned earlier is centered at the carrier frequency of the desired signal or the interferer. Another approach to achieving very narrowband frequency response at RF is to translate the frequency characteristic of a low-frequency filter, for example, a baseband filter, to the desired carrier frequency [48]. Architectures that achieve the frequency selectivity of baseband filters at RF are described in the following sections.

3.3.2.1 Feedforward

Consider the basic receiver front end shown in Figure 3.1, where a small desired signal (p_{sig}) and a large blocker (p_{int}) are indicated. We assume that the LNA is sufficiently broadband such that close-in interferers appear unmitigated at its output. We further assume that the LNA has sufficiently high gain such that the key linearity bottleneck appears at the input of the mixers.

A feedforward technique based on the use of an auxiliary receiver path to attenuate the interferer level at the LNA output, thus relaxing its linearity and that of the mixers, is reported in Ref. [49–51] (Figure 3.12a).

The auxiliary path employs a quadrature downconverter to translate the incident spectrum consisting of the desired signal and interferer to baseband, using the same LO as that used in the main receiver path. The downconverted signal and interferer are applied to baseband amplifiers (not shown in Figure 3.12a; see [49,50]) and passive high-pass filters* (HPF), translated back to the LO frequency using a passive linear I/Q upconverter, and then combined with the output of the main path LNA.

---

* In a strict sense, this needs to be a band-pass response. The filter must provide significant attenuation near the harmonics of the LO. Otherwise, noise at the output of the baseband amplifier can fold in-band.

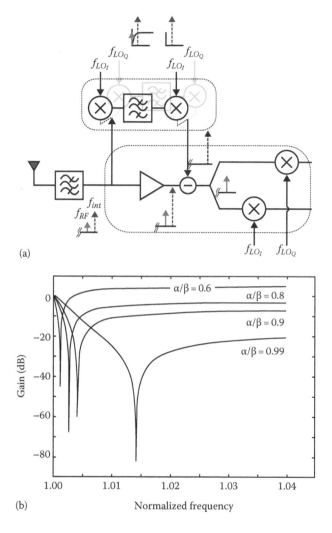

FIGURE 3.12 Feedforward interference cancellation. (a) Architecture. (b) Frequency response.

The desired signal in the auxiliary path, after downconversion, falls within the stopband of the HPF and is hence rejected. The output of the auxiliary path thus consists only of the interferer. The baseband amplifiers are used to equalize the gains of the main path LNA and the auxiliary path at the interferer frequency. If the phase of the auxiliary path output at the frequency of the interferer is set to 180° relative to the phase at the LNA output,* the interferer is rejected at the LNA output node after combining. Approaches to vary the phase in the auxiliary path are described in Ref. [49].

---

* This is indicated by the subtraction of the two paths in Figure 3.12a.

If a first-order high-pass filter is used in the auxiliary path, then the effective transfer function of the combined main and auxiliary paths at the output of the LNA is given by Ref. [50]:

$$H(j\omega) = \alpha + \beta \frac{j\omega/p}{1+j\omega/p} e^{j\phi} \tag{3.2}$$

where
   α is the gain of the main path
   β is the gain of the auxiliary path
   ω is the frequency $\omega_{in} - \omega_{LO}$, which is the offset frequency relative to the LO frequency
   p is the pole frequency of the HPF at baseband
   ϕ is the phase shift of the auxiliary path relative to the main path

For $\phi = \cos^{-1}(-\alpha/\beta)$, Equation 3.2 can be shown to provide a null at

$$\omega_n = \left\{ \frac{\alpha}{\sqrt{\beta^2 - \alpha^2}} \right\} p \tag{3.3}$$

Further, for $\omega \gg p$, the magnitude of the response is given by $\sqrt{\beta^2 - \alpha^2}$. For $\beta = \alpha$, we have $\phi = 180°$ and an asymptotic gain of 0 with the location of the null at infinite offset. For $\beta = 1.05\alpha$, the location of the null is at $3.12 \times p$ Hz and the asymptotic gain is -10 dB. The simulated location of the null and distant rejection as a function of normalized frequency (ω_{in}/ω_{LO}), with α/β as a parameter, is shown in Figure 3.12b.

Since the location of the null from Equation 3.3 is of the order of p, this approach is seen to provide baseband selectivity at RF. Equation 3.3 also indicates that the exact location of the null can be tuned by setting the values of the relative gain and phase of the main and auxiliary paths. Further, the transfer function of Equation 3.2 tracks ω_{LO}. Thus, a change in ω_{LO} will cause the value of the null location at RF to be modified by the same amount.

In Figure 3.12a, any in-band distortion or noise terms generated by the first down-converter and the baseband amplifiers preceding the HPF in the auxiliary path are not critical, since these terms are filtered by the high-pass filter. The most critical auxiliary path circuit is the I/Q upconversion mixer, which needs to have sufficient input linearity for the largest interferer that is being canceled. Additionally, in-band noise of this mixer must be small in comparison to the noise at the output of the LNA, in order to minimize noise figure degradation. On the other hand, the mixer does not need to exhibit gain, since the baseband amplifiers can be used for equalizing the gains of the main and auxiliary paths. Thus, a passive design is ideally suited for this purpose. The output linearity required in the baseband amplifier is also of the same order as that of the passive mixer. However, it can be expected that the dynamic range per-unit power performance at baseband is significantly larger than that at RF, thus contributing to the efficiency of the architecture.

It should be noted that while significant enhancement of blocker-induced compression can be achieved, the improvement can be smaller than the small-signal attenuation at the blocker frequency predicted by Equation 3.2 [50].

The aforementioned approach essentially partitions the dynamic range requirement between the main path and the auxiliary path. By relaxing the linearity requirement of the main path, the margin for sensitivity that is available in the presence of the interferer (e.g., 3 dB) can be applied toward relaxing other sources of degradation such as the LO phase noise.

Feedforward interference cancellation (FIC) can be employed for improving linearity in low-voltage designs, as demonstrated in Ref. [52]. FIC is employed here at the output of the main path downconverter, to relax the linearity constraint at this node. Using this technique, 13.4 dB IIP3 improvement is achieved in a 0.6 V receiver front end implemented in a 65 nm CMOS process. The auxiliary path requires 11 mW out of a total receiver power of 26 mW. Cancellation of close-in interferers is demonstrated.

A significant issue that needs to be considered in the design of feedforward architectures is the requirement for gain matching and phase alignment between the auxiliary and main paths, which can be achieved through calibration.

3.3.2.2 Feedback

The use of frequency translation in a feedback path has been demonstrated in Ref. [53,54] for rejecting transmitter leakage in CDMA receivers. The approach uses a downconverter and upconverter with a low-pass filtering baseband section between the two stages (Figure 3.13).

Unlike the feedforward architecture of Figure 3.12a, the auxiliary path uses an LO that is coincident with the frequency of the interferer. Thus, the *interferer* is downconverted to baseband, selected in the LPF, and subsequently upconverted, at which point it can be subtracted from the critical node. An integrator is employed at baseband in Ref. [53], and interference cancellation is achieved at the output of an LNA. Nearly 23 dB improvement in TX rejection is reported, with an excess current of 11 mA and 0.3 dB NF degradation. In Ref. [54], a low-pass filter is employed at baseband, and the interferer is attenuated at the input of the LNA using the auxiliary

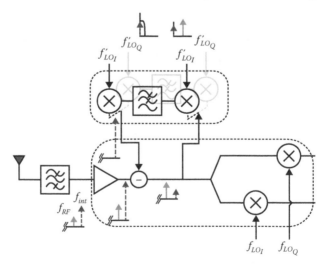

FIGURE 3.13 Feedback auxiliary path for interference cancellation at the LNA output.

feedback path. Nearly 20 dB of TX rejection is achieved through use of the auxiliary path with 0.22 dB noise figure degradation, with an excess bias current requirement of 4 mA. The designs of Refs. [53,54] utilize 0.18 μm CMOS.

A feedback architecture, with an auxiliary path that employs the same LO as the main path and an HPF for isolating the interferer at baseband, is reported in Ref. [55]. A 9 dB improvement in gain reduction caused by a blocker at 1.88 GHz is achieved by using interference cancellation.

Feedback-based approaches can show reduced sensitivity to mismatches of gain and phase between the main and auxiliary paths compared to feedforward [53]. On the other hand, a key consideration in the design of feedback-based auxiliary paths arises from stability. The order of the filter used in the auxiliary path determines the phase margin of the closed loop, and increasing the filter order degrades the phase margin [54]. Feedforward systems do not suffer from stability concerns and can allow for the use of higher-order filters in the auxiliary path.

3.3.3 Frequency Translation of Impedance in Passive Mixers for Filtering

Passive mixers (Section 3.2.6.2) transfer low-frequency impedance at their outputs to their RF inputs (e.g., [38,41,42,44,56]). This is due to the bilateral nature of the mixer switches when they are *on*. A consequence of this impedance transfer is that the selectivity provided by a baseband filter at the output of the mixer can be achieved at RF, at the output of the RF transconductor that is used to drive the mixer switches. In a voltage-mode mixer, for instance, if the output is loaded with a low-pass filter that presents a high impedance to the downconverted desired signal, and a low-impedance to close-in interferers, a narrow band-pass response is observed at RF at the output of the transconductor that mirrors the baseband response. This limits the voltage excursion at the output node of the transconductor that is caused by close-in blockers, thereby improving the linearity of the transconductor as well as the switches within the passive mixer. A simulated frequency response at RF of a passive I/Q mixer loaded with various values of capacitance assuming an input RF transconductor stage with 500 Ω output resistance and devices modeled by ideal switches is shown in Figure 3.14. A band-pass response can be observed at the transconductor output.

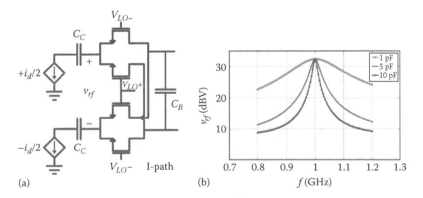

FIGURE 3.14 RF voltage at input of passive voltage-mode I/Q mixer. (a) I-path of mixer. (b) Frequency response at RF input.

In the case of current-mode mixers, the impedance at baseband is designed to be significantly lower at baseband than the impedance of the transconductor at RF. As such, the transconductor output impedance is lowered over the bandwidth where the baseband impedance is small.

A band-pass filter based on the aforementioned effect is demonstrated in Ref. [56]. A resistor R followed by a passive switching I/Q mixer with 25% duty-cycle LO, which is loaded by capacitors C, is used to achieve a frequency transferred impedance of a low-pass RC filter, centered around the LO frequency, at the input of the mixer. The mixer is driven by an external balun. The low-pass response itself is seen at baseband at the outputs of the mixer. The input-referred impedance is modeled as an equivalent shunt R–L–C network. It is shown that the bandwidth observed at RF due to the upconversion of the baseband response is set by the equivalent baseband RC filter. Using this approach, a tunable filter with a bandwidth of 35 MHz is demonstrated, with a tuning range from 100 MHz to 1 GHz. The design exhibits high linearity of 14 dBm IIP3, which is related to the use of the passive mixer, and also a low-noise figure of 3–5 dB. In Ref. [41], a capacitively loaded voltage-mode I/Q mixer with 25% duty-cycle LO is employed to implement a narrowband RF filter at the output of the RF LNA. This filter allows for rejection of interferers at the RF LNA output, thus relaxing its output linearity in the presence of blockers. The design is shown to reject a blocker at 20 MHz offset by nearly 15 dB. An interesting aspect to the impedance transfer property in such mixers, as discussed in Refs. [38,56], is that the impedance seen looking into the mixer is itself a function of the source impedance that is used to drive it.

An important consideration in the use of these filters is that since the filter is composed of switching elements, the filter can exhibit folding of signals from one LO harmonic to another; for instance, signals from around the third harmonic of the LO can fold around the fundamental frequency. This is a concern in systems with distant interferers and necessitates the use of prefiltering to reject the undesirable interferer energy near the harmonics of the LO [56].

3.4 HARMONIC REJECTION MIXERS AND DOWNCONVERTERS

A switching mixer amplitude-limits the LO waveform, in order to desensitize the mixer to LO amplitude variation and noise. Consequently, the mixer provides significant gain at not only the fundamental LO frequency but also the harmonics of the LO. For example, a switching mixer that employs an LO at frequency f_{LO} will also provide a finite conversion gain at all harmonics of f_{LO}, which is at mf_{LO}, where m is an integer. For an LO duty cycle of 50%, the harmonic response at even frequencies is ideally rejected. In systems where there is no signal of interest at the LO harmonics, sufficient filtering can be employed at the input of the radio receiver, in order to reject signals beyond a certain offset from the LO frequency, and avoid downconversion from LO harmonics. The level of rejection required is set by the standard and the anticipated interferer content at the LO harmonics and determines the complexity and cost of the front-end filter. In theory, given a sufficiently sharp filter, the response from LO harmonics can always be completely rejected.

In broadband applications, however, the system bandwidth can be sufficiently wide that the harmonics of the LO coincide with other desired signal channels.

This is the case, for example, in digital TV systems, for example, [7], where signals occupy 6 MHz wide channels in the frequency range from 48 to 860 MHz. In a receiver for this system, when channels toward the lower end of the signal band are downconverted, channels that appear at the LO harmonics are also downconverted.

One option for reducing the impact of undesirable harmonic responses in a direct downconversion radio is to place a tunable band-pass filter at the input, with a pass band that coincides with the desired signal. Achieving broadband tunability implies that active devices must be used in the filter. The filter can limit the dynamic range of the receiver in this case. Indeed implementation of such a broadband active tunable filter poses a significant design challenge to itself.

A very effective alternative in such cases is a harmonic rejection mixer (HRM), first proposed in Ref. [57] for transmitter applications. These mixers use linear combinations of phase-shifted two-level periodic switching waveforms, in order to synthesize a discrete approximation to a sine-wave LO (e.g., [58–61]). Each of the periodic switching waveforms can be employed in a unique downconversion path and scaled with an appropriate gain coefficient. While the waveforms operating on each of the paths contain the harmonics of the LO, after combining the phase-shifted LOs, specific harmonics can be rejected. By using two-level waveforms in the multiple paths, HRMs retain the advantage of switching mixers with regard to insensitivity to LO amplitude, unlike multipliers with sinusoidal LOs.

In this section, we will describe principle of operation and architectures for HRMs, primarily in the context of receivers. Limitations in harmonic rejection performance, as set by relative gain and phase matching of different paths, will be outlined. Previously reported approaches for desensitizing the design to gain and phase errors will be described. It will be shown that an HRM can also be used to internally synthesize harmonics of the fundamental. Harmonic rejection in such operating modes will be outlined.

3.4.1 HARMONIC REJECTION BASED ON SYNTHESIS OF A DISCRETE APPROXIMATION TO A SINE WAVE

In order to eliminate all LO harmonic content, a mixer needs to internally multiply the input with a sinusoidal LO. However, switching mixers inherently amplitude limit the LO, which introduces significant harmonic content, even if the harmonic content in the externally applied LO is small. If a mixer employs a hard-limiting switching stage, and uses an external LO that has a 50% dutycycle, the signal experiences an amplitude-limited LO that approximates a square wave that switches between −1 and +1. Such an LO waveform can be represented by a Fourier series that is given by

$$x_{LO}(t) = \frac{4}{\pi} \sum_{k=1}^{\infty} \frac{sin\{2\pi(2k-1)(t/T_{LO})\}}{2k-1} \tag{3.4}$$

where T_{LO} is the time period of the waveform. The mth odd harmonic of the LO is attenuated relative to the fundamental at $f_{LO} = 1/T_{LO}$ by a scaling factor of m, while the even harmonics are ideally absent.

By combining appropriately phase-shifted and amplitude-scaled pulse trains of the previous form, it is possible to synthesize LO waveforms that can reject specific harmonics within the LO. One approach to accomplishing this in mixers is to use phase shifting in the LO path while applying amplitude scaling in the RF or baseband paths using specific gain coefficients. Gain scaling is easier to perform in the signal paths rather than within mixer switches, while broadband phase shifting can be performed easily in the LO path using time delays in the digital logic used to synthesize the LOs.

The previous approach to harmonic rejection was first demonstrated in Ref. [57] for a transmitter application. The design used three mixing paths that had gains in the ratio of $1:\sqrt{2}:1$. The LOs applied on the three paths employed relative phase shifts of $-\pi/4{:}0{:}\pi/4$. An equivalent LO with harmonics at $(8m \pm 1)f_{LO}$, where $f_{LO} = 1/T_{LO}$ is the fundamental frequency and m is a positive integer, was obtained from the summation of the three mixing paths. Ideally, the magnitude of all other harmonics was zero.

The HRM architecture as applied to a receiver downconverter is shown in Figure 3.15a. The effective LO that operates on the input in the time domain and its frequency spectrum are shown in Figure 3.15b. The waveform can be seen to be a better approximation to a sine wave than a square wave, which implies a reduced harmonic content. Mathematically, the amplitude of the fundamental LO component in the waveform is scaled by a relative factor of $\left(e^{-j\pi/4} + \sqrt{2} + e^{j\pi/4}\right)$, which evaluates to $2\sqrt{2}$. The amplitude of the third harmonic at $3f_{LO}$ is scaled by $(1/3)\left(e^{-j3\pi/4} + \sqrt{2} + e^{j3\pi/4}\right)$, while that of the fifth harmonic is scaled by $(1/5)\left(e^{-j5\pi/4} + \sqrt{2} + e^{j5\pi/4}\right)$. These two terms can each be seen to evaluate to 0, implying a cancellation of the respective harmonic.

Two key sources of degradation need to be considered in the previous design. The first arises from the requirement for irrational transconductor gain ratio of $1:\sqrt{2}:1$. In practice, this ratio is implemented using device scaling, and thus only rational ratios are feasible. This implies that a practical implementation of the design will necessarily only approximate this ratio. This leads to a finite residual term at the harmonic. For instance, if a rational approximation to $1:\sqrt{2}:1$ results in an effective ratio of $1:\sqrt{2}(1+\alpha):1$, we will observe residual terms of relative magnitude $\alpha/6$ and $\alpha/10$ at the third and fifth harmonics, respectively. A related source of degradation appears from mismatch. While the ideal rational approximation to the previous gain ratios is $1:\sqrt{2}(1+\alpha):1$, in practice, each of the three gain terms may be different from this ratio, owing to device mismatches caused by statistical variation.

A second source of error arises from nonideal relative phases compared to the ideal shifts of $-\pi/4{:}0{:}\pi/4$. This too leads to nonideal rejection of harmonics. For instance, consider the case where the path with the nominal $0°$ phase shift has a phase of θ at the fundamental. The third harmonic term, with an ideal gain ratio of $1:\sqrt{2}:1$, will then be scaled by $(1/3)\left(e^{-j\pi/4} + \sqrt{2}e^{j3\theta} + e^{j\pi/4}\right)$, which results in a residual harmonic component of magnitude $\sqrt{2}\sin(3\theta)/3$. Since the fundamental term, for small values of θ, is still nearly $2\sqrt{2}$, the relative third harmonic is given by $\theta/2$. In general, the mth harmonic, which should ideally be rejected, will instead be at a relative level of $\theta/2$. The impact of the previous nonidealities on the LO spectrum is shown in Figure 3.15c.

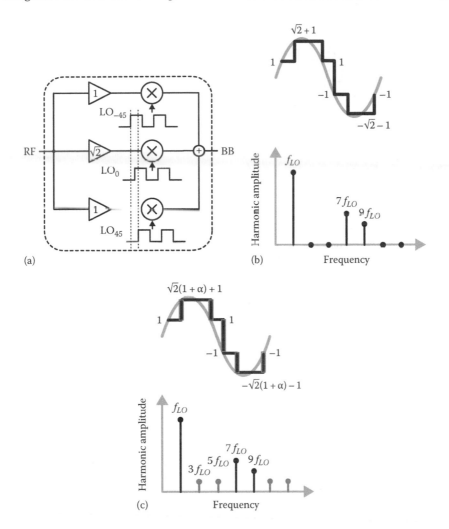

FIGURE 3.15 (a) 50% duty-cycle HRM, (b) equivalent LO and spectrum with ideal coefficients and LO phases, and (c) equivalent LO and spectrum with nonideal coefficients and LO phases.

We note that using three paths as shown here degrades noise performance compared to the use of a single transconductor without the three phase-shifted paths. The reason for this is that while only part of the signal response at the desired harmonic results in the net conversion gain, $\left(e^{-j\pi/4} + \sqrt{2} + e^{j\pi/4}\right)$, the noise contributed by each of the three transconductors adds in power with magnitude scaling of $1 : \sqrt{2} : 1$.

The design of Ref. [57] uses active transconductors prior to the switching stage with the approximate ratio of $1 : \sqrt{2} : 1$. An alternative design (Ref. [58]) implements the gain coefficients at baseband for improved matching performance and employs

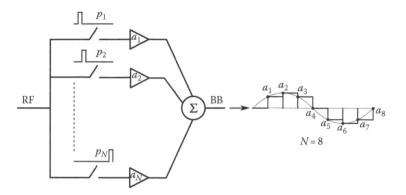

FIGURE 3.16   HRM using 1/N duty-cycle clocks and baseband gain coefficients.

1/8 duty-cycle clocks, instead of clocks with 50% duty cycle, to generate the same effective LO as that in Ref. [57] within an eight-phase HRM.

An N-phase HRM with baseband gain coefficients is shown in Figure 3.16. The effective LO waveform that multiplies the input is shown for $N = 8$. By using phase-shifted clocks with duty cycle of $1/N$ the input is sequentially applied to a set of gain coefficients $[a_1:a_N]$. These coefficients are derived by sampling a sinusoid of frequency f_{LO} N times per period, with time steps $1/(Nf_{LO})$ and a sample clock of frequency Nf_{LO}. Thus

$$a_k = G \sin\left\{\frac{2\pi k}{N}\right\} \tag{3.5}$$

where $k \in [1:N]$ and G is the largest scaling coefficient. In practice, the gain coefficients will be quantized approximations of those in Equation 3.5. By applying gains a_k to the input in a specific time sequence, an effective downconversion frequency of f_{LO} is obtained. For an N-phase HRM with ideal gain coefficients, it can be shown that all harmonics except for $(Nm \pm 1)f_{LO}$ are rejected, where m is a positive integer. It is thus theoretically possible to reject an arbitrarily large number of harmonics by increasing N, although this comes at the expense of a higher sample clock frequency.*

3.4.2   ENHANCING HARMONIC REJECTION PERFORMANCE

As was discussed earlier, harmonic rejection performance is ultimately limited by gain and phase errors. Designs without any special enhancements that rely solely on native matching performance of silicon devices have typically been observed to provide harmonic rejection in the range of 30–40 dB. Improving performance beyond this level requires the use of additional techniques.

One approach is to improve performance through calibration. This can be performed in the analog or digital domain. For instance, by measuring the harmonic rejection performance, the gains of the multiple paths and phases of the LOs used in

---

* N paths are depicted in Figure 3.16 for clarity, However, it is noted that many HRMs employ $N/2$ BB paths using differential circuits to change the sign of gain coefficients during the second half period of the equivalent LO (e.g., [58–60]).

each path can be calibrated to remove residual harmonics. In a practical design, it is important to ensure that the calibration is robust to voltage supply and temperature variations.

Approaches that employ specific design techniques termed two-stage harmonic rejection and clock retiming have been proposed in the recent past and have been shown to provide harmonic rejection in excess of 55 dB. These are described in the following sections.

3.4.2.1 Reducing Sensitivity to Gain Coefficient Errors

In Section 3.4.1, a discrete approximation to a sinusoidal LO was implemented using three square-wave LOs, with relative phase shifts of $-\pi/4{:}0{:}\pi/4$ and respective gain scaling of $1:\sqrt{2}:1$. As we noted previously, a nonideal gain ratio leads to a residual third and fifth harmonic. Thus, if the practical gain ratio is $1:\sqrt{2}(1+\alpha):1$, then the residual relative third and fifth harmonics are scaled by $\alpha/6$ and $\alpha/10$.

The concept of two-stage harmonic rejection for reducing the impact of gain-coefficient errors was demonstrated in Ref. [59]. In an eight-phase implementation of this approach, the RF was first downconverted to baseband in three harmonic-reject downconversion paths, with relative mutual phase shifts of $-\pi/4{:}0{:}\pi/4$. The first HRM stages used a nominal gain ratio of $1:\sqrt{2}:1$. The baseband outputs were then combined using a second stage of gain coefficients with the same nominal gain ratio. As shown in Ref. [59], if the first stage employed a nonideal gain ratio of $1:\sqrt{2}(1+\alpha):1$ and the second stage $1:\sqrt{2}(1+\beta):1$, then the net error at baseband was proportional to $\alpha\beta$.

This approach for achieving harmonic rejection was shown in Ref. [59] for a broadband receiver application. The design used passive mixers, with nonoverlapping waveforms of the form shown in Figure 3.16. The design in 65 nm CMOS spanned a bandwidth from 400 to 900 MHz, provided a gain of 34 dB, DSBNF of 4 dB, and IIP3 of 3.5 dBm, with power dissipation of 60 mW. Harmonic rejection of over 60 dB was achieved at the third and fifth harmonics, without filtering or calibration.

3.4.2.2 Reducing Sensitivity to LO Phase Errors

While the previous approach of cascading two harmonic rejection stages significantly improves the sensitivity to gain errors, the impact of phase errors is not mitigated. Thus, if the three LO paths have a relative phase shift of $-\pi/4{:}\theta{:}\pi/4$, instead of $-\pi/4{:}0{:}\pi/4$, the mth odd harmonic would be at a level of $\theta/2$, even with ideal gain ratios.

To mitigate this source of error, a multiclocking approach can be used where all phase clocks are synchronized to a single master clock. Multiclocking approaches have been used for reducing the sensitivity to phase errors in several contexts. In Ref. [62], sensitivity of image rejection to phase error is reduced by employing two switches connected in series in an I/Q mixer, both clocked at a 50% duty cycle. Another clock at frequency $4f_{LO}$ is used to desensitize the design to phase errors arising from non-ideal clock duty-cycle. As described earlier, in Ref. [36], a multiclocking approach is used to significantly reduce flicker noise in an active mixer. Insensitivity of harmonic rejection to phase errors by using multiclocking is shown in Ref. [60] in a single-stage active HRM. In this design, a primary clock of frequency Nf_{LO} is employed to generate N clock phases, and the source of error arising from the nonideality of the primary clock duty cycle is removed by combining the two polarities of the primary clock to create each mixing phase.

We describe the operation of a retiming circuit [60] in the context of a passive HRM with eight phases, as shown in Figure 3.17. The RF signal is sequentially steered to the gain units a_1–a_8 using nonoverlapping pulses p_1–p_8, respectively. Nonideal edge timing in these pulses, for example, p_1 in Figure 3.17, acts a source of phase error and leads to degradation in harmonic rejection. For instance, if p_1 extends into the duration of p_2, the RF signal would be split between gain paths a_1 and a_2 for a brief duration of time.

In order to avoid this, instead of using one path to apply the RF to a gain block, we can employ two paths that are retimed using the complementary primary clock pulses CLK_p and CLK_n, as shown in Figure 3.18. Each of the pulses p_i, $i \in [1{:}8]$, is now replaced by two pulses, p_{i+}, $i \in [1{:}8]$, and p_{i-}, $i \in [1{:}8]$. Further, these pulses are designed to overlap the primary clock pulses CLK_p and CLK_n, as shown in the figure. The RF signal is connected the baseband gains a_1–a_8 using effective pulses $p_{eff,i} = (p_{i+} \wedge CLK_p) \vee (p_{i-} \wedge CLK_n)$. The pulses $p_{eff,i}$ transition only at the edges of CLK_p and CLK_n for all values of i. The edges of the pulses $p_{i\pm}$ can now tolerate an uncertainty of $\pm T_{CLK}/4$, where T_{CLK} is the time period of the primary clock, for each of their transitions, without an impact on $p_{eff,i}$. This greatly desensitizes the design to phase errors.

The previous approach was demonstrated in a single-stage active HRM with a signal bandwidth of 100–300 MHz in Ref. [60], with gain of 12 dB, DSB NF of 11 dB, IIP3 of 12 dBm, and power of 69.4 mW. The harmonic rejection ratio was in the range of 52–55 dB.

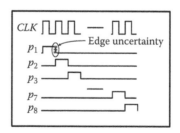

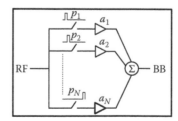

FIGURE 3.17   Passive HRM with potential timing error on edge of p_1.

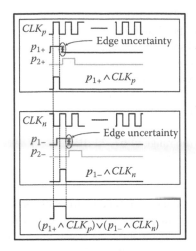

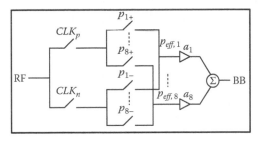

FIGURE 3.18 Clock retiming in passive HRM.

3.4.3 DESENSITIZING A TWO-STAGE DESIGN TO GAIN AND PHASE ERRORS

An HRM can be desensitized to both gain and phase errors as shown in Ref. [61]. A 16-phase two-stage HRM is employed with clock retiming in this design. In the design of Figure 3.18, a single RF gain path is steered through multiple baseband gains using the effective clocks $p_{eff,i}$, which are implemented using clock retiming. The impact of phase errors in the individual clocks p_i is mitigated using this approach. In a two-stage design, multiple RF gain coefficients are used in parallel to synthesize the effective discrete-level sinusoidal LOs, with phase offsets. Each RF gain coefficient has associated with it a clock-retiming network similar to that shown in Figure 3.18.

Since the same master clock is routed through all the clock-retiming networks, phase errors in the master clocks caused by routing can cause phase errors between the downconversion paths. As described in Ref. [61], the two-stage design with retiming ensures that relative phase errors between the multiple downconversion paths are also mitigated. The design of Ref. [61] is able to achieve over 65 dB harmonic rejection without gain or phase calibration or harmonic filtering.

Another key feature described in Ref. [61] is the capability of an HRM to synthesize multiple downconversion frequencies within the HRM, through a reconfiguration in the way in which gain coefficients are accessed. This employs the observation that the coefficients $[a_1{:}a_N]$, where $a_k = \sin(2\pi k/N)$, can be used to generate all harmonics of f_{LO} up to $(N/2)f_{LO}$ through a simple permutation of the sequence in which they are applied to the signal. If the nth harmonic is synthesized in this way where $n \in [1:N/2]$, the LO spectral content is observed only at nf_{LO} and $(Nm \pm n)f_{LO}$, for positive integer values of m, while all other harmonics of f_{LO} are ideally zero. An HRM based on this approach was demonstrated in 130 nm CMOS technology for the frequency band from 50–830 MHz in Ref. [61]. The HRM had a gain of 12 dB, DSBNF of 11 dB, and IIP3 of 5.4 dBm, with a power dissipation of 67 mW. In the fundamental LO mode, the harmonic rejection at the third, fifth, and seventh harmonics was measured in the range of 67–72 dB.

3.5 CONCLUSION

The appearance of energy from out-of-band sources into a desired channel can lead to significant degradation in signal quality. In this chapter, we identified key mechanisms responsible for this potential degradation. As noted, these can be linear time varying or nonlinear in nature. Several circuit and architectural techniques were described in CMOS and BJT technologies to mitigate the impact of these mechanisms.

REFERENCES

1. T.H. Lee, *The Design of CMOS Radio Frequency Integrated Circuits*, 2nd edn. Cambridge, U.K.: Cambridge University Press, 2004.
2. D. Kaczman, M. Shah, M. Alam, M. Rachedine, D. Cashen, L. Han, and A. Raghavan, A single-chip 10-Band WCDMA/HSDPA 4-Band GSM/EDGE SAW-less CMOS receiver with DigRF 3G interface and 90 dBm IIP2. *IEEE Journal of Solid-State Circuits* 44(3), 718–739, 2009.
3. N.K. Yanduru, D. Griffith, K.-M. Low, and P.T. Balsara, RF receiver front-end with +3dBm out-of-band IIP3 and 3.4dB NF in 45 nm CMOS for 3G and beyond. *Proceedings of IEEE Radio Frequency Integrated Circuits Symposium, 2009*, June 7–9, 2009. pp. 9–12.
4. R. Gharpurey, N. Yanduru, F. Dantoni, P. Litmanen, G. Sirna, T. Mayhugh, C. Lin, I. Deng, P. Fontaine, and F. Lin, A direct-conversion receiver for the 3G WCDMA standard. *IEEE Journal of Solid-State Circuits* 38(3), 556–560, 2003.
5. Technical Specification Group Radio Access Network, User equipment (UE) radio transmission and reception (FDD) (release 9), 2009, 3rd generation partnership project std. [online]. Available: http://www.3gpp.org/ftp/Specs/archive/25_series/25.101.
6. M. Brandolini, P. Rossi, D. Manstretta, and F. Svelto, Toward multistandard mobile terminals—Fully integrated receivers requirements and architectures. *IEEE Transactions on Microwave Theory and Techniques* 53(3), 1026–1038, 2005.
7. D. Im, H. Kim, and K. Lee, A broadband CMOS RF front-end for universal tuners supporting multi-standard terrestrial and cable broadcasts. *IEEE Journal of Solid-State Circuits* 47(2), 392–406, 2012.
8. R. Gharpurey, Linearity enhancement techniques in radio receiver front-ends. *IEEE Transactions on Circuits and Systems I: Regular Papers* 59(8), 1667–1679, 2012.
9. Y. Tsividis, K. Suyama, and K. Vavelidis, Simple 'reconciliation' MOSFET model valid in all regions. *Electronics Letters* 31(6), 506–508, 1995.
10. S. Kang, B. Choi, and B. Kim, Linearity analysis of CMOS for RF application. *IEEE Transactions on Microwave Theory and Techniques* 51(3), 972–977, 2003.
11. B. Toole, C. Plett, and M. Cloutier, RF circuit implications of moderate inversion enhanced linear region in MOSFETs. *IEEE Transactions on Circuits and Systems I: Regular Papers* 51(2), 319–328, 2004.

12. P.R. Gray, P.J. Hurst, S.H. Lewis, and R.G. Meyer, *Analysis and Design of Analog Integrated Circuits*. New York: Wiley, 2009.
13. S. Narayanan, Transistor distortion analysis using Volterra series representation. *Bell System Technical Journal* 46(3), 991–1024, 1967.
14. S. Narayanan, Application of Volterra series to intermodulation distortion analysis of transistor feedback amplifiers. *IEEE Transactions on Circuit Theory* 17(4), 518–527, 1970.
15. K.L. Fong and R.G. Meyer, High-frequency nonlinearity analysis of common-emitter and differential-pair transconductance stages. *IEEE Journal of Solid-State Circuits* 33(4), 548–555, 1998.
16. J. Long and M. Copeland, A 1.9 GHz low-voltage silicon bipolar receiver front-end for wireless personal communications systems. *IEEE Journal of Solid-State Circuits* 30(12), 1438–1448, 1995.
17. V. Aparin and C. Persico, Effect of out-of-band terminations on intermodulation distortion in common-emitter circuits. *Proceedings of IEEE MTT-S*, vol. 3, 1999, pp. 977–980.
18. K.L. Fong, High-frequency analysis of linearity improvement technique of common-emitter transconductance stage using a low-frequency-trap network. *IEEE Journal of Solid-State Circuits* 35(8), 1249–1252, 2000.
19. V. Aparin and L.E. Larson, Linearization of monolithic LNAs using low-frequency low-impedance input termination. *Proceedings of IEEE ESSCIRC*, 2003, pp. 137–140.
20. R.G. Meyer and A.K. Wong, Blocking and desensitization in RF amplifiers. *IEEE Journal of Solid-State Circuits* 30(8), 944–946, 1995.
21. K.L. Fong, C.D. Hull, and R.G. Meyer, A class AB monolithic mixer for 900-MHz applications. *IEEE Journal of Solid-State Circuits* 32(8), 1166–1172, 1997.
22. C.S. Aitchison, M. Mbabele, M.R. Moazzam, D. Budimir, and F. Ali, Improvement of third-order intermodulation product of RF and microwave amplifiers by injection. *IEEE Transactions on Microwave Theory and Techniques* 49(6), 1148–1154, 2001.
23. S. Lou and H.C. Luong, A linearization technique for RF receiver front-end using second-order-intermodulation injection. *IEEE Journal of Solid-State Circuits* 43(11), 2404–2412, 2008.
24. Y. Ding and R. Harjani, A +18 dBm IIP3 LNA in 0.35 μm CMOS. *Proceedings of IEEE ISSCC*, February 2001, pp. 162–163.
25. T.W. Kim and B. Kim, A 13-dB IIP3 improved low-power CMOS RF programmable gain amplifier using differential circuit transconductance linearization for various terrestrial mobile D-TV applications. *IEEE Journal of Solid-State Circuits* 41(4), 945–953, 2006.
26. M.T. Terrovitis and R.G. Meyer, Intermodulation distortion in current-commutating CMOS mixers. *IEEE Journal of Solid-State Circuits* 35(10), 1461–1473, 2000.
27. D. Webster, J. Scott, and D. Haigh, Control of circuit distortion by the derivative superposition method. *IEEE Microwave and Guided Wave Letters* 6(3), 123–125, 1996.
28. B. Kim, J.-S. Ko, and K. Lee, Highly linear CMOS RF MMIC amplifier using multiple gated transistors and its Volterra series analysis. *Proceedings of IEEE MTT-S*, vol. 1, 2001, pp. 515–518.
29. T.W. Kim, B. Kim, and K. Lee, Highly linear receiver front-end adopting MOSFET transconductance linearization by multiple gated transistors. *IEEE Journal of Solid-State Circuits* 39(1), 223–229, 2004.
30. V. Aparin and L.E. Larson, Modified derivative superposition method for linearizing FET low-noise amplifiers. *IEEE Transactions on Microwave Theory and Techniques* 53(2), 571–581, 2005.
31. S.C. Blaakmeer, E.A.M. Klumperink, D.M.W. Leenaerts, and B. Nauta, Wideband balun-LNA with simultaneous output balancing, noise-canceling and distortion-canceling. *IEEE Journal of Solid-State Circuits* 43(6), 1341–1350, 2008.

32. N.K. Yanduru and K.-M. Low, A highly integrated GPS front-end for cellular applications in 90 nm CMOS. *Proceedings of IEEE DCAS*, 2008, pp. 1–4.
33. D. Manstretta, M. Brandolini, and F. Svelto, Second-order intermodulation mechanisms in CMOS downconverters, *IEEE Journal of Solid-State Circuits* 38(3), 394–406, 2003.
34. M. Brandolini, P. Rossi, D. Sanzogni, and F. Svelto, A +78 dBm IIP2 CMOS direct downconversion mixer for fully integrated UMTS receivers. *IEEE Journal of Solid-State Circuits* 41(3), 552–559, 2006.
35. M.T. Terrovitis and R.G. Meyer, Noise in current-commutating CMOS mixers. *IEEE Journal of Solid-State Circuits* 34(6), 772–783, 1999.
36. R.S. Pullela, T. Sowlati, and D. Rozenblit, Low flicker-noise quadrature mixer topology. *Proceedings of IEEE ISSCC*, February 2006, pp. 1870–1879.
37. W. Redman-White and D.M.W. Leenaerts, 1/f noise in passive CMOS mixers for low and zero IF integrated receivers. *Proceedings of IEEE ESSCIRC 2001*, September 2001, pp. 41–44.
38. B.W. Cook, A. Berny, A. Molnar, S. Lanzisera, and K.S.J. Pister, Low-power 2.4-GHz transceiver with passive RX front-end and 400-mV supply. *IEEE Journal of Solid-State Circuits* 41(12), 2757–2766, 2006.
39. E. Sacchi, I. Bietti, S. Erba, L. Tee, P. Vilmercati, and R. Castello, A 15 mW, 70 kHz 1/f corner direct conversion CMOS receiver. *Proceedings of IEEE CICC*, September 2003, pp. 459–462.
40. Y. Feng, G. Takemura, S. Kawaguchi, N. Itoh, and P. Kinget, A low-power low-noise direct-conversion front-end with digitally assisted IIP2 background self calibration. *Proceedings of IEEE ISSCC*, February 2010, pp. 70–71.
41. J. Borremans, G. Mandal, V. Giannini, T. Sano, M. Ingels, B. Verbruggen, and J. Craninckx, A 40 nm CMOS highly linear 0.4–6 GHz receiver resilient to 0 dBm out-of-band blockers. *Proceedings of IEEE ISSCC*, February 2011, pp. 62–64.
42. D. Ghosh and R. Gharpurey, A power-efficient receiver architecture employing bias-current-shared RF and baseband with merged supply voltage domains and 1/f noise reduction. *IEEE Journal of Solid-State Circuits* 47(2), 381–391, 2012.
43. H. Khatri, P.S. Gudem, and L.E. Larson, Distortion in current commutating passive CMOS downconversion mixers. *IEEE Transactions on Microwave Theory and Techniques* 57(11), 2671–2681, 2009.
44. A. Mirzaei, H. Darabi, J.C. Leete, and Y. Chang, Analysis and optimization of direct-conversion receivers with 25% duty-cycle current-driven passive mixers. *IEEE Transactions on Circuits and Systems I* 57(9), 2353–2366, 2010.
45. H. Khatri, L. Liu, T. Chang, P. Gudem, and L. Larson, A SAW-less CDMA receiver front-end with single-ended LNA and single-balanced mixer with 25% duty-cycle LO in 65 nm CMOS. *Proceedings of IEEE RFIC*, June 2009, pp. 13–16.
46. W.B. Kuhn, N.K. Yanduru, and A.S. Wyszynski, Q-enhanced LC bandpass filters for integrated wireless applications. *IEEE Transactions on Microwave Theory and Techniques* 46(12), 2577–2586, 1998.
47. B. Tenbroek, J. Strange, D. Nalbantis, C. Jones, P. Fowers, S. Brett, C. Beghein, and F. Beffa, Single-chip tri-band WCDMA/HSDPA transceiver without external SAW filters and with integrated TX power control. *Proceedings of IEEE ISSCC*, February 2008, pp. 202–607.
48. L. Franks and I. Sandberg, An alternative approach to the realization of network transfer functions: The N-path filter. *Bell System Technical Journal* 39(5), 1321–1350, 1960.
49. R. Gharpurey and S. Ayazian, Feedforward interference cancellation in narrow-band receivers. *Proceedings of IEEE DCAS*, 2006, pp. 67–70.
50. S. Ayazian and R. Gharpurey, Feedforward interference cancellation in radio receiver front-ends. *IEEE Transactions on Circuits and Systems II* 54(10), 902–906, 2007.
51. H. Darabi, A blocker filtering technique for SAW-less wireless receivers. *IEEE Journal of Solid-State Circuits* 42(12), 2766–2773, 2007.

52. A. Balankutty and P. Kinget, 0.6 V, 5 dB NF, −9.8 dBm IIP3, 900 MHz receiver with interference cancellation. *Proceedings of IEEE VLSI*, 2010, pp. 183–184.
53. A. Safarian, A. Shameli, A. Rofougaran, M. Rofougaran, and F. De Flaviis, Integrated blocker filtering RF front ends. *Proceedings of IEEE RFIC*, June 2007, pp. 13–16.
54. V. Aparin, A new method of TX leakage cancelation in W/CDMA and GPS receivers. *Proceedings of IEEE RFIC*, June 2008, pp. 87–90.
55. T.D. Werth, C. Schmits, R. Wunderlich, and S. Heinen, An active feedback interference cancellation technique for blocker filtering in RF receiver front-ends. *IEEE Journal of Solid-State Circuits* 45(5), 989–997, 2010.
56. A. Ghaffari, E.A.M. Klumperink, M.C.M. Soer, and B. Nauta, Tunable high-Q N-path band-pass filters: Modeling and verification. *IEEE Journal of Solid-State Circuits* 46(5), 998–1010, 2011.
57. J. Weldon, R. Narayanaswami, J. Rudell, L. Lin, M. Otsuka, S. Dedieu, L. Tee, K.-C. Tsai, C.-W. Lee, and P. Gray, A 1.75-GHz highly integrated narrow-band CMOS transmitter with harmonic-rejection mixers. *IEEE Journal of Solid-State Circuits* 36(12), 2003–2015, 2001.
58. A. Molnar, B. Lu, S. Lanzisera, B. Cook, and K. Pister, An ultra-low power 900 MHz RF transceiver for wireless sensor networks. *Proceedings of IEEE Custom Integrated Circuits Conference*, October 2004, pp. 401–404.
59. Z. Ru, N. Moseley, E. Klumperink, and B. Nauta, Digitally enhanced software-defined radio receiver robust to out-of-band interference. *IEEE Journal of Solid-State Circuits* 44(12), 3359–3375, 2009.
60. A. Rafi, A. Piovaccari, P. Vancorenland, and T. Tuttle, A harmonic rejection mixer robust to RF device mismatches. *IEEE ISSCC Digest of Technical Papers*, February 2011, pp. 66–68.
61. T. Forbes, W.-G. Ho, and R. Gharpurey, Design and analysis of harmonic rejection mixers with programmable LO frequency. *IEEE Journal of Solid-State Circuits* 48(10), 2363–2374, 2013.
62. T. Hornak, K. Knudsen, A. Grzegorek, K. Nishimura, and W. McFarland, An image-rejecting mixer and vector filter with 55-dB image rejection over process, temperature, and transistor mismatch. *IEEE Journal of Solid-State Circuits* 36(1), 23–33, 2001.

4 Interference Filtering with On-Chip Frequency-Translated Filters in Wireless Receivers

Hooman Darabi and Ahmad Mirzaei

CONTENTS

4.1 INTRODUCTION

This chapter introduces frequency-translated filters that are used in wireless receivers to deal with interferers. These filters frequency-translate baseband impedances to synthesize high-Q band-pass filters (BPFs) with center frequencies precisely controlled by a clock frequency [1–6]. These filters are composed of only MOS switches and capacitors, making them ideal for integration while they follow the technology scaling.

The chapter is organized as follows. First, we briefly discuss that why wireless receivers need external surface acoustic wave (SAW) filters and why these filters must be removed and and replaced with integrated counterparts. Then, the simplest form of a frequency-translated filter is introduced, which is the conventional passive mixer driven by 50% duty-cycle local oscillator (LO) clocks. It is revealed that while this filter offers a high-Q filtering, it suffers from the image problem, preventing it from being useful as on-chip high-Q filters. Moving forward, four-phase filters are presented to eliminate the image issue while preserving the high-Q filtering. Then, the design of a SAW-less fully integrated quad-band 2.5G receiver is described, which uses these on-chip high-Q filters to deal with blockers. Next, transfer functions of the four-phase filters are derived, and effects of various imperfections such as LO phase noise, second-order nonlinearity, and thermal noise of switches are discussed.

4.2 GENERAL CONSIDERATIONS IN WIRELESS RECEIVERS

In any wireless standard, the receiver must satisfy a certain blocking template defined at various blocker frequencies and levels. For instance, as shown in Figure 4.1, in the Global System for Mobile Communications (GSM) standard, a desired signal only 3 dB above the sensitivity could be accompanied by an out-of-band blocker that can be as large as 0 dBm and as close as only 80 MHz to the edge of the Personal Communication Services (PCS) band. Since the desired signal is weak, the gain of the low-noise amplifier (LNA) must be adequately high. Thus, the blocker must be filtered out prior to reaching the amplifier's input in order not to compress it.

Due to the modest quality factor (Q) of on-chip inductors, conventionally, out-of-band blockers are attenuated by an off-chip SAW filter placed prior to the LNA. Being expensive and bulky, it is desirable to remove the SAW filter especially in multiband applications such as cellular. In addition to the cost implications, SAW filters not only degrade the receiver's sensitivity due to their inevitable insertion loss, they also remove the flexibility of sharing LNAs in multiband applications.

In the conventional quad-band 2.5G receiver, the use of four external SAW filters is inevitable due to the stringent out-of-band blocking requirements of GSM (Figure 4.2a). Also, each band requires three matching components, which amounts to a total of 16 external RF components. Using on-chip four-phase filters, Ref. [7] was able to remove the entire external filters including their matching components, saving a considerable amount of bill of material (Figure 4.2b). Moreover, the receiver in Ref. [7] features two on-chip baluns with only two single-ended inputs, one for the

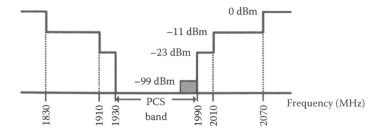

FIGURE 4.1 GSM out-of-band blocking profile for the PCS band.

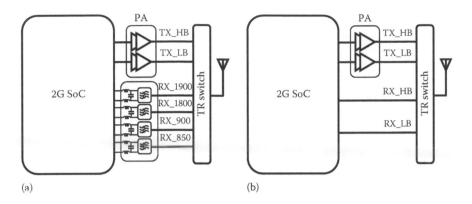

(a)   (b)

FIGURE 4.2   (a) Conventional quad-band 2.5G SoC with SAW filters. (b) Saw-less quad-band 2.5G SoC proposed in Ref. [7] using integrated four-phase filters.

high band and one for the low band. Consequently, the number of receiver RF inputs from the typical of eight is reduced to two, which substantially simplifies the design of the package and the PCB, especially for SoC where a large portion of the package is dedicated to the baseband routing.

4.3   TWO-PHASE HIGH-Q FILTERING

As shown in Figure 4.3, an RF current I_{RF} is commutated by mixer switches clocked at a rate f_{LO} with rail-to-rail square-wave differential LOs. The differential downconverted currents flow into baseband impedances Z_{BB} and become baseband voltages across them. Due to the lack of reverse isolation of passive mixers, these baseband voltages are upconverted back to the RF, becoming an RF voltage at the RF side of the switches. For simplicity, as shown in Figure 4.3, assume that the RF current is a single tone at $f_{LO} + f_m$ and this current is commutated by the switches clocked at a rate f_{LO}. It can be shown that around f_{LO} the resulting RF voltage has two frequency components: one at the incident frequency, $f_{LO} + f_m$, called the main frequency, and one

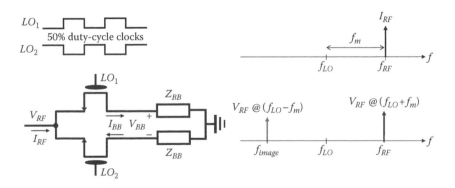

FIGURE 4.3   Operation principals of a 50% passive mixer.

at the image frequency, $f_{LO} - f_m$. Phasors of the two main and image components of the RF voltage are given by these expressions [8]:

$$V_{RF} @ (f_{LO} + f_m) = \left[R_{SW} + \frac{2}{\pi^2} Z_{BB}(jf_m) \right] I_{RF} \qquad (4.1)$$

$$V_{RF} @ (f_{LO} - f_m) = \frac{2}{\pi^2} Z_{BB}(-jf_m) I_{RF}^* \qquad (4.2)$$

and the significance of these expressions will be discussed soon.

If we focus on only the main component of the RF voltage, assuming that the excitation is an RF current with infinite output impedance, the input impedance would be given by this expression:

$$Z_{in} \cong R_{SW} + \frac{2}{\pi^2} \left\{ Z_{BB} \left(s - j\omega_{LO} \right) + Z_{BB} \left(s + j\omega_{LO} \right) \right\} \qquad (4.3)$$

which is equal to the switch resistance in series with a BPF. In fact, this BPF is the same as the low-Q low-pass filter that is frequency-shifted to the RF to become a high-Q band pass. So, with two MOS switches and two 180° out-of-phase LO clocks, we were able to synthesize a high-Q BPF with a tunable center frequency precisely controlled by the LO clock.

Phasors of the two main and image components of the RF voltage are given by (4.1) and (4.2). According to (4.1) and (4.2), if the distance between the incident frequency and the LO is larger than the baseband filter's bandwidth, the resulting RF voltage will be lowered by the same amount at both the main and the image frequencies. In other words, for both frequency components, the baseband impedance is shifted to f_{LO} to construct a built-in high-Q BPF.

We surmise that this impedance transformation property can be utilized to attenuate unwanted blockers. As shown in Figure 4.4, assume that the high-Q two-phase

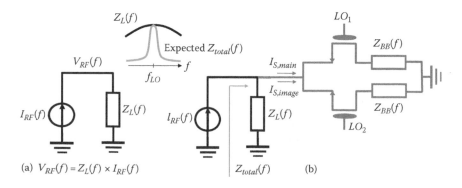

FIGURE 4.4 High-Q filtering of a two-phase filter. (a) LTS impedance with modest Q. (b) High Q filtering by adding a two-phase filter.

band-pass impedance is placed at any point inside a receiver front end. We expect that this arrangement would exhibit a low-impedance path to those blockers whose distance to ω_{LO} is greater than the bandwidth of the low-pass filter while maintaining a large input impedance for frequency components of interest close to ω_{LO}. To understand the concept, let us turn our attention to Figure 4.4a where an RF current $I_{RF}(f)$ flows into an RF impedance $Z_L(f)$. This arrangement can be the Norton equivalent of an RF node inside the receiver front end. Owing to the limited Q of on-chip inductors, $Z_L(f)$ would exhibit either a low-Q band-pass response centered at the desired RF or a low-pass response with a bandwidth large enough to pass the desired signal without a significant attenuation. The transfer function from the RF current to the RF voltage across the RF impedance is simply equal to $Z_L(f)$ plotted in Figure 4.4, and the resulting filtering is generally not sharp.

Now, as shown in Figure 4.4b, let us assume that a two-phase switching system is connected to the same RF node. To understand the effect, we need to evaluate the resulting RF voltage across Z_L. Ideally, we would like the desired signal around the LO not to experience much attenuation with respect to the original RF system, while far-out blockers are significantly attenuated by the switching system.

To evaluate the filtering performance of the proposed high-Q filter, it must be noted that based on our previous discussions the voltage across Z_L shall contain components at the main and image frequencies.* So, after some math, these two RF voltage components are found to be

$$V_{RF,main} =$$

$$\left[R_{SW} \parallel Z_L(f) + \frac{\left(\dfrac{Z_L(f)}{Z_L(f)+R_{SW}} \right)^2 \dfrac{2}{\pi^2} Z_{BB}\left(f-f_{LO}\right)}{1+\dfrac{2}{\pi^2} Z_{BB}\left(f-f_{LO}\right) \sum_{k=-\infty}^{+\infty} \dfrac{1}{(2k+1)^2 [Z_L(2kf_{LO}+f)+R_{SW}]}} \right] \times I_{RF}(f)$$

$$(4.4)$$

$$V_{RF,image} = \frac{\left(\dfrac{Z_L(f)}{Z_L(f)+R_{SW}} \right)^2 \dfrac{2}{\pi^2} Z_{BB}(f-f_{LO})}{1+\dfrac{2}{\pi^2} Z_{BB}\left(f-f_{LO}\right) \sum_{k=-\infty}^{+\infty} \dfrac{1}{(2k+1)^2 [Z_L(2kf_{LO}+f)+R_{SW}]}} \times I_{RF}^{*}(f)$$

$$(4.5)$$

Equations 4.4 and 4.5 are not friendly at all, so we use a graphical approach presented in Figure 4.5 to explain their implications. Assume that the input frequency is swept from f_{LO} to upper frequencies. The transfer function for the RF voltage at the main frequency is plotted in Figure 4.5a, which is rightfully a high-Q band pass. However, around f_{LO} where the desired signal is located, the gain drops by approximately 6 dB

---

* Plus frequency components around $3\omega_{LO}$, $5\omega_{LO}$, and beyond, which are ignored for now.

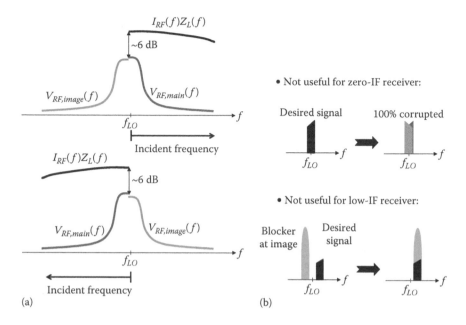

FIGURE 4.5 Image problem of a two-phase high-Q filter. (a) Transfer functions. (b) Explanation of image problem in zero and low-IF receivers.

with respect to the original transfer function with no two-phase high-Q filter attached. The image component created on the left side of f_{LO} is also high-Q band pass and is roughly the mirror of the main transfer function. Similarly, if the excitation frequency is swept over frequencies below f_{LO}, we will have the main and the image transfer functions where both are high-Q band pass with 6 dB gain loss around f_{LO}.

Because of the image component, the two-phase filter is useless for both zero- and low-IF receivers. For a zero-IF receiver, the resulting RF voltage would be composed of the desired RF signal plus its own image aliased and both would have equal strengths (Figure 4.5b). Having the same strengths for the image and the main components is the worst possible situation eliminating the possibility of removing the image and extracting the desired signal. For a low-IF receiver, a blocker at the image frequency, which can be orders of magnitude stronger than the desired signal, would be folded on top of the desired signal and would completely destroy it (Figure 4.5b). Therefore, because of the image issue, the two-phase high-Q filter is useless as is and needs to be modified.

4.4 FOUR-PHASE HIGH-Q FILTERING

To get rid of the image, we can increase the number of clock phases to four, leading to the four-phase filter presented in Figure 4.6. In the four-phase filter, four MOS switches are attached to four baseband low-pass impedances on one side and the switches are tied together on the other side. The switches are clocked by four-phase 25% duty-cycle LO clocks with an angular frequency of ω_{LO}, and they are

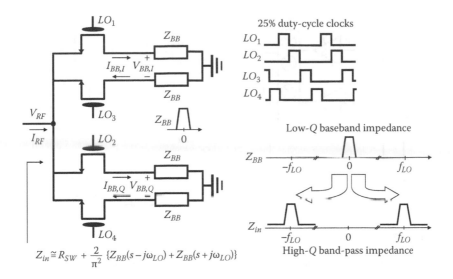

The impedance transformation formula shown in the figure:

$$Z_{in} \cong R_{SW} + \frac{2}{\pi^2}\{Z_{BB}(s - j\omega_{LO}) + Z_{BB}(s + j\omega_{LO})\}$$

FIGURE 4.6   Impedance transformation of a four-phase filter.

progressively phase-shifted by 90°. It can be shown that the input impedance is given by the same expression as that of the two-phase filter [9]:

$$Z_{in} \cong R_{SW} + \frac{2}{\pi^2}\left\{Z_{BB}\left(s - j\omega_{LO}\right) + Z_{BB}\left(s + j\omega_{LO}\right)\right\} \tag{4.6}$$

meaning that the impedance seen from the RF side of the switches is that of the baseband impedance shifted in frequency to ω_{LO} (in series with the switch resistance). Therefore, a simple low-Q low-pass filter is translated to a high-Q band pass whose center is very well controlled by the frequency of the clock ω_{LO}. More importantly, the image problem is solved.

If we place this four-phase filter at any point inside a receiver front end, this arrangement would exhibit a low-impedance path to those blockers whose distance to LO is greater than the bandwidth of the low-pass filter while maintaining a relatively large input impedance for the frequency components of interest nearby ω_{LO}. In other words, we have now achieved a high-Q BPF with no image problem. We will discuss about the practical issues of this four-phase filter and its robustness against some implementation-related imperfections. But first, let us discuss the design of a 2.5G SAW-less receiver, which makes use of four-phase filters to deal with out-of-band blockers.

4.5   SAW-LESS QUAD-BAND 2.5G RECEIVER USING FOUR-PHASE FILTERS

Figure 4.7 shows the overall receiver block diagram integrated as a part of 2.5G SoC. It uses a low-IF architecture with active RC filters providing partial channel selection. The low-pass filter is a third-order Butterworth design with the real pole in the

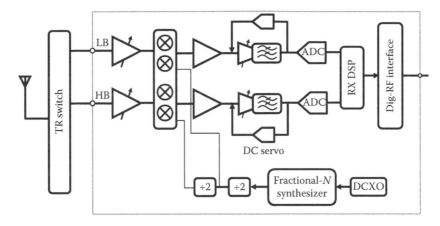

FIGURE 4.7   Block diagram of the quad-band SAW-less 2.5G receiver.

first stage and a biquad stage next. Analog DC offset cancelation is provided by HPF servo loops and the residue of DC offset is corrected in the digital signal processing (DSP). The LPF is followed by a 14-bit ADC that provides the digital spectrum to the receiver (RX) DSP unit, which is responsible for several functionalities such as full channel selection, equalization, and the image rejection.

For receiving GSM/GPRS/EDGE signals with 200 kHz channel bandwidth, the baseband impedance Z_{BB} in the high-Q BPF is chosen to be simply a capacitor, C_{BB}, which is shown in Figure 4.8. The BPF is more suitable to be implemented differentially in order to make it robust to common-mode noise sources. To receive GSM signals, since the out-of-band blockers can be as close as 20 MHz to the desired signal, the impedance of the baseband capacitor C_{BB} at 20 MHz must be low enough

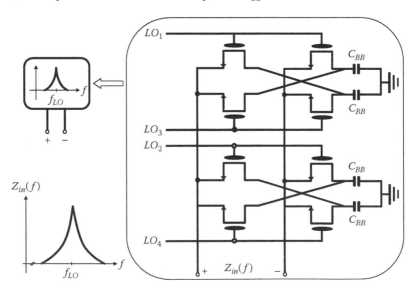

FIGURE 4.8   Utilized on-chip high-Q BPF.

to adequately attenuate the 0 dBm blocker without allowing any gain compression. This determines the minimum size for C_{BB}. Depending on the availability of space on the chip, a larger C_{BB} results in a more attenuation even for blockers that are closer to the desired signal. The switch resistance eventually limits the attenuation roll-off at far-out frequencies.

Illustrated in Figure 4.9, an on-chip balun converts the single-ended receiver input from the antenna switch to a differential signal to drive the inputs of a differential LNA. The secondary of the transformer is tuned to the desired band, to ensure an S_{11} of less than −10 dB seen from the input of the RF IC. This single-ended to differential conversion provides a voltage gain of close to 9 dB, resulting in a simulated receiver noise figure of 2.8 dB. The LNA is a common-source cascode amplifier loaded by a differential tuned inductor. To prevent the saturation of input devices of the LNA by the 0 dBm blocker, an on-chip four-phase high-Q BPF is differentially placed at the LNA inputs. This BPF provides a low input impedance for far-out blockers without impacting the desired signal much, yet causing out-of-band blockers to be significantly attenuated at this stage. The blockers now have experienced enough attenuation to the level not to cause any significant gain compression or intermodulation at the input devices of the LNA. However, this attenuation is not sufficient for the 0 dBm blocker at the 20 MHz offset, which is required for low band. Therefore, the second four-phase high-Q BPF is utilized at the inputs of the cascode devices. This additional filtering guarantees that the attenuated blockers do not cause large voltage swings at the LNA output, while the desired signal is experiencing the intended amplification by the LNA. Furthermore, the downconversion

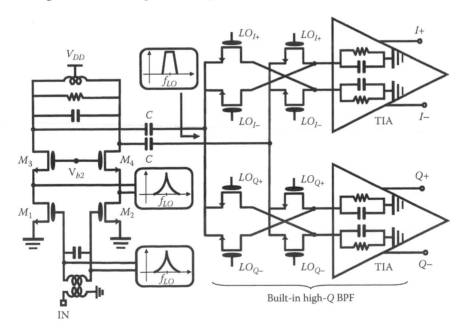

FIGURE 4.9 Front-end circuits of the quad-band 2.5G SAW-less receiver with high-Q four-phase filters.

current-driven passive mixer along with the low-pass response of the transimpedance amplifier can be utilized as the third high-Q BPF. For this purpose, the switches
of the passive mixer need to be driven by the same 25% duty-cycle signal that clocks
the two other high-Q BPFs.

There is an on-chip synthesizer operating at double frequency and an on-chip
divide-by-two circuit provides quadrature 50% duty-cycle rail-to-rail clocks at the
desired LO frequency. Using four AND gates, these 50% duty-cycle quadrature signals are utilized to generate the required 25% duty-cycle quadrature clocks, and after
being buffered, they drive switches of the high-Q BPFs and the downconversion
passive mixer. The clock buffers are slightly skewed to make the generated 25%
clocks nonoverlapped and the amount of nonoverlapping is kept very small. It can be
shown that a small nonoverlapping doesn't alter the nature of the input impedance.
On the contrary, the overlapped clocks can potentially make the two switches turn
on simultaneously, causing unwanted charge sharing between the two corresponding
capacitors.

The chip was fabricated in 65 nm digital CMOS and is verified both as a stand-
alone receiver and with the baseband running in the call mode in the platform level.
The die photo of the receiver is shown in Figure 4.10. It occupies an active area of
2.4 mm^2.

When the on-chip BPFs are disabled, as plotted in Figure 4.11a, the measured S_{11}
is below −10 dB across both the low and high bands. With the high-Q BPF at the
secondary of the transformer enabled, the measured S_{11} is less than −10 dB only at
a very narrow frequency range around the RX-LO and sharply becomes larger than
−5 dB at other frequencies (Figure 4.11b). The narrow frequency band where S_{11}

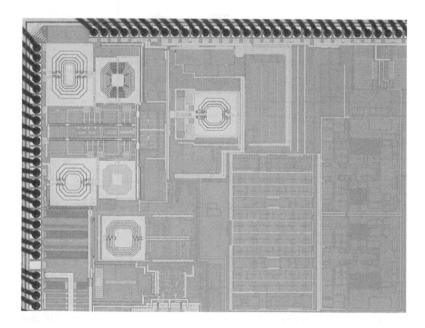

FIGURE 4.10 Receiver die photo.

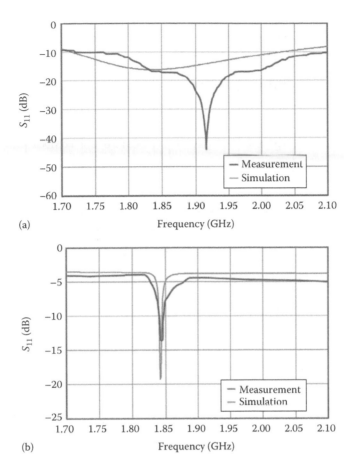

(a)

(b)

FIGURE 4.11 Measured S_{11}: (a) when the high-Q four-phase BPF is disabled; (b) when the high-Q filter is enabled.

is less than −10 dB moves with the RX-LO. This phenomenon can be explained as follows: when the four-phase BPF at the secondary of the transformer is activated, it exhibits high input impedance at a very narrow frequency range around the LO; hence, it is expected not to load the secondary of the transformer much around the RX-LO. That's why around the LO the measured S_{11} is not degraded. However, at higher offset frequencies, the four-phase filter exhibits a large capacitive input impedance, which increases the return loss as a result. This can potentially offer even more filtering through mismatching the antenna impedance at frequencies away from the desired channel.

The receiver noise figure measured across all receiver bands and channels is about 3.1 dB (Figure 4.12a). This is about 0.5 dB higher than a typical cellular receiver noise figure. However, given that the SAW filter adds an additional loss of about 1.5 dB, the overall noise figure of the system is 1 dB better. The corresponding voice sensitivity in the PCS band is better than −110 dBm shown in Figure 4.12b, which is

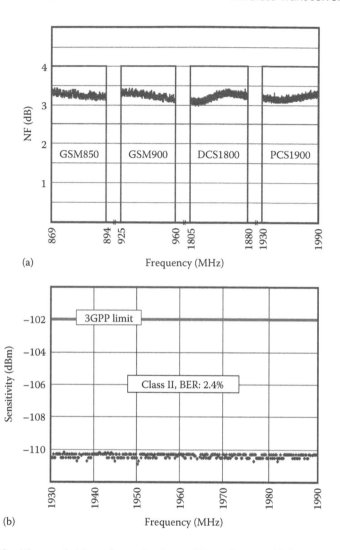

FIGURE 4.12 Measured: (a) receiver noise figure; (b) receiver sensitivity.

about 1 dB better than the similar radios with external SAW filters especially around the edges of the PCS band. Note that the four-phase BPFs are nominally disabled and only need to turn on at infrequent events of receiving a large out-of-band blocker detected through a wideband RSSI circuit.

The NF of the receiver versus the blocker power at ±80 and ±20 MHz offsets for the PCS band is shown in Figure 4.13a. For a 0 dBm blocker at ±80 MHz, the measured blocker NF is 11.4 dB, which is well below the 15 dB estimated requirement of 3GPP. For a blocker at an offset of ±20 MHz, a −11 dBm blocker results in a measured NF of 10.9 dB, which is again well below the target.

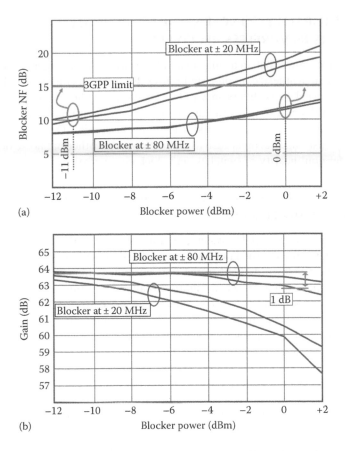

FIGURE 4.13   Measured: (a) receiver blocker noise figure; (b) receiver gain versus blocker power.

All other bands satisfy the requirements as well, and only the more challenging PCS band is shown as an example. As shown in Figure 4.13b, a 0 dBm blocker at an offset of ±80 MHz compresses the receiver gain by only 0.8. Also, for a blocker at an offset of ±20 MHz, a −11 dBm blocker compresses the RX gain by only 0.3 dB.

At the maximum gain, the measured in-band IIP3 is better than −12.4 dBm and IIP2 is measured better than +45 dBm for the high band and +50 dBm for the low band. The measured 600 kHz blocker NF is 6.9/4.6 dB for the high/low bands, respectively. The 3 MHz blocker NF is 8.6/7.3 dB. The receiver adjacent and in-band blocking performance is measured at the platform level, which agrees well with the device measurements. In all cases, the 3GPP requirement is met with margin. The receiver dissipates 55 mA, which includes the RX-PLL and clock generation circuitry.

4.6 NONIDEALITY CONCERNS OF FOUR-PHASE FILTERS IN WIRELESS RECEIVERS

So far, we have shown that an arrangement of four MOS switches and four baseband impedances can synthesize an on-chip high-Q BPF.* The technique can be utilized to replace external SAW filters in wireless receivers. Although the employed on-chip high-Q BPF has a simple structure, numerous shortcomings need to be studied. This section analyzes the robustness of the four-phase high-Q filters in SAW-less receivers against imperfections such as clock phase noise, thermal noise of switches, second-order nonlinearity of switches, and clock phase error.

4.6.1 IMPACT OF RF HARMONIC UP- AND DOWNCONVERSIONS ON TRANSFER FUNCTION

In a linear time-invariant (LTI) system, regardless of the type of the excitation voltage or current, the input impedance, which is defined as the Fourier or Laplace transform of the port voltage divided to the port current, is independent of the output impedance of the excitation (Z_L in Figure 4.14a) and is a function of only the LTI network and its constituent elements. This is not the case in a linear time-variant (LTV) system. As shown in Figure 4.14b, let us assume that a four-phase filter is connected to an RF node, which is modeled by its Norton equivalent system, that is, an RF current with an output impedance equal to Z_L. The RF current is commutated by the switches of the four-phase filter, downconverted, integrated into the baseband impedances, and eventually low-pass filtered. Due to the lack of a reverse isolation, the baseband voltages across the baseband impedances are upconverted back to the RF to become an RF voltage that contains frequency components not only at the incident frequency but also at the third, fifth, and all other odd harmonics. These high-frequency components of the RF voltage appear across Z_L, initiating RF

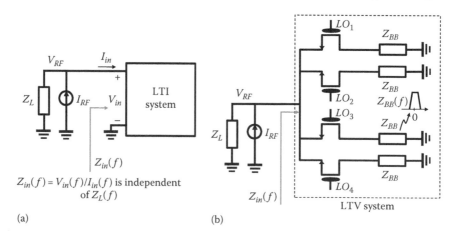

$Z_{in}(f) = V_{in}(f)/I_{in}(f)$ is independent of $Z_L(f)$

(a)   (b)

FIGURE 4.14 (a) LTI versus (b) LTV systems.

---

* In a differential implementation of the four-phase filter, eight switches are needed.

currents at these harmonics. Due to the harmonics of the LO clocks, these frequency components of the RF current are downconverted back to the baseband, modifying the original baseband voltages. In the steady state, once we satisfy KCL and KVL at all nodes and branches, it can be proved that the input impedance around the LO frequency is given by the following expression:

$$Z_{in}(f) \| Z_L(f) = R_{SW} \| Z_L(f)$$

$$+ \frac{\dfrac{2}{\pi^2} Z_{BB}(f - f_{LO})}{1 + \dfrac{2}{\pi^2} Z_{BB}(f - f_{LO}) \displaystyle\sum_{k=-\infty}^{+\infty} \dfrac{1}{(2k+1)^2 [Z_L(2kf_{LO} + f) + R_{SW}]}} \tag{4.7}$$

which is coupled to Z_L through the previous complicated format. In fact, Z_{in} depends not only on Z_L at frequencies around the LO but also on the values of Z_L at other odd harmonics of the LO clock. Because of this effect, the input impedance seen from the RF side is not infinite even though the baseband impedances are infinite at DC.

Similarly, we can find the RF voltage versus the excitation current I_{RF} (Figure 4.15), given by this expression:

$$\frac{V_{RF}(f)}{I_{RF}(f)} = R_{SW} \| Z_L(f) + \frac{\dfrac{2}{\pi^2} Z_{BB}(f - f_{LO})}{1 + \dfrac{2}{\pi^2} Z_{BB}(f - f_{LO}) \displaystyle\sum_{k=-\infty}^{+\infty} \dfrac{1}{(2k+1)^2 [Z_L(2kf_{LO} + f) + R_{SW}]}}$$

$$\tag{4.8}$$

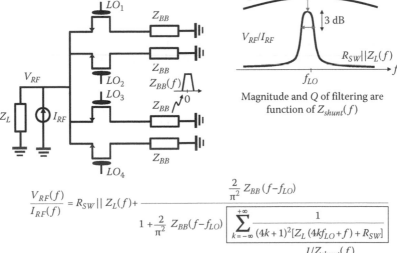

FIGURE 4.15   Impact of harmonic up- and downconversions on gain and Q of transfer function.

If the four-phase filter were not connected to this node, the transfer function would simply be Z_L. When the four-phase filter is connected, even if the baseband impedances are capacitors that are infinite at DC, the input impedance becomes finite. As a result, the transfer function at the LO frequency drops below the ideal target. This infinite summation is defined as the shunt admittance (Figure 4.15). The Q of the filtering and the drop at the center frequency both are functions of this shunt admittance. Also, another important thing is the attenuation at far-out offset frequencies where the baseband impedance diminishes to zero. This far-out attenuation is simply equal to the switch resistance in parallel with the RF impedance Z_L. Thus, in order to increase this attenuation, the switch resistance must be adequately lowered.

For example, in the special case where the RF impedance is a resistor of size R_L and the baseband impedances are simply capacitors of size C_{BB}, this drop in the transfer function is equal to 1.9 dB, and the 3 dB bandwidth is given by this expression:

$$f_{-3dB} = \frac{1}{2\pi R_L C_{BB}} \tag{4.9}$$

which is a function of only R_L and C_{BB}. That is why by increasing the LO frequency a higher Q can be achieved. The transfer function of the RF voltage versus the input current is given by this equation:

$$\frac{V_{RF}(f)}{I_{RF}(f)} = \frac{R_L}{R_L + R_{SW}} \left[R_{SW} + \frac{\frac{8}{\pi^2} R_L}{1 + j4(R_L + R_{SW})C_{BB}(f - f_{LO})} \right] \tag{4.10}$$

In reality, we cannot have a resistor with infinite RF bandwidth. As shown in Figure 4.16, we can model this finite bandwidth by adding the capacitor C_L in parallel with Z_L. It is interesting to study the effect of this capacitor. From the forgoing discussions, we realize that the harmonic effect lowers the Q and leads to a gain drop at the LO frequency. Due to the presence of the capacitor C_L, the overall impedance of Z_L at the harmonics decreases, increasing the shunt admittance. In other words, the harmonic effect becomes even much stronger, leading to further drop at the peak frequency and lower Q as the capacitor C_L increases (Figure 4.16). Also, since the shunt admittance has a positive imaginary part, the center frequency is slightly shifted toward the left and this shift continues as C_L increases.

4.6.2 IMPACT OF LO PHASE NOISE

As shown in Figure 4.17, assume that a weak desired signal around f_{LO} is accompanied by a strong blocker at f_b, where $|f_b - f_{LO}| \ll f_{LO}$. The LO is not clean and is noisy. Each of the baseband capacitors holds three types of voltage [1]: (1) the

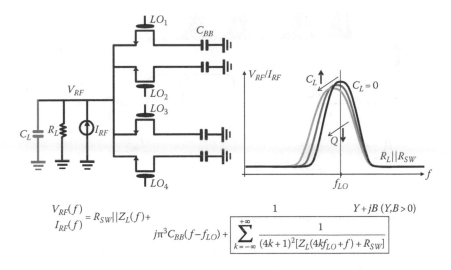

$$\frac{V_{RF}(f)}{I_{RF}(f)} = R_{SW}||Z_L(f) + \cfrac{1}{j\pi^3 C_{BB}(f-f_{LO}) + \displaystyle\sum_{k=-\infty}^{+\infty} \cfrac{1}{(4k+1)^2[Z_L(4kf_{LO}+f) + R_{SW}]}} \qquad Y+jB \; (Y,B>0)$$

FIGURE 4.16   Impact of harmonic up- and downconversion on transfer function.

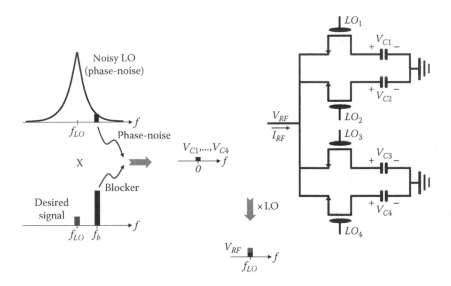

FIGURE 4.17   Impact of LO phase noise.

downconverted desired signal around DC, (2) the downconverted and attenuated blocker at $f_b - f_{LO}$, and (3) the reciprocally downconverted blocker by the LO phase noise, which is around DC as well. Now, looking back from the RF side, the first two types of voltages are upconverted back to the RF, creating components of the RF voltage at the desired and the blocker frequencies, which is the intended functionality of the high-Q filter. However, the third type of the capacitor voltages that are emerged from the LO phase noise are also upconverted back to the RF and appear as an RF noise voltage around the LO, where the desired signal is located.

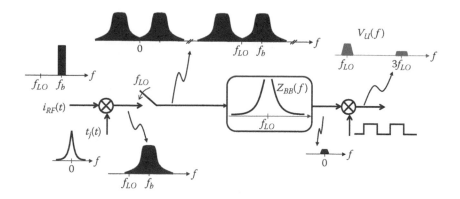

FIGURE 4.18 Modeling impact of LO phase noise.

We can perform all the math to find the unwanted series voltage that models the phase-noise impact. The final equations turn out to be complicated and unfriendly [1] but can graphically be described as shown in Figure 4.18. The incoming RF current $i_{RF}(t)$ is multiplied to the LO jitter. The blocker is assumed to be a narrowband located at f_b, and the jitter has a low-pass spectrum. Thus, the power spectral density of the jitter times the RF current is concentrated at the blocker frequency as well. Now, the result is impulse-sampled uniformly at the rate of f_{LO}, which as shown in Figure 4.18 replicates the spectrum at integer multiples of f_{LO}. Consequently, the sampled signal would have frequency components around DC. The impulse samples go through baseband filtering defined by the baseband impedance of the four-phase filter. Hence, the baseband impedance Z_{BB} filters out the high-frequency components including those at the clock frequency and its harmonics. The resulting low-pass signal is multiplied by the 50% LO clock and is frequency-translated to f_{LO}, its harmonics increasing the receiver noise floor.

For example, when the four-phase filter is used for a 2.5G receiver [7], a 0 dBm blocker at 20 MHz offset or above and a phase noise of −160 dBc/Hz at the blocker frequency result in an input-referred noise floor of −167.5 dBm/Hz. This noise floor is translated to a blocker NF of 6.5 dB, well below the 15 dB target specified by the 3GPP.

4.6.3 IMPACT OF SECOND-ORDER NONLINEARITY

Let us assume that there is an amplitude-modulated blocker in the RF current, which is represented as $i_b(t) = I_b(t)\cos(2\pi f_b t)$ in Figure 4.19. The voltage–current relationship of each switch in the high-Q BPF can be modeled as $v_{SW} = R_{SW}i_{SW} + \beta i_{SW}^2$, in which v_{SW} represents the voltage drop across the switch, i_{SW} is the current flowing through the switch, R_{SW} is the switch resistance, and β is the coefficient of the second-order nonlinearity. It can be shown that if the switches are perfectly matched, due to the second-order nonlinear term, the blocker current would appear as $\beta i_b^2(t)$ at the RF side of the switches. This unwanted term would pose no harm as it would not create any frequency components at the desired signal frequency. $\beta i_b^2(t)$ would generate

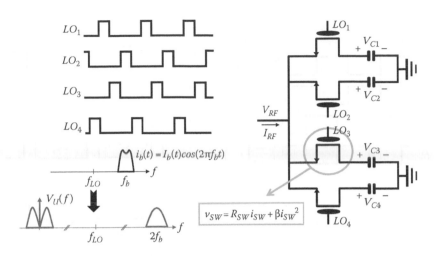

FIGURE 4.19   Impact of second-order nonlinearity.

components either at DC or around $2f_b$, none of which is harmful. However, the mismatches are inevitable, and it can be shown that due to the presence of mismatches low-frequency components of the AM blocker are upconverted to around f_{LO}, increasing the receiver noise floor. Since the switches are sized to be relatively large for a lower *on* resistance, the second-order nonlinearity typically poses no threat.

4.6.4   IMPACT OF QUADRATURE ERROR

As shown in Figure 4.20, assume that there is a phase error of $\Delta\varphi$ in the original quadrature 50% duty-cycle clocks. First, we must find out how this phase error shapes outputs of the AND gates in the 25% clock generator circuit. As demonstrated in Figure 4.20, as a result of the phase error $\Delta\varphi$ in the 50% clocks, only the duty cycles of the 25% clocks change and no phase error among them occurs. In other words, fundamental harmonics of LO_1, ..., LO_4 are still phase-shifted by 0°, 90°, 180°, and 270°, respectively. The duty cycles of one pair of LOs, LO_1 and LO_3, increase and those of the other pair, LO_2 and LO_4, decrease. Now, assume that the incident RF current is a tone at $f_{LO} + f_m$ whose phasor is denoted by I_{sig}, in which $f_m \ll f_{LO}$ and without loss of generality f_m is assumed to be positive. It can be shown that the resulting RF voltage will have two frequency components. The first component is at the incident frequency, $f_{LO} + f_m$, whose phasor is calculated to be $\{R_{SW} + (2/\pi^2)Z_{BB}(f_m)\}I_{sig}$ and is the desired component. The second component, which we call the image component, appears at the image frequency, $f_{LO} - f_m$, which is derived to be $\Delta\varphi \times \{R_{SW} + (2/\pi^2)Z_{BB}(f_m)\}I_{sig}$. This component is undesired.

For example, in a low-IF receiver, this phenomenon along with the downconversion mixer of the receiver can fold the image components to on top of the desired signal. In a zero-IF receiver, the desired signal will be flipped over its center and is added to itself. It should be mentioned that because the mismatch is typically weak, the image folding can be corrected in the digital domain.

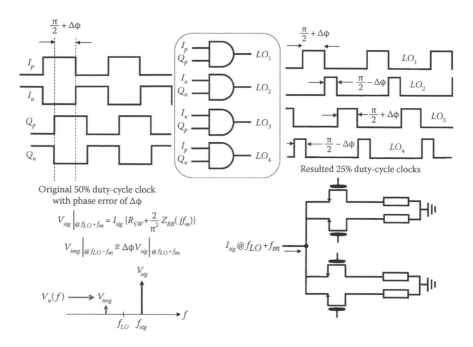

FIGURE 4.20   Impact of quadrature phase error in LO clocks.

4.6.5 THERMAL NOISE OF SWITCHES

Since the switches of the four-phase BPF carry no DC currents, they do not contribute any flicker noise. However, like any switched capacitor circuits, thermal noise of the switches needs to be carefully studied. In a MOS device, white noise originates from thermal noise due to the conductance of the channel (conductance of the inversion layer). Since the proposed four-phase BPF is an LTV system, it is simpler to utilize the Thevenin theorem (Figure 4.21). According to this theorem, any linear one-port circuit can be replaced with an equivalent circuit called the Thevenin

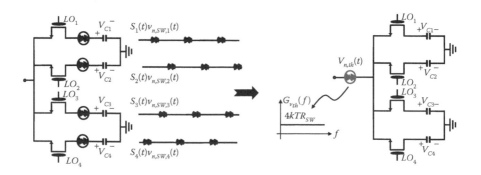

FIGURE 4.21   Modeling thermal noise of switches.

equivalent, where the branch voltages and branch currents of other circuits attached to this circuit do not change. The Thevenin equivalent of a linear one-port circuit is the same circuit but with zeroed independent voltage and current sources in series with the open-circuit voltage called the Thevenin voltage. In fact, the zeroed system is the same four-phase BPF but with noiseless switches.

The Thevenin voltage, $v_{n,th}(t)$, in Figure 4.21 is equal to: $S_1(t)v_{n,sw,1}(t) + S_2(t)v_{n,sw,2}(t) + S_3(t)v_{n,sw,3}(t) + S_4(t)v_{n,sw,4}(t)$, where $S_i(t)v_{n,sw,i}(t)$ is the ith, $i = 1, ..., 4$, cyclostationary noise contributed by the ith switch. $S_i(t)$ is equal to 1, when the corresponding LO is high and is zero otherwise. The white noise sources from the four switches are statistically independent, and all hold identical statistical properties. Knowing that at two different moments a white noise source has statistically independent values, $v_{n,th}(t)$ also turns out to be a white noise voltage with a power spectral density of $4kTR_{SW}$. Thus, the four phase BPF with noisy switches can be equivalently replaced with a noiseless BPF in series with a white noise voltage whose power spectral density is equal to that of the thermal-noise voltage of a resistor of size R_{SW}.

Since the BPF exhibits large input impedance and hence large Thevenin impedance at frequencies around the LO and its harmonics, the transfer function by which those noise components appear at the desired signal band would be weak. As shown in Figure 4.22, in a special case when the RF impedance is a resistor equal to R_L and the baseband impedances are capacitors of size C_{BB}, the power spectral density coming from switch noises across R_L around the LO is found to be given by $4kTR_{SW}(1 - 8/\pi^2)$. We can calculate the power spectral density of noise from R_L across R_L too, which is equal to $4kTR_L(8/\pi^2)$. Comparing the previous two quantities, one can readily observe that the switch noise contribution is negligible compared to the noise of R_L.

4.6.6 HARMONIC FOLDING

Let us assume that there is a blocker around $3f_{LO}$ received by the receiver antenna. Passing through the antenna switch and the package (no SAW filter exists), this blocker

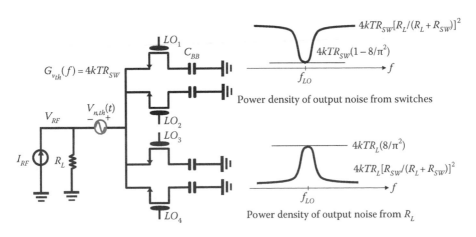

FIGURE 4.22 Impact of thermal noise of switches.

experiences some attenuation but is never fully suppressed. Eventually, this blocker is harmonically downconverted to on top of the desired signal that is around f_{LO}. As a result, the receiver noise floor may increase. Similarly, a blocker around fifth harmonic can be harmonically downconverted and aliased to on top of the desired signal. Since the attenuation from the package and the antenna switch is not sufficient, an exception must be taken for out-of-band blockers at $3f_{LO}$ and $5f_{LO}$. For example, in the SAW-less receiver of Ref. [7], the blocker level at these frequencies is relaxed from 0 to −43 dBm.

4.7 CONCLUSIONS

We realize that one can synthesize high-Q filters by frequency translation of low-Q baseband filters. Being constructed from only MOS switches and capacitors makes these filters very friendly for integration and they follow the technology scaling. These filters offer high-Q BPFs with a Q proportional to the size of the constituent baseband capacitor, and their center frequencies are precisely controlled and tuned by the clock frequency. These filters are very linear and of low noise, which makes them perfect candidates for SAW-less receivers. We also described the design and architecture of a fully integrated SAW-less quad-band 2.5G receiver, which uses on-chip four-phase filters to deal with out-of-band blockers.

REFERENCES

1. A. Mirzaei and H. Darabi, Analysis of imperfections on performance of 4-phase passive-mixer-based high-Q bandpass filters in SAW-less receivers, *IEEE Transactions on Circuits and Systems I: Regular Papers* 58(5), 879–892, 2011.
2. A. Ghaffari, E. Klumperink, and B. Nauta, A differential 4-path highly linear widely tunable on-chip bandpass filter, in *IEEE Radio Frequency Integrated Circuits Symposium (RFIC)*, Anaheim, CA, pp. 299–302, 2010.
3. M. Soer, E. Klumperink, P.-T. de Boer, F.E. van Vliet, and B. Nauta, Unified frequency-domain analysis of switched-series—RC passive mixers and samplers, *IEEE Transactions on Circuits and Systems I: Regular Papers* 57(10), 2618–2631, 2010.
4. C. Andrews and A. Molnar, Implications of passive mixer transparency for impedance matching and noise figure in passive mixer-first receivers, *IEEE Transactions on Circuits and Systems I: Regular Papers* 57(12), 3092–3103, 2010.
5. B. Cook, A. Berny, A. Molnar, S. Lanzisera, and K. Pister, Low-power 2.4-GHz transceiver with passive RX front-end and 400-mV supply, *IEEE Journal of Solid-State Circuits* 41(12), 2757–2766, 2006.
6. C. Andrews and A. Molnar, A passive-mixer-first receiver with baseband-controlled RF impedance matching, ≪6dB nf, and ≫27 dBm wideband IIP3, in *IEEE International Solid-State Circuits Conference*, San Francisco, CA, pp. 46–47, 2010.
7. A. Mirzaei, H. Darabi, A. Yazdi, Z. Zhou, E. Chang, and P. Suri, A 65 nm CMOS quad-band SAW-less receiver SoC for GSM/GPRS/EDGE, *IEEE Journal of Solid-State Circuits* 46(4), 950–964, 2011.
8. A. Mirzaei, H. Darabi, J. Leete, X. Chen, K. Juan, and A. Yazdi, Analysis and optimization of current-driven passive mixers in narrowband direct-conversion receivers, *IEEE Journal of Solid-State Circuits* 44(10), 2678–2688, 2009.
9. A. Mirzaei, H. Darabi, J. Leete, and Y. Chang, Analysis and optimization of direct-conversion receivers with 25% duty-cycle current-driven passive mixers, *IEEE Transactions on Circuits and Systems I: Regular Papers* 57(9), 2353–2366, 2010.

5 Analog Signal Processing for Reconfigurable Receiver Front Ends

Bodhisatwa Sadhu and Ramesh Harjani

CONTENTS

5.1 INTRODUCTION

The wireless industry continues to flourish at breakneck speeds. As a result, spectral congestion caused by wireless user traffic has already become a significant concern that threatens further growth of the technology [1,2]. However, this congestion is primarily due to suboptimal frequency usage arising from the inflexibility of the spectrum licensing process. This suboptimal spectrum allocation and its consequent inefficiency can be solved by utilizing the concept of dynamic spectrum access. This allows other users to utilize temporarily unused spectrum for communications. Typically, a software-defined radio (SDR) hardware is envisioned for exploiting dynamic spectrum access using cognitive radio (CR) capability. In this chapter, we define the CR in the narrow sense of an intelligent device that is able to dynamically adapt and negotiate wireless frequencies and communication protocols for efficient communications. To enable this capability, each participating device needs to have many features such as geolocation, analysis of the external communications environment, sensing the dynamic spectrum usage, and the ability to change the frequency and bandwidth of transmission, adjust the output power level, and even alter transmission parameters and protocols [3].

Figure 5.1 provides an indication of the growth of CRs as a topic of research in the recent past. The figure plots the number of publications that show up in IEEE search for a given keyword across different years. Some of the keywords represent growing research areas in wireless, while other keywords such as *VLSI* and *DSP* represent popular areas of research in electronics and have also been included for comparison. The first CR paper was published in 1999 by J. Mitola [4]; research in this area was relatively dormant, however, till 2004. Since then, with technology maturing to meet

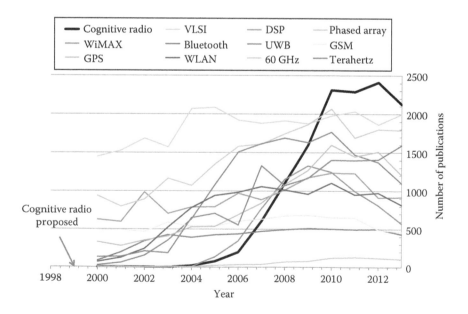

FIGURE 5.1   Number of publications in different research areas in the last decade.

the requirements of the CR hardware, and the rising need for reduced spectrum congestion, CRs have seen a tremendous growth in research activity and are now among the most researched areas in the wireless industry.

A CR can be divided into (1) an SDR hardware unit and (2) an intelligence software unit to provide the required software-based intelligence (cognition) to the radio [5]. In this chapter, we will discuss the hardware SDR unit and, more specifically, the spectrum-sensing receiver front end of the SDR.

The CR concept has been based on the use of dynamic spectrum access [6,7]. Dynamic spectrum access relies on dynamic spectrum monitoring using a spectrum sensor. Combined with spatial and temporal information, it can be used to perform dynamic spatial [8] or spatiospectral beamforming [9] to exploit temporal, spatial, and spectral degrees of freedom. In this chapter, we focus on the spectrum-sensing aspect of the CR.

Among other features, this continuous frequency monitoring makes the CR unique in its hardware. From the perspective of the hardware design, the spectrum sensor remains a challenging aspect of CR design. Even for narrowband (small frequency range, <100 MHz) spectrum sensors, limiting the power consumption is a challenge. The CR spectrum sensor needs to detect signals at all frequencies of interest instantaneously. In addition, very high detection sensitivity is desired (perhaps 100 times better than a conventional narrowband radio) to overcome the hidden-terminal problem, shadowing, channel fading, multipath, etc., lest it causes interference to other users due to incorrect sensing [6].

In this chapter, we discuss the suitability of passive switched capacitor signal processing techniques for spectrum-sensing applications. We present various techniques in passive switched capacitors that allow them to be used in high-speed, low-power RF applications. As an example, we present a prototype passive charge-based FFT design, first presented in Ref. [10], which can instantaneously analyze wideband signals (5 GHz bandwidth) with very low power consumption using these techniques.

5.2   REVIEW OF SDR SPECTRUM SENSORS

The architecture design for the SDR analog/RF is significantly different from that of traditional narrowband radios. In the original software radio proposal by Joseph Mitola in 1992 [4], he envisioned an architecture that digitized the RF bandwidth (no downconversion) and performed spectrum analysis and demodulation in the digital domain. While providing the maximum amount of flexibility through increased software capability in the digital domain, this architecture imposes impractical requirements on the analog-to-digital and digital-to-analog converters. For example, as discussed in Ref. [11], a 12 GHz, 12-bit analog-to-digital converter (ADC) that might be used for this purpose would dissipate 500 W of power! As a result, the ideal goal of being able to communicate at any desirable frequency, bandwidth, modulation, and data rate by simply digitizing the input and invoking the appropriate software remains far from realizable.

Subsequent proposals for spectrum-sensing architectures can be divided into two fundamental categories: a scanner type and a wide bandwidth instantaneous digitizer type.

5.2.1 SCANNER ARCHITECTURE

In this scheme, a narrowband, wide-tuning receiver scans and digitizes the entire bandwidth (similar to a benchtop spectrum analyzer) for analysis. The digital back end processes each band sequentially and stitches the frequency domain outputs to obtain a spectral map of the environment. A sample architecture is shown in Figure 5.2. Note, however, that in order to overcome issues such as multipath, fading, hidden nodes, and interference problems [6,12], the sensitivity and dynamic range requirements of the architecture are more challenging than a traditional communications receiver. Moreover, sensing may be a blind detection problem, as opposed to traditional reception where *a priori* knowledge of the transmitted signal is available.

Although the scanning architecture is able to reuse some features of a traditional receiver architecture, this detection technique suffers from multiple shortcomings. These systems lack the agility to be able to detect any fast-hopping signals. Frequency domain stitching is power hungry in the digital domain due to the need to correct for phase distortion introduced by the analog filters. Moreover, stitching the frequency domain information from several scans is imperfect in the face of multipath; consequently, signals spanning across multiple scan bandwidths are imperfectly reconstructed. Due to these and other reasons, it is desirable to construct a real-time instantaneous bandwidth digitizer (similar to J. Mitola's original software radio idea) in the spectrum sensor.

5.2.2 WIDEBAND DIGITIZER ARCHITECTURE

Unlike the scanning-type architecture, a wide instantaneous band digitizer is expected to digitize the entire RF bandwidth simultaneously. Understandably, the wideband digitizer has widely been considered as the bottleneck to the realization of the SDR-based CR. A number of efforts in recent years have focused on wider bandwidths, broadband matching, higher front-end linearity, and, most importantly, wideband ADCs.

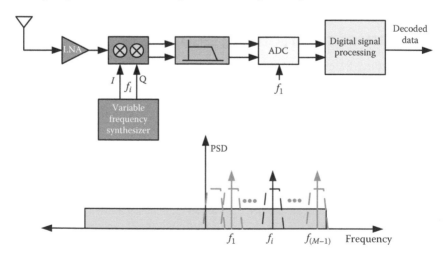

FIGURE 5.2 A narrowband, wide-tuning scanning approach for signal sensing. (From Sadhu, B. et al., *Hindawi Int. J. Antennas Propag.*, 2014.)

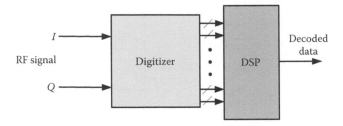

FIGURE 5.3   A wideband R-to-D architecture for spectrum sensing. (From Sadhu, B. et al., *Hindawi Int. J. Antennas Propag.*, 2014.)

Several architectures have been proposed for the RF front end. Of these, the most popular is the extension of the traditional receiver architecture as shown in Figure 5.3 effectively performing an RF to digital conversion (R-to-D). Typically, the front end also requires a wideband low-noise amplifier (LNA) prior to the digitizer (not shown). Moreover, the front end needs to handle a very large dynamic range due to the generally large peak-to-average power ratio (PAPR) of wideband signals. The increase in PAPR for wide bandwidths is described in Figure 5.4. As shown, the PAPR for the narrowband signals is only 2, while that for the wideband signal (five times the bandwidth) with multiple signals, all having similar powers, is 10. As a result of the large PAPR of the wideband inputs, a very linear front end is required. The linearity requirements of the LNA have been addressed in Ref. [13–15]. Another approach using a low-noise transconductance amplifier (LNTA) followed by mixers is discussed in Ref. [16]. Moreover, passive mixer-first topologies have been proposed for high IIP_3 performance [17].

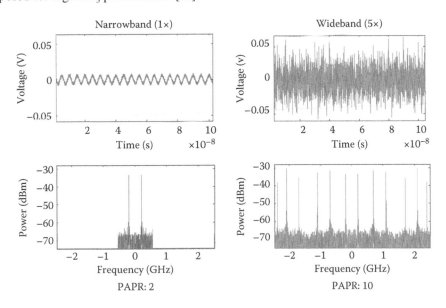

FIGURE 5.4   PAPR increase in wideband signals compared to narrowband signals. (From Sadhu, B. et al., *Hindawi Int. J. Antennas Propag.*, 2014.)

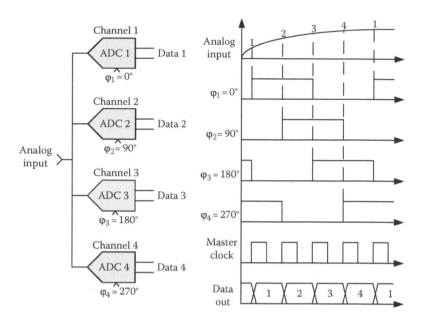

FIGURE 5.5 Time-interleaved ADCs for broadband channelization. (From Sadhu, B. et al., *Hindawi Int. J. Antennas Propag.*, 2014.)

The digitizer block shown in the figure is essentially an ADC with performance specifications beyond the capability of state-of-the-art converters. This wideband digitizer can be implemented in multiple ways, all based on some forms of multiplexing in order to ease the requirements on the ADCs. A multiplexed broadband approach using time-interleaving as shown in Figure 5.5 was proposed in Ref. [18]. This scheme reduces the sampling rate of ADCs. However, all the ADCs still see the full bandwidth and, therefore, still require high dynamic range capability.

In order to reduce the dynamic range requirements on the ADCs, the signal can be transformed to a different domain prior to digitization [19]. Specifically, a frequency domain transform is particularly attractive [20]. A frequency domain transform can be approximated in practice using band-pass filters for channelization, as proposed in Ref. [21]. This reduces the dynamic range requirements of the ADCs but introduces the problem of designing impractically sharp band-pass filters. In Ref. [22], replacing sharp band-pass filters by frequency downconverters followed by sharp low-pass filters eliminates this problem as shown in Figure 5.6. However, these are based on PLLs, mixers and low-pass filters [23], or on injection-locked oscillators [24]* and can be power hungry. Moreover, harmonic mixing of signals within the SDR input bandwidth severely corrupts the channelized baseband signals. Additionally, due to overlap between bands and phase issues, signal reconstruction from the digitized filter-bank outputs is challenging.

---

* Note that injection-locked oscillators have the advantage of a larger noise suppression bandwidth (≈lock range) [25] and provide better reciprocal mixing robustness compared to PLLs (assuming the reference phase noise is better than the VCO phase noise).

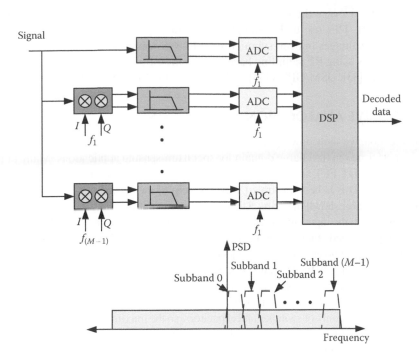

FIGURE 5.6 Low-pass filter-bank approach for channelization. (From Sadhu, B. et al., *Hindawi Int. J. Antennas Propag.*, 2014.)

In this chapter, we propose a digitizer approach based on analog signal processing, using passive switched capacitors to condition the signal prior to digitization by ADCs (Figure 5.7). The RF discrete time (DT) signal processing, as shown in the second block in Figure 5.7, eases the dynamic range requirements on the ADCs by prefiltering the signal.

For RF-sampled processors, an RF sampler has historically been an inherent bottleneck. However, with the scaling of technology and subsequent improvements in switch performance, RF sampling has become feasible in modern silicon processes. Moreover, it is possible to use charge domain sampling to leverage the inherent benefits of including a built-in antialias filter into the sampler [26], robustness to jitter [27],

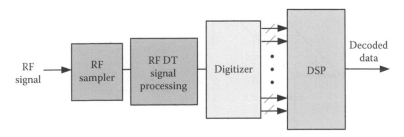

FIGURE 5.7 An envisioned SDR architecture enabled by passive analog signal processing. (From Sadhu, B. et al., *Hindawi Int. J. Antennas Propag.*, 2014.)

and the ability to vary the resulting filter notches by simply varying the integration period [27–31]. This use of RF samplers and subsequent DT processing provide a number of advantages in deep submicron CMOS processes [32]. Recently, other DT radio receivers using RF sampling have been demonstrated using CMOS technology for Bluetooth [33], GSM/GPRS [34], WLAN [35], and SDR-type applications [11,36].

5.3   PASSIVE ANALOG SIGNAL PROCESSING

In this section, we show how signal sampling and variable-rate analog signal processing can be performed in the charge domain for spectrum-sensing applications. Many of the benefits of the DT FFT architecture are based on the use of passive DT charge-based computations. This is best illustrated with the help of an example design. The passive switched capacitor shown in Figure 5.8 is able to operate at RF sampling speeds [28–31].

In this circuit, the input signal is sampled progressively in time ($\varphi_1 - \varphi_n$). After N clock periods, the averaged output is sampled onto the capacitor C_s, which has previously been discharged. The complete circuit implements an N-tap finite impulse response (FIR) filter that is decimated by N. Interestingly, if the capacitor C_s is not discharged between each rotation, then the circuit implements an N-tap FIR filter combined with a first-order IIR filter that is decimated by N. Note there is no active element (i.e., amplifier) in this circuit. The circuit consists only of switches and capacitors, so the maximum sampling rate is only dependent on the RC settling times. Additionally, the only power dissipation, other than that required for sampling the signal from the input, is due to the charging and discharging of the switch-transistor gate capacitors in a very digital-like way. As a result, a variety of functions on the sampled signal can be computed very fast and using minimal power.

5.3.1   PASSIVE COMPUTATIONS

For performing any linear function, addition and multiplication operations need to be performed. Note that all passive switched capacitor operations are destructive in nature. Therefore, once an operation is performed, the input values are lost. For

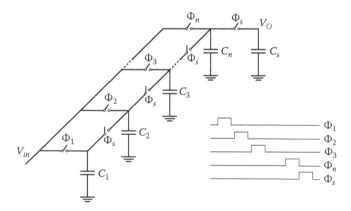

FIGURE 5.8   Switched capacitor implementation of a passive N-tap FIR with a decimation by N. (From Sadhu, B. et al., *Hindawi Int. J. Antennas Propag.*, 2014.)

performing multiple operations on a single input, multiple copies of the input need to be maintained. Here, we present techniques to perform these operations using passive switched capacitor circuits. In order to select a suitable technique for implementation, it is necessary to compare these techniques based on their robustness to nonidealities, ease of implementation, power consumption, speed, etc.

5.3.1.1 Addition Operation

1. *Parallel connection*: Using passive switched capacitors, voltages may be added by sharing the charges on two participating capacitors by connecting them in parallel as shown in Figure 5.9. The result of this operation (for capacitors with equal capacitances) is the average value $(V_1 + V_2)/2$ of the input voltages V_1 and V_2, which is a scaled version of their sum operation. Also note that two copies of the output are obtained, and these can be used for two subsequent independent operations as desired. However, the operation inherently attenuates the output by half. From an implementation perspective, use of parallel capacitors allows the sharing of one plate (ground plate) for all the capacitors. This can greatly reduce the parasitic capacitance and resistance of the capacitor and the area of the overall implementation.

2. *Series connection*: An alternative technique is to connect the capacitors in series. The result of this operation is the sum $(V_1 + V_2)$ of the input voltages V_1 and V_2. In this scheme, it is possible to use slightly delayed clock phases for the top and bottom plate switches in order to make the charge injection independent of the input voltage [37]. However, in the latter technique, switches are required both on the top and bottom plate, thereby increasing the power consumption in this circuit. The two switches placed in series halve the speed of this circuit for identical switch sizes. Moreover, only one output (which can be used for exactly one subsequent operation) is obtained. Also, both the top and bottom plate parasitics are problematic.

5.3.1.2 Multiplication

1. *Charge stealing*: Multiplication in the charge domain can be performed by scaling the voltage on a capacitor using a share operation with another known capacitor (stealing capacitor). The charge on the stealing capacitor is not utilized later. The overall operation causes a subunity scaling on the original value. The scaling factor for a capacitor of value C and a stealing capacitor of value C_s is given by $m = C/(C + C_s)$.

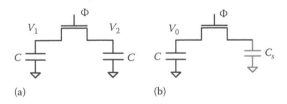

(a)   (b)

FIGURE 5.9 Techniques for charge domain addition and multiplication operations. (a) Add (share). (b) Multiply (steal). (From Sadhu, B. et al., *Hindawi Int. J. Antennas Propag.*, 2014.)

Figure 5.9b shows a scaling operation using a stealing capacitor of size C_s with no initial voltage on it. After the sharing operation, the final value on the capacitor with initial value V_0 becomes $V_0 \, C/(C + C_s)$. C_s can be chosen appropriately to obtain a particular scaling factor. Note that although this technique is capable of performing both subunity scaling and multiplication with a known attenuation, at least one of the operands needs to be known in advance for this implementation. In case voltage-dependent variable capacitors (i.e., capacitor DACs) are utilized, dynamic operands can also be used.

2. *Pulse-width modulation (PWM)*: Another technique to perform multiplication using passive switched capacitors is to modulate the turn on time of the switch and perform an incomplete share operation with a fixed stealing capacitor. The duration of the operation determines the multiplication factor. It is possible to multiply two unknown operands using this technique. Unfortunately, considering the nonlinearity in the resistance and the share operation, the errors caused by this technique make it unusable. However, the concept can be used to devise another PWM scheme, which allows complete settling, thereby making it more reliable. In this modified technique, the switch can be turned on using a sequence of randomly placed pulses and sharing the capacitor charge using a small stealing capacitor for each clock cycle. The stealing capacitor is discharged at the end of each cycle. Complete settling is allowed in each cycle. The total number of on-pulses determines the amount of scaling. Maximum scaling is obtained when all the clock cycles have on-pulses, while no scaling is obtained when all the clock cycles have off-pulses. Although this technique is relatively accurate, and is able to handle dynamic operands, it is slow and consumes more power than the charge-stealing technique. Also, depending on the accuracy required, the attenuation is considerable.

3. *Current domain*: If the charge is converted to the current domain, a single, variable-duration PWM scheme can be used to perform multiplication. Also, multiplication would not entail an inherent attenuation. However, the technique is very power hungry, and the accuracy of the transconductance amplifier that translates from charge to current domain needs to be very high.

Due to their low-power, high-speed characteristics, we have focused on the parallel connection scheme for addition and the charge-stealing scheme for multiplications in our designs. For many relevant linear algebra problems, multiplication using fixed coefficients is sufficient, and this technique lends itself easily to such applications.

5.3.2 Switching Schemes

To implement these addition and multiplication schemes, a variety of switched capacitor topologies can be used. Note that complex multiplication can be performed using a combination of scalar multiplication operations as discussed in Ref. [38]. In this subsection, we discuss the various topologies and their trade-offs. For the addition operation, two capacitors can be shared as shown in Figure 5.9a and represented

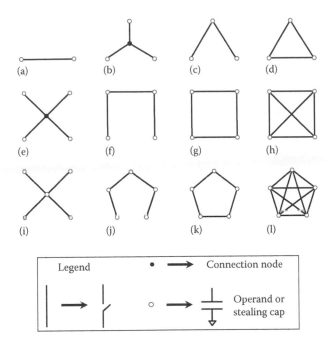

FIGURE 5.10 Different switching topologies for charge domain operations. (a) Two capacitor sharing. (b, c, d) Share and scale with 3 capacitors and 2 or 3 switches. (e, f, g, h) Multiplication by $(c + c \cdot j)$ using 4 capacitors. (i, j, k, l) Multiplication by $(c + c \cdot j)$ using 5 capacitors. (e, f, g, h, i, j, k, l) 4 input radix-4 operations. (From Sadhu, B. et al., *Hindawi Int. J. Antennas Propag.*, 2014.)

by Figure 5.10a. We can combine a share followed by scaling into a single operation by connecting three capacitors (two with input samples and one empty) and sharing their charges. This can be performed in different ways using two or three switches as shown in Figure 5.10b–d. It can be shown that three appropriately sized switches in the scheme of Figure 5.10d minimize the settling error [4]. Multiplication by a factor $c + c j$ is a special case scaling operation that can be performed using a single step operation [4]. Depending on the normalization of the scaling factor, this may be performed using four capacitors (Figure 5.10e–h), or using five capacitors (Figure 5.10i–l). Moreover, in the case of four input operations (radix-4 operations), these schemes (Figure 5.10e–l) are useful.

While many schemes (Figure 5.10a,b,d,e,h,i,l) ensure settling symmetry, others (Figure 5.10c,f,g,j,k) use fewer switches for lower power at the expense of settling performance and mismatch. Some variants (Figure 5.10d,h,l with equal size switches) provide both settling speed and symmetry at the cost of larger power. When the switches between the operand capacitors are sized differently from those connecting to the stealing capacitor, in (d) and (l), these same configurations can be optimized for an enhanced settling-per-power performance. Finally, when comparing the different schemes with their appropriate switch sizes, different trade-offs with regard to charge injection, clock feedthrough error, etc., should be considered.

For our design, we chose to use (a) and (d) to perform radix-2 scalar operations, while complex operations are performed by cascading to sets of operations.

Configurations (h) and (l) were used to perform single-phase complex multiplications in special cases. In the case of (d) and (l), optimized switch sizing was used to mitigate their extra power demands while still realizing their enhanced settling performance for a net settling-per-power gain versus (b,c) and (i–k), respectively.

5.3.3 NONIDEALITIES

Several nonidealities haunt passive switched capacitor circuits. The problem of nonidealities is aggravated by the absence of a virtual ground node unlike in op-amp-based active switched capacitor circuits. The effect of sampling clock jitter in passive switched capacitor circuits has been analyzed [27]. Two important nonidealities, clock feedthrough and charge injection, become a nuisance in the absence of a virtual ground node. Consequently, traditional circuit techniques such as bottom plate sampling are difficult to implement. Also, poor matching between nMOS and pMOS switches, and the reducing difference between V_{dd} and V_{th} in scaled technologies, makes the use of transmission gate switches less effective for mitigating these nonidealities. The noise in the system is dominated by the kT/C noise of the RC filter formed by the switch–capacitor combination. Moreover, for a multistage switched capacitor operation, the sampled noise voltages from one stage recombine in the later stages. These combining noise samples in a particular stage are correlated, and therefore, the final noise becomes a complicated function of the noise sampled at each stage of the switched capacitor operation. The switch resistance (along with the capacitance of the capacitor) determines the settling time constant. However, the switch resistance is inherently nonlinear and input signal dependent. Consequently, in the case of high speeds of operation, incomplete settling can cause significant signal-dependent errors in computations.

Since switched capacitor circuits utilize a clock signal, the accuracy of the clock is critical to performance. Specifically, jitters in clocks reduce the accuracy of the switched capacitor computations by translating timing uncertainty to charge and voltage uncertainty. Fortunately, new techniques based on transconductance linearization can be used to achieve low-phase noise clocks in SiGe bipolar [39] and even in scaled CMOS circuits [40]. For increased frequency flexibility, highly optimized switched inductor- [41] and switched capacitor-based [42] LC VCOs can be utilized to obtain a wide range of frequencies without sacrificing noise performance. Moreover, on-chip self-healing techniques [43] utilizing a digital back end can be used for healing the switched capacitor circuits as well as improving the clock jitter [44].

For high-speed designs, it is necessary to accurately model these nonidealities in the circuit simulator. It is also useful to have the ability to individually turn off these nonidealities to trace the effect of each nonideality on the output error. For our designs, we model the nonidealities in MATLAB® and include them in system level simulations using MATLAB or Simulink® [4]. This allows us to effectively capture the nonidealities and optimize the designs in their presence.

5.4 ANALOG DOMAIN FFT

In this section, we discuss an RF sampler followed by a DT Fourier transform engine to perform channelization of the wideband RF input. The use of RF samplers and subsequent DT processing, prior to digitization, provides a number of advantages in deep submicron CMOS processes including high linearity, programmability, large bandwidth, robustness to jammers, immunity to clock jitter, and low-power ADCs [32,45,46]. Recently, DT radio receivers using RF sampling have been demonstrated using CMOS technology for Bluetooth [33], GSM/GPRS [34], WLAN [35,47], and SDR-type applications [11,36].

In this technique, the DT DFT is used as a functionally equivalent linear phase N-path filter to perform channelization [48]. The output of each bin is effectively filtered using frequency-shifted complex *sinc* filters, followed by downconversion. This is exactly equivalent to a mixer followed by a low-pass filter scheme [48]. This scheme reduces both the speed and dynamic range of the ADCs and, by virtue of being linear phase, allows for simple reconstruction in the digital domain using an IFFT without any loss of information. A few current-based analog DFT filters have been designed [49,50]. However, these designs are speed limited and consume significant power (Table 5.3), minimizing the overall gains. Additionally, they use active devices for signal processing and are therefore expected to be more nonlinear at high operating speeds. In this work, we describe the design of a charge domain DFT filter, the charge re-use analog Fourier transform (CRAFT), based on passive switched capacitors. It performs an analog domain 16-point DFT running at input rates as high as 5 GS/s and uses only 3.8 mW, or 12.2 pJ per 16-point DFT conversion. By virtue of its *I/Q* input processing capability, it is able to transform a 5 GHz asymmetrically modulated bandwidth instantaneously. The design was first presented in Ref. [10]; this chapter includes further details of the implementation, modeling, and mitigation of nonidealities and additional measurements.

Figure 5.11 shows a sample architecture with a CRAFT RF front end. We will show that this architecture reduces both the ADC speed (by N) and the required ADC dynamic range by removing out-of-band signals per ADC, at a negligible power overhead. The impact of the CRAFT front end on the ADC input bandwidth and dynamic range for well-spread signals is shown in Figure 5.12 where the expected ADC requirements for a 5 GS/s input are plotted among measured ADC

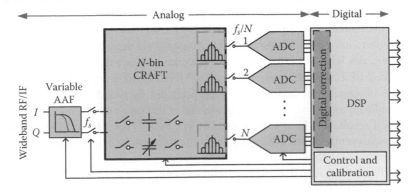

FIGURE 5.11   An SDR architecture enabled by CRAFT. (From Sadhu, B. et al., *Hindawi Int. J. Antennas Propag.*, 2014.)

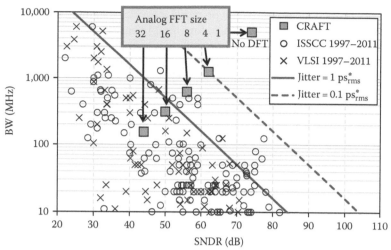

* Assumes white noise (see [9] for details on the assumption)

FIGURE 5.12 Feasibility of ADCs versus FFT size of the CRAFT front end. (From Sadhu, B. et al., *Hindawi Int. J. Antennas Propag.*, 2014.)

implementations [51]. As seen in Figure 5.12, the CRAFT front end reduces the required speed of each ADC as well as their input dynamic range through channelization. This brings the ADC requirements from being infeasible (top right corner) toward being achievable (bottom left half), thereby solving the critical wideband digitizing problem in SDRs. In contrast, a time-interleaved ADC approach only reduces the speed requirement of each ADC without reducing the dynamic range as shown by the vertical arrow in Figure 5.12; consequently, the total ADC power remains approximately unchanged.

5.5   CRAFT DESIGN CONCEPT

CRAFT operations are based on charge reuse. Once sampled, the charge on a capacitor is shared and reshared with other charge samples such that the resulting mathematical manipulation is an in-place DFT. By basing the design only on toggling switches (transistor gates), low power and high speeds are made feasible. Additionally, the power scales with frequency, supply, and technology in a digital-like fashion.

A radix-2 FFT algorithm was used in CRAFT. The FFT computation uses only two types of operations: addition and multiplication by twiddle factors. Note that these twiddle factors $W = e^{-(2\pi j/16)}$ are equispaced points on a unit circle in the complex plane. As a result, for these scaling factors, W^k, $\Re\{W^k\} \leq 1, \forall k$, and $\Im\{W^k\} \leq 1, \forall k$. Since passive computations inherently attenuate the signal, these operations are particularly suited for subunity scaling.

To perform the CRAFT operations, the following charge domain computations are used. Addition is performed by sharing the charges on two capacitors as shown in Figure 5.9a. Subunity multiplication is performed by stealing charge away from a capacitor using a suitably sized stealing capacitor (C_s) as shown in Figure 5.9b. These two simple operations form the basis of all operations performed in CRAFT.

The 16-point, radix-2 CRAFT operation can be represented as a linear matrix transform*:

$$\mathbf{X} = \mathbf{F}\mathbf{x} \qquad (5.1)$$

These operations are utilized in the CRAFT processor as shown in Figure 5.13. Row 1 shows the *share* operation. Subsequent subunity twiddle factor multiplication, if required, is performed by charge stealing as shown in row 2: *share and multiply* operation in Figure 5.13. Negation is performed by swapping the positive and negative wires as shown in row 3, while multiplication by $j = \sqrt{-1}$ is performed by swapping the appropriate wires of the real and imaginary components as shown in row 4 of Figure 5.13. These techniques are extended to perform complex multiplication, as shown in row 5. A sample butterfly operation using multiple complex multiplications is also shown in row 6. Note that each butterfly operation requires

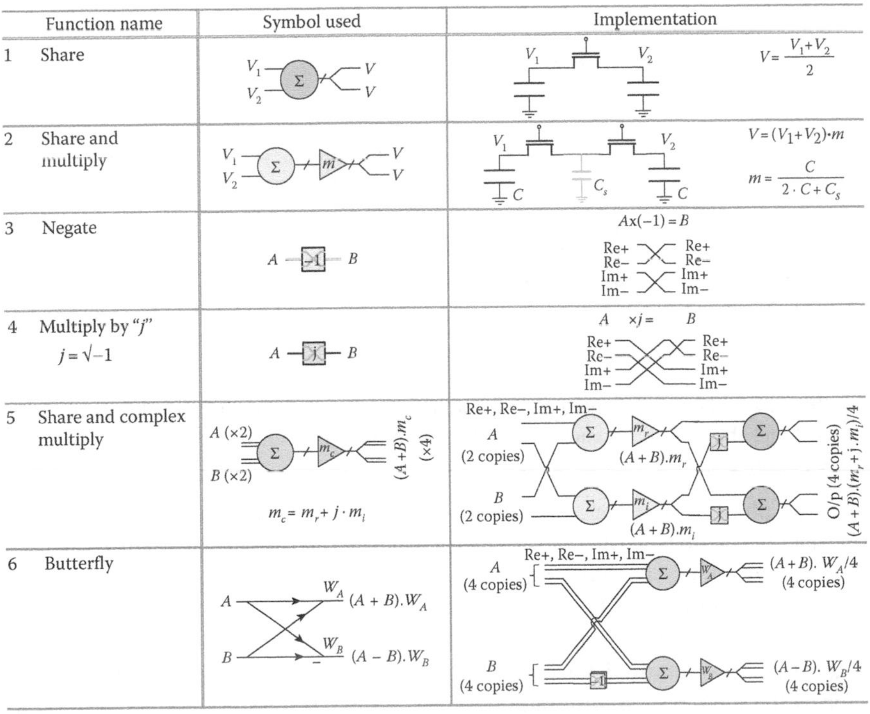

	Function name	Symbol used	Implementation
1	Share		$V = \dfrac{V_1 + V_2}{2}$
2	Share and multiply		$V = (V_1 + V_2) \cdot m$ $m = \dfrac{C}{2 \cdot C + C_s}$
3	Negate		$Ax(-1) = B$
4	Multiply by "j" $j = \sqrt{-1}$		$A \quad \times j = \quad B$
5	Share and complex multiply	$m_c = m_r + j \cdot m_i$	
6	Butterfly		

FIGURE 5.13 Mathematical operations in the CRAFT processor using charge sharing. (From Sadhu, B. et al., *Hindawi Int. J. Antennas Propag.*, 2014.)

---

* Note that any linear transform with a fixed matrix can be performed using the addition and multiplication techniques outlined earlier. Due to the inherent attenuation in charge domain operations, the result is a scaled version of the desired transform.

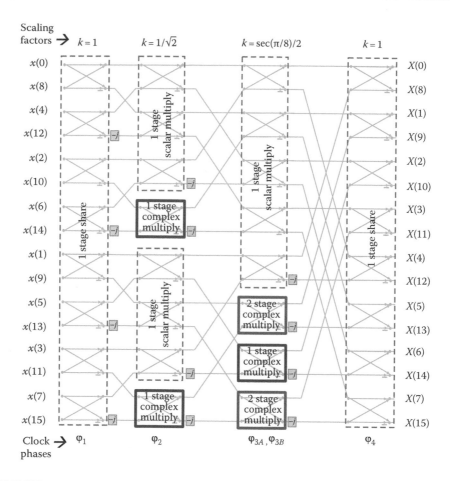

FIGURE 5.14 The implemented CRAFT algorithm showing the different kinds of butter-flies used and the scaling in each stage. (From Sadhu, B. et al., *Hindawi Int. J. Antennas Propag.*, 2014.)

two clock phases. However, in the CRAFT engine, we optimize the operations such that only 5, instead of 8, clock phases are utilized for the four butterfly stages. The optimized butterfly blocks are then used to construct the 16-point CRAFT engine shown in Figure 5.14.

The in-place CRAFT computations are destructive; therefore, multiple copies of each value are required for multiple operations. Since a radix-2 FFT algorithm performs the DFT with a minimum number of operations per operand (less copies required), it is selected for CRAFT.

Conceptually, the FFT is computed as follows: The input signal is sampled onto capacitors. Since each input is operated on twice in an FFT butterfly, and twice for complex operations, four copies of each sample are required. Also, considering I, Q (=2), and differential (=2) inputs, the 16-point FFT requires 16 × 2 (complex math) × 2 (butterfly branches) × 2 (I, Q) × 2 (differential) = 256 sampling capacitors. Details on the CRAFT implementation can be found in Ref. [4].

5.6   CIRCUIT NONIDEALITIES AND MITIGATION

As expected, the design relies heavily on digital state machines. Also, the design's regularity and complexity approach that of digital designs. However, despite these similarities, the CRAFT engine remains vulnerable to analog circuit nonidealities including noise, matching, and nonlinearity. Consequently, accurate modeling of nonidealities is critical. Switched capacitor circuit noise, charge injection, and charge accumulation were not adequately modeled in the Spectre models available. Moreover, in order to isolate the impact of individual nonidealities per stage, it was necessary to enable/disable them independently in simulation. For this design, CRAFT-specific models of dominant nonidealities were developed in MATLAB. Each nonideality is modeled as an independent error source that can be enabled in simulation to isolate the impact of each error source.

The different operations in CRAFT can be divided into the initial sampling phase and the subsequent processing phases. In this section, the effect of noise on the sampler as well as the core is discussed. This is followed by a discussion on incomplete settling due to the high speed of operation. Other nonlinearities such as clock feedthrough and charge injection are also modeled and mitigated [52] using techniques outlined in Ref. [51,53,54] but have been omitted from this discussion in the interest of space.

5.6.1   NOISE

5.6.1.1   Sampling Noise

Motivation: During a voltage sampling operation, noise presents itself as a final-value disturbance with a power kT/C. In this design, 4×2 capacitors, $200\,f$F each, are used to sample four copies of each input pseudodifferentially. The full-scale input is $V_{pp,\text{diff}} = 1.2V$. This sampler yields a sampled noise voltage of $\sqrt{V_{n,rms}^2} = V_{n,rms} = 144\,\mu V$ on each of the four single-ended copies (-63.4 dBFS). As the noise is uncorrelated, averaging these copies at the output and forming a differential output give a total noise of $V_{n,rms} = 144 \cdot \dfrac{1}{\sqrt{4}} \cdot \sqrt{2} = 102\,\mu V$ (-72.4 dBc) for a full-scale 1-tone input.

Modeling: Sampling noise effects are included in our MATLAB system simulation model. After sampling, a Gaussian random variable with $\sigma = V_{n,rms} = \sqrt{kT/C}$ is added to each capacitor's final value.

Mitigation: The sampling capacitor size sets both the sampler's noise as well as the baseline for the processing noise because the same capacitors are used for computations. The capacitor size is selected to be $200\,f$F based on the simulated total output-referred noise.

5.6.1.2   Processing Noise

Motivation: Similar to the sampling operation, noise from the CRAFT core switches corrupts the computations. At the end of every share operation, kT/C noise power is added to each output copy. Interestingly, this instantaneous sampled noise on the two capacitors arising from a single switch is completely correlated (equal magnitude, opposite signs). Similarly, the noise sampled on the capacitors during a share and multiply operation arises from the same switches and is therefore partially correlated. Also, any noise in the input operands (e.g., from the previous stage of operations) is averaged during an operation.

Modeling: For modeling, Gaussian random variables are used that are distributed in magnitude and sign on the output copies in exactly the manner the noise-transfer functions dictate for both reset and processing operations. This generates the proper noise correlation that, averaged over multiple simulations, provides the expected output-referred noise power.

Mitigation: Noise in the later stages of the CRAFT engine is reduced due to copy averaging among four copies of each output to reduce the noise power by 4. Moreover, correlation ρ among copies is exploited for noise reduction by averaging before the latching operation.

5.6.1.3 Total Output Noise

The noise contribution of each stage is computed analytically and tabulated in Table 5.1. The attenuation ($A_{v,out}$) reduces the output-referred noise by $A_{v,out}^2$. The single-ended, copy-averaged noise, including the residual noise from the stealing capacitor reset operation, is computed as shown in the last column, yielding a total output-referred noise of $0.30 \cdot kT/C$ (−63.3 dBFS per differential $\Re$, $\Im$ output for a noise EVM of −61.3 dBFS).

5.6.2 Incomplete Settling

Motivation: For an RF DT signal processor, very high speeds are mandated. Increasing the switch size to allow better settling not only increases the power consumption

---

TABLE 5.1
Summary of Noise Contribution in CRAFT

Stage	Noise Sources	$\overline{V_n^2}$ per Copy	$A_{v,out}$	ρ_{out}	$P_{n,out} = \left(\dfrac{1}{4}\right) \cdot \overline{V_n^2} \cdot A_{v,out}^2 \cdot (1 + \rho_{out})$
	Sampler (4-Copy)	$1.000 \cdot \dfrac{kT}{C}$	0.38	0	$0.0366 \cdot \dfrac{kT}{C}$
1	2 pt. share	$0.500 \cdot \dfrac{kT}{C}$	0.38	0	$0.0183 \cdot \dfrac{kT}{C}$
2	2 pt. sh./scale ($m = 0.707$)	$0.750 \cdot \dfrac{kT}{C}$	0.54	0	$0.0549 \cdot \dfrac{kT}{C}$
3	2 pt. sh./scale ($m = 0.541$)	$0.854 \cdot \dfrac{kT}{C}$	1	−0.11	$0.1899 \cdot \dfrac{kT}{C}$
4	2 pt. share	$0.500 \cdot \dfrac{kT}{C}$	1	−1	0
	Total output noise				$0.2997 \cdot \dfrac{kT}{C}$

Source: Sadhu, B. et al., *Hindawi Int. J. Antennas Propag.*, 2014.

but also causes larger charge injection and clock feedthrough errors. For example, in CRAFT, stages 1–4 have average simulated settling time to time constant ratios of $\Gamma_{S1} = 7$, $\Gamma_{S2} = 4$, $\Gamma_{S3} = 5$, and $\Gamma_{S4} = 4$, respectively [55]. For the modeling and mitigation sections, first, a single-ended settling scenario is considered. This is then extended to the pseudodifferential operation in CRAFT.

Modeling:

1. *Single-ended settling:* A two-point share and multiply operation, as shown in Figure 5.13, has the settling response of Equation 5.2. The input capacitors (C), with initial voltages v_{a0} and v_{b0}, are connected to the stealing capacitor (C_s, with no initial voltage: $v_{s0} = 0$) by switches of resistance R_{sw}, where the scaling factor $m = 2/(2 + C_s/C)$. Therefore,

$$
V_a(t) = \overbrace{\frac{1}{2}(v_{a0} + v_{b0})m\left[1 + \left(\frac{1}{m} - 1\right)e^{-t/\tau_s}\right]}^{\text{scaled-sum term}} + \overbrace{\frac{1}{2}(v_{a0} - v_{b0})e^{-t/\tau_d}}^{\text{difference term}}
$$

$$
V_b(t) = \underbrace{\frac{1}{2}(v_{a0} + v_{b0})}_{\text{ideal result}}m\underbrace{\left[1 + \left(\frac{1}{m} - 1\right)e^{-t/\tau_s}\right]}_{\text{scaling error factor}} - \underbrace{\frac{1}{2}(v_{a0} - v_{b0})e^{-t/\tau_d}}_{\text{difference settling error}}
$$

(5.2)

The difference and sum-settling time constants, τ_d and τ_s, respectively, are

$$
\tau_d = R_{sw}C \quad \tau_s = R_{sw} \cdot (C \parallel C_s/2) = R_{sw}C \cdot (1 - m)
$$

2. *Pseudodifferential settling:* Pseudodifferential operands are described by the relationships in the following, where V^+ and V^- represent the positive and negative components of the operand and are stored on separate capacitors:

$$
V_A(t) = V_a^+(t) - V_a^-(t) \quad V_B(t) = V_b^+(t) - V_b^-(t)
$$

(5.3)

The time-domain settling response of these pseudodifferential operands is shown in Figure 5.15 for $m = 1$ (sharing operation). In general, for a share and multiply operation, using $m' = m\left[1 + ((1/m) - 1)e^{-t/\tau_s}\right]$ and $\epsilon = \frac{1}{2}(v_{a0} - v_{b0})e^{-t/\tau_d}$, we can write

$$
V_A(t) = \overbrace{\left[\frac{1}{2}(v_{a0}^+ + v_{b0}^+)m'^+ + \epsilon^+\right]}^{V_a^+(t)} - \overbrace{\left[\frac{1}{2}(v_{a0}^- + v_{b0}^-)m'^- + \epsilon^-\right]}^{V_a^-(t)}
$$

$$
= (v_{a0}^+ + v_{b0}^+)\underbrace{\frac{1}{2}(m'^+ + m'^-)}_{m_d'} + \underbrace{(\epsilon^+ - \epsilon^-)}_{\epsilon_d} = (v_{a0}^+ + v_{b0}^+)m_d' + \epsilon_d
$$

(5.4)

where

$$
m_d' = m\left[1 + \left(\frac{1}{m} - 1\right)\frac{1}{2}\left(e^{-t/\tau_s^+} + e^{-t/\tau_s^-}\right)\right]
$$

$$
\epsilon_d = (v_{a0}^+ - v_{b0}^+)(1/2)\left(e^{-t/\tau_d^+} + e^{-t/\tau_d^-}\right)
$$

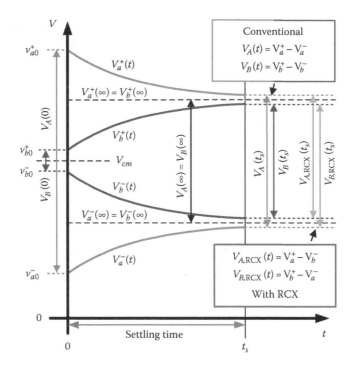

FIGURE 5.15 Reduction in differential settling error using RCX ($m = 1$). (From Sadhu, B. et al., *Hindawi Int. J. Antennas Propag.*, 2014.)

Mitigation:

1. *Single-ended settling*: The settling accuracy can be improved using a third switch ($R_{sw,3}$) for the share and multiply operation. This improves the differential settling error by providing an alternate settling path. Note that the widths of the main sharing switches (having a resistance, R_{sw}) effectively occur in series for the differential settling equation. As a result, increasing the third switch width (having a resistance, $R_{sw,3}$) improves the differential settling with twice the power efficiency as compared to increasing the main switch width: $\tau'_d = \left(2R_{sw} \parallel R_{sw,3}\right) \cdot (C/2)$.

2. *Pseudodifferential settling*: Considering the pseudodifferential settling expressions under small operand swings, a method for RC settling error cancellation (RCX) is developed. Recognizing that V_a^- and V_b^- settle to the same value but have ε error of opposite sign, they can be interchanged to form the new differential operands as follows (compare with Equation 5.3):

$$V_{A,RCX}(t) = V_a^+(t) - V_b^-(t) \quad V_{B,RCX}(t) = V_b^+(t) - V_a^-(t)$$

This is implemented using only wire swapping as shown in Figure 5.15, greatly reducing differential settling error. While the pseudodifferential *inputs* (+,−) to the operation are (V_a^+, V_a^-) and (V_b^+, V_b^-), the capacitors holding the *output* results (two copies) are

$\left(V_a^+, V_b^-\right)$ and $\left(V_b^+, V_a^-\right)$. This changes the $\left(e^{-t/\tau_d^+} + e^{-t/\tau_d^-}\right)$ term in ε_d to $\left(e^{-t/\tau_d^+} - e^{-t/\tau_d^-}\right)$, allowing the differential settling error to be canceled when $\tau_d^+ \approx \tau_d^-$. Also,

$$\epsilon_{d,RCX} = \left(\epsilon^+ + \epsilon^-\right) = \left(\epsilon^+ - \epsilon^-\right)\left[\frac{\epsilon^+ + \epsilon^-}{\epsilon^+ - \epsilon^-}\right] = \epsilon_d \cdot \tanh\left[t\left(\frac{\tau_d^+ - \tau_d^-}{2\tau_d^+\tau_d^-}\right)\right] < \epsilon_d = \left(\epsilon^+ - \epsilon^-\right)$$

shows that RCX always yields a net improvement in intercopy error and is used in all the CRAFT butterflies. Simulations show a net improvement of about 10 dB in the FFT settling error. It is important to understand that the intercopy error is not actually being canceled; it is just partially translated into a common-mode component. The common-mode rejection of the stages that follow rejects this settling error.

Note that the switch resistance is voltage dependent causing a spread in the realized time constants. Consequently, effective time constants based on simulations are calculated and utilized in the incomplete settling and RCX equations to ensure our computation accuracy requirements.

5.7 MEASUREMENT RESULTS

The design has been implemented in the IBM 65 nm CMOS process. Measurement results are shown in Figures 5.17 through 5.19a. The measurements shown are after calibration to compensate for systematic offsets due to parasitics. The calibration process used is detailed in the following.

5.7.1 CALIBRATION

The accuracy of the CRAFT operations is dependent on the matching between the capacitors used to realize it. Any systematic mismatch between the capacitors causes computation errors that reduce the dynamic range of the system. However, since these errors are systematic, and input independent, it is possible to calibrate for them in the digital domain. For measurements, a simple calibration technique can be used and is described as follows.

First, the nonidealities in the CRAFT operation are represented using a modified FFT matrix, $\mathbf{F}'$, which includes the effects of mismatch and represents the nonideal CRAFT operation. The resultant outputs are represented by $\mathbf{X}':\mathbf{X}' = (1/k)\ \mathbf{F}'\mathbf{x}$. A calibration scheme is constructed by observing that the FFT matrix $\mathbf{F}$ comprises 256 elements. Consequently, a set of 16 mutually independent inputs, vectors $\mathbf{x}_i$ with entries $x_i(t)$, constitutes 256 independent equations. Using the measured results $\mathbf{X}'_i$, with entries $X'_i(n)$, the nonideal $\mathbf{F}'$ matrix can be determined.

For convenience, we generate 16 orthogonal inputs comprising 16 separate 1-tone signals, each centered on a bin i. Note that in an on-chip implementation, perfect tones are not easily available. However, preloaded samples of linearly independent (not necessarily orthogonal) inputs that require a low-resolution DAC can be used instead to provide a similar calibration accuracy.

After determining an estimate of $\mathbf{F}'$, a one-time correction matrix $\widehat{\mathbf{H}}$ is computed:

$$\widehat{\mathbf{F}'} = \mathbf{X}'\widehat{\mathbf{x}}^{-1} \quad \Rightarrow \widehat{\mathbf{H}} = \mathbf{F}(\widehat{\mathbf{F}'})^{-1} = \mathbf{F}\widehat{\mathbf{x}}(\mathbf{X}')^{-1}$$

where
$\widehat{\mathbf{x}}$ is the calibration input
$\mathbf{X}'$ is the calibration output

All subsequent measurements across different magnitude and frequency inputs are then corrected by applying $\widehat{\mathbf{H}}$ to $\mathbf{X}' \Rightarrow \widehat{\mathbf{X}} = \widehat{\mathbf{F}}\mathbf{x} = \widehat{\mathbf{H}}\mathbf{F}'\mathbf{x} = \widehat{\mathbf{H}}\mathbf{X}'$.

5.7.2 TEST SETUP

The test setup is shown in Figure 5.16. I and Q inputs are generated using the Tektronix AWG-7122B arbitrary waveform generator and input to the CRAFT engine using 50 Ω terminated probe pads, which feed the sampler array. The latched outputs are externally buffered and captured by external ADCs controlled by an FPGA (NI-7811R) programmed using LabVIEW®.

Note that the CRAFT processor runs at an input/output rate of 5 GS/s per I and Q channel. Additionally, the outputs are analog values. Due to the very high speed of operation, and the limitations in the number of I/O pins, the outputs are first latched using the OTA-based analog latches and multiplexed out at a slower rate limited to 40 MS/s. The CRAFT speed is not compromised due to the output readout limitation. This output multiplexing and readout is controlled by an FPGA (NI-7811R) programmed using LabVIEW®. The outputs are first buffered, then digitized by external ADCs, and finally captured by the FPGA. Using this scheme, RF inputs that are synchronous as well as asynchronous to the clock can be captured and phase aligned. The latter is particularly important to emulate practical scenarios.

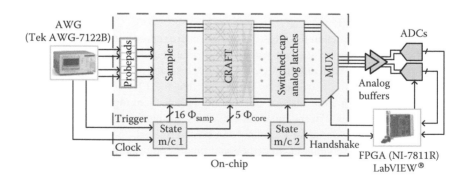

FIGURE 5.16 Test setup for the CRAFT processor (on and off chip). (From Sadhu, B. et al., *Hindawi Int. J. Antennas Propag.*, 2014.)

5.7.3 On-Bin 1-Tone Measurements

Figure 5.17 shows the CRAFT output magnitudes $\sqrt{\Re^2 + \Im^2}$ with a 312.5 MHz = 5 GHz/16 1-tone input at 5 GS/s. Curve I shows the measured uncalibrated output magnitude across 16 bins. Curve II in Figure 5.17 shows the calibrated plot depicting the circuit noise floor (including the integrated noise of the analog latches) at −46 dB. To explore the nonlinearity floor, a synchronous average over 500 measurements is used. The resultant Curve III shows the nonlinearity floor with 43 dB SFDR* and 48 dB SNDR.† Note that the SNDR and SFDR also signify the out-of-band rejection of the FFT as a filter. The nonlinearity predicted by simulations of the stand-alone CRAFT engine is shown in Curve IV. Note that Curve III not only includes the nonidealities in CRAFT but, unlike Curve IV, also includes the nonidealities of the 8-bit resolution AWG inputs, the sampler, and systematic and random sampling jitter. Unfortunately, despite the use of

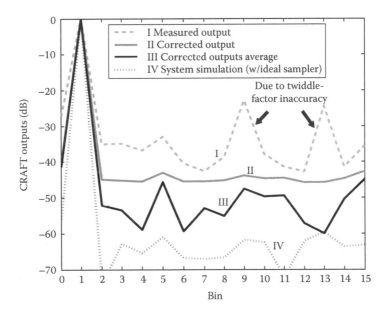

FIGURE 5.17 CRAFT outputs with a 362.5 MHz 1-tone input sampled at 5 GS/s. (From Sadhu, B. et al., *Hindawi Int. J. Antennas Propag.*, 2014.)

---

* SFDR for a 1-tone test is calculated as the difference between a full-scale on-bin signal and the largest off-bin output.

† $SNDR = 20 \times \log_{10}\left(\dfrac{\sqrt{\sum_{k=1}^{N} V_{ideal}^2(k)}}{\sqrt{\dfrac{1}{N}\sum_{k=1}^{N} \{V_{meas}(k) - V_{ideal}(k)\}^2}} \right)$

where N is the number of FFT bins. Since the outputs will be observed/digitized on a per bin basis, the total (noise + distortion) in the denominator is averaged over the N bins to yield the average (noise + distortion) per bin.

state-of-the-art test equipment, the limitations in the input and output test setup severely limit the observable nonidealities in CRAFT.

5.7.3.1 SNDR Variation

One-tone measurements for all bins were performed as shown in Figure 5.18a; the resulting SNDR at 1, 3, and 5 GS/s for a 1-tone input frequency placed at different bins is shown in Figure 5.18b. The average SNDR across bins is ≈50 dB at 1 and

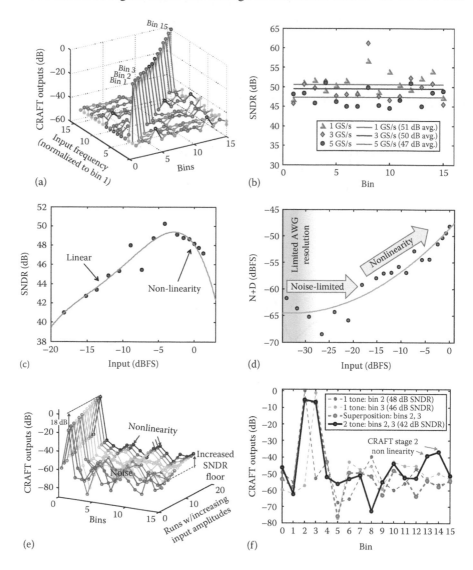

FIGURE 5.18 Measurement results including 1-tone measurements (a), SNDR variation across bins and frequencies (b), SNDR and noise and distortion across input amplitudes (c–e), and results of a 2-tone test (f). (From Sadhu, B. et al., *Hindawi Int. J. Antennas Propag.*, 2014.)

TABLE 5.2

CRAFT SNDR and SFDR with Single-Tone, On-Bin, 0 dBFS Inputs

	dB	Min.	Max.	Mean	Sigma
1 GS/s	SNDR	46.6	56.5	50.6	2.7
	SFDR	37.7	47.8	43.2	4.1
3 GS/s	SNDR	56.5	61.2	49.5	3.8
	SFDR	35.8	45.5	42.1	5.0
5 GS/s	SNDR	44.5	51.1	47.3	2.3
	SFDR	35.6	43.3	40.5	2.3

Source: Sadhu, B. and Harjani, R., *Cognitive Radio Receiver Front-Ends: RF/Analog Circuit Techniques*, Springer, 2014.

3 GHz and degrades to 47 dB at 5 GHz; SNDR better than 45 dB is maintained across all frequencies. The SNDR and SFDR measurements are tabulated in Table 5.2. This provides 7–8 bits of spectrum-sensing resolution over a 5 GHz (2.5 GHz × 2 due to I, Q inputs) frequency range in the digital back end.

Figure 5.18c plots the output SNDR versus the input amplitude. A fourth-order fit shows a linear SNDR improvement with increasing amplitude before being limited by a combination of the circuit nonlinearities and the AWG resolution (8 bits).

The CRAFT output magnitudes with varying input amplitudes for a 1-tone, on-bin 312.5 MHz input at 5 GS/s are shown in Figure 5.18c. The input is varied over an 18 dB range. As seen in Figure 5.18d, the circuit is limited by the noise floor for low-input amplitudes, while it becomes nonlinearity-limited at larger input amplitudes. Note that Curve III in Figure 5.17 represents a cross section of Figure 5.18e.

5.7.4　On-Bin 2-Tone Measurements

Results from a 2-tone test with tones on adjacent bins are shown in Figure 5.18f. Assuming a preceding AGC, the maximum (time domain) amplitude of the 2-tone signal is normalized to that of a single tone in the 1-tone test as shown. Two 2-tone measurements and their superposition (normalized for the same maximum amplitude) are also shown. The difference between the superposition and the measured 2-tone output is indicative of the additional nonlinearity introduced. The relative increase in bins 13 and 14 is likely to be due to uncorrected twiddle factor errors in stage 2 of the CRAFT engine. Note that the apparent effect of these twiddle factor errors, leaking signal onto the negative of the signal frequency, is identical to the effect of I–Q mismatch errors in a homodyne receiver.

5.7.5　Power Consumption

The CRAFT core consumes 12.2 pJ/conv. and uses 3.8 mW of power when interleaved by 2 for a 5 GS/s input and 5 GS/s output. Measurements of the energy consumption

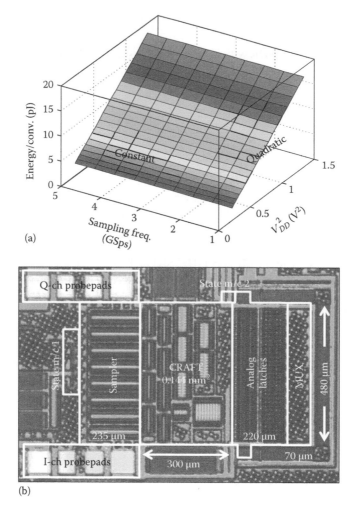

(a)

(b)

FIGURE 5.19 The energy consumption relation with frequency and supply (a) and a die photograph (b). (From Sadhu, B. et al., *Hindawi Int. J. Antennas Propag.*, 2014.)

versus supply voltage and frequency are shown in Figure 5.19a. These measurements clearly show the expected digital-like relationship of the CRAFT energy with frequency and supply voltage. This further corroborates our premise that CRAFT is expected to respond favorably to technology scaling.

Figure 5.19b shows a die photograph of the CRAFT chip. The core occupies an area of 300 μm × 480 μm = 0.144 mm² as shown.

5.8   CONCLUSION

This chapter describes the different kinds of wideband spectrum-sensing architectures and the suitability of switched capacitor circuits for designing such front ends. As an example, a wideband ultralow-power RF front-end channelizer based on a

TABLE 5.3
Comparison with Other FFT Implementations

Ref.	Domain	Bins	SNDR (dB)	Power (mW)	Speed (GS/s)	E/Conv. (pJ/Conv.)[a]	E/Conv. (Ratio[a])
[49]	Current	64	—	389	1.2	345.8	28×
[50]	Current	8	36	19	1.0	405.3	33×
[56]	Digital	128	51[b]	175	1.0	1600	131×
This work	Charge	16	47[c]	3.8	5.0	12.2	1×

Source: Sadhu, B. et al., *Hindawi Int. J. Antennas Propag.*, 2014.
[a] Scaled for complexity similar to the scaling used in Ref. [50].
[b] 8.5-bit ENOB assumed for a 10-bit internal word length.
[c] After twiddle factor correction.

16-point FFT is described. The design is based on a charge reuse technique that enables it to run at 5 GS/s with a 47 dB SNDR capable of transforming a 5 GHz signal (I/Q) while consuming only 3.8 mW (12.2 pJ/conv.).

The current chapter details the design of the CRAFT engine. It also describes the interface circuitry and the test setup required to test such a high-speed, high–dynamic range system. Nonidealities in the CRAFT computations are discussed, analytical models derived, and new circuit techniques developed for mitigating these. These techniques can be easily extended to other passive switched capacitor circuits to improve their performance. Measurement results are then presented.

Table 5.3 compares the CRAFT performance with one digital and two analog domain FFT implementations. As shown, CRAFT operates at speeds 5× faster than previous state-of-the-art designs. Additionally, it consumes better than 28× lower energy. As an RF channelizer, it is expected to reduce digitization requirements enabling wideband digital spectrum sensing. As a result, it advances the state of the art for wideband SDR architectures.

ACKNOWLEDGMENTS

The authors are grateful to Brian M. Sadler and Martin Sturm for their help.

REFERENCES

1. B. Fette, *Cognitive Radio Technology*. Newnes: Oxford, U.K., 2006.
2. B. Sadhu and R. Harjani, *Cognitive Radio Receiver Front-Ends: RF/Analog Circuit Techniques*. Springer: New York City, 2014.
3. T. Ulversoy, Software defined radio: Challenges and opportunities, *IEEE Communications Surveys Tutorials* 12(4), 531–550, 2010.
4. J. Mitola III, Software radios—Survey, critical evaluation and future directions, in *NTC-92. National Telesystems Conference, 1992.* Washington, DC, pp. 13/15–13/23, May 1992.
5. B. Sadhu, M. Sturm, B.M. Sadler, and R. Harjani, Passive switched capacitor RF front-ends for spectrum sensing in cognitive radios, *International Journals of Antennas and Propagation* 2014, Article ID 947373, 20 pages, 2014. doi:10.1155/2014/947373.

6. D. Cabric, I. O'Donnell, M.-W. Chen, and R. Brodersen, Spectrum sharing radios, *IEEE Circuits and Systems Magazine* 6(2), 30–45, 2006.

7. Q. Zhao and B. Sadler, A survey of dynamic spectrum access, *IEEE Signal Processing Magazine* 24(3), 79–89, 2007.

8. A. Valdes-Garcia, A. Natarajan, D. Liu, M. Sanduleanu, X. Gu, M. Ferriss, B. Parker et al., A fully-integrated dual-polarization 16-element W-band phased-array transceiver in SiGe BiCMOS, in *Radio Frequency Integrated Circuits Symposium (RFIC), 2013 IEEE*. IEEE, Seattle, Washington, 2013, pp. 375–378.

9. S. Kalia, S.A. Patnaik, B. Sadhu, M. Sturm, M. Elbadry, and R. Harjani, Multi-beam spatio-spectral beamforming receiver for wideband phased arrays, *IEEE Transactions on Circuits and Systems I: Regular Papers* 60(8), 2018–2029, 2013.

10. B. Sadhu, M. Sturm, B.M. Sadler, and R. Harjani, A 5 GS/s 12.2 pJ/conv. analog charge-domain FFT for a software defined radio receiver front-end in 65 nm CMOS, *IEEE Radio Frequency Integrated Circuits Symposium*, Montreal, Canada, June 2012.

11. R. Bagheri, A. Mirzaei, S. Chehrazi, M.E. Heydari, M. Lee, M. Mikhemar, W. Tang, and A.A. Abidi, An 800-MHz to 6-GHz software-defined wireless receiver in 90-nm CMOS, *IEEE Journal of Solid-State Circuits* 41(12), 2860–2876, 2006.

12. S.M. Mishra, A. Sahai, and R.W. Brodersen, Cooperative sensing among cognitive radios, *IEEE International Conference on Communications*, Istanbul, Turkey, pp. 1658–1663, June 2006.

13. Y. Ding and R. Harjani, A +18 dBm IIP3 LNA in 0.35 μm CMOS, in *IEEE International Solid-State Circuits Conference*, San Francisco, CA, pp. 162–163, 443, 2001.

14. V. Aparin and L. Larson, Modified derivative superposition method for linearizing FET low-noise amplifiers, *IEEE Transactions on Microwave Theory and Techniques* 53(2), 571–581, 2005.

15. W.-H. Chen, G. Liu, B. Zdravko, and A. Niknejad, A highly linear broadband CMOS LNA employing noise and distortion cancellation, *IEEE Journal of Solid-State Circuits* 43(5), 1164–1176, 2008.

16. Z. Ru, N. Moseley, E. Klumperink, and B. Nauta, Digitally enhanced software-defined radio receiver robust to out-of-band interference, *IEEE Journal of Solid-State Circuits* 44(12), 3359–3375, 2009.

17. M. Soer, E. Klumperink, Z. Ru, F. van Vliet, and B. Nauta, A 0.2-to-2.0 GHz 65 nm CMOS receiver without LNA achieving >11 dBm IIP3 and <6.5 dB NF, in *IEEE International Solid-State Circuits Conference*, San Francisco, CA, pp. 222–223, 223a, February 2009.

18. I.D. O'Donnel and R.W. Brodersen, An ultra-wideband transceiver architecture for low power, low rate, wireless systems, *IEEE Transactions on Vehicular Technology* 54(5), 1623–1631, 2005.

19. S. Hoyos, B. Sadler, and G. Arce, Analog to digital conversion of ultra-wideband signals in orthogonal spaces, in *IEEE Conference on Ultra Wideband Systems and Technologies*, pp. 47–51, November 2003.

20. S. Hoyos, B. Sadler, and G. Arce, Broadband multicarrier communication receiver based on analog to digital conversion in the frequency domain, *IEEE Transactions on Wireless Communications* 5(3), 652–661, 2006.

21. S.R. Valezquez, T.Q. Nguyen, S.R. Broadstone, and J.K. Roberge, A hybrid filter bank approach to analog-to-digital conversion, *Proceedings of the IEEE-SP International Symposium on Time-Frequency and Time-Scale Analysis*, Philadelphia, PA, pp. 116–119, October 1994.

22. W. Namgoong, A channelized digital ultrawideband receiver, *IEEE Transactions on Wireless Communications*, 2(3), 502–510, May 2003.

23. T.-L. Hsieh, P. Kinget, and R. Gharpurey, A rapid interference detector for ultra wide-band radio systems in 0.13 μm CMOS, *IEEE Radio Frequency Integrated Circuits Symposium*, Atlanta, GA, pp. 347–350, April 2008.

24. M. Elbadry, B. Sadhu, J. Qiu, and R. Harjani, Dual channel injection-locked quadrature LO generation for a 4 GHz instantaneous bandwidth receiver at 21 GHz center frequency, *IEEE Radio Frequency Integrated Circuits Symposium*, Montreal, Quebec, Canada, pp. 333–336, June 2012.

25. S. Kalia, M. Elbadry, B. Sadhu, S. Patnaik, J. Qiu, and R. Harjani, A simple, unified phase noise model for injection-locked oscillators, *IEEE Radio Frequency Integrated Circuits Symposium*, Baltimore, MD, pp. 1–4, June 2011.

26. L. Carley and T. Mukherjee, High-speed low-power integrating CMOS sample-and-hold amplifier architecture, *Proceedings of the IEEE 1995 Custom Integrated Circuits Conference, 1995*, Santa Clara, CA, pp. 543–546, May 1995.

27. A. Mirzaei, S. Chehrazi, R. Bagheri, and A. Abidi, Analysis of first-order anti-aliasing integration sampler, *IEEE Transactions on Circuits and Systems I* 55(10), 2994–3005, 2008.

28. G. Xu and J. Yuan, Comparison of charge sampling and voltage sampling, in *IEEE Midwest Symposium on Circuits and Systems*, vol. 1, Lansing, MI, pp. 440–443 vol. 1, 2000.

29. S. Karvonen, T. Riley, and J. Kostamovaara, Charge-domain FIR sampler with programmable filtering coefficients, *IEEE Transactions on Circuits and Systems II* 53(3), 192–196, 2006.

30. J. Yuan, A charge sampling mixer with embedded filter function for wireless applications, in *International Conference on Microwave and Millimeter Wave Technology, 2000*, Beijing, China, pp. 315–318, 2000.

31. R. Bagheri, A. Mirzaei, M. Heidari, S. Chehrazi, M. Lee, M. Mikhemar, W. Tang, and A. Abidi, Software-Defined Radio receiver: Dream to reality, *IEEE Communications Magazine* 44(8), 111–118, 2006.

32. Z. Ru, E. Klumperink, and B. Nauta, On the suitability of discrete-time receivers for Software-Defined Radio, in *IEEE International Symposium on Circuits and Systems*, New Orleans, LA, pp. 2522–2525, May 2007.

33. K. Muhammad, D. Leipold, B. Staszewski, Y.-C. Ho, C. Hung, K. Maggio, C. Fernando et al., A discrete-time bluetooth receiver in a 0.13 μm digital CMOS process, in *IEEE International Solid-State Circuits Conference*, San Francisco, CA, pp. 268–527 vol. 1, February 2004.

34. K. Muhammad, Y. Ho, T. Mayhugh, C. Hung, T. Jung, I. Elahi, C. Lin et al., A discrete time quad-band GSM/GPRS receiver in a 90 nm digital CMOS process, in *IEEE Custom Integrated Circuits Conference*, San Jose, CA, pp. 809–812, September 2005.

35. D. Jakonis, K. Folkesson, J. Dbrowski, P. Eriksson, and C. Svensson, A 2.4-GHz RF sampling receiver front-end in 0.18-μm CMOS, *IEEE Journal of Solid-State Circuits* 40(6), 1265–1277, 2005.

36. Z. Ru, E. Klumperink, and B. Nauta, Discrete-time mixing receiver architecture for RF-sampling Software-Defined Radio, *IEEE Journal of Solid-State Circuits* 45(9), 1732–1745, 2010.

37. N.J. Guilar, F. Lau, P.J. Hurst, and S.H. Lewis, A passive switched-capacitor finite-impulse-response equalizer, *IEEE Journal of Solid-State Circuits* 42(2), 400–409, 2007.

38. K. Martin, Complex signal processing is not complex, *IEEE Transactions on Circuits and Systems I* 51(9), 1823–1836, 2004.

39. J.-O. Plouchart, M. Ferriss, B. Sadhu, M. Sanduleanu, B. Parker, and S. Reynolds, A 73.9–83.5 GHz synthesizer with −111 dBc/Hz phase noise at 10 MHz offset in a 130 nm

SiGe BiCMOS technology, in *Radio Frequency Integrated Circuits Symposium (RFIC), 2013 IEEE*, IEEE, Seattle, WA, pp. 123–126, 2013.

40. B. Sadhu, M.A. Ferriss, J.-O. Plouchart, A.S. Natarajan, A.V. Rylyakov, A. Valdes-Garcia, B.D. Parker et al., A 21.8–27.5 GHz PLL in 32 nm SOI using G_m linearization to achieve −130 dBc/Hz phase noise at 10 MHz offset from a 22 GHz carrier, *IEEE Radio Frequency Integrated Circuits Symposium*, Montreal, Canada, June 2012.

41. B. Sadhu, J. Kim, and R. Harjani, A CMOS 3.3–8.4 GHz wide tuning range, low phase noise LC VCO, *Proceedings of IEEE Custom Integrated Circuits Conference*, San Jose, CA, pp. 559–562, September 2009.

42. B. Sadhu and R. Harjani, Capacitor bank design for wide tuning range LC VCOs: 850 MHz-7.1 GHz (157%), *Proceedings of IEEE International Symposium on Circuits and Systems*, Paris, France, pp. 1975–1978, May 2010.

43. S. Sun, F. Wang, S. Yaldiz, X. Li, L. Pileggi, A. Natarajan, M. Ferriss et al., Indirect performance sensing for on-chip analog self-healing via Bayesian model fusion, in *Custom Integrated Circuits Conference (CICC), 2013 IEEE*, IEEE, San Jose, CA, pp. 1–4, 2013.

44. B. Sadhu, M. Ferriss, A. Natarajan, S. Yaldiz, J.-O. Plouchart, A. Rylyakov, A. Valdes-Garcia et al., A linearized, low-phase-noise VCO-based 25-GHz PLL with autonomic biasing, *IEEE Journal of Solid-State Circuits* 48(5), 1138–1150, 2013.

45. A. Abidi, The path to the Software-Defined Radio receiver, *IEEE Journal of Solid-State Circuits* 42(5), 954–966, 2007.

46. V. Arkesteijn, E. Klumperink, and B. Nauta, Jitter requirements of the sampling clock in software radio receivers, *IEEE Transactions on Circuits and Systems II: Express Briefs* 53(2), 90–94, 2006.

47. F. Montaudon, R. Mina, S. Le Tual, L. Joet, D. Saias, R. Hossain, F. Sibille et al., A scalable 2.4-to-2.7 GHz Wi-Fi/WiMAX discrete-time receiver in 65 nm CMOS, in *IEEE International Solid-State Circuits Conference, 2008. ISSCC 2008. Digest of Technical Papers*, San Francisco, CA, pp. 362–619, February 2008.

48. F. Harris, C. Dick, and M. Rice, Digital receivers and transmitters using polyphase filter banks for wireless communications, *IEEE Transactions on Microwave Theory and Techniques* 51(4), 1395–1412, 2003.

49. F. Rivet, Y. Deval, J.-B. Begueret, D. Dallet, P. Cathelin, and D. Belot, The experimental demonstration of a SASP-based full software radio receiver, *IEEE Journal of Solid-State Circuits* 45(5), 979–988, 2010.

50. M. Lehne and S. Raman, A 0.13-μm 1-GS/s CMOS discrete-time FFT processor for ultra-wideband OFDM wireless receivers, *IEEE Transactions on Microwave Theory and Techniques* 59(6), 1639–1650, 2000.

51. B. Murmann, ADC performance survey 1997–2012, 2013. Available at: http://www.stanford.edu/~murmann/adcsurvey.html

52. B. Sadhu, *Circuit Techniques for Cognitive Radio Receiver Front-Ends*, Dissertation, University of Minnesota, Minneapolis, MN, 2012.

53. Y. Ding and R. Harjani, A universal analytic charge injection model, in *IEEE International Symposium for Circuits and Systems* 1, 144–147, 2000.

54. G. Wegmann, E. Vittoz, and F. Rahali, Charge injection in analog MOS switches, *IEEE Journal of Solid-State Circuits* 22(6), 1091–1097, 1987.

55. M. Soer, E. Klumperink, P.-T. de Boer, F. van Vliet, and B. Nauta, Unified frequency-domain analysis of switched-series passive mixers and samplers, *IEEE Transactions on Circuits and Systems I* 57(10), 2618–2631, 2010.

56. Y.-W. Lin, H.-Y. Liu, and C.-Y. Lee, A 1-GS/s FFT/IFFT processor for UWB applications, *IEEE Journal of Solid-State Circuits* 40(8), 1726–1735, 2005.

6 Design Considerations for Direct Delta-Sigma Receivers

Jussi Ryynänen, Kimmo Koli, Kim Östman,
Mikko Englund, Olli Viitala, and Kari Stadius

CONTENTS

6.1 INTRODUCTION

Modern CMOS technology development is driven by the needs of digital circuitry. Higher density, speed, and energy efficiency are obtained as minimum transistor sizes scale down with each new technology node. This trend also drives RF designs toward digital-type circuits that benefit from the scaling. All-digital phase-locked loops have been widely used in cellular synthesizers, and even more aggressive direct digital synthesis methods have been studied in academia. Likewise, transmitter demonstrations of power digital-to-analog converters (DACs) have pushed the boundary of digital signal processing to the power amplifier output node.

On the receiver side, we have seen similar development. The passive mixer-first structures that originate from traditional RF designs can be considered as sampling structures that are placed at the receiver input. Moreover, the analog-to-digital converter (ADC) design community has developed delta-sigma-type receiver concepts that could be utilized as conventional receivers [1–4]. However, on the receiver side, all of the aforementioned structures still need further performance development to override the direct-conversion architecture that is the current industry standard.

Another trend that has been visible in cellular standardization for a couple of years is the expanding number of utilized frequency bands. The current long time evolution (LTE) standards cover tens of bands under 4 GHz. From an RFIC design perspective, this has led to a situation where the circuits should cover a wide range of frequencies. At the same time, there should be signal rejection of unwanted blockers already at the RF nodes, indicating that programmable RF filters are a must.

This chapter focuses on the direct delta-sigma receiver (DDSR) concept, which was introduced in 2010 [5]. The original idea of this architecture is to extend the operation of a continuous-time (CT) delta-sigma A/D converter to RF by introducing an upconverted feedback loop to the RF front-end low noise amplifier (LNA) output. The fundamental difference between a conventional direct-conversion receiver and the DDSR is illustrated in Figure 6.1. As can been seen, the DDSR RF front end receives feedback from an N-stage baseband delta-sigma converter output. This transforms a traditional direct-conversion front end and a baseband delta-sigma converter into a complete RF-to-digital converter. The feedback loop places the LNA into a dual role. It still acts as a traditional amplifier. However, it is also part of the first integrator stage of the DDSR, where the discrete-time (DT) signal is fed back to the LNA output.

This chapter is organized as follows. The first section focuses on the essentials of delta-sigma converters and is targeted to a reader who is not familiar with them. It points out the most essential design parameters for understanding the DDSR and suggests further reading on the fundamental theory of delta-sigma converters. The second section further extends the delta-sigma concept to the DDSR. It focuses on the parameters that are specific to the DDSR and that are not typically considered in traditional converter design. The following section focuses on RF design issues: what does the DDSR entail for an RF designer and, especially, what happens when we extend the DDSR concept to wideband operation. This is followed by a discussion on quantization noise feedback trade-offs, which is an inherent design aspect of the DDSR. Finally, we show case study examples of implemented DDSR architectures to provide an understanding of the current state of the art.

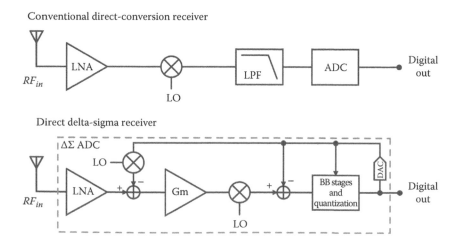

FIGURE 6.1 Conceptual comparison of a conventional receiver and a DDSR.

6.2   BASEBAND DELTA-SIGMA MODULATORS

Delta-sigma modulation is a noise-shaping signal-processing technique that has been applied in different contexts to modify the spectral shape of noise. Applications include, but are not limited to, ADC and DAC. In this section, we concentrate on delta-sigma modulation in the ADC context, since familiarity with this subject is crucial for understanding DDSR operation. This section is not meant to be a comprehensive discussion, and thus the interested reader is encouraged to refer to Refs. [6,7] for deeper insight into delta-sigma modulator (DSM) data converters.

An essential part of any ADC is the quantizer. An N-bit quantizer discretizes an analog signal by comparing it to $2^N - 1$ predetermined threshold levels. In the simplest case, an ADC consists of only a quantizer with sample-and-hold capability. The quantizer input is sampled and held at a sampling interval determined by the sampling clock and then discretized. The output consists of the input signal and an error signal whose power is determined by the number of quantizer bits N. A higher N reduces the quantization error, and thus the signal-to-noise ratio (SNR) is higher. If the quantizer input is *busy*, that is, the state of the output changes almost every sampling interval, the quantization error can be approximated as a white noise whose power is evenly distributed between dc and the sampling frequency. Thus, in addition to increasing N, we can also increase the SNR by sampling at a higher frequency than necessary and thus reduce the quantization noise at frequencies of interest. This is referred to as oversampling. The term oversampling ratio (OSR) is often used to tie the wanted frequency bandwidth f_{BW} to the sampling frequency f_S so that

$$OSR = \frac{f_S}{2 f_{BW}}.$$
(6.1)

A DSM provides an additional way of reducing the quantization noise within f_{BW}. In a DSM, we add a high-gain filter, commonly referred to as the loop filter, before the quantizer. If we then bring a feedback through a DAC from the quantizer output to the filter input, the quantization noise is shaped away from frequencies where the filter has gain. For example, the loop filter in a low-pass DSM is usually implemented with integrators, and so the quantization noise is shaped away from low frequencies. The loop filter can be either DT or CT. In a DT DSM, the input is sampled before filtering, and thus an antialias filter is often required before the actual ADC. A DT DSM is commonly implemented with switched-capacitor techniques. In a CT DSM, the input is sampled after filtering, which means that the loop filter also acts as an antialias filter. A CT DSM is typically implemented with op-amp-RC or $g_m C$ integrators. However, it should be noted that the CT DSM is also ultimately DT due to the sampling at the quantizer input. This must be taken into account when designing the loop filter. For example, while the s-plane can be used to approximate system behavior at low frequencies, the actual response is periodic and repeats itself at integer multiples of the sampling clock frequency f_S.

There exist various loop filter architectures. One common architecture is a cascade of integrators in feedback (CIFB). The signal flow graph in Figure 6.2 illustrates a traditional third-order CIFB DSM. The main parts of the converter are

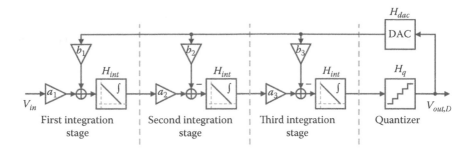

FIGURE 6.2   Signal flow graph of a third-order DSM based on a CIFB architecture.

the integrators that form the loop filter, the quantizer, and the feedback DAC. The input signal is passed through a cascade of integrators and then discretized by the quantizer. The DAC is used to convert the digital output stream of the quantizer back to the analog domain, and this signal is then weighted and subtracted in each integration stage. To evaluate how the DSM ADC affects the input signal and the quantization noise of the quantizer, we can define two transfer functions. The signal transfer function (*STF*) is a linearized model that can be used to evaluate the signal response from the ADC input to its output. The noise transfer function (*NTF*) is a linearized model that can be used to evaluate how the quantization noise is shaped by the ADC.

In order to obtain useful versions of the *STF* and *NTF*, the quantizer and DAC have to be replaced with their linear models. The quantizer model should take into account the sample-and-hold functionality, the effective gain of the quantizer, and the quantization error due to finite resolution. As mentioned earlier, white noise can be used to approximate the quantization error. Thus, quantization error can be accounted by summing a noise voltage V_{noise} to the output of the quantizer. For the DAC, we need to take into account its pulse shape, which in the simplest case is merely a delayed version of the quantizer output. After linearization, the *STF* and *NTF* in Figure 6.2 can be derived as follows:

$$STF = \frac{V_{out,D}}{V_{in}} = \frac{a_1 a_2 a_3}{\dfrac{1}{H_q H_{int}^3} + b_3 \dfrac{H_{dac}}{H_{int}^2} + b_2 a_3 \dfrac{H_{dac}}{H_{int}} + b_1 a_2 a_3 H_{dac}}, \tag{6.2}$$

$$NTF = \frac{V_{out,D}}{V_{noise}} = \frac{\dfrac{1}{H_q H_{int}^3}}{\dfrac{1}{H_q H_{int}^3} + b_3 \dfrac{H_{dac}}{H_{int}^2} + b_2 a_3 \dfrac{H_{dac}}{H_{int}} + b_1 a_2 a_3 H_{dac}}, \tag{6.3}$$

where H_{int} is the integrator transfer function, H_q is the linearized quantizer model, and H_{dac} is the linearized DAC transfer function. The reason why we have decided to leave the specific transfer functions undefined at this point is that one may prefer to use either s- or z-plane models depending on the application. For example, if we choose to analyze the behavior of a CT DSM in the s-plane, we can use the following functions for H_{int}, H_q, and H_{dac}:

$$H_{int}(s) = \frac{\omega_0}{s}, \tag{6.4}$$

$$H_q(s) = q\frac{1 - e^{-(s/f_S)}}{(s/f_S)}, \tag{6.5}$$

$$H_{dac}(s) = e^{-sD}. \tag{6.6}$$

H_{int} is now an ideal s-plane integrator with the natural frequency ω_0. H_q approximates the zero-order-hold functionality of the quantizer at a specific sampling frequency f_S, and q is the effective gain of the quantizer. H_{dac} consists of a simple delay of time D.

By using the parameter values listed in Table 6.1 for our third-order CT CIFB DSM, we arrive at the *STF* and *NTF* plotted in Figure 6.3. The loop filter is designed by using the *cookbook design procedure* explained in Ref. [6]. The chosen filter family is Butterworth, and the parameter ω_0 is chosen such that the *NTF* exhibits high-frequency gain without rendering the modulator unstable. The architecture results in

TABLE 6.1

Parameter Values Used for the Example in Figure 6.3

Parameter	a_1	a_2	a_3	b_1	b_2	b_3	ω_0	Q	D	f_S
Value	1	1	1	1	2	2	$2\pi120e6$	2.6	$0.01/f_S$	2e9

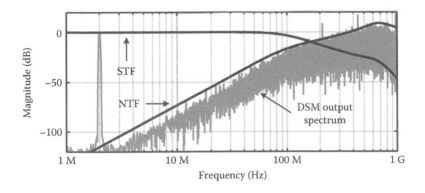

FIGURE 6.3 *STF*, *NTF*, and output spectrum of a third-order CT CIFB DSM. The chosen architecture results in a low-pass *STF* and a high-pass *NTF*. The spectrum obtained from a transient simulation follows the trends set by the analytical *STF* and *NTF*.

a low-pass *STF* and a high-pass *NTF*. Because the DSM is a highly nonlinear system, the responses have been validated in a transient simulation. Figure 6.3 also displays the output spectrum from this simulation. The input signal applied to the DSM consists of three tones and noise. The input tones are 0 dB at 2 MHz, 20 dB at 210 MHz, and 40 dB at 440 MHz and can be observed in the output spectrum. We can see that the analytical *STF* and *NTF* approximate the behavior of the DSM output spectrum. The high-frequency bump in both the *STF* and the *NTF* is caused by the feedback delay and the nonunity effective gain of the used 1-bit quantizer (here, q is approximated to be 2.6). This high-frequency behavior degrades DSM performance and may even lead to instability. However, different compensation techniques such as a direct feedback around the quantizer can be used to compensate for the effects of the feedback delay [8].

From Figure 6.3, it is easy to see that if we are only interested in the signals below 10 MHz, the quantization noise inside this bandwidth is attenuated significantly by the *NTF*, in turn providing a high SNR. Thus, it can be seen that we do not necessarily require a high-resolution quantizer in order to achieve high SNR in a DSM ADC. In fact, single-bit quantization is commonly used, since it is inherently linear and simple to implement. When combined with a high oversampling rate and noise shaping, the resulting SNR can still be very high. However, a low-resolution quantizer has its shortcomings. Perhaps the most severe drawback of single-bit quantization, especially in a CT DSM, is its susceptibility to sampling clock jitter. A higher-resolution quantizer can be used to alleviate this problem. Increased resolution lowers the quantization noise level and improves resilience against sampling clock jitter, but this trades off with increased design complexity. Recently, jitter sensitivity has also been reduced by using finite impulse response (FIR) filter techniques in the first feedbacks [9–11]. For convenience, Table 6.2 summarizes how different nonidealities affect the performance of the DSM.

TABLE 6.2
Summary of the Most Important DSM Nonidealities

Nonideality	Cause	Effect
Excess loop delay (CT DSM)	Nonzero switching time of transistors	Increases *NTF* numerator order, may limit performance
Clock jitter	Nonideal sampling clock	Increases quantization noise floor
Nonlinearity in quantizer and feedback DACs	Component mismatch, limited transistor output impedance	Increases quantization noise, distorts feedback signals
Loop filter gain error	Component mismatch	Alters *NTF*, may decrease SNR as a result
Low integrator DC gain	Limited output impedance and transconductance of transistors	Degrades noise shaping
Insufficient op-amp GBW and SR	Insufficient op-amp current, poor frequency compensation	Increases odd-order harmonics that decrease the SNR

6.3 FROM BASEBAND DSM TO RF-TO-DIGITAL CONVERSION

The preceding section concentrated on the basic concepts of quantization and the delta-sigma noise-shaping principle. We were interested in removing as much of the quantization noise from the desired bandwidth as possible in order to achieve a high SNR. In this section, we move from the DC-centered baseband DSM to the local oscillator (LO)–centered DDSR. We will see that in a DDSR, the STF becomes as important as or even more important than the NTF.

The key idea of the DDSR is to combine typical receiver signal-processing tasks into one functional block. The signal-processing tasks include the amplification, filtering, and downconversion performed by the RF front end and the digitalization performed by the baseband chain. The most significant difference between a DDSR and a conventional DSM is that in a conventional DSM, the input signal is usually well defined. The input is scaled so that the available dynamic range is fully utilized. Also, the signal bandwidth is typically limited before A/D conversion. In contrast, the input of a receiver can be far from optimal. The desired signal can be barely distinguishable from the thermal noise floor. Even worse, the neighboring channels and bands may contain undesired signals that are several decades larger in magnitude. In the worst case, the undesired signals can saturate the receiver and thus prevent reception of the desired signal. We have two alternative ways to cope with the challenges brought by the incorporation of the RF front end inside a DSM. One way is to design for a very high dynamic range, so that both the weak desired signal and strong undesired signal can be converted simultaneously. However, obtaining the needed dynamic range is challenging even in the case of a conventional DSM. The other way is to filter out as much of the undesired signal as possible, so that ideally only the desired signal is converted. In this case, the filtering should be applied as early as possible in the signal chain in order to prevent strong undesired signals from overloading the input stages. A feedback DSM architecture is chosen for the DDSR, because it offers superior filtering of the input signal.

Consider the DDSR signal flow graph in Figure 6.4. Compared to a general feedback-type DSM, the most notable differences are the addition of mixers to

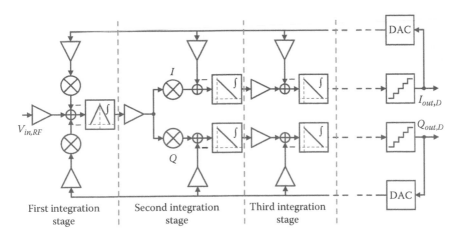

FIGURE 6.4 DDSR signal flow graph.

perform the frequency translation between RF and BB and that the flow diagram divides into I/Q quadrature branches after the RF stages in the same manner as in a conventional direct-conversion receiver. From a direct-conversion receiver viewpoint, an important difference is that the quantized baseband output is now fed back to the RF stages.

The first and second integration stages of the DDSR operate partly at RF and partly at baseband. Figure 6.5a illustrates the circuits that form the first stage. The RF input signal is first amplified by an LNA. Passive mixers are then employed at the LNA output node to translate the frequencies between the RF and baseband I and Q branches. The use of passive mixers has additional benefits, as they have the ability to translate impedances between their RF and baseband ports. When the two passive mixers are combined with the baseband capacitors C_{Npath}, the circuit functions as an N-path filter. The low-pass response of C_{Npath} is translated into a band-pass response at the LNA output node. The center frequency of this band-pass response follows the LO frequency. Thus, band-pass filtering is applied to the RF input, while the baseband feedbacks are low-pass filtered. Another way to interpret this is that the feedback is integrated by a conventional $g_m C$ integrator, while frequency-shifted $g_m C$ integration is applied to the RF input. Although this may appear to be an optimal solution, there are several nonidealities that must be taken into account. These nonidealities will be discussed in more detail later.

The second integration stage is illustrated in Figure 6.5b. A buffer drives the RF input signal through passive I/Q mixers to the baseband Miller integrator inputs. The feedback is brought to the same Miller integrator inputs and is integrated and subtracted from the input signal. In this integrator, the output is on the baseband side and the nonidealities are not as significant as in the first stage. Since this circuit structure has an RF input and an integrated baseband output, we will refer to it as a downconverting g_mOPA-C integrator. The capacitors C_{mix} are used to apply additional attenuation outside the wanted channel. They have no significant effect on the DDSR transfer function, because the Miller effect causes the integration capacitors C_{int} to dominate. As a practical note, the RF input signal is divided

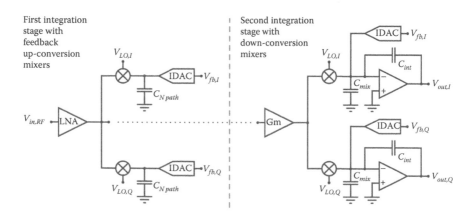

FIGURE 6.5 The first and second integration stages of the DDSR.

equally between the I/Q branches in a symmetric case, and thus the corresponding feed-forward DSM coefficient has to be divided by 2.

The design guidelines of a conventional DSM can be applied when designing the DDSR loop filter. However, these guidelines may not be optimal for a DDSR, since we have to fulfill also the role of a direct-conversion receiver. In order to further demonstrate the differences between a conventional DSM and the DDSR, consider the *STF* and *NTF* plots in Figure 6.6a and b. The *STF* and *NTF* in Figure 6.6a are examples of what one could see in the case of conventional second-order feedback CT DSM with f_S = 2 GHz. The cutoff frequency of the loop filter is relatively high, and about 3 dB of high-frequency gain is applied to the *NTF* to trade off the noise-shaping performance with the stability of the modulator. The *STF* gain is 0 dB, and some filtering is inherently applied due to the feedback architecture. Depending on the needed SNR, the actual signal bandwidth f_{BW} could be, for example, 10 MHz, which is significantly less than the cutoff frequency of the loop filter.

Figure 6.6b is an example of a second-order DDSR *STF* and *NTF* for a 20 MHz RF bandwidth (f_{LO} ± 10 MHz) when using f_{LO} = 1 GHz and f_S = 2 GHz. The *STF* now has gain in the passband between −20 MHz and 20 MHz offsets from the f_{LO}. The gain is applied in the first stage in order to relax the noise performance requirements of the following stages. Note that the loop filter cutoff frequency is now only about two times the actual bandwidth, which means that while the filtering of the unwanted signals outside the bandwidth is enhanced, we pay the price in the form of nonoptimal noise shaping. We have to rely on the oversampling and the number of quantizer bits in addition to the noise shaping by the loop filter to ensure that the quantization noise will not dominate the noise performance of the entire receiver.

It is clear that a trade-off will have to be made in loop filter design between input signal filtering and noise shaping. If the cutoff frequency of the loop filter is high, we need to have high dynamic range to prevent any high-power out-of-band signals from overloading the receiver. On the other hand, if we push the filter cutoff frequency too low, the noise shaping might not be sufficient and the quantization noise could dominate and desensitize the receiver. Additionally, if the cutoff frequency of the loop filter is very low compared to the sampling frequency of the quantizer, the quantizer input is no longer *busy*. This means that the white noise model of the

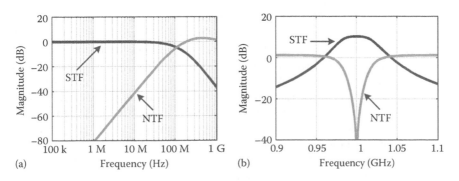

FIGURE 6.6   *STF* and *NTF* examples for (a) a DSM and (b) a DDSR.

quantization noise no longer applies and we may see unwanted noise concentration spikes in the output spectrum. The usable ratio of cutoff frequency to sampling frequency depends on loop filter design and the threshold levels of the quantizer. This problem can be alleviated by injecting dither to the quantizer input.

6.4   RF DESIGN CONSIDERATIONS

When comparing DDSR blocks to a regular baseband DSM, the RF section stands out as a significant difference. This is not only due to the frequency translations that are involved but also due to the nonideal properties of the RF stages themselves. Accordingly, in this section, we discuss the roles of these stages, the need for modeling their nonidealities, and the most important areas of codesign between the RF section and the rest of the DDSR.

Because the DDSR embeds a direct-conversion receiver front end into the DSM structure, the front-end LNA and downconversion blocks naturally take on dual roles. On the one hand, they function as conventional blocks that amplify, filter, and downconvert incoming RF signals, and as such, they play a significant role in determining receiver linearity and sensitivity. On the other hand, the LNA and the downconversion block act as the first two integrators in the integrator cascade making up the DDSR. In this second role, they participate in forming the transfer function of the DDSR loop filter, thus impacting the desired channel filtering and noise-shaping properties of the receiver.

The first stage, comprised of the LNA and the N-path filter as shown in Figure 6.5, is called an *N-path $g_m C$ integrator*. From a conventional viewpoint, the LNA provides voltage amplification to the RF input signal and is the most important block in terms of noise and linearity. The N-path filter at the LNA output provides a band-pass-type frequency response around f_{LO}, which is beneficial for attenuating unwanted blocker signals. This enhances the out-of-band linearity of the receiver significantly. The response is dependent on the LNA output impedance.

From the DSM viewpoint, the LNA acts as a transconductor that drives the integration capacitors C_{Npath}. The response at the LNA output is centered on f_{LO} due to the frequency translation by the N-path filter. In similar fashion, the signal from the baseband current-mode DAC (IDAC) in the feedback path is integrated by C_{Npath} and then upconverted to the LNA output through passive mixing.

The second stage is comprised of an RF amplifier, a current-commutating passive downconversion mixer, and a baseband Miller integrator. Again, from a conventional viewpoint, this arrangement resembles a current-mode downconversion system that is loaded by a transimpedance amplifier. However, the DSM viewpoint entails looking at the arrangement as a g_mOPA-C-based integrator. In contrast to the baseband realization, the voltage-to-current conversion is performed at RF around f_{LO}, after which the current is frequency-translated to baseband. The feedback path operates entirely at baseband, and the integrated baseband output signal voltage is processed further by subsequent baseband stages in conventional DSM fashion.

In comparing the two RF stages to their baseband prototype counterparts, certain nonidealities of RF implementation become evident. For example, frequency translation in both stages causes effects that must be accounted for in order to obtain

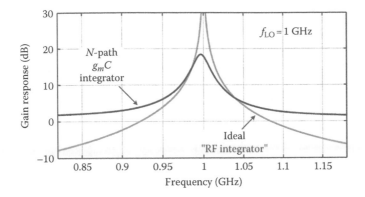

FIGURE 6.7 Example responses of an ideal RF integrator and the N-path $g_m C$ integrator when $f_{LO} = 1$ GHz.

the desired response. The most important nonidealities revolve around the first integrator. We will now detail the nonidealities related to this integrator and discuss a method for modeling them in a receiver design.

The main nonideal features of the integrator response are depicted in Figure 6.7. First, the integrator gain is relatively low at f_{LO} (i.e., at dc in a conventional DSM). Whereas the output impedance of the baseband transconductor in a $g_m C$ integrator is usually several kΩ, the LNA in the N-path $g_m C$ integrator has an intrinsic maximum output impedance of only a few hundred ohms. The gain at f_{LO} is thus most often limited to values around 20 dB, whereas baseband integrator design usually targets a dc gain of at least 40 dB, thanks to high output impedance. The most significant consequence of this nonideality is a reduced contribution to noise-shaping operation by the outermost feedback loop. Narrowband LC loads usually provide the highest output impedance and allow for simpler techniques to boost the impedance further.

Second, the N-path $g_m C$ integrator has an out-of-band attenuation floor that depends on the passive mixer switch resistance R_{SW}. In the example of Figure 6.7, the gain response reaches a floor at 2 dB, whereas the gain of the ideal integrator has no lower limit at faraway offsets from f_{LO}.

Third, the response of the first integrator is not exactly centered on f_{LO}. This causes an imbalance for the two sidebands of the incoming RF signal, especially in wideband implementations. In such circuits, the frequency offset is caused by parasitic capacitance at the LNA output node, increasing with frequency and capacitance. In addition to these three nonidealities, the basic N-path $g_m C$ integrator is unable to distinguish between signals at f_{LO} and its odd harmonics.

A modeling method is needed to account for the most important nonidealities during receiver design. Our proposed approach is to partially replace the ideal coefficient-based representation of the DDSR with transfer functions that entail nonideal integration. This approach is illustrated in Figure 6.8 where the CT coefficients a_1 and b_1, both related to the first integrator, are replaced by the nonideal integration functions H_{rf} and H_{fb}. For example, if we assume that there are no further baseband

RF-related section

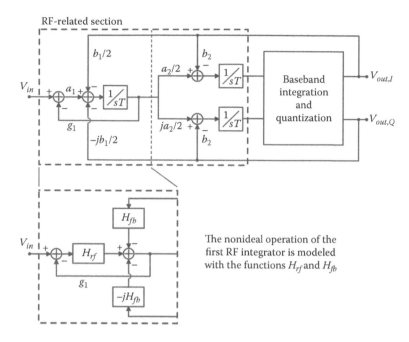

FIGURE 6.8 RF-centric signal flow graph of the DDSR, using (a) ideal and (b) nonideal integration transfer functions to model the N-path g_mC integrator.

integration stages beyond the second integrator and that $g_1 = 0$ for simplicity, we obtain these STF and NTF approximations [12] for the I branch:

$$STF_I = \frac{1}{2} \frac{a_2 H_{rf}}{sT + b_2 + a_2 H_{fb}}, \quad NTF_I = \frac{sT}{sT + b_2 + a_2 H_{fb}}.$$

Proper circuit models are then required to obtain the more precise form of these transfer functions. The passive mixer–based work in Ref. [13] provides a useful lumped-element starting point for linear time-invariant modeling. Approximate CT DDSR transfer functions for the first integrator are derived in Ref. [12] and account for the three nonidealities listed earlier. The entire DDSR model, including the nonidealities, can then be used in a variety of mixed-signal simulators for complete receiver design. As a result, the design process produces a significantly improved understanding of the actual signal filtering and noise-shaping capability to be expected from the receiver. This approach has been an essential part of designing the receivers discussed later in the case study section.

The second DDSR integrator also has a number of nonidealities when compared to an ideal baseband integrator prototype. These nonidealities are related to the output impedance of the RF transconductor and to the effect of frequency translation on the exact frequency response. Although they can be modeled in similar fashion to those of the N-path g_mC integrator, it can also be shown that they are much less significant.

Still, the RF designer should be aware of the nonidealities and deliver information on block properties to the system designer.

In conventional direct-conversion receiver design, the only point of contact between the RF and baseband section designers is the downconversion mixing interface. Apart from proper joint design of this interface, the section designers can mostly work independently by adhering to the gain, noise, and linearity partitioning requirements provided by the system designer. The DDSR is different in that there is one additional RF–baseband interface due to the upconverted delta-sigma feedback and in that the properties of the RF stages affect not only traditional analog receiver performance metrics but also the properties of the DSM as a whole. Here, we will briefly discuss two of the most important codesign issues, namely, the gain control and the trade-off between noise figure (NF), noise shaping, and blocker filtering.

Both RF stages can be designed to include gain control functionality. The LNA gain (related to coefficient a_1) can be adjusted without any effect on the receiver STF and NTF. Gain tuning in the second stage (coefficient a_2) should be balanced by simultaneous tuning of the outermost feedback loop (coefficient b_1) in the opposite direction. This preserves the shape and bandwidth of the STF while at the same time changing the gain level. In practice, this means that once the required extent of gain tuning and the initial properties of the blocks related to a_2 and b_1 are known, the system design process should give the section designers specifications on how much tuning should be included in each block. Continued iterations between all designers are then needed to ensure a proper tuning range.

Moreover, the design of the N-path g_mC integrator governs a system-wide trade-off between receiver NF, noise shaping, and the amount of blocker filtering. This is related to the closed-loop behavior of the DDSR, causing the gain at the LNA output to be suppressed inside the loop filter bandwidth [12] so that the bandwidth of the LNA gain response is similar to the loop filter bandwidth. In particular, a higher value of C_{Npath} provides a sharpened bandpass filtering response and improved noise shaping in the NTF, but it also leads to stronger gain suppression and thus an increased NF. The acceptable amount of trade-off between these three receiver metrics should be decided at the system level and then translated to N-path g_mC integrator component requirements. In Ref. [14], we propose a circuit-level design method for navigating this trade-off in light of target specifications.

6.5   QUANTIZATION NOISE FEEDBACK

The effective noise shaping of a high-order low-pass DSM enables a high signal-to-quantization-noise ratio (SQNR). This is achieved with a relatively high OSR, that is, the desired signal bandwidth is narrow compared to the sampling frequency f_S. As a result, the quantization noise level is low not only around DC but also around the sampling frequency f_S and its multiples. Elsewhere, the quantization noise is stronger than it would be without the noise shaping. The sinc function inherent in the quantization mainly helps to attenuate quantization noise around the multiples of f_S, where the quantization noise level is low anyway.

In frequency-translating DSMs and the DDSR, the delta-sigma-modulated feedback signal is upconverted to the desired RF frequency range by f_{LO}. Strongly nonlinear frequency upconversion is very challenging when using a very wideband feedback signal: choosing the wrong ratio f_S/f_{LO} of the sampling and LO frequencies leads to completely losing the intended feedback signal under undesired quantization noise mixing products.

When frequency mixing is performed at the sampling frequency, that is, $f_{LO} = f_S$, upconversion is successful since the desired feedback signal is shifted in frequency to the next quantization noise minimum around the sampling frequency f_S. Similarly, all other quantization noise minima are shifted upward on the frequency axis by the same amount, thus keeping the output spectrum intact. It is easy to see that successful frequency upconversion is obtained when the LO frequency is an integer multiple of the sampling frequency, that is, $f_{LO} = Nf_S$.

A more complex example is shown in Figure 6.9, where the LO frequency equals the DSM Nyquist frequency, that is, $f_{LO} = f_S/2$. Figure 6.9a shows the original delta-sigma-modulated baseband signal, repeated at intervals of f_S with gradual high-frequency attenuation provided by the sinc function. The spectrum in Figure 6.9b is the result of upconversion with an ideal sinusoidal LO signal, and in a similar manner Figure 6.9c is the sinusoidal upconversion with the third harmonic of the LO. Figure 6.9d depicts the upconversion with a squarewave LO signal, showing the individual mixing products of the fundamental as well as the third and the

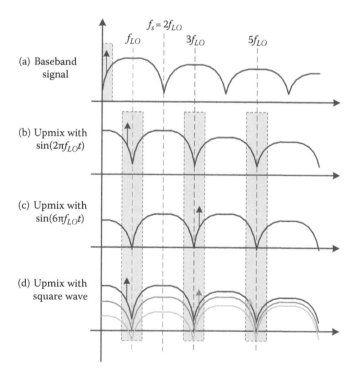

FIGURE 6.9 The DDSR feedback signal (a) before and (b) after upconversion with sinusoidal LO-signal $f_{LO} = f_S/2$, (c) with the third harmonic of f_{LO} and (d) upconversion with a squarewave LO signal with the fundamental, third, and fifth harmonic mixing products shown separately.

fifth harmonics of f_{LO}. The fundamental mixing product and the harmonic mixing products fall nicely together, and the harmonic products are attenuated sufficiently such that they do not decrease the total SQNR. The three gray areas in the three bottom spectrums describe the frequency ranges that will be downconverted to baseband. The downconversion products from the harmonics are slightly attenuated with respect to the fundamental downconversion product, and thus the SQNR is not degraded.

Another example is shown in Figure 6.10 with $f_{LO} = f_S/3$. With this f_S/f_{LO} ratio, quantization noise levels are drastically increased in the upmixed spectrum, even when mixing with an ideal sinusoidal LO signal. This is caused by quantization noise folding from negative frequencies (the dashed line spectrum). Upconversion with the third harmonic of f_{LO} in Figure 6.10c is symmetric around DC such that folding from negative frequencies does not degrade spectral purity. The spectral energy from mixing with the third LO harmonic is strong around the LO frequency, but as Figure 6.10d shows, the dominating quantization noise is still produced by folding from negative frequencies. When the feedback signal is downconverted back to baseband, also downconversion from the third LO harmonic slightly raises the quantization noise level at the desired frequency band.

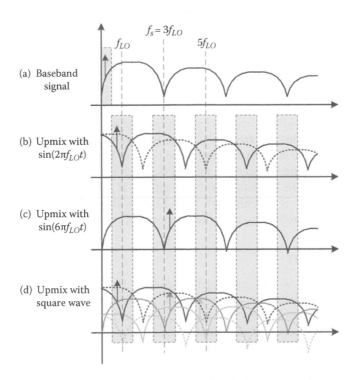

FIGURE 6.10 The DDSR feedback signal (a) before and (b) after upconversion with a sinusoidal LO signal $f_{LO} = f_S/3$, (c) with the third harmonic of f_{LO}, and (d) upconversion with a squarewave LO signal with fundamental, third, and fifth harmonic mixing products shown separately.

Based on these two examples, the quality of the original feedback signal is retained when the relation between f_{LO} and f_S is chosen so that the upconverted spectrum is symmetric around DC. The LO frequency should therefore be an integer multiple of the Nyquist frequency $f_S/2$, that is, $f_{LO} = Nf_S/2$. This condition is relatively strict. For example, if f_{LO} and f_S are both set to 1 GHz, even a 5 % deviation for the LO frequency can degrade ADC bandwidth from 10 to 5 MHz for the same SNR. Moreover, when f_{LO} is not a multiple of the Nyquist frequency, neither feedback signal upconversion from the LO harmonics nor downconversion from the LO harmonics back to baseband significantly degrades the SQNR. This is because the dominant performance degradation is caused in the feedback upconversion of the fundamental LO frequency. Therefore, resorting to harmonic rejection techniques in the feedback upmixing does not affect the quantization noise level.

Additional filtering of the DDSR feedback signal is required prior to frequency upconversion if greater freedom for choosing f_{LO} and f_S is required. The most straightforward filtering implementation would be a simple FIR filter. In the case of our second example where $f_{LO} = f_S/3$, an FIR filter with the transfer function $H(z) = (1 + z^{-1} + z^{-2})/3$ would already significantly improve on the obtainable SNR. However, the notches provided by FIR filtering are very narrow, and therefore this technique does not solve the problem for wideband transmissions such as LTE 20.

The N-path technique significantly improves the situation with the DDSR feedback signal upconversion. The differential I/Q feedback signals are fed with I/Q differential-output IDACs to the N-path capacitors, thus providing first-order low-pass filtering of the DDSR feedback signal before upconversion. Since the feedback signal is now filtered before upconversion, f_{LO} can be chosen freely, provided that f_{LO} is equal to or higher than f_S. When f_{LO} is decreased well below f_S, the low-pass filtering provided by the N-path technique may not be sufficient to prevent the mixing products from folding from the negative frequencies to the desired signal band. In order to extend this LO frequency range to lower RF bands, a combination of the N-path technique and FIR filtering could be used. Determining when additional FIR filtering is required depends on the DDSR order, the architecture, and the coefficients, as well as the resolution of the quantizer. If a 1-bit quantizer is used, the quantization noise level is significantly higher than with multibit quantizers, and thus a different level of prefiltering is required before upconversion.

6.6   CASE STUDY

In this final section, we present implementation and measurement result details by using DDSRs that have been published to date. This provides a realistic understanding of the performance levels that can be achieved by using the DDSR concept. The original introduction of the DDSR concept was accompanied by a 900 MHz prototype in a 65 nm CMOS, reported in Ref. [5]. The subsequent work in Ref. [15] presents two implementations in a 40 nm CMOS, with one targeted for narrowband (2.5 GHz) operation and the other for wideband (0.7–2.7 GHz) operation. We will highlight the most important implementation aspects by focusing on the latter designs, which emphasize programmability in both the coefficients and the bandwidth of the loop filter.

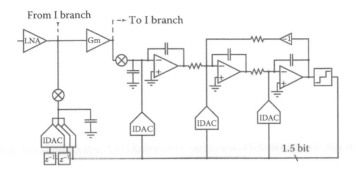

FIGURE 6.11   Block-level schematic of the four-stage DDSR implementation in Ref. [15].

As shown in Figure 6.11, the receivers in Ref. [15] are based on a four-stage feedback modulator structure, comprised of a cascade of integrators. The number of stages is essentially a trade-off between the achievable level of out-of-band blocker filtering on the one hand and the complexity and stability of the design on the other. The first two stages operate partly at RF and partly at baseband, whereas the final two stages operate at baseband.

The loop filter baseband bandwidth can be programmed to either 1.5 or 15 MHz, allowing for RF signal channels up to about 20 MHz. Furthermore, programmability has been implemented in many of the RF and baseband coefficients (a_i, b_i) of the prototype DSM signal flow graph. For example, all feedback coefficients b_i and the forward coefficient a_2 are programmable in the wideband receiver, and the narrowband receiver adds programmability in a_1 as an additional level of freedom as explained previously.

In the RF section, the narrowband receiver contains an inductively degenerated common-source LNA with a Q-boosted LC load that is tuned to 2.5 GHz. The effective transconductance of the LNA can be programmed, thus providing tunability in the modulator coefficient a_1. The wideband 0.7–2.7 GHz receiver uses a common-source LNA with a PMOS load, together with tunable common-drain RC feedback for proper input matching. It is important to note the lower output impedance of the wideband LNA as the major difference between the receivers. As discussed before, this is a nonideal feature that requires modeling, and it also leads to a degraded ability for noise shaping by the outermost feedback loop.

The second stage in both receivers is a g_m-boosted source follower that drives a current-commutating downconversion interface as a transconductor. The three stages operate in parallel to provide programmability in the transconductance and thus in the modulator coefficient a_2. The LO signal is generated externally, and an on-chip divider together with a custom buffer arrangement [16] creates the 25% duty cycle used for both the RF feedback and downconversion mixers. To ensure proper operation and correct LO phase polarity throughout the receiver, it is important that the same divider is used to generate the LO signal for both mixers.

In the baseband section, the integrators are built using regular operational amplifiers. The loop filter bandwidth is programmed by changing the size of the integration capacitor. Feedback from the digital receiver output to each integrator stage is provided by D/A converters (IDAC), where current-mode operation has been chosen for proper performance in terms of speed and power consumption. Coefficient programmability is implemented by 5-bit (two-decade) tuning of the IDAC bias currents.

Three important design choices have been made at baseband to increase the achievable signal-to-noise-and-distortion ratio (SNDR). First, 1.5-bit operation is used throughout the digital parts. This is done by designing the quantizer with two parallel comparators, each of which uses its own reference voltage level. Moreover, the levels can be reprogrammed depending on operating conditions. As discussed previously, the high-order loop filter also exhibits excess loop delay, and a programmable delay compensation circuit is implemented at the input of the comparators. Second, an internal feedback is used from the input of the comparators to the input of the third integrator. This creates a notch in the *NTF*, with the purpose of maximizing the in-channel noise shaping by placing the notch at the edge of the desired baseband channel. Finally, a 2-tap FIR filter is placed in the outermost feedback loop. As discussed previously, the purpose of such a filter is to improve SNDR by reducing the transfer of quantization noise to the LNA output node. Especially for the wideband case, it is important to note that the placement of the notches is programmable, thus allowing for a wider range of f_{LO}/f_S ratios.

The 2.5 GHz and 0.7–2.7 GHz receivers are implemented on the same chip and share the baseband circuitry. Only the native $V_{DD} = 1.1$ V of the utilized 40 nm CMOS technology is used. For the wideband receiver, Figure 6.12 displays the intended functionality of the receiver through an example output spectrum plot using $f_{LO} = 2.5$ GHz. First, a –68 dBm RF input signal at 2.5001 GHz is downconverted to 100 kHz as expected. Second, the quantization noise is shaped away from the desired channel bandwidth according to the designed *NTF*. Finally, a –43 dBm out-of-band RF blocker signal is inserted at 2.5875 GHz and is filtered as indicated by the overlaid *STF*, disappearing into the quantization noise floor. Similar operation is observed throughout the full operating range.

The measurement results of the three published receivers are summarized in Table 6.3. Analog receiver metrics and A/D converter metrics are both included to facilitate comparison within both domains.

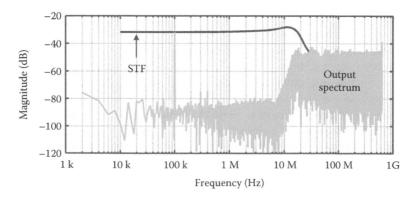

FIGURE 6.12 Example measured output spectrum for $f_{IN} = 2.5001$ GHz, $f_{LO} = 2.5$ GHz, $f_{BW} = 15/2$ MHz, and $f_{clk} = 1.25$ GHz. The measured *STF* is overlaid.

TABLE 6.3

Summary of Measurement Results

Case	NB RX [5]	WB RX [15]		NB RX [15]
Frequency (GHz)	0.9	0.7–2.7	0.7–2.7	2.5
f_S^a (GHz)	1	1.25	1.25	1
Signal BW (MHz)	18	15	1.4	15
Gain (dB)	40	37	37	41
NF (dB)	6.2	5.9–8.8	5.9–8.2	4.2
OB IIP3[b] (dBm)	+4	–2	–2	0
Non-cal. IIP2[c] (dBm)	>+45	>+38	>+35	>+31
BCP 1 dB (dBm)	18[c]	12[d]	14[d]	15[d]
Peak SNDR (dB)	54	43	40	50
Power (mW)	80 at 1.2 V		90 at 1.1 V	
Active area (mm²)	1.2	1.0		1.1

[a] Maximum used f_S
[b] Closest interferer at f_{LO} + 95 MHz.
[c] Blocker or closest interferer at f_{LO} + 80 MHz.
[d] Blocker at f_{LO} + BW/2 + 85 MHz.

Contrasting the wideband and narrowband cases provides some interesting insights. First, the lower NF of the narrowband receiver shows the benefit of the chosen LNA architecture and a Q-boosted on-chip load resonator. This benefit is apparent also in the achievable maximum SNDR, where the outermost feedback loop is able to provide better noise shaping as compared to using a wideband LNA with lower output impedance in the first stage.

Nevertheless, also the wideband receiver provides the expected functionality with satisfactory performance and thus demonstrates the feasibility of the DDSR concept for operation in a wide LO frequency range. For all receiver versions, the out-of-band IIP3 is ultimately limited by the RF front end, and the power consumption is divided approximately evenly between the RF and baseband sections.

REFERENCES

1. R. Winoto and B. Nikolic, A highly reconfigurable 400–1700 MHz receiver using a down-converting sigma-delta A/D with 59-dB SNR and 57-dB SFDR over 4-MHz bandwidth, *Proceedings of the 2009 Symposium on VLSI Circuits*, Kyoto, Japan, pp. 142–143, June 2009.
2. E. Martens, A. Bourdoux, A. Couvreur, R. Fasthuber, P. Van Wesemael, G. Van der Plas, J. Craninckx, and J. Ryckaert, RF-to-baseband digitization in 40 nm CMOS with RF bandpass ΔΣ modulator and polyphase decimation filter, *IEEE Journal of Solid-State Circuits* 47(4), 990–1002, 2012.
3. H. Shibata, R. Schreier, W. Yang, A. Shaikh, D. Paterson, T.C. Caldwell, D. Alldred, and P.W. Lai, A DC-to-1 GHz tunable RF ΔΣ ADC achieving DR = 74 dB and BW = 150 MHz at f_0 = 450 MHz using 550 mW, *IEEE Journal of Solid-State Circuits* 47(12), 2888–2897, 2012.

4. C. Wu and B. Nikolic, A 0.4 GHz–4 GHz direct RF-to-digital ΣΔ multi-mode receiver, *Proceedings of the 2013 European Solid-State Circuits Conference*, Bucharest, Romania, pp. 275–278, September 2013.

5. K. Koli, S. Kallioinen, J. Jussila, P. Sivonen, and A. Pärssinen, A 900-MHz direct delta-sigma receiver in 65-nm CMOS, *IEEE Journal of Solid-State Circuits* 45(12), 2807–2818, 2010.

6. S. Norsworthy, R. Schreier, and G. Temes, *Delta-Sigma Data Converters: Theory, Design, and Simulation*. New York: IEEE Press, 1997.

7. R. Schreier and G. Temes, *Understanding Delta-Sigma Data Converters*. New York: IEEE Press, 2004.

8. M. Keller, A. Buhmann, J. Sauerbrey, M. Ortmanns, and Y. Manoli, A comparative study on excess-loop-delay compensation techniques for continuous-time sigma-delta modulators, *IEEE Transactions on Circuits and Systems–I: Regular Papers* 55(11), 3480–3487.

9. O. Oliaei, Sigma-delta modulator with spectrally shaped feedback, *IEEE Transactions on Circuits and Systems–II: Analog and Digital Signal Processing* 50(9), 518–530, 2003.

10. B. Putter, ΣΔ ADC with finite impulse response feedback DAC, *IEEE International Solid-State Circuits Conference Digest of Technical Papers*, San Francisco, CA, pp. 76–77, February 2004.

11. P. Shettigar and S. Pavan, Design techniques for wideband single-bit continuous-time ΔΣ modulators with FIR feedback DACs, *IEEE Journal of Solid-State Circuits* 47(12), 2865–2879, 2012.

12. K.B. Östman, M. Englund, O. Viitala, K. Stadius, K. Koli, and J. Ryynanen, Characteristics of LNA operation in direct delta-sigma receivers, *IEEE Transactions on Circuits and Systems—II: Express Briefs* 61(2), 70–74, 2014.

13. C. Andrews and A.C. Molnar, Implications of passive mixer transparency for impedance matching and noise figure in passive mixer-first receivers, *IEEE Transactions on Circuits and Systems—I: Regular Papers* 57(12), 3092–3103, 2010.

14. K.B. Östman, M. Englund, O. Viitala, K. Stadius, J. Ryynanen, and K. Koli, Design tradeoffs in N-path G_mC integrators for direct delta-sigma receivers, *Proceedings of the 2013 European Conference on Circuit Theory and Design*, Dresden, Germany, pp. 1–4, September 2013.

15. M. Englund, K. B. Östman, O. Viitala, M. Kaltiokallio, K. Stadius, K. Koli, and J. Ryynänen, A programmable 0.7-to-2.7 GHz direct ΔΣ receiver in 40 nm CMOS, *IEEE International Solid-State Circuits Conference Digest of Technical Papers*, San Francisco, CA, pp. 470–471, February 2014.

16. M. Kaltiokallio, R. Valkonen, K. Stadius, and J. Ryynanen, A 0.7–2.7-GHz blocker-tolerant compact-size single-antenna receiver for wideband mobile applications, *IEEE Transactions on Microwave Theory and Techniques* 61(9), 3339–3349, 2013.

Section II

Millimeter-Wave Transceivers

7 60 GHz CMOS Transceiver Chipset for Mobile Devices

Koji Takinami, Takayuki Tsukizawa,
Naganori Shirakata, and Noriaki Saito

CONTENTS

7.1 INTRODUCTION

In recent times, 60 GHz millimeter-wave systems have become increasingly attractive due to the escalating demand for multi-Gb/s wireless communication. The 9 GHz of unlicensed bandwidth that has been allocated at 60 GHz enables extremely high data rates as compared to those offered by 2.4 and 5 GHz wireless LAN standards. Traditionally, expensive III–V semiconductors such as gallium arsenide (GaAs) are required for this frequency. However, recent works have demonstrated the ability to realize a 60 GHz transceiver by means of a cost-effective CMOS process. Such regulatory changes and technological advances have driven not only academia but also industry bodies to explore the 60 GHz frequency band targeting high-volume and low-cost consumer applications.

This chapter aims to give the most up-to-date status of the 60 GHz wireless transceiver development, with an emphasis on realizing low power consumption and small form factor that is applicable for mobile terminals. Section 7.2 gives a brief overview of the system and industrial standards. Section 7.3 discusses choices of transceiver architectures. Section 7.4 presents the low-power transceiver design including built-in self-calibration (BiSC) techniques that are essential to improve yield and robustness in mass production. The transceiver is integrated into a miniaturized radio-frequency (RF) module with antennas that can be implemented in mobile handsets. Finally, concluding remarks are given in Section 7.5.

7.2   60 GHZ WIRELESS SYSTEM

7.2.1   USE CASE SCENARIOS

The use of high frequencies comes with both advantages and disadvantages. Large path loss, as well as high attenuation due to obstacles such as a human body, limits 60 GHz applications to those suitable for short-range wireless communication within 10 m. On the other hand, higher frequencies lead to smaller sizes of RF components including antennas, enabling compact realization of an array structure that offers improved antenna gain with high directivity [1,2]. These properties unique to 60 GHz help reduce interference between terminals, offering opportunities to maintain multi-Gb/s throughput even in high-density environments.

The first commercially available consumer products using 60 GHz appeared in the early 2010s and were based on the WirelessHD standard, which supports uncompressed high-definition (HD) video links between fixed devices such as a set-top box (STB) and a TV. The chipset employed sophisticated beamforming technology to extend communication distance [3], but this resulted in high power consumption. Even though substantial progress has been made since then, employing a 60 GHz system in mobile terminals such as smartphones and tablets remains a difficult challenge since it requires extensive reduction in power consumption as well as small form factor.

Figure 7.1 shows use cases that are made possible by the 60 GHz mobile solution. One example application is HD video streaming from mobile devices. This is good for gaming, for instance, because the required latency can be in the order of a few tens of milliseconds to enable a natural responsiveness between the mobile device and the large screen TV as illustrated in Figure 7.1a. Another example application is fast file transfer. In order to download 30 min worth of HD contents, which is roughly 2.4 GB, it takes several minutes with the traditional IEEE 802.11 at 2.4/5 GHz, whereas it can take less than 10 s by using the 60 GHz wireless system. Therefore, one can download videos from the PC to the mobile device or even share the downloaded contents with others instantaneously.

7.2.2   FREQUENCY ALLOCATION

The most recent global frequency allocation at 60 GHz is illustrated in Figure 7.2 [4]. Frequency bands around 60 GHz are available worldwide. The ITU-R recommended channelization comprises four channels, each 2.16 GHz wide, centered on 58.32,

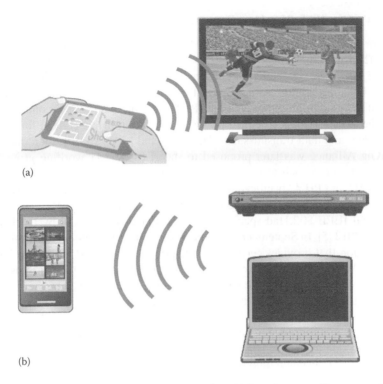

(a)

(b)

FIGURE 7.1 60 GHz mobile use case examples. (a) Low-latency video streaming and (b) fast file transfer.

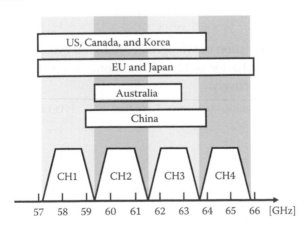

FIGURE 7.2 Frequency allocations by region and channel in WiGig/IEEE 802.11ad.

60.48, 62.64, and 64.80 GHz, respectively. The frequency allocations in each region do not match exactly, but there is substantial overlap of at least 3.5 GHz of contiguous spectrum in all regions. This allows a single device to operate worldwide without hardware modifications just like the traditional wireless LAN based on the IEEE 802.11 at 2.4/5 GHz.

7.2.3   WiGig/IEEE 802.11AD STANDARD

There are many different standards for 60 GHz wireless communication. Within the IEEE 802.15 working group, Task Group 3 (TG3) had led the standardization and the IEEE 802.15.3c was issued in 2009 by Task Group 3c (TG3c). Soon after, the Wireless Gigabit (WiGig) Alliance was formed to develop an industrial standard for 60 GHz. In addition to the physical (PHY) and medium access control (MAC) layers, the WiGig Alliance defined protocol adaptation layers (PALs) for effective audio, video, and data transmissions. The PHY and MAC specification defined by the WiGig Alliance was later proposed to the IEEE 802.11 working group, Task Group *ad* (TGad), which was the group tasked with defining modifications to the 802.11 MAC and PHY specifications to enable operation in the 60 GHz frequency band capable of a maximum throughput of at least 1 Gb/s.

The final IEEE 802.11ad specification was officially approved by the IEEE in December 2012 [5]. In September 2013, the Wi-Fi Alliance, a global nonprofit industry association that provides certification programs, announced successful integration of the WiGig Alliance, placing WiGig/IEEE 802.11ad in the lead position to become the de facto standard for wireless communication at 60 GHz.

In WiGig/IEEE 802.11ad, both single carrier (SC) modulation and orthogonal frequency division multiplexing (OFDM) modulation have been adopted, considering various use case scenarios. In general, SC modulation is suitable for reduced power consumption due to its low peak-to-average power ratio (PAPR), whereas OFDM modulation offers better multipath tolerance. Table 7.1 shows the modulation and coding schemes (MCSs) for SC modulation, where MCS 0 to MCS 4 are mandatory. MCS 0 is exclusively used to transmit control channel messages employing DBPSK

TABLE 7.1

MCS for SC Modulation in WiGig/IEEE 802.11ad (MCS 0 and MCS 1 Employ Spreading Factor of 32 with $\pi/2$ Rotation and Spreading Factor of 2, Respectively)

MCS Index	Modulation	Code Rate	PHY Payload Rate (Mb/s)
0	DBPSK	1/2	27.5
1	$\pi/2$-BPSK	1/2	385
2	$\pi/2$-BPSK	1/2	770
3	$\pi/2$-BPSK	5/8	962.5
4	$\pi/2$-BPSK	3/4	1155
5	$\pi/2$-BPSK	13/16	1251.25
6	$\pi/2$-QPSK	1/2	1540
7	$\pi/2$-QPSK	5/8	1925
8	$\pi/2$-QPSK	3/4	2310
9	$\pi/2$-QPSK	13/16	2502.5
10	$\pi/2$–16QAM	1/2	3080
11	$\pi/2$–16QAM	5/8	3850
12	$\pi/2$–16QAM	3/4	4620

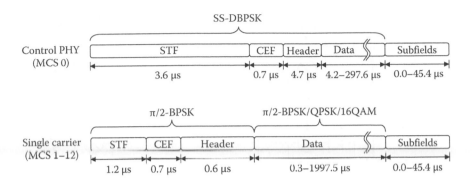

FIGURE 7.3   Packet structure examples.

modulation with code spreading to ensure better robustness. Figure 7.3 shows the packet structure examples. They consist of a short training field (STF) and a channel estimation field (CEF), followed by a header, data, and subfields. The subfields can be used for fine beamforming training.

To achieve multi-Gb/s throughput, the MAC layer requires many modifications from the one in the traditional IEEE 802.11. For example, a minimum interval time between transmission packets, called short inter-frame space (SIFS), is shortened from 16 μs in the IEEE 802.11a/g to 3 μs in the WiGig/IEEE 802.11ad. Many timing parameters of random backoff in carrier sense multiple access with collision avoidance (CSMA/CA) are also shortened. On the other hand, the maximum payload size is extended from 2304 to 7920 octets. These modifications reduce packet overhead and improve transmission efficiency to achieve multi-Gb/s throughput.

The standard also specifies fast session transfer (FST), which enables wireless devices to seamlessly transit between the 60 GHz frequency band and the legacy 2.4 and 5 GHz bands, to complement optimal performance with wider communication range.

7.3   RADIO ARCHITECTURE

Two of the most popular choices of radio architecture are super heterodyne and direct conversion [6]. In the super heterodyne architecture, the frequency conversion is performed in two steps. The use of an intermediate frequency (IF) that is lower than the original carrier frequency makes radio design easier, but it requires two phase-locked loop (PLL) frequency synthesizers. In addition to occupying large chip area, oscillators fabricated on the same chip suffer from unwanted coupling. To avoid these issues, a sliding IF architecture is commonly used in modern integrated transceivers [3,7,8]. As shown in Figure 7.4a, two local oscillator (LO) frequencies are generated from one frequency synthesizer by using frequency multipliers or dividers. The IF is often chosen relatively high so that it provides large channel separation between the carrier frequency and its image frequency, relaxing the requirement for image rejection and eliminating external IF filtering.

Direct conversion, on the other hand, performs the frequency conversion in one step, making the radio architecture very simple as illustrated in Figure 7.4b, since

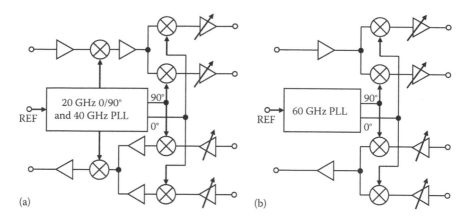

FIGURE 7.4   (a) Sliding IF and (b) direct conversion architectures.

the LO frequency is the same as the carrier frequency; special care must be taken to avoid frequency pulling from the transmitter (TX) output to the PLL. Also, the LO leakage in the receiver (RX) path makes the LO self-mixing, resulting in a dynamic dc offset. In spite of these technical difficulties, recent works [9–16] have realized a 60 GHz transceiver with direct conversion architecture. Compared to the super heterodyne or the sliding IF, the direct conversion architecture typically achieves smaller die area and lower power consumption.

In some 60 GHz applications, it is preferable to place an RF module and a base-band IC (BBIC) in separate locations. Figure 7.5 is an example where the RF module and the BBIC are interfaced with a single coaxial cable [17–19]. This can be viewed as the super heterodyne architecture split across two chips. The IF and the LO as well as some control signals are fed through the coaxial cable, which is suitable for PC platforms. For example, in the laptop PC, the RF module should be mounted on the display frame to ensure better signal radiation from the antenna, whereas the BBIC must be placed close to the processors to minimize signal routings.

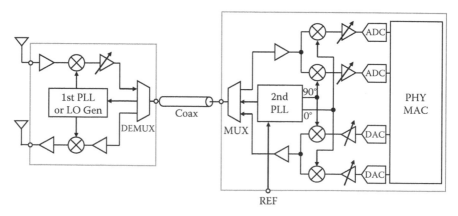

FIGURE 7.5   Heterodyne architecture using coaxial cable.

7.4 LOW-POWER CMOS TRANSCEIVER BASED ON WiGig/IEEE 802.11AD

This section presents a fully integrated transceiver chipset based on the WiGig/IEEE 802.11ad for mobile usage [15,20]. The chipset is developed for SC modulation, which is advantageous for reducing power consumption. However, SC modulation is sensitive to in-band amplitude variations, which are primarily a result of gain variations of analog circuits and multipath delay spread.

Figure 7.6 illustrates the distortion mechanisms. Since the WiGig/IEEE 802.11ad employs a wide modulation bandwidth of 1.76 GHz, it is challenging to maintain a flat frequency response over the entire modulation bandwidth. In general, the TX signal suffers large attenuation at higher offset frequencies as a result of baseband analog circuits, with additional asymmetric distortion due to the nonideal frequency response in the 60 GHz signal path. Moreover, the multipath environment causes frequency-dependent distortion in the received signal, which is further distorted by the RX analog circuits.

To reduce performance degradations due to these signal distortions, the proposed chipset employs built-in TX in-band calibration and an RX frequency domain equalizer (FDE). In order to minimize the increase of hardware size, the chipset utilizes existing circuit blocks and also employs a unique signal processing algorithm of the FDE. These techniques relax the requirement of the gain flatness and process variations for high-speed analog circuits, leading to less power consumption with minimum hardware overhead.

7.4.1 TRANSCEIVER SYSTEM

Figure 7.7 shows the block diagram of the proposed 60 GHz CMOS transceiver chipset. It employs two separate antennas, one for the TX and the other for the RX, to minimize the insertion loss from the RF front end to antennas. The radio-frequency IC (RFIC) includes the RF front end and the PLL synthesizer, which support all four channels allocated at 60 GHz. The BBIC includes PHY and MAC layers as well as high-speed interfaces.

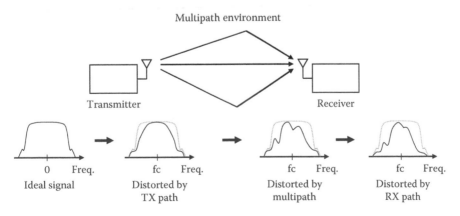

FIGURE 7.6 Illustration of distortion mechanisms.

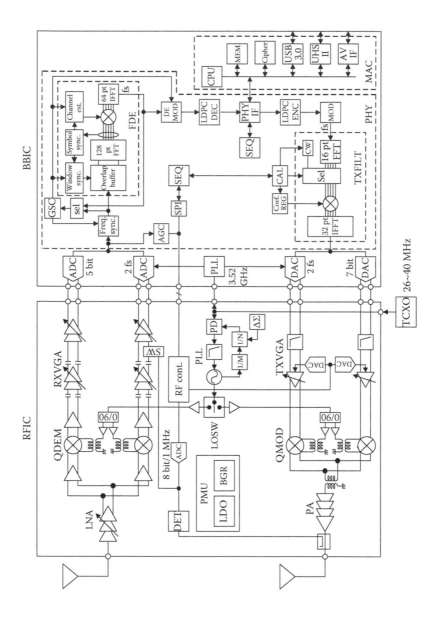

FIGURE 7.7 Block diagram of the RFIC and the BBIC. (From Saito, N., Tsukizawa, T., Shirakata, N. et al., A fully integrated 60-GHz CMOS transceiver chipset based on WiGig/IEEE 802.11ad with built-in self calibration for mobile usage, *IEEE J. Solid-State Circuits*, 48(12), 3146, 2013. Copyright 2013 IEEE.)

The RFIC employs direct conversion architecture in both TX and RX paths, which is advantageous for reducing power consumption and die area as mentioned in Section 7.3. Integrated low-dropout regulators (LDOs) provide regulated 1.25 and 1.4 V dc voltages from the 1.8 V power supply to improve external noise tolerance. A dedicated LDO is used for the PLL to achieve better noise immunity. The TX consists of a four-stage power amplifier (PA), a quadrature modulator (QMOD), and TX variable gain amplifiers (TXVGAs). TXVGAs employ the fourth-order passive low-pass filters (LPFs) for anti-aliasing. The automatic power control (APC) is performed by using an envelope detector (DET) and an 8-bit 1 MHz analog-to-digital converter (ADC). The RX consists of a four-stage low noise amplifier (LNA), a quadrature demodulator (QDEM), and RX variable-gain amplifiers (RXVGAs). In order to cancel the dc offset, the transceiver adopts the ac coupling in the RXVGAs. Since the WiGig/IEEE 802.11ad standard requires a stringent settling time for achieving high-speed data rates, a two-step gain control algorithm with high-pass filter (HPF) cutoff frequency switching is employed [21]. In the proposed algorithm, the cutoff frequency of the HPF is set to be 160 MHz at start-up to make the convergence time faster. The variable gain amplifier (VGA) gain is adjusted by using a binary search algorithm to achieve ±3 dB accuracy. After 0.6 µs, the cutoff frequency is switched to a lower value for fine adjustment, and the VGA gain is tuned to within ±1 dB accuracy by using a linear search algorithm with 1 dB VGA gain steps.

The LO generation consists of a 30 GHz push–push voltage-controlled oscillator (VCO) to eliminate a power hungry 60 GHz frequency divider. The PLL synthesizer incorporates fast frequency calibrations for the VCO subband selection and the injection-locked 30 GHz frequency divider [22]. The quadrature hybrid generates 0°/90° LO signals to drive the QMOD and the QDEM. In the direct conversion architecture, it is essential to have accurate quadrature generation at 60 GHz. In this design, the magnetic coupling hybrid is used to achieve small amplitude imbalance and phase error over wide frequency range [23]. The measured insertion loss is less than 1 dB. The measured amplitude imbalance and the phase error are less than 0.3 dB and 3 deg, respectively, across CH1–CH4.

The BBIC is interfaced with the RFIC through differential IQ analog signals and digital control signals. The TX path of the BBIC incorporates 7-bit current steering IQ digital-to-analog converters (IQ-DACs) with a 3.52 GHz sampling clock. The current mode output is directly fed to the RFIC in order to eliminate current consumption of the voltage level shifter and the I-V converter. The RX signal from the RFIC is fed to the 5-bit IQ-ADCs, which are interfaced with the RFIC via 100 ohm differential microstrip lines with the full scale of 200 mVpp differential voltage. The BBIC incorporates the FDE, the TX filter (TXFILT), and the sequencer (SEQ) as well as the low-density parity check (LDPC) decoder and encoder, which support MCS 0 to MCS 9. Even though the standard specifies higher MCSs, the chipset does not support them to reduce power consumption. USB 3.0 and UHS-II (SDIO 4.0) provide external high-speed interfaces. The AV interface is used for video streaming.

7.4.2 TX CALIBRATION

As mentioned earlier, it is challenging to maintain flat frequency response over the wide modulation bandwidth. Although TX amplitude variations can be minimized

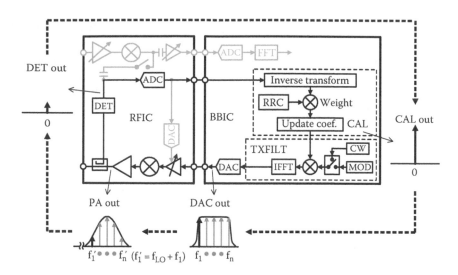

FIGURE 7.8 Block diagram of the proposed TX in-band amplitude calibration. (From Saito, N., Tsukizawa, T., Shirakata, N. et al., A fully integrated 60-GHz CMOS transceiver chipset based on WiGig/IEEE 802.11ad with built-in self calibration for mobile usage, *IEEE J. Solid-State Circuits*, 48(12), 3146, 2013. Copyright 2013 IEEE.)

by a factory calibration, it requires additional testing costs in mass production. So the chipset employs BiSC to compensate for TX in-band amplitude variations. Figure 7.8 shows the block diagram of the proposed TX in-band amplitude calibration. Even though 5-bit 3.52 GHz IQ-ADCs in the BBIC can be used to sample the 1.76 GHz bandwidth TX signal, a loopback calibration suffers from additional amplitude variations in the RX path. To minimize the estimation error, the proposed calibration detects the power level right after the PA output through the wide-bandwidth 10 dB coupler by using the DET and the 8-bit 1 MHz ADC, which are readily available from the APC feedback loop, thus requiring no additional hardware.

As illustrated in Figure 7.8, the BBIC generates continuous wave (CW) signals with 110 MHz steps from −880 MHz (f_1) to +880 MHz (f_n) in the digital domain, which are converted to analog waveforms. At the DAC output, each signal has constant amplitude level, which is distorted due to frequency variations in the subsequent analog circuits. At the PA output, the signal is extracted through the coupler followed by the DET. The DET generates the dc signal in proportion to the PA output level. The output is digitized by the ADC and then fed back to the calibration algorithm in the BBIC. The calibration (CAL) block in Figure 7.8 performs inverse transformation and updates filter coefficients for each frequency. The root-raised cosine (RRC) filtering is added to the filter coefficients, which are multiplied with the modulation signal in the TXFILT to compensate for the entire frequency distortion in the TX path.

7.4.3 RX Calibration

In general, the time domain equalizer (TDE) is widely used for the SC modulation. However, when the propagation delay is much longer than the symbol duration,

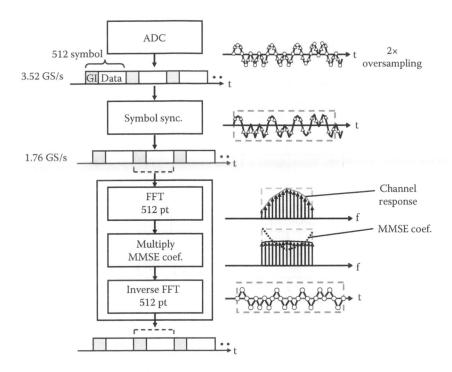

FIGURE 7.9 Flow chart of the conventional FDE. (From Saito, N., Tsukizawa, T., Shirakata, N. et al., A fully integrated 60-GHz CMOS transceiver chipset based on WiGig/ IEEE 802.11ad with built-in self calibration for mobile usage, *IEEE J. Solid-State Circuits*, 48(12), 3146, 2013. Copyright 2013 IEEE.)

which is the case for the WiGig/IEEE 802.11ad standard, the TDE results in long-tap equalization, making the FDE more preferable, especially for smaller hardware size. Figure 7.9 shows the flow chart of the conventional FDE. The payload consists of 512 symbols having the 64-symbol guard interval (GI) and 448-symbol data field, which are concatenated periodically. The total 512-symbol data are first synchronized and decimated by 2 to reduce sampling speed, and then it is equalized by using 512 pt FFT and inverse fast Fourier transform (IFFT). This is equivalent to the 64-tap TDE that can equalize up to 64-symbol delay [24]. However, since the sampling speed is in gigahertz, this symbol synchronization process normally requires parallel signal processing [25], resulting in large hardware and higher power dissipation due to large size FFT and IFFT.

Figure 7.10 shows the flow chart of the proposed oversampling/overlapping FDE. The incoming signal is sampled with a 50% overlap window using 64 symbols, which is the same length as the GI. After being converted into the frequency domain, decimation is performed by extracting 64 pt out of 128 pt, which corresponds to the desired signal. Then, it is equalized based on the minimum mean square error (MMSE) coefficients, which are calculated from the channel impulse response estimated by the CEF of the packet. Because the proposed overlap FDE does not discard the GI, the equalized GI can be used for estimating the residual frequency offset

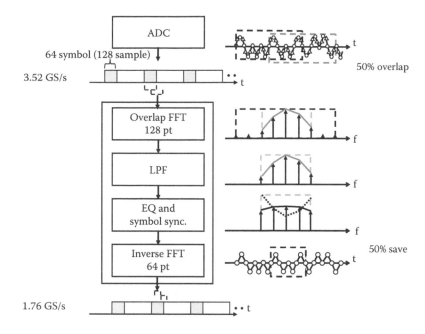

FIGURE 7.10 Flow chart of the proposed oversampling/overlapping FDE. (From Saito, N., Tsukizawa, T., Shirakata, N. et al., A fully integrated 60-GHz CMOS transceiver chipset based on WiGig/IEEE 802.11ad with built-in self calibration for mobile usage, *IEEE J. Solid-State Circuits*, 48(12), 3146, 2013. Copyright 2013 IEEE.)

and the phase error. Therefore, carrier tracking and equalization can be performed simultaneously in frequency domain. These phase errors due to synchronization error are estimated at every 512-symbol interval and compensated by every equalizing block in the frequency domain. The signals are then transformed into the time domain by IFFT. After concatenation with 50% overlaps, the equalized 1.76 GS/s data are generated.

Due to 50% overlap, the proposed FDE is equivalent to the 32-tap TDE in theory [24]. Thus, the maximum excess delay equalized by the proposed FDE is 32-symbol delay, which is half of the conventional FDE. This is acceptable for a line-of-sight (LOS) short-range communication up to a few meters of distance in a living room, where the time of arrival of a main reflected wave is estimated within about 13 ns (23 symbols) [26]. The proposed FDE achieves a low power consumption of 115.7 mW including the power hungry 3.52 GS/s decimation filtering, whereas the publications [25] and [27] dissipate more than 200 mW in the data path alone.

7.4.4 MODULE AND ANTENNA DESIGN

Figure 7.11 shows the simplified structure of the RF module [28]. It employs the multilayer polyphenylene ether (PPE) resin substrate, which is less expensive than the low-temperature cofired ceramic (LTCC) substrate, which had been widely used at millimeter-wave frequencies.

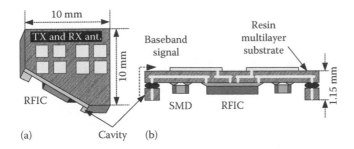

FIGURE 7.11   Module structure. (a) Perspective view and (b) cross-sectional view.

The TX/RX antennas are printed on the top plane, whereas the RFIC is placed on the bottom plane right under the antennas. This structure shortens line length from the RFIC to antennas to minimize signal loss at 60 GHz. Each TX/RX antenna consists of a four-element patch array. The baseband signals and the control signals are fed from the I/O pins on the cavity frame to the RFIC through internal layers. This structure provides more than 50 I/O pins with narrow I/O pad pitch of 500 μm. The substrate material has relative dielectric constant ε_r = 3.6 and loss tangent tan δ = 0.016 at 60 GHz, offering less than 1 dB loss from the RFIC to the antennas. Figure 7.12 shows measured antenna radiation patterns. The antenna gain is 6.7 dBi and the beamwidth is 51° for the XZ plane and 55° for the YZ plane.

Figure 7.13 illustrates the layer structure of the module. Line and GND layers (c) provide solid ac ground and also help suppress back radiation from the antennas to the RFIC. However, if the distance from the RFIC to the ground plane is too close, the frequency characteristics of the RFIC may be shifted, which deteriorates transceiver performance. To avoid this, the GND pattern of the RFIC mounted layer (d) is removed to maintain 160 μm spacing from the RFIC.

7.4.5   MEASUREMENT RESULTS

Figure 7.14a and b shows die photos of the RFIC and the BBIC, which are fabricated in 90 and 40 nm CMOS processes, respectively. The chip sizes of the RFIC and the

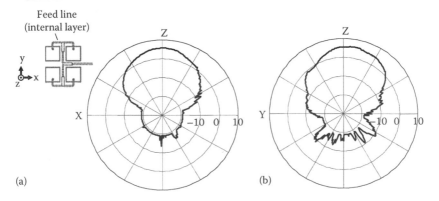

FIGURE 7.12   Measured antenna radiation patterns (CH2). (a) XZ plane and (b) YZ plane.

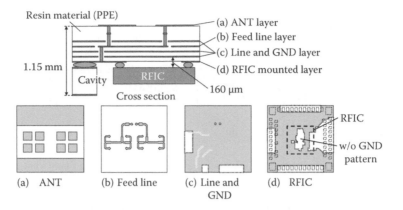

FIGURE 7.13 Layer structure of the module.

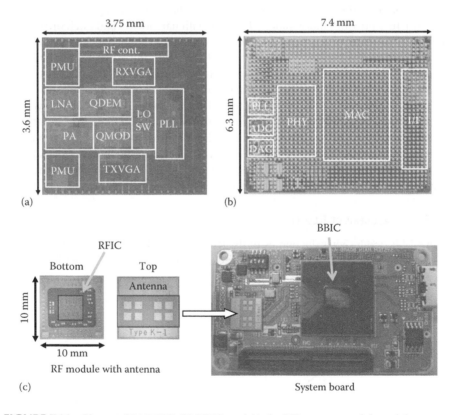

FIGURE 7.14 Photos of (a) RFIC, (b) BBIC, and (c) the RF antenna module and the system board. (From Saito, N., Tsukizawa, T., Shirakata, N. et al., A fully integrated 60-GHz CMOS transceiver chipset based on WiGig/IEEE 802.11ad with built-in self calibration for mobile usage, *IEEE J. Solid-State Circuits*, 48(12), 3146, 2013. Copyright 2013 IEEE.)

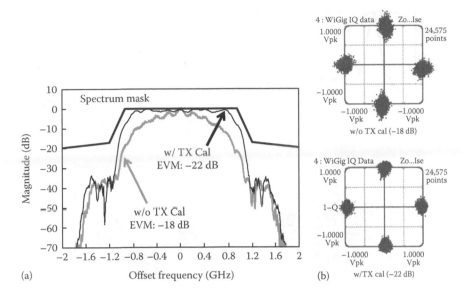

(a)   Offset frequency (GHz)   (b)

FIGURE 7.15 Improvement of the proposed built-in self-calibration at 2 dBm output power. (a) Measured output spectrum and (b) measured constellations at CH2 with MCS 9. (From Saito, N., Tsukizawa, T., Shirakata, N. et al., A fully integrated 60-GHz CMOS transceiver chipset based on WiGig/IEEE 802.11ad with built-in self calibration for mobile usage, *IEEE J. Solid-State Circuits*, 48(12), 3146, 2013. Copyright 2013 IEEE.)

BBIC are 3.6 mm × 3.75 mm and 6.3 mm × 7.4 mm, respectively. The supply voltages are 1.8 and 2.5 V for the RFIC and 1.1, 1.8, and 3.3 V for the BBIC. Figure 7.14c shows the photograph of the RF module and the system board.

Figure 7.15a and b compares the measured output spectrum and constellations with and without the proposed TX in-band amplitude calibration. Without calibration, it shows 17 dB passband droop at 880 MHz. The modulation signal itself has 3 dB droop and the antialiasing filter introduces 3 dB attenuation at 880 MHz. The remaining 11 dB bass-band droop is mainly caused by the analog baseband circuits. The proposed calibration reduces the passband droop to 6 dB, resulting in the error vector magnitude (EVM) improvement of 4 dB from −18 to −22 dB.

The FDE performance is measured by using two system boards on the desk, TX and RX, at the height of 1 m. As shown in Figure 7.16a, a large reflected wave is observed at 8 ns delay in this specific condition, which results in the asymmetric output spectrum as shown in Figure 7.16b. Using the proposed FDE, the MMSE coefficients effectively equalize the in-band amplitude variations. As a result, the IQ constellation is converged to the reference points as shown in Figure 7.16c.

Figure 7.17 shows the measured MAC throughput from one station (STA1) to the other (STA2), using MCS 5 to MCS 9 with different communication distances. The chipset achieves 1.8 Gb/s up to 40 cm and 1.5 Gb/s up to 1 m. The signal waveform is also captured over the air by an external horn antenna during bidirectional transfer as shown in Figure 7.18. The data from the STA2 have the preamble of 1.9 μs and the MCS 9 data of 14 μs. After TX signal transmission from the STA2, the acknowledge

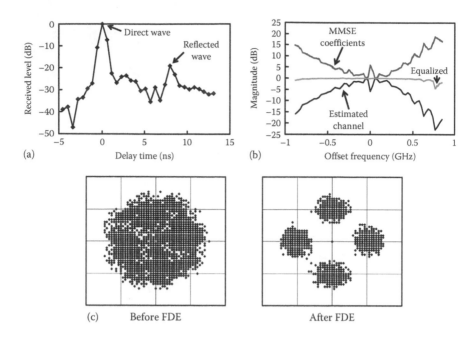

FIGURE 7.16 Measured RX FDE results. (a) Received signal against delay time, (b) magnitude compensation of the FDE, and (c) comparison of constellations at −60 dBm received power. (From Saito, N., Tsukizawa, T., Shirakata, N. et al., A fully integrated 60-GHz CMOS transceiver chipset based on WiGig/IEEE 802.11ad with built-in self calibration for mobile usage, *IEEE J. Solid-State Circuits*, 48(12), 3146, 2013. Copyright 2013 IEEE.)

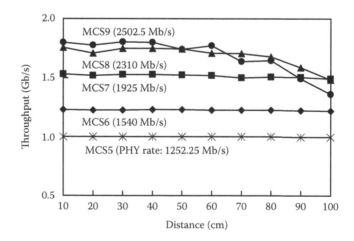

FIGURE 7.17 Measured MAC throughput over the air from one station (STA1) to the other (STA2). (From Saito, N., Tsukizawa, T., Shirakata, N. et al., A fully integrated 60-GHz CMOS transceiver chipset based on WiGig/IEEE 802.11ad with built-in self calibration for mobile usage, *IEEE J. Solid-State Circuits*, 48(12), 3146, 2013. Copyright 2013 IEEE.)

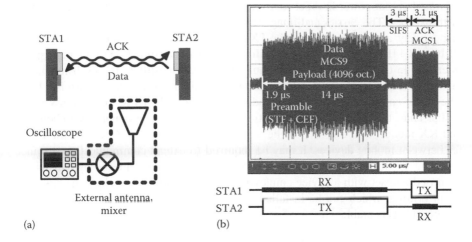

(a)

(b)

FIGURE 7.18 Measured signal waveform captured over the air. (a) Measurement setup and (b) measured signal waveform. (From Saito, N., Tsukizawa, T., Shirakata, N. et al., A fully integrated 60-GHz CMOS transceiver chipsct based on WiGig/IEEE 802.11ad with built-in self calibration for mobile usage, *IEEE J. Solid-State Circuits*, 48(12), 3146, 2013. Copyright 2013 IEEE.)

frame (ACK) of MCS 1 from the STA1 is transmitted after the SIFS period of 3 µs, based on the WiGig/IEEE 802.11ad frame format.

The performance is summarized in Table 7.2. The TX equivalent isotropic radiated power (EIRP) is +8.5 dBm and the EVM of MCS 9 is −22 dB. The maximum PA output power is set to be +2 dBm, which corresponds to 6 dB backoff from the

TABLE 7.2
Performance Summary

	RFIC	BBIC
Process (nm)	90	40
Supply voltage (V)	1.8/2.5	1.1/1.8/3.3
Chip size (mm²)	3.75 × 3.6	7.4 × 6.3
TX performance		
EIRP (dBm)	8.5	
EVM (dB)	−22	
Carrier/image leakage (dBc)	−39.0/−37.7	
Power consumption (mW)	347 (RFIC)/441 (BBIC)	
RX performance		
Received range (dBm)	−78 ~ −25	
NF at maximum gain (dB)	7.1	
Power consumption (mW)	274 (RFIC)/710[a] (BBIC)	

[a] Excluding interface.

saturated output power in order to avoid the hot carrier injection (HCI) degradation of transistors. The carrier and image leakages are lower than −39.0 and −37.7 dBc, respectively, after calibration. In the TX mode, the chipset consumes 347 mW in the RFIC and 441 mW in the BBIC, excluding the power consumption of the high-speed interface. In the RX mode, the minimum input level sensitivity is −78 dBm and the maximum input level is −25 dBm, and the noise figure at maximum gain is 7.1 dB. The power consumption in the RX mode is 274 mW in the RFIC and 710 mW in the BBIC. Thus, the total power consumption of the chipset is less than 1 W in both TX and RX modes. The chipset will be useful for realizing multi-Gb/s fast file transfer between mobile devices. It may be required to extend communication distance for some applications such as HD video streaming. This is achieved by pairing the proposed chipset with a device employing high antenna gain, such as the one with beamforming, which can be installed in a STB or a TV where larger power consumption is acceptable.

7.5   CONCLUSIONS

This chapter has presented a 60 GHz transceiver chipset based on the WiGig/IEEE 802.11ad that integrates RF, PHY, and MAC layers. In order to compensate for the amplitude variations both in the TX and the RX, the proposed chipset employs built-in TX in-band amplitude calibration and the RX FDE. The measurement shows that the proposed TX in-band amplitude calibration improves the TX EVM by 4 dB. In addition, the RX FDE effectively equalizes the in-band gain variation in the RX signal due to multipath environment. The chipset achieves the MAC throughput of 1.8 Gb/s for up to 40 cm and 1.5 Gb/s for up to 1 m while consuming less than 1 W, demonstrating the feasibility of adopting the 60 GHz wireless system in mobile devices.

ACKNOWLEDGMENT

This work was partly supported by "The research and development project for expansion of radio spectrum resources" of The Ministry of Internal Affairs and Communications, Japan. The authors would like to thank the members of Panasonic Corporation for their valuable advice and technical support.

REFERENCES

1. A.M. Niknejad and H. Hashemi, *mm-Wave Silicon Technology, 60 GHz and Beyond*. Springer, New York, 2008.
2. N. Guo, R.C. Qiu, S.S. Mo, and K. Takahashi, 60-GHz millimeter-wave radio: Principle, technology, and new results. *Hindawi Publishing Corporation EURASIP Journal on Wireless Communications and Networking*, 2007, 1, 19–26, 2007.
3. S. Emami, R.F. Wiser, E. Ali et al., A 60 GHz CMOS phased-array transceiver pair for multi-Gb/s wireless communications. In *IEEE International Solid-State Circuits Conference (ISSCC) Digest of Technical Papers*, San Francisco, CA, February 2011, pp. 164–165.

4. Agilent Technologies Application Note, Wireless LAN at 60 GHz—IEEE 802.11ad explained, May 2013.
5. IEEE Std. 802.11ad, Part 11: Wireless LAN medium access control (MAC) and physical layer (PHY) specifications, amendment 3: Enhancements for very high throughput in the 60 GHz band, December 2012.
6. B. Razavi, *RF Microelectronics*, 2nd edn. Prentice Hall, New Jersey, 2011.
7. A. Siligaris, O. Richard, B. Martineau et al., A 65 nm CMOS fully integrated transceiver module for 60 GHz wireless HD applications. In *IEEE International Solid-State Circuits Conference (ISSCC) Digest of Technical Papers*, , San Francisco, CA, February 2011, pp. 162–163.
8. T. Mitomo, Y. Tsutsumi, H. Hoshino et al., A 2 Gb/s-throughput CMOS transceiver chipset with in-package antenna for 60 GHz short-range wireless communication. In *IEEE International Solid-State Circuits Conference (ISSCC) Digest of Technical Papers*, San Francisco, CA, February 2012, pp. 266–267.
9. M. Tanomura, Y. Hamada, S. Kishimoto et al., TX and RX front-ends for 60 GHz band in 90 nm standard bulk CMOS. In *IEEE International Solid-State Circuits Conference (ISSCC) Digest of Technical Papers*, February 2008, pp. 558–559.
10. C. Marcu, D. Chowdhury, C. Thakkar et al., A 90 nm CMOS low-power 60 GHz transceiver with integrated baseband circuitry. In *IEEE International Solid-State Circuits Conference (ISSCC) Digest of Technical Papers*, San Francisco, CA, February 2009, pp. 314–315.
11. K. Okada, K. Matsushita, K. Bunsen et al., A 60 GHz 16QAM/8PSK/QPSK/BPSK direct-conversion transceiver for IEEE 802.15.3c. In *IEEE International Solid-State Circuits Conference (ISSCC) Digest of Technical Papers*, , San Francisco, CA, February 2011, pp. 160–161.
12. K. Okada, K. Kondou, M. Miyahara et al., A full 4-channel 6.3 Gb/s 60 GHz direct-conversion transceiver with low-power analog and digital baseband circuitry. In *IEEE International Solid-State Circuits Conference (ISSCC) Digest of Technical Papers*, San Francisco, CA, February 2012, pp. 218–219.
13. V. Vidojkovic, G. Mangraviti, K. Khalaf et al., Low-power 57-to-66 GHz transceiver in 40 nm LP CMOS with −17 dB EVM at 7 Gb/s. In *IEEE International Solid-State Circuits Conference (ISSCC) Digest of Technical Papers*, San Francisco, CA, February 2012, pp. 268–269.
14. K. Takinami, J. Sato, T. Shima et al., A 60 GHz CMOS transceiver IC for a short-range wireless system with amplitude/phase imbalance cancellation technique. *IEICE Transactions on Electronics*, 2012, E95-C(10), 1598–1609.
15. T. Tsukizawa, N. Shirakata, T. Morita et al., A fully integrated 60 GHz CMOS transceiver chipset based on WiGig/IEEE802.11ad with built-in self calibration for mobile applications. In *IEEE International Solid-State Circuits Conference (ISSCC) Digest of Technical Papers*, San Francisco, CA, February 2013, pp. 230–231.
16. B. Razavi, Z. Soe, A. Tham et al., A low-power 60-GHz CMOS transceiver for WiGig applications. In *Proceedings of IEEE Symposium on VLSI Circuits (VLSIC)*, Kyoto, Japan, June 2013, pp. 300–301.
17. A. Yehezkely and O. Sasson, Single transmission line for connecting radio frequency modules in an electronic device. U.S. Patent 0,307,695 A1, December 6, 2012.
18. E. Cohen, M. Ruberto, M. Cohen, O. Degani, S. Ravid, and D. Ritter, A CMOS bidirectional 32-element phased-array transceiver at 60 GHz with LTCC antenna. *IEEE Transactions on Microwave Theory and Techniques*, 2013, 61(3), 1359–1375.
19. M. Boers, I. Vassiliou, S. Sarkar et al., A 16TX/16RX 60 GHz 802.11ad chipset with single coaxial interface and polarization diversity. In *IEEE International Solid-State Circuits Conference (ISSCC) Digest of Technical Papers*, San Francisco, CA, February 2014, pp. 344–345.

20. N. Saito, T. Tsukizawa, N. Shirakata et al., A fully integrated 60-GHz CMOS transceiver chipset based on WiGig/IEEE 802.11ad with built-in self calibration for mobile usage. *IEEE Journal of Solid-State Circuits*, 2013, 48(12), 3146–3159.
21. R. Kitamura, K. Tanaka, T. Morita, T. Tsukizawa, K. Takinami, and N. Saito, A 1 μs settling time fully digital AGC system with a 1 GHz bandwidth variable gain amplifier for WiGig/IEEE 802.11ad multi-Gigabit wireless transceivers. *IEICE Trans. on Electronics*, 2013, E96-C(10), 1301–1310.
22. T. Shima, K. Miyanaga, and K. Takinami, A 60 GHz PLL synthesizer with an injection locked frequency divider using a fast VCO frequency calibration algorithm. In *Proceedings of IEEE Asia-Pacific Microwave Conference (APMC)*, Kaohsiung, Taiwan, December 2012, pp. 646–648.
23. T. Nakatani, T. Shima, and J. Sato, Small and low-loss quadrature hybrid and T/R local signal selection switch for 60 GHz direct conversion transceivers. In *Proceedings of IEEE Topical Meetings on Silicon Monolithic Integrated Circuits in RF Systems (SiRF)*, Austin, TX, January 2013, pp. 3–5.
24. F. Hsiao, A. Tang, D. Yang, M. Pham, and M. Chang, A 7 Gb/s SC-FDE/OFDM MMSE equalizer for 60 GHz wireless communications. In *Proceedings of IEEE Asian Solid-State Circuits Conference (A-SSCC)*, Jeju, South Korea, October 2011, pp. 293–296.
25. D.D. Falconer and S.L. Ariyavisitakul, Broadband wireless using single carrier and frequency domain equalization. In *Proceedings of Sixth International Symposium on Wireless Personal Multimedia Communications (WPMC)*, Hawaii, October 2002, pp. 27–36.
26. A. Maltsev, V. Erceg, E. Perahia et al., Channel models for 60 GHz WLAN systems. IEEE, document 802.11-09/0334r8, May 2010.
27. F.C. Yeh, T.Y. Liu, T.C.Wei, W.C. Liu, and S.J. Jou, A SC/OFDM dual mode frequency-domain equalizer for 60 GHz multi-Gbps wireless transmission. In *Proceedings of IEEE International Symposium on VLSI Design, Automation and Test (VLSI-DAT)*, Hsinchu, Taiwan, April 2011, pp. 1–4.
28. M. Nakamura, R. Shiozaki, Y. Kashino, H. Uno, and S. Fujita, A cavity structure miniaturized 60 GHz transceiver module with built-in antennas for WiGig/IEEE 802.11ad based mobile devices. In *Proceedings of IEEE Global Symposium on Millimeter Waves (GSMM)*, Sendai, Japan, April 2013.

8 Ultrahigh-Speed Wireless Communication with Short-Millimeter-Wave CMOS Circuits

Minoru Fujishima

CONTENTS

8.1 INTRODUCTION

The short millimeter wave has the frequency range of 100 to 300 GHz. One of the motivations for studying this range is to realize ultrahigh-speed communication. Data rates in both wireless and wireline communications are increasing every year, as shown in Figure 8.1.[1] In particular, the progress in wireless communication has been much faster than that in wireline communication until recently. Currently, 60 GHz wireless communication,[2–14] which is the most focused communication band in the millimeter-wave range, has been adopted as the high-speed wireless communication standard such as IEEE 802.11ad, which enables data rates of 6–7 Gbps. Since the increase in speed in wireless communication has been almost 10-fold every 4 years until now, 100 Gbps wireless will appear in around 2020, assuming the development is extrapolated into the future. To realize wireless communication at a higher speed than that of the 60 GHz one, we have to study wireless communication with higher carrier frequency than that with 60 GHz. This is one of the motivations of studying wireless communication with D-band, which is allocated in the lowest frequency band in short millimeter wave.

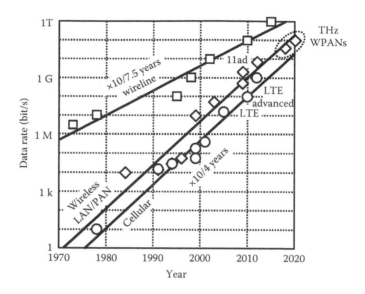

FIGURE 8.1 Evolution of data rates in wireline and wireless communications. (From Kürne, T., IEEE 802.15-10-0320-02-0000-Tutorial_Igthz.)

Transmitter and receiver integrated circuits based on complementary metal oxide semiconductor (CMOS), which has the ability to realize such circuits with low-power consumption at a low cost compared with those based on compound semiconductors,[15,16] are expected to become widespread. Although high-frequency characteristics of silicon devices are inferior to those of compound semiconductor devices such as GaAs MESFETs and InP HBT, the transceiver system utilizing a compound semiconductor also dissipates a large amount of power. On the other hand, owing to the miniaturization of gate length, the maximum operation frequency (f_{max}) of N-channel metal oxide semiconductor field effect transistor (NMOSFET) in CMOS technology exceeds 300 GHz even in mass production technology. The f_{max} of NMOSFET also continues to increase and is expected to be 1 THz by around 2020 according to the prediction based on the International Technology Roadmap for Semiconductors (ITRS) 2012 RFAMS,[17] as shown in Figure 8.2. Considering system integration, particularly mobile application, CMOS circuits should be a strong candidate even in the D-band.

In this chapter, the design of promising D-band CMOS circuits for future mobile communication is described from device to system layers, and finally, ultrahigh-speed wireless communication is demonstrated.

8.2 DEVICE DESIGN AND MODELING IN D-BAND

As shown in Figure 8.3, generally, there are three layers for realizing the CMOS chip, namely, device, circuit, and system layers. In each layer, there are three tasks to accomplish, namely, design, measurement, and modeling. The starting point for general analog and radiofrequency (RF) CMOS designers is the circuit layer. The process design kit (PDK) is usually prepared by a foundry, and the designers do not have to design and characterize devices and establish the model, which is considered

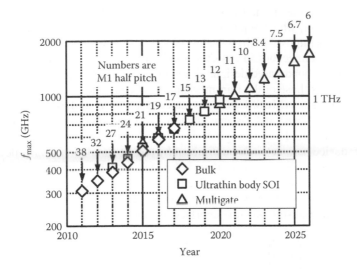

FIGURE 8.2 Evolution of maximum operation frequency (f_{max}) of NMOSFET.[17] The f_{max} improves independently of the transistor structure.

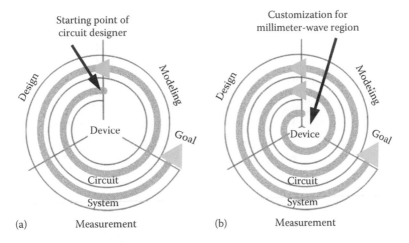

FIGURE 8.3 Chip development processes in the cases of (a) general analog/RF circuits and (b) millimeter-wave circuits.

to be device engineers' duty. However, on the other hand, millimeter-wave designers sometimes have to start from the design of the device layout since the PDK for millimeter wave, particularly for the D-band, is not necessarily provided. Therefore, customization of the millimeter-wave region from the original PDK is generally necessary even though foundry PDK is also utilized. As a result, the most different layer between analog/RF and millimeter-wave designs is the device layer. In the case of layout for a millimeter wave, it is not easy to apply layout parasitic extraction (LPE) since not only parasitic capacitance and resistance affect the performance, but inductance as well.

Moreover, the parasitic inductance is not easy to extract without using a 3D electromagnetic simulator, which requires a high computational cost. So, applying LPE for the whole chip is impossible. To solve this problem, we adopt the layout without using a parasitic wire connect or LPE, as shown in Figure 8.4, which is named bond-based design.[18,19] In the bond-based design, all the devices used are in tile layout, with the same interface as those of the transmission lines. Thus, metal oxide semiconductor field effect transistors (MOSFETs), pads, and capacitors as well as transmission lines including L-shape bend and T-shape junction have the same ground–signal–ground interface when a coplanar waveguide is adopted in the transmission line. In this case, after device parameters are optimized by circuit simulation, all the device tiles according to the optimization results are bonded with the same transmission line interface. This design style can minimize the discrepancy between the circuit results and measurement results. However, when the bond-based design is used, the device model, including transmission line interface, is necessary.

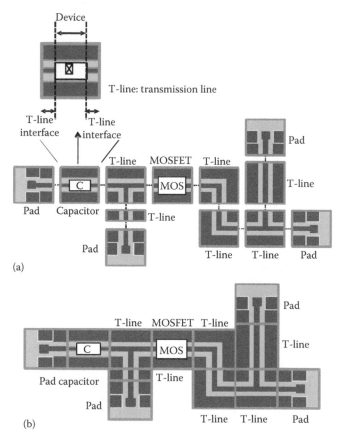

FIGURE 8.4   (a) Precharacterized elements and (b) complete layout diagram in a millimeter-wave CMOS circuit. All the devices in the millimeter-wave CMOS circuits have the same interface as transmission lines have, so that all the devices can be bonded without using wires. This layout is called bond-based design. (From Fujimoto, R. et al., *IEICE Trans Fundam.*, E96-A(2), 486, February 2013; Pengg, F.X., *IEEE MTT-S*, 1, 271, June 2002.)

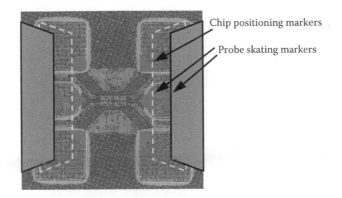

Chip positioning markers

Probe skating markers

FIGURE 8.5 Scotch-tape markers on the display to control the positions of probes.

Precise measurement is also important for robust device modeling. One of the techniques for precise measurement is position control of the high-frequency probe. Since the wavelength decreases with increasing frequency, even a small fluctuation of the probing position on a pad can affect measurement precision.

We use scotch-tape markers on the display, monitoring the silicon chip surface with a microscope, as shown in Figure 8.5. By this technique, the probing position and scratch shapes look almost the same even if two probes are moved manually for different distances between probes, as shown in Figure 8.6. This is because both start and end points of the probe skating are well controlled approximately within 1 μm. Note that the lateral resolution of landing the probes shown in the y-direction is also controlled using chip-positioning markers on the display.

After the devices are measured, modeling for the D-band is established. Here, although the MOSFET model provided from the foundry may not meet the short-millimeter-wave band, it is still helpful since valuable information, such as dc characteristics with threshold variation, is included. Thus, we focus on the error between the provided model and short-millimeter-wave characteristics, and try to establish the error model. Conventionally, studies on the extraction of MOSFET parameters in the millimeter-wave region have been reported.[20-24] In these studies,

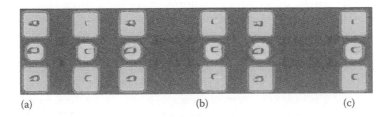

(a)　　　　　　　　　　(b)　　　　　　　　　　(c)

FIGURE 8.6 Chip micrograph after probing of on-chip devices: (a) through dummy, (b) transmission line of 80 μm length, and (c) transmission line of 120 μm length. Positions as well as shapes of scratches made by probing are almost the same on pads even though the distances of the two probes are different among (a–c). Positions of probes and pads are aligned by scotch-tape markers attached on a display monitoring a chip surface.

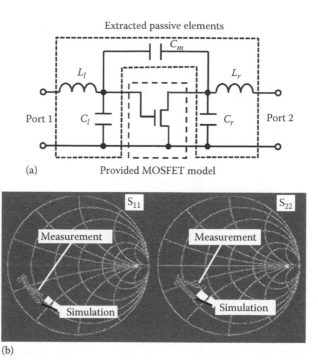

(a)

(b)

FIGURE 8.7   (a) Basic RF device model using subcircuit extension comprising series inductances and parallel capacitances. (b) Comparison of reflection characteristics obtained by
measurement and simulation using a single-stage common-source amplifier using the subcircuit extension model. Note that transmission coefficients S_{21} and S_{12} show good agreement
between measurement and simulation results.

the subcircuit extension model, comprising series inductors and parallel capacitors,
was used for a simple RF model, such as that shown in Figure 8.7a. The reported
method of parameter extraction can be used to explain the behavior of a MOSFET
up to microwave frequency range and at a fixed biasing point. However, the method
cannot be used to explain the behavior over a broad frequency range, nor at the
desired biasing points, as shown in Figure 8.7b. In the conventional subcircuit
model, the topology is determined first, and a non-quasi-static (NQS) effect caused
by the delay from the stimulation of gate voltage to the reaction of drain current
cannot be considered unless the core compact FET model takes it into account. As
a result, it is difficult to match all the scattering parameters between measurement
and simulation results as shown in Figure 8.7b. Although the NQS delay is typically
0.1–0.2 ps[25] and it is insignificant in microwave, it affects the device performance
significantly in the terahertz region. Measured NQS delays of a MOSFET with the
40 nm process after removing parasitic resistances are shown in Figure 8.8. Note
that not all the foundries provide the compact model with sufficiently precise NQS
effect in the terahertz region, although the NQS is sometimes considered in the
original compact model. To address this issue, we modeled the wrapper admittance matrix Y_{wrap} derived by subtracting admittance matrices obtained from the

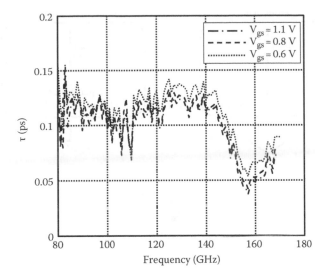

FIGURE 8.8 Measured NQS delay of a MOSFET with gate width and length of 32 μm and 40 nm, respectively. Drain voltage is 1.1 V.

simulation results using the provided PDK model YPDK from the matrices obtained from the measured results Y_{meas}, as shown in Figure 8.9a.[26] This error term has four elements of the 2 × 2 admittance matrix. Note that y_{12} and y_{21} of Y_{wrap} are different if the NQS is not precisely considered. Each element of the 2 × 2 error matrix is modeled by four equivalent circuit branches, and y_{12} and y_{21} branches are connected to voltage-controlled voltage sources. Admittances in four branches are derived from four elements of Y_{wrap} as

$$y_{in} = y_{11} + y_{12},$$

$$y_{reverse} = -y_{12},$$

$$y_{forward} = -y_{21}, \text{ and} \qquad (8.1)$$

$$y_{out} = y_{22} + y_{21},$$

where y_{in}, $y_{reverse}$, $y_{forward}$, and y_{out} are shown in Figure 8.9b. This modeling with the wrapper matrix is named the Y-wrapper model. The Y-wrapper model can cover the dedicated effects in the D-band even if ultrahigh-frequency effects are not considered in the original core FET model. The Y-wrapper model is valid across all bias conditions by using voltage-dependent passive devices in each branch shown in Figure 8.9b. Figure 8.10 shows comparison of reflection characteristics between measured and simulated results using the Y-wrapper model. As shown in Figure 8.10, the Y-wrapper model shows better agreement with the measured results than the subcircuit model shown in Figure 8.7b. By using a single-stage common-source

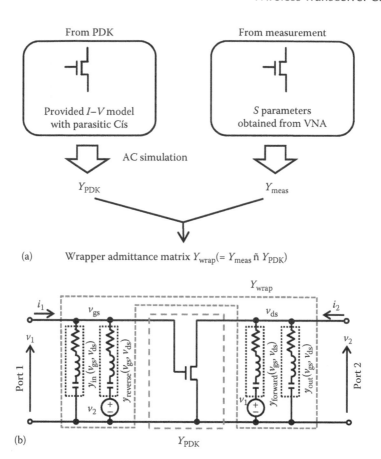

(a)   Wrapper admittance matrix $Y_{\text{wrap}}(= Y_{\text{meas}} \tilde{\text{n}} Y_{\text{PDK}})$

(b)

FIGURE 8.9   (a) In the Y-wrapper model, wrapper admittance matrix Y_{wrap} is calculated by subtracting the admittance matrix obtained from the provided PDK model YPDK from that obtained from measurement Y_{meas}. Four elements of Y_{wrap} are replaced by four branches shown in (b). Note that voltage sources shown in (b) are controlled by the voltage of the port on the other side.

amplifier as shown in Figure 8.11, measured results and simulation results are compared. Comparison results are shown in Figure 8.12 at a variety of gate and drain voltages. As shown in figures, from off- to on-states, all the S-parameters show good agreement between those obtained from the measurement and simulation with the Y-wrapper model. Note that since the error term modeled by the Y-wrapper is not precisely based on the physical model, the Y-wrapper model have to be extracted for each transistor size when different transistor sizes are used in a circuit. To extract the values of the components in the Y_{wrap}, characteristics of each branch in frequency domain are approximated by rational polynomial Laplace functions, orders of which should be optimized to fit the measured results. Additionally, although temperature dependency can be implemented in the Y-wrapper model by using temperature-dependent passive branches, it is not considered yet in the current Y-wrapper model.

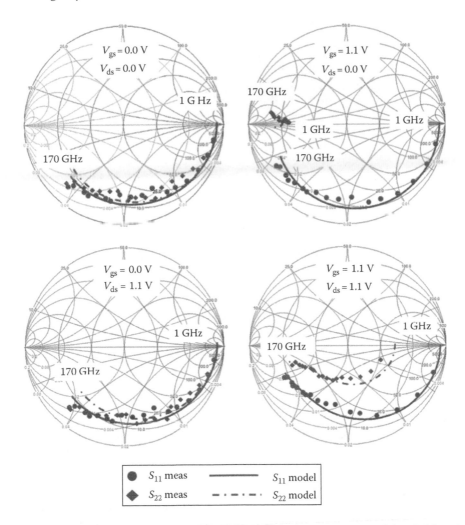

FIGURE 8.10 Comparison of reflection characteristics between measured (symbols) and Y-wrapper model (lines) scattering parameters of a MOSFET at a variety of gate and drain voltages. Note that transmission coefficients S_{21} and S_{12} show good agreement between measurement and simulation results.

8.3 ULTRAWIDE-FREQUENCY D-BAND AMPLIFIER

Now, we can identify the starting point of the circuit design. One of the advantages of utilizing the D-band is potential availability of a wide-frequency band. If we can utilize a wide-frequency band, simple modulation such as amplitude-shift keying (ASK) can be adopted to realize low-power consumption. Note that since the envelope of a communicating signal is formed with the amplitude variation, demodulation fails when the envelope of the signal collapses. As a result, the ASK requires a flat frequency response of both gain and group delay to preserve

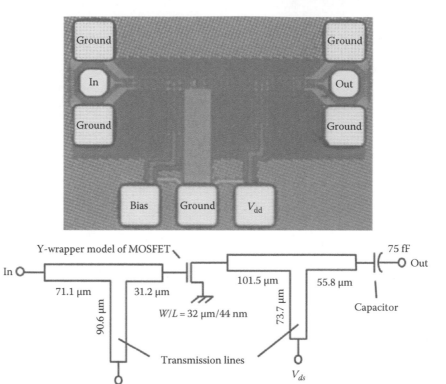

FIGURE 8.11 One-stage common-source amplifier utilized for confirmation of reliability of the Y-wrapper model. A chip micrograph and its circuit are shown.

the envelope. For example, as shown in Figure 8.13, a large variation of either gain or group delay degrades the ASK envelope at the output of the amplifier. Here, in the microwave ultrawide band (UWB), a technique using a group delay equalizer has been proposed to keep the group delay constant in the target frequency band.[27] However, this technique increases the power consumption and decreases the communication speed owing to the equalizer, which adjusts the group delay using passive devices with a considerable insertion loss. On the other hand, when an amplifier with a constant group delay is achieved, no gain degradation due to the equalizer occurs, and the receiver can obtain sufficient sensitivity. Flat gain and group delay of the amplifier can be achieved by properly distributing the poles of interstage matching networks. Design details are described in Ref. [28]. We have fabricated the D-band wideband amplifier using standard 1P12M 65 nm CMOS technology. Figure 8.14 shows a six-stage wideband 140 GHz CMOS amplifier, which comprises one cascade stage and five common-source stages. As shown in the chip micrograph, this amplifier is based on the bond-based design. In the amplifier micrograph, dedicated decoupling capacitors, named zero-ohm

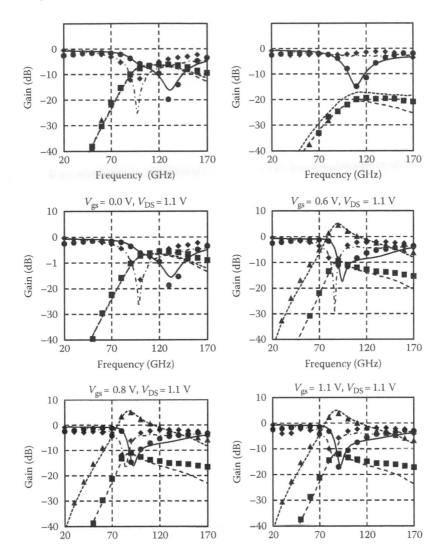

FIGURE 8.12 Comparison between measured (symbols) and Y-wrapper model (lines) scattering parameters at a variety of gate and drain voltages. These results confirm that the proposed model matches the measured results at any frequency and bias.

transmission line, are shown at the end of the short stub.[29] Measured gain and group-delay performance are shown in Figure 8.15a, where simulated results are also shown for comparison. The performances are compared among conventional millimeter-wave amplifiers,[30–35] as shown in Figure 8.15b. With careful optimization of device parameters, we have successfully demonstrated a gain bandwidth of 12 GHz within 0.1 dB ripple and ±13.1 ps group delay within the 3 dB bandwidth of 27 GHz.

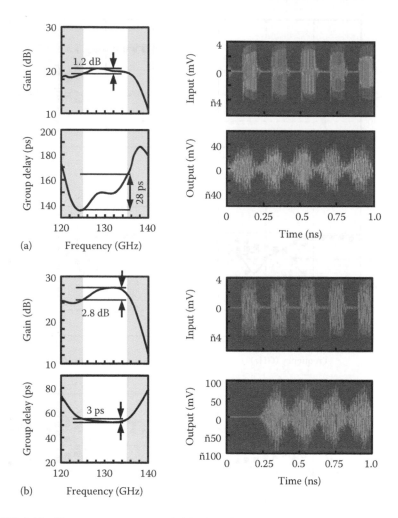

FIGURE 8.13 Frequency response and time domain response are compared in the case of two amplifiers with (a) flat gain and large group delay variation and (b) flat group delay and large gain variation. On and off periodical ASK signal with 10 Gbps data rate is applied to both amplifiers. In both cases, the envelope of the output is distorted and the on–off ratio is degraded.

8.4 135 GHZ LOW-POWER AND HIGH-SPEED WIRELESS COMMUNICATION CHIPSET

To realize high-speed wireless communication with low-power consumption, we fabricated an ASK transceiver chipset. To achieve an ASK transceiver in sub-terahertz region, frequency multipliers in a transmitter and sub-harmonic mixers in a receiver were used in Ref. [40]. However, multiplier-based architecture consumes a large amount of power due to considerable insertion loss in frequency-multiplying operation. The building block without using multiplier-based architecture used in

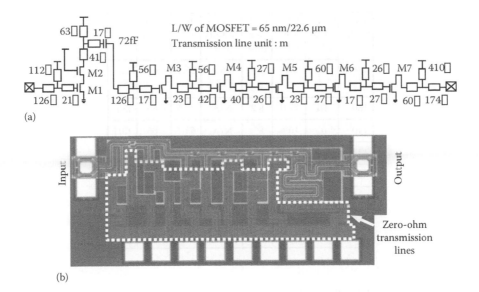

(a)

(b)

FIGURE 8.14 (a) Schematic and (b) chip micrograph of a 140 GHz CMOS amplifier fabricated with a 65 nm CMOS process. It comprises a single-stage cascade amplifier and five-stage common-source amplifiers.

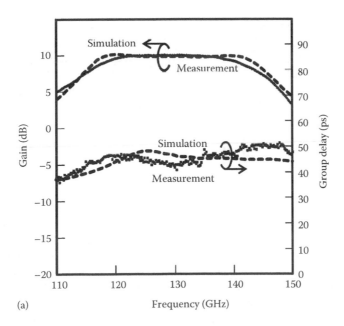

(a)

FIGURE 8.15 (a) Measurement and simulation results of frequency responses of gain and group delay.   (*Continued*)

Freq. (GHz)	0.1 dB –BW (GHz)	3 dB–BW (GHz)	Peak gain (dB)	GD [ps] @ 3 dB-BW	P_{DC} (mM)	Techn ology (nm)	Ref.
136.1	12	27.6	9.9	46.2 ± 13.1	57.1	65	This work
150	1.1	27.0	8.2	N/A	25.5	65	[30]
140	0	10	8	N/A	63	65	[31]
144	1.9	10	9	N/A	75.6	65	[32]
95	2.8	19	10	N/A	35.0	65	[33]
63	2.3	14	17	N/A	36	90	[34]
60	1.0	8.5	12.2	120.7 ± 26.3	29.1	130	[35]

(b)

FIGURE 8.15 (CONTINUED)   (b) Performance comparison between wideband millimeter-wave CMOS amplifiers.

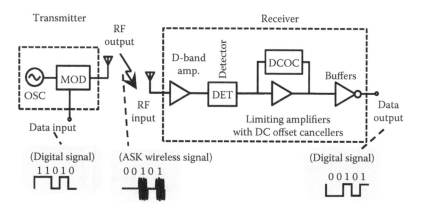

FIGURE 8.16   Block diagram of a 135 GHz CMOS transceiver chipset.

this study is shown in Figure 8.16. As shown in Figure 8.16, simple architecture is realized utilizing ASK modulation, where power-hungry PLL, ADC/DCA, and DSP are not necessary.[5,13] As a result, low-power consumption is realized. On the other hand, a wide-frequency band is required for ASK modulation. For example, 10 Gbps ASK modulation requires at least a 10 GHz bandwidth. It is not sufficient to use a 60 GHz band, where at most a 9 GHz total bandwidth is available. On the other hand, a wider frequency band is potentially available above 100 GHz,[36] where mainly passive applications such as radio astronomy and earth exploration satellite are allocated. We chose the 135 GHz band to avoid the conflict of frequency bands allocated exclusively to passive applications, as shown in Figure 8.17.[37]

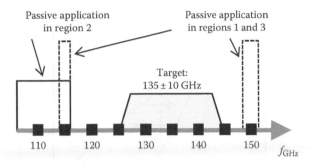

FIGURE 8.17 Target operation frequency for a D-band CMOS transceiver. Target frequency is selected while avoiding frequency bands in the D-band allocated exclusively to passive applications. Here, region 1 is Europe, Africa, and Russia. Region 2 is North and South America. Region 3 is Asia and Oceania.

8.4.1 Transmitter Chip

The ASK transmitter module has only an oscillator and an ASK modulator. A building block of a general ASK transmitter is shown in Figure 8.18a, where a power amplifier (PA) is required to realize sufficient output power for wireless transmission. Since the PA consumes typically half of the transmitter power, we can reduce the power consumption by half if the PA is omitted. Figure 8.18b shows the proposed transmitter. To increase the output power of the oscillator, a push–push oscillator, generating a second-harmonic signal, is adopted instead of using a fundamental

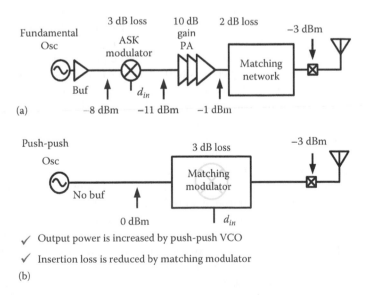

FIGURE 8.18 Block diagrams of (a) conventional ASK transmitter and (b) proposed PA-free ASK transmitter. By using a push–push oscillator and a matching modulator, comparable output power is obtained without using a PA.

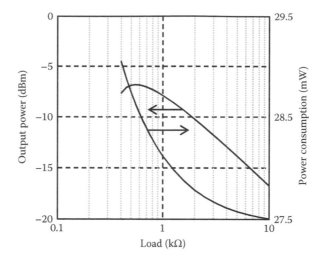

FIGURE 8.19 Simulated output power and power consumption of a fundamental oscillator as a function of load impedance.

oscillator. Although the fundamental signal is generally stronger than the second-harmonic signal, the output power of the buffer is small. Figure 8.19 shows the simulated output power and power consumption of a fundamental oscillator as a function of load impedance. To achieve a high frequency, the core gate width should be smaller than that of push–push one and the output power may be improved up to −6.8 dBm at the optimized load of 500 Ω. However, 500 Ω is hardly achieved by the gate of a MOSFET buffer since the input resistance increases up to only 132 Ω by decreasing the gate width up to 1 μm at the frequency of the voltage controlled oscillator (VCO). On the other hand, in the case of the push–push oscillator, since the fundamental oscillation frequency is half of 135 GHz and no buffer MOSFET is required, the core MOSFET can be large, which can generate a larger amount of output power. When the parameter is optimized, −0.7 dBm output power can be obtained. Additionally, when the PA is omitted, the ASK modulator and output matching network can be merged, which reduces the total insertion loss. As a result, sufficient output power can be obtained even if the PA is omitted.

The output power of the push–push oscillator varies depending on the load impedance. Output-power contour as a function of load impedance is shown in Figure 8.20a, which is obtained from load-pull simulation based on periodic steady-state (PSS) analysis. As shown in Figure 8.20a, the capacitive load can increase the output power of the push–push oscillator. Figure 8.20b shows schematics of a transmitter. The modulator comprises two shunt switches and a series transmission line (TL2). TL1 is a power feed for an oscillator. By changing the parameters of TL1, TL2, and the coupling capacitor, the load of the oscillator is adjusted. Note that the shorter length of TL2 degrades the on–off ratio of the ASK modulator as shown in Figure 8.21a. Trajectories of the load impedance of the oscillator with

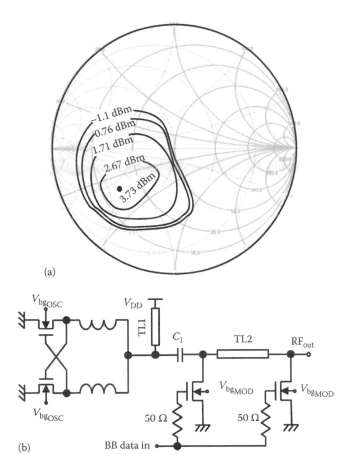

FIGURE 8.20 Output power of the push–push oscillator is dependent on the load. (a) Power contour plot is shown as a function of load impedance. (b) Schematic of a 135 GHz CMOS amplifier-free transmitter. The load of the push–push oscillator is adjusted by TL1, TL2, and C1.

changing TL1, TL2, and C1 are shown in Figure 8.21b. The load impedance should be optimized within the sufficient on–off ratio. Measured output spectrums of the modulated and unmodulated transmitter output is shown in Figure 8.22, where the on–off ratio of the modulator is 19 dB. Note that the output frequency slightly changes from 135.49 to 135.32 GHz when the ASK modulator is turned on to off. The measured output power and the power consumption of the transmitter are shown in Figure 8.23. Measured output power from the transmitter is smaller than the expected one obtained from the simulation results even though the insertion loss by the modulator is considered. One of the reasons for the discrepancy is that nonlinear characteristics of the D-band MOSFET model is not sufficiently accurate. Phase noise of the transmitter is shown in Figure 8.24 when the modulator is on. Phase noise at 1 MHz offset is $-80\,\text{dBc}/\sqrt{\text{Hz}}$.

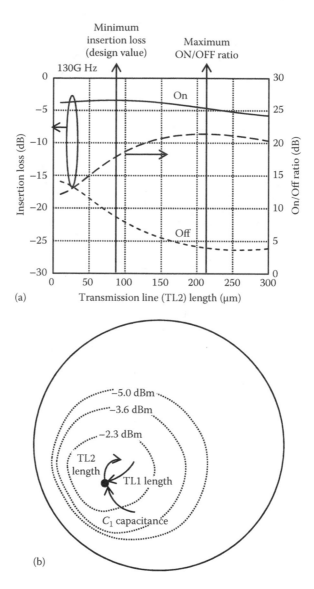

(a)

(b)

FIGURE 8.21 (a) Simulated results of insertion loss and on–off ratio of the modulator as a function of the length of transmission line TL2 in Figure 8.16. Insertion loss is evaluated by S21 when the port 1 is oscillator output and the port 2 is 50 Ω with a pad. (b) Trajectories of the load impedance of the oscillator with changing TL1, TL2, and C1 in Figure 8.16.

8.4.2 Receiver Chip

The receiver comprises the D-band amplifier, detector, and limiting amplifier with dc offset cancellers. The envelope detector is realized by a common-source amplifier with gate bias of near threshold voltage to maximize nonlinearity.[38] As described in Section 8.3, the flatness of gain and group-delay frequency response are important

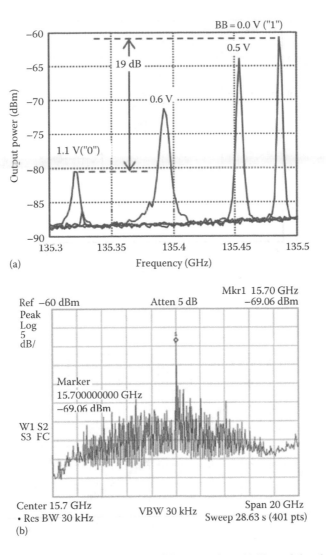

FIGURE 8.22 Measured output spectrums of the transmitter. (a) Unmodulated output spectrums with the gate voltages of the modulator (BB) of 1.1 V (*off*), 0.6 V, 0.5 V, and 0 V (*on*), and (b) modulated output spectrum.

for high-speed ASK modulation. Note that the detector and D-band amplifier operate in the 135 GHz band, as shown in Figure 8.25a. As a result, the total frequency response of the D-band amplifier and detector should be flat with regard to gain and group delay to achieve a high data rate. Here, since the detector converts 135 GHz to the baseband, the input frequency of the D-band amplifier and the output frequency of the detector are different, where normal S-parameter analysis cannot be applied. To check the frequency flatness with regard to gain and group delay, the ideal mixer and oscillator module are inserted in the circuit simulation as shown in Figure 8.25b. In this case, the baseband signal is applied at the simulation input,

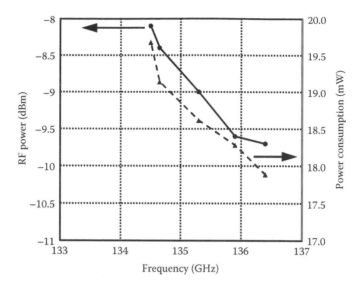

FIGURE 8.23 Measured output power and power consumption of the transmitter as a function of oscillation frequency, which can be changed by back-gate voltage V_{bgosc} in Figure 8.20.

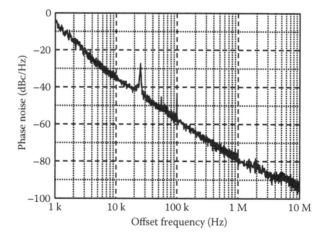

FIGURE 8.24 Measured phase noise of the transmitter when the modulator is on. Back-gate voltage of the oscillator is 0 V and oscillation frequency is 135 GHz.

which is upconverted at the input of the D-band amplifier and downconverted by the detector. As a result, gain and frequency flatness can be checked in the baseband frequency. Since the input signal of the detector is the data modulated by the D-band carrier and downconverted to the baseband, we first evaluated the frequency response of the detector utilizing the same method shown in Figure 8.25b. Then, the frequency response of the D-band amplifier is designed to compensate the frequency response of the detector to achieve a total flat response. Figure 8.26 shows

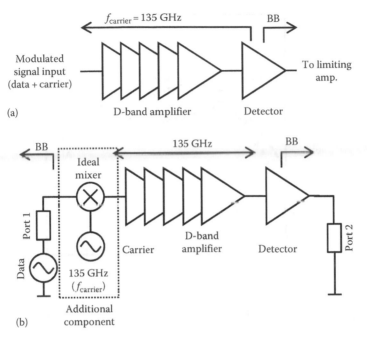

(a)

(b)

FIGURE 8.25 (a) Co-design of the D-band amplifier and detector is required for a high-speed ASK signal in order to maintain the envelop of high-speed ASK modulation. (b) Frequency responses of gain and group delay are verified by the baseband signal. The test signal in circuit simulation is upconverted at the input of a receiver.

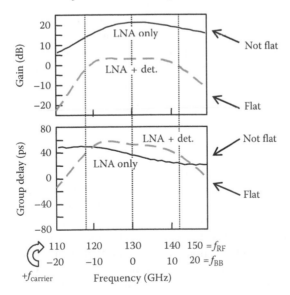

FIGURE 8.26 Simulated frequency responses of gain and group delay. Although frequency responses of the D-band amplifier alone are not flat, total frequency responses of the D-band amplifier and detector display flat frequency responses.

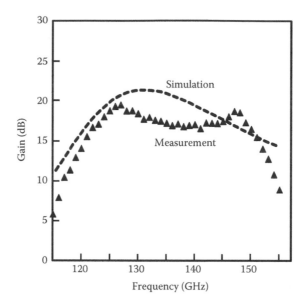

FIGURE 8.27 Measured and simulated gain responses of the small-signal amplifier used in the receiver.

the frequency responses of gain and group delay obtained from the circuit simulation. As shown in Figure 8.26, the total frequency response of the D-band amplifier and the detector is flat, although those of the D-band amplifier alone are not flat. The measured S-parameters of the D-band amplifier are shown in Figure 8.27. Although no measurement results of the total frequency response of the D-band amplifier and the detector, as shown in Ref. [39], are available, sufficient frequency characteristics are indirectly confirmed by the measurement results of the communication test described in the next subsection.

8.4.3 MEASUREMENT

This transceiver chipset is fabricated with 40 nm CMOS technology. Chip micrographs of the transmitter and receiver are shown in Figure 8.28. The core area is 0.32 mm² for a transmitter, where the chip area is reduced by the PA-free architecture. The chip size of the receiver is 0.8 mm × 2.1 mm. Figure 8.29 shows the measurement setup of the wireless test using the transceiver chipset. The wireless test is realized using two probe stations, and two horn antennas are directly connected to the waveguide probes. A $2^{31}-1$ pseudorandom-bit signal (PRBS) is applied for the digital input of the transceiver. The distance between two horn antennas is 10 cm. Note that the PA will be necessary in the case of longer communication distance. The eye diagram obtained at the receiver output in the case of 10 Gbps data rate is shown in Figure 8.30a. Figure 8.30b shows bit error rate as a function

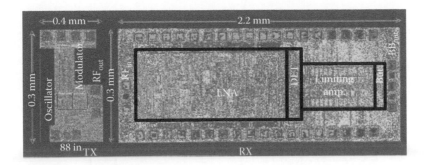

FIGURE 8.28   Chip micrographs of a transmitter (TX) and a receiver (RX). TX and RX chip areas are 0.32 mm², and 0.8 mm × 2.1 mm, respectively, including pads.

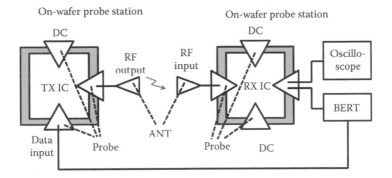

FIGURE 8.29   Measurement setup of the transmitter and receiver with the wireless signal.

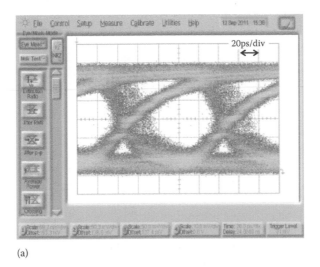

(a)

FIGURE 8.30   Measured response of the transmitter and receiver with the wireless signal. $2^{31}-1$ PRBS is applied for digital input. (a) The eye diagram obtained at the receiver output.

(*Continued*)

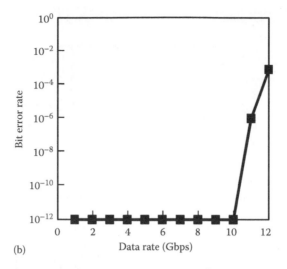

(b)

FIGURE 8.30 (CONTINUED)   Measured response of the transmitter and receiver with the wireless signal. $2^{31}-1$ PRBS is applied for digital input. (b) Bit error rate as a function of data rate. Wireless distance is 0.1 m. The maximum measured data rate with a BER of less than 10^{-11} is 10 Gbps.

TABLE 8.1
Measured Power Consumption of the Transmitter and Receiver

TX	Oscillator	17.9
	Modulator	0
RX	LNA	59.8
	Detector	0.7
	BB amp	16.8
	Buffer	3.1
	Total	98.4

The power consumption is 17.9 mW for the transmitter and 80.5 mW for the receiver. The total power consumption is 98.4 mW.

of data rate. The bit error rate is less than 10^{-11} below 10 Gbps. It is found that the error-free maximum data rate is 10 Gbps. Table 8.1 shows the measured power consumption of the transmitter and receiver. Table 8.2 shows the comparison of short-millimeter-wave transceivers.[14,15,40–43] The power consumption of our transceiver is 98 mW, where the power consumptions of a transmitter and a receiver are 18 and 80 mW, respectively. This is the lowest power consumption among short-millimeter-wave transceivers.

TABLE 8.2

Comparison of Short-Millimeter-Wave Transceivers

	[14]	[15]	[40]	[41]	[42]	[43]	This work
Technology	InP HEMT	SiGe BiCMOS	65 nm CMOS	65 nm CMOS	65 nm CMOS	65 nm CMOS	40 nm CMOS
Frequency (GHz)	125	140	120	120/140	116	260	135
TX, RX blocks	TX, RX	TX, RX	TX, RX	RX only	TX only	Tx, Rx	TX, RX
Wireless propagation	Yes	Yes	No	Yes	No	Yes	Yes
Power consumption (mW)	750	1500	80.9 *1	85.7 (120 GHz) 111.7 (140 GHz)	200	1173	98
Maximum data rate (Gbps)	10	4	9	3.0 (120 GHz) 3.6 (140 GHz)	>10	10	10
BER	10^{-12}	N/A	10^{-9}	10^{-11}	N/A	N/A	10^{-11}

*1 PA and LNA are not included.

8.5 CONCLUSIONS

In this chapter, an overview of chip development for the D-band is described. To realize the D-band circuits, device optimization, as well as circuit optimization, is generally necessary. By carefully optimizing the device layout and model, we realized a 135 GHz 98 mW 10 Gbps ASK transmitter and receiver chipset fabricated with 40 nm CMOS technology. In the chipset, a PA-free architecture is adopted in the transmitter to realize low-power operation, and codesign of the D-band amplifier and detector is applied in the receiver to realize a high data rate. The chipset was verified with a wireless propagation test for 10 cm distance. As a result, the CMOS transceiver operating over 100 GHz will be one of the candidates for high-speed and low-power wireless applications.

REFERENCES

1. T. Kürne, IEEE 802.15-10-0320-02-0000-Tutorial_Igthz. https://mentor.ieee.org/802.15/dcn/12/15-12-0320-02-0thz-what-s-next-wireless-communication-beyond-60-ghz-tutorial-ig-thz.pdf.
2. K. Raczkowski, W.D. Raedt, B. Nauwelaers et al., A wideband beamformer for a phased-array 60 GHz receiver in 40 nm digital CMOS, *IEEE International Solid-State Circuit Conference*, pp. 40–41, February 2010.
3. W.L. Chan, J.R. Long, M. Spirito et al., A 60 GHz-band 2 × 2 phased-array transmitter in 65 nm CMOS, *IEEE International Solid-State Circuit Conference*, pp. 42–43, February 2010.

4. F. Vecchi, S. Bozzola, M. Pozzoni et al., A wideband mm-wave CMOS receiver for Gb/s communications employing interstage coupled resonators, *IEEE International Solid-State Circuit Conference*, pp. 220–221, February 2010.

5. K. Kawasaki, Y. Akiyama, K. Komori et al., A millimeter-wave intra-connect solution, *IEEE International Solid-State Circuit Conference*, pp. 414–415, February 2010.

6. K. Okada, K. Matsushita, K. Bunsen et al., A 60-GHz 16QAM/8PSK/QPSK/BPSK direct-conversion transceiver for IEEE802.15.3c, *IEEE International Solid-State Circuit Conference,* pp. 160–161, February 2011.

7. A. Siligaris, O. Richard, B. Martineau et al., A 65 nm CMOS fully integrated transceiver module for 60 GHz wireless HD applications, *IEEE International Solid-State Circuit Conference,* pp. 162–163, February 2011.

8. S. Emami, R.F. Wiser, E. Ali et al., A 60 GHz CMOS phased-array transceiver pair for multi-Gb/s wireless communications, *IEEE International Solid-State Circuit Conference*, pp. 164–165, February 2011.

9. M. Tabesh, J. Chen, C. Marcu et al., A 65 nm CMOS 4-element sub-34 mW/element 60 GHz phased-array transceiver, *IEEE International Solid-State Circuit Conference,* pp. 166–167, February 2011.

10. T. Mitomo, Y. Tsutsumi, H. Hoshino et al., A 2 Gb/s-throughput CMOS transceiver chipset with in-package antenna for 60 GHz short-range wireless communication, *IEEE International Solid-State Circuit Conference*, pp. 266–267, February 2012.

11. K. Okada, K. Kondou, M. Miyahara et al., A full 4-channel 6.3 Gb/s 60 GHz direct-conversion transceiver with low-power analog and digital baseband circuitry, *IEEE International Solid-State Circuit Conference*, pp. 218–219, February 2012.

12. A. Oncu and M. Fujishima, 19.2 mW 2 Gbps CMOS pulse receiver for 60 GHz band wireless communication, *Symposium on VLSI Circuits*, pp. 158–159, June 2008.

13. A. Oncu, S. Ohashi, K. Takano et al., 1 Gbps/ch 60 GHz CMOS multichannel millimeter-wave repeater, *Symposium on VLSI Circuits*, pp. 92–94, June 2010.

14. T. Kosugi, M. Tokumitsu, T. Enoki et al., 120-GHz Tx/Rx chipset for 10-Gbit/s wireless applications using 0.1 μm-gate InP HEMTs, *IEEE Compound Semiconductor Integrated Circuit Symposium*, pp. 171–174, October 2004.

15. E. Laskin, P. Chevalier, B. Sautreuil et al., A 140-GHz double-sideband transceiver with amplitude and frequency modulation operating over a few meters, *IEEE Bipolar/BiCMOS Circuits and Technology Meeting*, pp. 178–181, October 2009.

16. N. Deferm and P. Reynaert, A 120 GHz 10 Gb/s phase-modulating transmitter in 65 nm LP CMOS, *IEEE International Solid-State Circuit Conference,* pp. 290–292, February 2011.

17. RFAMS table in International Technology Roadmap for Semiconductors (ITRS) 2012. http://www.itrs.net/links/2012itrs/home2012.htm

18. Y. Manzawa, Y. Goto, and M. Fujishima, Bond-based design for MMW CMOS circuit optimization, *Asia-Pacific Microwave Conference,* pp. 1–4, December 2008.

19. R. Fujimoto, M. Motoyoshi, K. Takano et al., A 120 GHz/140 GHz dual-channel OOK receiver using 65 nm CMOS technology, *IEICE Transactions on Fundamentals*, E96-A(2), 486–493, February 2013.

20. F.X. Pengg, Direct parameter extraction on RF-CMOS, *IEEE MTT-S*, 1, 271–274, June 2002.

21. C.H. Doan, S. Emami, A.M. Niknejad et al., Millimeter-wave CMOS design, *IEEE JSSC*, 40, 144–155, January 2005.

22. R. Fujimoto, K. Takano, M. Motoyoshi et al., Device modeling techniques for high-frequency circuits design using bond-based design at over 100 GHz, *IEICE Transactions on Electronics*, 94, 589–597, April 2011.

23. B. Heydari, M. Bohsali, E. Adabi et al., Millimeter-wave devices and circuit blocks up to 104 GHz in 90 nm CMOS, *IEEE JSSC*, 42, 2893–2903, December 2007.

24. M.T. Yang, P.P.C. Ho, Y.J. Wang et al., Broadband small-signal model and parameter extraction for deep sub-micron MOSFETs valid up to 110 GHz, *IEEE RFIC*, 269–372, June 2003.
25. K.H.K. Yau, E. Dacquay, I. Sarkas et al., Device and IC characterization above 100 GHz, *IEEE Microwave Magazine*, 13, 30–54, February 2012.
26. K. Katayama, M. Motoyoshi, K. Takano et al., Bias-voltage-dependent subcircuit model for millimeter-wave CMOS circuit, *IEICE Transactions on Electronics*, 95, 1077–1085, June 2012.
27. K. Murase, R. Ishikawa, and K. Honjo, Group delay equalized monolithic microwave integrated circuit amplifier for ultra-wideband based on right-left-handed transmission line design approach, *IEEE Transactions on IET*, 3, 967–973, August 2009.
28. M. Motoyoshi, R. Fujimoto, K. Takano et al., 140 GHz CMOS amplifier with group delay variation of 10.2 ps and 0.1 dB bandwidth of 12 GHz, *IEICE Electronics Express*, 8(4), 1192–1197, July 2011.
29. Y. Manzawa, M. Sasaki, and M. Fujishima, High-attenuation power line for wideband decoupling, *IEICE Transactions on Electronics*, E92-C(6), 792–797, June 2009.
30. M. Seo, B. Jagannathan, C. Carta et al., A 1.1 V 150 GHz amplifier with 8 dB gain and +6 dBm saturated output power in standard digital 65 nm CMOS using dummy-prefilled microstrip lines, *IEEE International Solid-State Circuit Conference*, pp. 484–485, February 2009.
31. S.T. Nicolson, A. Tomkins, K.W. Tang et al., A 1.2 V 140 GHz receiver with on-die. Antenna in 65 nm CMOS, *IEEE RFIC Symposium*, pp. 229–232, June 2008.
32. E. Laskin, M. Khanpour, S.T. Nicolson et al., Nanoscale CMOS transceiver design in the 90–170-GHz range, *IEEE Transactions on MTT*, 57, 3477–3490, November 2009.
33. E. Laskin, M. Khanpour, R. Aroca et al., A 95 GHz receiver with fundamental-frequency VCO and static frequency divider in 65 nm digital CMOS, *IEEE International Solid-State Circuit Conference*, pp. 180–181, February 2008.
34. Y. Natsukari and M. Fujishima, 36 mW 63 GHz CMOS differential low-noise amplifier with 14 GHz bandwidth, *Symposium on VLSI Circuits*, pp. 252–253, June 2009.
35. C.C. Chen, Y.S. Lin, P.L. Huang et al., A 4.9-dB NF 53.5–62-GHz. Micro-machined CMOS wideband amplifier with small group-delay-variation, *IEEE MTT-S*, pp. 489–492, May 2010.
36. A. Hirata, R. Yamaguchi, T. Kosugi et al., 10-Gbit/s wireless link using InP HEMT MMICs for generating 120-GHz-band millimeter-wave signal, *IEEE Transactions on Microwave and Technology*, 57(5), 1102–1109, May 2009.
37. N. Ono, M. Motoyoshi, K. Takano et al., 135 GHz 98 mW 10 Gbps ASK transmitter and receiver chipset in 40 nm CMOS, *Symposium on VLSI Circuits*, pp. 50–51, June 2012.
38. A. Oncu, B.B.M. Badalawa, and M. Fujishima, 60 GHz-pulse detector based on CMOS nonlinear amplifier, *IEEE SiRF*, pp. 1–4, January 2009.
39. E. Dacquay, A. Tomkins, K.H.K. Yau et al., D-band total power radiometer performance optimization in an SiGe HBT technology, *IEEE Transactions on MTT*, 60, 813–826, March 2012.
40. R. Fujimoto, M. Motoyoshi, K. Takano et al., A 120-GHz transmitter and receiver chipset with 9-Gbps data rate using 65-nm CMOS technology, *Asian Solid-State Circuit Conference*, pp. 281–284, November 2010.
41. R. Fujimoto, M. Motoyoshi, K. Takano et al., A 120-GHz/140-GHz dual-channel ASK receiver using standard 65 nm CMOS technology, *European Microwave Integrated Circuits Conference*, pp. 628–631, October 2011.
42. N. Deferm and P. Reynaert, A 120 GHz 10 Gb/s phase-modulating transmitter in 65 nm LP CMOS, *IEEE International Solid-State Circuit Conference*, pp. 290–292, February 2011.
43. J. Park, S. Kang, S. Thyagarajan, E. Alon et al., A 260 GHz fully integrated CMOS transceiver for wireless chip-to-chip communication, *Symposium on VLSI Circuits*, pp. 48–49, June 2012.

9 Photonics-Enabled Millimeter-Wave Wireless Systems

Jeffrey A. Nanzer, Timothy P. McKenna, and Thomas R. Clark, Jr.

CONTENTS

9.1 MILLIMETER-WAVE WIRELESS SYSTEMS

The term *millimeter-wave*, commonly abbreviated MMW, mmW, mm-wave, or mmWave, nominally refers to the frequency range of 30–300 GHz (wavelengths of the range 1–10 mm) and corresponds to the extremely high frequency (EHF) band of the International Telecommunication Union (ITU) band designations and extends from the K_a-band through the mm band of the IEEE standard letter band designations (see Table 9.1). The millimeter-wave region of the electromagnetic spectrum has long been an area of significant interest in the remote-sensing community due to the ability to achieve finer spatial resolution with smaller physical apertures than can be achieved at lower frequencies. Applications in missile guidance radar systems and radio astronomy extend back decades [1,2], where technologies focused on W-band (75–110 GHz) frequencies, in particular near 94 GHz where atmospheric absorption is low, allowing long propagation distances. Reasonable fractional bandwidths at millimeter-wave frequencies translate to very large relative baseband bandwidths, thereby allowing for significantly better resolution in time and space. Many devices were developed for these applications; however, compared to lower-frequency

TABLE 9.1

IEEE Microwave and Millimeter-Wave Band Designations

Band Designation	Frequency Range (GHz)
L	1–2
S	2–4
C	4–8
X	8–12
K_u	12–18
K	18–26.5
K_a	26.5–40
V	40–75
W	75–110
mm	110–300

devices, the performance of millimeter-wave components and systems was lacking in terms of noise, efficiency, and cost.

The drive for increased signal bandwidth—and thus increased data rate—in current and future communications systems coupled with the need for improved spatial resolution in remote-sensing systems has driven millimeter-wave technologies to the point where greater availability, improved performance, and lower cost of the devices has made the development and production of millimeter-wave applications much more feasible. Wireless communications systems today generally operate at lower microwave frequencies where components have better performance at lower cost. WiFi (IEEE 802.11) compliant networks operate in the 2.4–5 GHz band, while GSM cellular telephone systems operate below 2.0 GHz. The bandwidth at these frequencies is, however, constrained compared to millimeter-wave frequencies, and future systems will require bandwidth exceeding what is possible at microwave frequencies. A high bandwidth wireless personal area network (WPAN) standard was recently defined (IEEE802.15.3c) specifying wireless data rates up to 5.28 Gbps operating between 58.3 and 64.8 GHz, and the recent IEEE 802.11ad is specified to operate in the 60 GHz band. Communications systems operating in the V-band (40–75 GHz) are thus actively being pursued, as well as systems in the W-band. Meanwhile, millimeter-wave security sensors achieving spatial imaging accuracy on the order of millimeters are already deployed in airports and other checkpoints, with future developments seeking to enable longer ranges and improved resolution. While many advances in millimeter-wave technologies have been made, future applications drive requirements that necessitate further technological development. For example, communications systems require more bandwidth and improved spectral efficiency to achieve higher data rates, while radar and imaging systems require increased transmit power and improved receiver sensitivity along with increased bandwidth.

Electronic millimeter-wave devices are able to support some of the requirements of developing systems; however, limitations exist that are extremely challenging to overcome. For example, generating sufficiently low-noise millimeter-wave signals to enable high-modulation format communications signals has proven difficult using electronic techniques. Electrical oscillators generate stable tones with low-phase noise at MHz frequencies that can then be multiplied up to GHz frequencies; however, the phase noise of the output frequency is increased by 6 dB with every doubling in frequency, severely reducing the noise performance at high millimeter-wave frequencies. Millimeter-wave diode oscillators such as IMPact ionization avalanche transit-time (IMPATT) and Gunn oscillators can generate high powers across a wide frequency range; however, they have generally poor noise performance. Upconversion and downconversion of waveforms to and from the millimeter-wave carrier frequency using electronic heterodyne architectures are inherently limited by the finite bandwidth that mixers can support, and the amplitude and phase variation across the passband can vary significantly. Millimeter-wave transmission lines are lossy, placing severe restrictions on the distance that millimeter-wave signals can be transported.

Photonic techniques have made significant advances in recent years and offer approaches to significantly reduce or eliminate the drawbacks associated with millimeter-wave technologies. Techniques such as photonic signal generation, signal upconversion and downconversion, and optical remoting offer significant advantages in performance over electronic techniques. Combined with the contemporary advances in millimeter-wave electronic technology, photonic technology enables significant advances in communications and remote sensing. This chapter will discuss several recent advances focused on millimeter-wave wireless systems, with a focus on millimeter-wave communications systems.

9.2   MOTIVATIONS FOR MILLIMETER-WAVE TECHNOLOGY

Relative to systems operating at lower frequencies, millimeter-wave wireless systems have important benefits. Due to the shorter wavelengths of the radiated signals, components and systems can be made smaller and more compact than microwave systems. The antenna aperture in particular can be made physically much smaller while maintaining the same electrical performance such as gain and directivity. Conversely, physically increasing the size of a millimeter-wave antenna improves directivity and gain, while the size can still be relatively small compared to microwave frequencies. For example, a 10 GHz Cassegrain reflector antenna with 48 dBi gain requires a primary reflector with a diameter of approximately 3 m, whereas a 94 GHz Cassegrain with the same gain requires a primary reflector with a diameter of approximately 0.3 m.

A significant benefit of operating wireless systems at millimeter-wave frequencies is the large bandwidth that is available. Because the fractional bandwidth of standard electronic designs is relatively constant over frequency, wider bandwidths, and thus increased data rates, can be achieved at millimeter-wave frequencies. The primary detriment to propagation is the absorption of electromagnetic energy (or, equivalently, attenuation of the signal) incurred by propagation through the atmosphere. Figure 9.1 shows the atmospheric absorption at sea level as a function of frequency with 15°C temperature and 50% humidity, where it can be seen that there exist frequency bands where the absorption is low as well as some bands where the absorption is very high. The overall trend is an increase in absorption as the frequency increases. The high absorption bands are due to absorption of water or

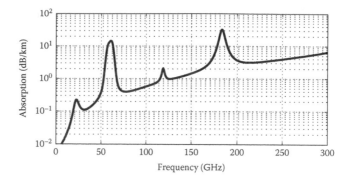

FIGURE 9.1   Sea-level atmospheric absorption at 15°C with 50% humidity.

oxygen molecules; the areas in between these bands are referred to as *atmospheric windows*, where long-range propagation is possible. The absorption plot changes with temperature, humidity, and weather conditions, generally with greater attenuation than is shown [3].

Atmospheric absorption is not the only factor to consider when designing a millimeter-wave wireless system. In particular, for communication systems, the optimal frequency band should also be determined by the required data rate that is to be realized. An optimal band can be determined by considering the channel capacity using Shannon's theorem $C = B \log_2(1 + \text{SNR})$, where C is the channel capacity, B is the bandwidth, and SNR is the signal-to-noise ratio. The size of the system is also an important consideration; thus, in determining the optimal frequency band, the antenna diameter is fixed in the following analysis to 0.3 m (12 in.), and a Cassegrain reflector is assumed. To evaluate the optimal frequency, a figure of merit used is the transmit power multiplied by the SNR required to achieve the specified capacity. Figure 9.2 shows the results of an evaluation to achieve capacities ranging from 1 to 100 Gbps for a 50 km link for a 10% fractional bandwidth assuming the atmospheric conditions described in Figure 9.1. For capacities 10 Gbps and higher, the optimal region for operation occurs in the W-band (75–110 GHz) due to the lower SNR requirements than at the lower frequencies. For lower capacities, the 30–50 GHz is preferable due to lower atmospheric attenuation.

9.3   MOTIVATIONS FOR PHOTONIC TECHNOLOGY

Photonics technology offers a number of benefits over traditional all-electronic implementations. Particularly advantageous is the broad bandwidth capability of photonics, which can be fully utilized with photonic frequency synthesis [4], and photonic upconversion [5–8] and downconversion [9–12], as shown in the photonics-enabled millimeter-wave transceiver of Figure 9.3. Low loss signal transport through photonic remoting and photonic interfacing of subsystems also allows preferential hardware distribution, packaging and efficient interconnections [13]. Additionally, photonic implementations support the linearity and SNR requirements needed for complex spectrally efficient modulation formats. A wide variety

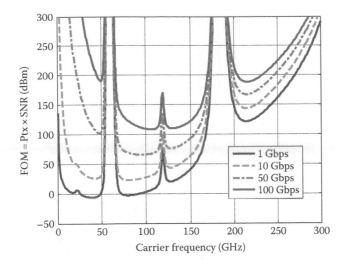

FIGURE 9.2 Optimal carrier frequency vs. capacity analysis for link range of 50 km, antenna diameter of 12 in., and 10% fractional bandwidth.

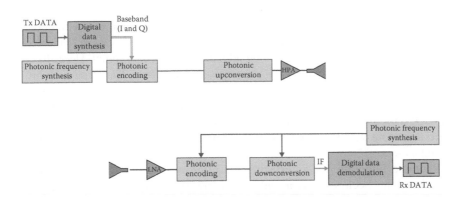

FIGURE 9.3 Photonics-enabled transmitter and receiver architecture.

of such formats and the supporting optical components have been developed for fiber-optic telecommunications over the past few years with data rates far exceeding those discussed earlier.

As discussed earlier, electronic mixer bandwidths are presently an impediment to achieving high data rate in millimeter-wave systems. In [13], a 9 Gbps W-band wireless communications link was presented with 2.8 bits/s/Hz spectral efficiency. In that work, photonic upconversion and photonic carrier generation were utilized; however, the bandwidth was limited by an electronic mixer in the receiver architecture. The total bandwidth and nonuniformity across the operating band of the electronic mixer required alteration of the modulation format at the spectral edges, as shown in Figure 9.4, to a more distortion- and noise-tolerant format, which also had lower spectral efficiency, quadrature phase shift keying (QPSK). The subcarrier

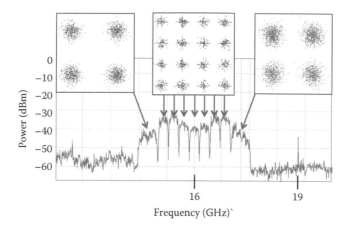

FIGURE 9.4 RF spectrum at intermediate frequency downconverted from 80 GHz carrier. Also shown are representative constellations for center seven 16-QAM encoded subcarriers and the electronic mixer edge subcarriers encoded with QPSK.

signals in the center of the band, where the mixer had less distortion, could maintain a high modulation of 16-QAM. Millimeter-wave amplifiers and antennas support much wider bandwidths than electronic mixers and, when combined with photonic upconversion and downconversion, offer a promising path forward. Figure 9.5a shows an experimental example of a broadband complex data signal. Here, 40 Gbps 16-QAM data have been encoded on a single optical carrier. Handling such broadband signals is not possible with today's electronic mixer technology. Figure 9.5b shows the experimental constellation and eye diagrams of the 40 Gbps 16-QAM encoded data output. An architecture employing photonic upconversion and photonic downconversion will enable full capitalization of the photonic bandwidth advantages.

Photonic remoting and interfacing for microwave and millimeter-wave systems has long been seen as an advantage for future systems. The attenuation over optical fiber is typically measured to be ~0.2 dB/km and the RF operating

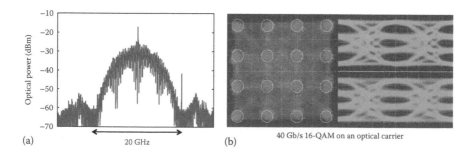

FIGURE 9.5 (a) Experimental optical spectrum of a broadband signal (40 Gbps 16-QAM) encoded onto the optical carrier. (b) Experimental received constellation and eye diagram of the 40 Gbps 16-QAM data on the optical carrier.

frequency is deemed largely irrelevant for millimeter-wave frequencies as the signals are transported on optical carriers (typically ~193 THz for the lowest fiber
loss region). Even short subsystem interfaces in waveguide (~3 dB/m in WR-10 at
80 GHz) and integrated circuit stripline (~0.09 dB/mm for 45 μm thick benzocy-
clobutene at 80 GHz) would require all desired circuit functions to be closely integrated. Employing fiber remoting and photonic generation, signal conditioning
and upconversion and downconversion functions will enable much more efficient
systems with necessary implementation options for numerous form factors. The
following sections discuss architectures and progress toward incorporating all of
the advantages and versatility inherent to photonic technologies into millimeter-
wave wireless systems.

9.4 PHOTONIC GENERATION OF MILLIMETER-WAVE SIGNALS

Photonics offer a number of ways to generate signals at millimeter-wave frequencies, including photomixing [14], quantum cascade lasers [15], and laser pulse techniques [16]. Through photomixing, continuous, stable, variable-frequency signals
can be generated at frequencies limited only by the photodiode bandwidth and the
modulating signal. In this technique, a dual-wavelength (dual-λ) signal consisting of
two optical signals separated in (optical) frequency by the desired millimeter-wave
frequency f_{mmw} are combined onto a single optical fiber and input to a high-speed
photodiode. The electrical signal at f_{mmw} is generated as the beat frequency between
the two optical fields through photomixing. The photodiode current is proportional
to the incident electric field through

$$i(t) = \eta E^2(t),\tag{9.1}$$

where η is the photodiode responsivity. The incident electric field contains components at the two frequencies separated by f_{mmw}, and is given by

$$E(t) = A_1 \cos(2\pi f_1 t) + A_2 \cos(2\pi f_2 t).\tag{9.2}$$

In general, a phase offset may be present between the two components; however, for
the generation of millimeter-wave frequencies, the phase offset is inconsequential.
The photodiode current due to the incident field is then given by

$$i(t) = \alpha \left\{ \begin{array}{l} A_1^2 + A_2^2 + A_1^2 \cos(4\pi f_1 t) + A_2^2 \cos(4\pi f_2 t) \\ +A_1 A_2 \cos\left[2\pi(f_1 - f_2)t\right] + A_1 A_2 \cos\left[2\pi(f_1 + f_2)t\right] \end{array} \right\}.\tag{9.3}$$

The frequency terms $f_1 + f_2$, $2f_1$, and $2f_2$ are optical frequencies and are filtered by
the finite bandwidth at the photodiode output. The constant terms exist at dc and are

filtered out by the electronics, such as the high-power amplifier (HPA) or antenna. Thus, the only component remaining is the millimeter-wave term of interest, given by

$$i(t) = i_0 \cos(2\pi f_{\text{mmw}} t), \tag{9.4}$$

where $f_{\text{mmw}} = |f_1 - f_2|$ and the responsivity and amplitudes have been folded into the peak photocurrent i_0. Thus, by generating dual-λ optical signals with signal components separated by the desired carrier frequency, millimeter-wave signals are generated at the output of the photodiode. These signals can then be fed to a front end consisting of an HPA and antenna.

Crucial to the carrier generation via the two optical fields mixing to form the millimeter-wave carrier is the quality and relation of the fields, and, in particular, the phase noise of the resulting millimeter-wave carrier is important for high-modulation formats. There are generally three methods to generate dual-λ optical signals for photomixing generation of the millimeter-wave carrier: a heterodyne carrier generation (HCG) architecture using two separate lasers (simple but generally too noisy), a double-sideband suppressed carrier (DSCS) architecture using a single laser (simple and low noise), and a multiwavelength laser generating the two optical tones (more complicated to implement but potentially has very low noise).

The benefits of an HCG architecture using separate lasers include unrestricted frequency tuning, provided the photodiode bandwidth is sufficient, as well as efficient use of the generated optical power, as 100% of the laser output power can be utilized to generate the millimeter-wave carrier. While it is simple, stringent requirements on both the optical wavelength stability and the optical lineshape are prohibitive for high-modulation formats. For two independent lasers, there is no correlation between the two fields and the linewidth, and therefore phase noise, of the resultant millimeter-wave carrier, is completely determined by the two optical lineshapes. In Figure 9.6, the optical linewidth limitation with the two optical fields mixing on an ideal photodiode is shown through simulation. The rms noise is quantified as an error vector magnitude (EVM). As an example, for a desired transmitter SNR of 25 dB, supporting a 64-QAM modulation format, a transmitter EVM better than 8% or a linewidth of both optical fields to be better than 6 kHz is required, as shown by the dotted line in Figure 9.6. While not an impossible requirement, such lasers are currently not widely available.

The simplest approach to dual-wavelength optical signal generation, one that also achieves relatively low noise, is a DSCS generation method, shown in Figure 9.7. Here, a single laser is input to a null-biased intensity modulator, resulting in multiple double sidebands at frequencies of integer multiples of the drive frequency and a suppressed optical carrier. The signal driving the modulator must be at most $\frac{1}{2} f_{\text{mmw}}$, as the resulting sidebands tones are separated from the optical carrier by this frequency. By appropriately null-biasing the modulator, the carrier can be suppressed, leaving only the two sidebands, separated by f_{mmw}. For example, a Mach–Zehnder modular biased at quadrature and driven with a 40 GHz electrical signal will generate sidebands separated by 80 GHz. After photomixing, the 80 GHz tone will remain at the output of the photodiode. The level of carrier suppression achieved depends on the

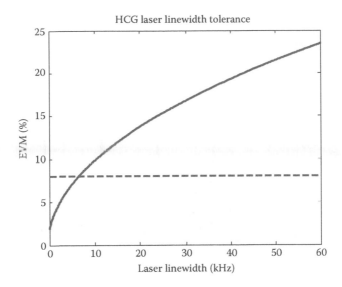

FIGURE 9.6 HCG tolerance to laser linewidth including target EVM of 8%.

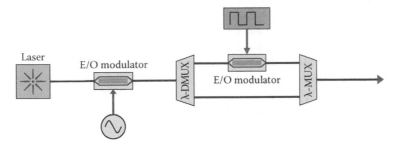

FIGURE 9.7 Double sideband carrier suppression (DSCS) carrier generation. The multiplexer and demultiplexer allow the modulation of data onto only one optical carrier, as discussed in Section 9.5 (see Figure 9.11).

accuracy of the modulator bias. While in this example the photomixing of the optical carrier will generate signals at 40 GHz, which are far enough separated from the desired 80 GHz to be easily filtered, these undesired signals take energy away from the desired carrier and thus reduce efficiency. For efficient generation of the millimeter-wave carrier, the carrier should therefore be suppressed.

Because the null-biased modulator output spectrum contains multiple harmonics of the sidebands, millimeter-wave frequencies can be generated by modulating signals at frequencies less than $\frac{1}{2}f_{mmw}$. Using this approach, lower-frequency microwave oscillators can be used in place of millimeter-wave oscillators as frequency multiplication can be performed in the optical domain with little penalty to the output power. Figure 9.8 illustrates this approach with a simulation of the optical spectrum at the output of the electro-optic modulator for a 40 GHz oscillator drive and a 13.33 GHz oscillator drive with 6 dB higher power. The power difference in the two 80 GHz

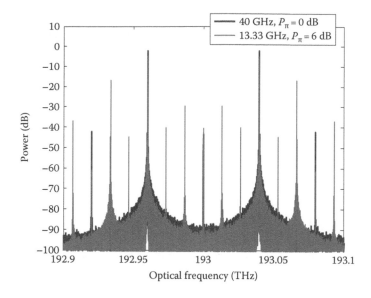

FIGURE 9.8 Double-sideband suppressed carrier (DSSC) optical spectrum of 40 GHz oscillator vs. 13.33 GHz oscillator drive.

separated sidebands is minimal. Another benefit of this architecture is the single laser nature, which all but removes the optical linewidth considerations, as illustrated in Figure 9.9. Here, the optical lineshape effect on the EVM is shown with delay mismatch up to 10 ns between the two paths after the modulator. Even a relatively poor linewidth distributed feedback laser, common for telecommunications, falls well within typical tolerances for transmitter and receiver SNR targets with very coarse delay matching.

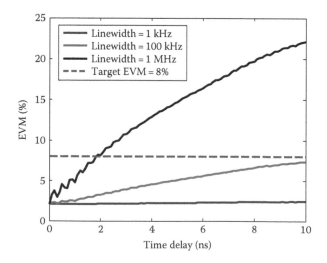

FIGURE 9.9 Linewidth and time delay tolerance of DSCS carrier generation.

Dual-wavelength optical signals can also be achieved using dual-wavelength lasers. These approaches typically result in lower noise signals due to the optical tones being generated in the same devices and thus having correlated noise that is removed in the photomixing process; however, such devices are also generally quite complicated. In one example [17], a dual-wavelength laser based on stimulated Brillouin scattering (SBS) was able to achieve sub-Hz linewidths. The SBS laser is a simple fiber ring resonator (FRR) with regularly spaced resonances that have a full-width at half-maximum of 0.5 MHz at a resonance spacing of 10.3 MHz. The resonator is pumped at two simultaneously resonant optical frequencies through the previously mentioned DSCS technique; these two optical tones are fed to the optical resonator. Each pump produces a gain band down-shifted in frequency by the Brillouin shift, which is 10.9 GHz for the selected fiber type and operating wavelength (~1550 nm). These gain bands have a full-width at half-maximum of ~15 MHz, so at least one resonance will fall under each band, and each gain band yields a lasing line at whichever cavity resonance is nearest its gain peak. Because SBS is a back-scattering phenomenon, the lasing lines propagate in the direction opposite the pumps, and they must be separated from the pumps with an optical circulator. Figure 9.10 illustrates a measured laser output spectrum using this technique with the two wavelengths separated by 60 GHz.

Because the two lasing lines of the SBS laser share a common cavity, most of the environmental noise sources, such as thermal drift and acoustic pickup, are present in both tones and thus correlated, and therefore the noise is canceled out when the two optical signals are photomixed, significantly lowering noise compared to the other signal generation techniques. Furthermore, the lasing frequencies are determined by the positions of the resonances, rather than the gain bands that are approximately 30 times broader than the resonances; this suppresses any transfer of noise from the pump source to the lasing lines [17], further reducing noise. The significant noise benefits of this type of laser are counter-balanced by the complexity involved

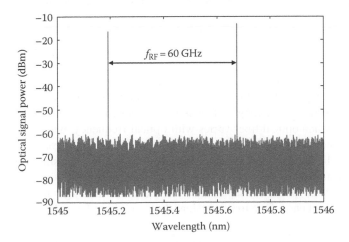

FIGURE 9.10   Measured optical spectrum of the SBS laser.

in designing and constructing the laser. SBS lasers and similar dual-wavelength lasers are currently laboratory-level instruments and are thus not optimized for size or power consumption.

9.5 PHOTONIC UPCONVERSION AND DATA ENCODING*

The photonic portions of millimeter-wave photonic transmitter systems consist of an electrical-to-optical (EO) conversion, a length of optical fiber to remote the signal, and an optical-to-electrical (OE) conversion. The EO conversion imposes the radio frequency (RF) signal of interest onto an optical carrier for transport over optical fiber to a desired location, which may be kilometers away. This architecture exploits the high bandwidth and low loss capabilities of optical fiber. At the remote antenna site, an OE conversion is performed, extracting the RF information for transmission to the receiver. The RF information may be encoded onto the intensity, the phase, or the frequency of a single-wavelength optical carrier, such as the output of a laser, by using corresponding EO and OE conversion technologies. The capability of systems utilizing photonic upconversion to transmit millimeter-wave signals with better spectral purity and thus higher data rates than electronic millimeter-wave transmitters has made photonic upconversion a popular approach to generate broadband millimeter-wave signals with data rates 10 Gbps and higher [19].

One of the simplest methods of digital data encoding is amplitude-shift keying (ASK), which represents binary data as two distinct signal amplitudes that are distinguished by a threshold operation. The amplitudes may be shifted between a maximum value representing 1 and zero amplitude representing 0, which is referred to as on–off keying (OOK). Multiple amplitude levels may be employed to represent multiple digital bit streams, thereby increasing the data throughput without increasing the signal bandwidth. Phase-shift keying (PSK) is an alternative method of digital encoding that represents digital signals as two different phases of the carrier signal. The two phases are generally spaced 180° apart so that mixing with a continuous-wave signal of the same frequency will produce constructive and destructive interference, resulting in a binary signal that can be decoded in the same manner as a simple ASK signal. Multiple phase levels may also be encoded, representing multiple digital bit streams.

It was noted earlier that the fiber span from the EO to the remote antenna may be kilometers in length. The primary limitation encountered when transporting broadband optical signals over long distances is chromatic dispersion [20]: the index of refraction of silica optical fiber varies with optical frequency, which causes the different frequency components comprising the encoded signal to propagate at different velocities in the optical fiber. Standard intensity modulation of a single optical wavelength with millimeter-wave data signal will generate optical sidebands on either side of the optical tone, each separated from the tone wavelength by the millimeter-wave center frequency. Due to dispersion, the two data sidebands will experience a

---

* Portions of this section are reproduced with permission from [18]. ©The Johns Hopkins University Applied Physics Laboratory.

differential time delay. Since the signal of interest is generated by photomixing of each sideband with the carrier, this time delay will cause periodic destructive interference between the two mixing products. In the millimeter-wave domain, this manifests as periodic fading of the millimeter-wave carrier as the fiber length is increased, placing limitations on the exact length of fiber that can be used for a given application. This dispersion problem is eliminated, however, by using a dual-wavelength optical carrier and by encoding only one wavelength of the optical signal with the baseband data to be transmitted rather than encoding a single optical carrier with the data at the millimeter-wave carrier. The process is illustrated in Figure 9.11: the two wavelengths output by the dual-wavelength laser are demultiplexed using an optical splitter and two optical filters, and the EO modulation is performed on one of the individual wavelengths before the optical signals are recombined. Figure 9.12 shows an example of the resulting optical spectrum of the recombined wavelengths; the spectrally wider tone at the higher optical frequency is carrying the encoded data. Because only one sideband is modulated with the baseband data, the dispersion affects only the bandwidth of the baseband data signal, rather than the bandwidth

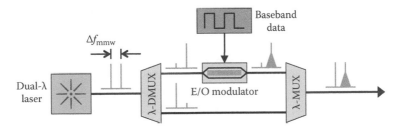

FIGURE 9.11 Diagram showing the single-sideband modulation technique. The light gray lines represent the spectral location of the optical tones at various points in the process and are separated by the desired millimeter-wave carrier frequency denoted by Δf_{mmw}. The two carriers are initially demultiplexed and one is modulated with the desired baseband data while the other is unaffected. After multiplexing, the optical signal contains modulation on one wavelength only.

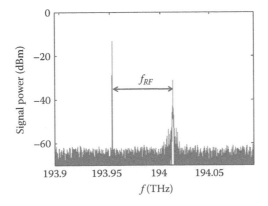

FIGURE 9.12 Measured optical spectrum with single-sideband modulation.

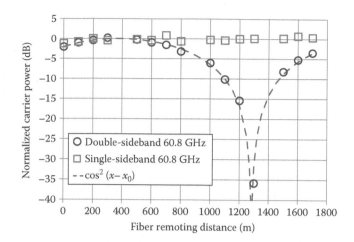

FIGURE 9.13 Received power at the 60 GHz carrier as a function of fiber length for different carrier generation methods. Circles denote the conventional method of modulating a single optical carrier, while squares denote a dual-wavelength generation method. The dashed line is a raised cosine fit to the conventional method data.

of the dual-λ optical signal (i.e., the separation of the two optical wavelengths). The optical signal is downconverted to millimeter-wave through photomixing in the photodiode. The unmodulated optical wavelength undergoes dispersion relative to the modulated optical signal; however, the phase of the unmodulated reference signal is arbitrary and thus does not affect the generation of the millimeter-wave signal at the photodiode.

To demonstrate that a dual-wavelength system is robust against dispersion-induced fading, the detected power of a 60 GHz RF carrier generated using an SBS laser was measured while varying the length of fiber in the system. For comparison, the experiment was repeated for the case of a conventional photonic link using intensity modulation of a single optical carrier to generate the 60 GHz signal. The results are shown in Figure 9.13, with the SBS system performance indicated by the squares. The conventional photonic link, indicated by the circles, experiences a null in signal power at a remoting distance of 1300 m. Shown in the dashed line is the best fit to a raised cosine function, which characterizes the dispersive fading. As is seen from the data, the interference pattern exhibits a horizontal offset of the peak normalized power, such that the peak occurs at ~370 m of remoting fiber rather than zero. This is due to chirp induced by the LiNbO$_3$ modulator [21]. In contrast with the single-optical-carrier architecture, the SBS-generated source does not suffer from signal fading due to dispersion.

9.6 PHOTONIC DOWNCONVERSION AND DOWNSAMPLING

Frequency downconversion is an essential part of all but the lowest frequency receivers, allowing for transmitted high-frequency information to be demodulated and processed by lower-speed precision electronics. Millimeter-wave photonic links, offering low loss signal transport over optical fiber and wideband operation, have the potential to solve many of the current technology deficiencies using simple, flexible hardware systems. A typical solution is to add an electronic mixer or set of mixers to perform the downconversion operation. It was seen previously that electronic mixers used for upconversion caused limitations in transmitters, and the same challenges exist when using mixers for downconversion in receivers. The disadvantages of this method include the addition of electronic mixer-induced signal distortions, bandwidth limitations, and the need for an electronic local oscillator. It is possible, however, to avoid these limitations by performing the frequency conversion in the optical domain through photonic downconversion [9,22], with broadband optical modulator technology, or photonic downsampling [23–25], with short pulse mode-locked laser technology. Both photonic downconversion and downsampling capitalize on the broadband nature of optical technology, while requiring only baseband or low intermediate frequency electronics for demodulation, digitization, and processing of the information contained in the signal. In this section, both downconversion and downsampling approaches are described, and their operation in millimeter-wave communication experiments is demonstrated.

Photonic downconversion can be performed in an analogous manner to the upconversion process described previously, where two optical carriers are generated as shown in Figure 9.14. Here the intended intermediate frequency (IF) is selected to support the wide bandwidth data demodulation in the digital domain. An electro-optic phase modulator, commercially available with response from less than 1 GHz to greater than 100 GHz, encodes the signal from an antenna onto the phase of an optical carrier separated from a second optical carrier by the λ-DEMUX filter. The encoded carrier (λ_1) is mixed onto a low-speed photodiode—the bandwidth of which is determined by the IF and data modulation format—with the second optical carrier (λ_2), which is separated from λ_1 by a frequency differing from the millimeter-wave carrier frequency by f_{IF}. This converts the millimeter-wave modulation of the encoded signal to a modulated electrical signal at f_{IF}.

In [26], this concept was experimentally demonstrated in a K_a-band receiver. The system was operated at 10 Gbps using 16-QAM modulation and root-raised cosine filtering for a spectral efficiency of 3.33 b/s/Hz. The sensitivity of the receiver is shown in Figure 9.15. Also shown is the constellation for a received power of −42.92 dBm, which results in an equivalent bit error ratio (BER) better than 10^{-9}. A minimum received power of −50.42 dBm resulted in a 13.8% EVM, which corresponds to a

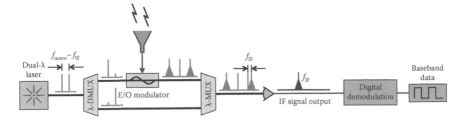

FIGURE 9.14 Photonic downconversion and digital demodulation.

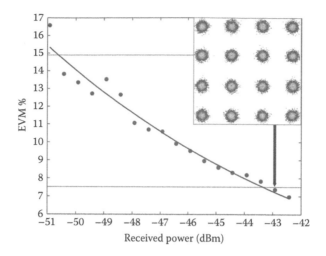

FIGURE 9.15 Plot of EVM as a function of received signal power. BER of 10^{-3} and 10^{-9} are indicated. Inset: signal constellation corresponds to a received power of −42.92 dBm.

BER of 4.6×10^{-4}, assuming white Gaussian noise as the primary impairment [27]. This BER would be correctable using standard forward error correction (FEC) codes, which are capable of correcting error rates up to 2×10^{-3}.

Another attractive alternative to downconversion exists where the millimeter-wave carrier is directly sampled to baseband. Photonic downsampling uses the precise timing of mode-locked lasers to sample with a sampling time and sampling jitter small compared to the millimeter-wave carrier period. This offers the ability to downsample microwave signals with very high resolution and with instantaneous bandwidth equal to the Nyquist sampling bandwidth [10,23]. The timing jitter of electronic clocks currently limits the electronic downsampling technique to lower frequencies, but mode-locked lasers with sub-10 fs jitter [28] and wideband electro-optic modulators allow for the photonic downsampling of signals at frequencies of 40 GHz and beyond [29]. As with all subsampling systems, strict filtering should be applied to avoid degraded system performance due to noise aliasing.

In [30], a 40 GHz wireless communication link was demonstrated that uses a photonic downsampling technique allowing demodulation of a multiple subcarrier 16-QAM signal with a total rate of 3 Gbps. Figure 9.16 shows the system setup; the transmitter concept is similar to the approach described previously and shown in Figure 9.11. Two optical sidebands with 40 GHz separation are generated with a laser-modulator system driven by a sinusoidal RF source at 20 GHz. The lower sideband is bandpass filtered and in-phase and quadrature (I/Q) data are single-sideband (SSB) modulated onto the optical carrier using a dual-parallel Mach–Zehnder modulator (DP-MZM). SSB modulation is used to prevent double-sideband signal copies from aliasing onto each other during the downsampling process at the receiver. The I/Q signals consist of one or more 16-QAM subcarriers each with a rate of 250 MSym/s. The data-encoded sideband was recombined with the unmodulated

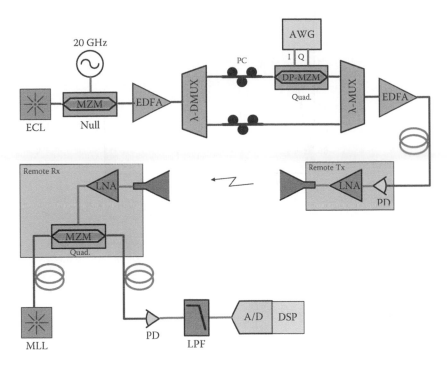

FIGURE 9.16 Schematic of fiber remote transmitter and receiver. ECL, external cavity laser; MZM, Mach–Zehnder modulator; EDFA, erbium-doped fiber amplifier; DP-MZM, dual-parallel Mach–Zehnder modulator; AWG, arbitrary waveform generator; PD, photodiode; LNA, low noise amplifier; MLL, mode-locked laser; LPF, lowpass filter; A/D, electronic analog-to-digital converter; DSP, digital signal processing.

reference sideband before mixing on a photodiode with a bandwidth of 50 GHz. The resulting 40 GHz electrical signal was amplified before radiating from a 24 dBi horn antenna. The wireless signal was received by a matching 24 dBi horn antenna. An actively harmonically mode-locked fiber laser (MLL) with a 10 GHz repetition rate operating at 1550 nm with 12.4 dBm of average output power was used to optically sample the received signal. The received signal was amplified by a low-noise amplifier (LNA) with 30 dB gain that directly drove a GaAs MZM with greater than 40 GHz bandwidth that amplitude-modulated the pulse train from the MLL. The modulated pulse train was detected by a photodiode with a 3 dB BW of 12 GHz and the resulting electrical signal was lowpass filtered and then digitized. Demodulation was performed in software on the signal content in the first Nyquist zone, as shown in Figure 9.17. The transmitted signal consisted of subcarriers at 40.2, 40.5, and 40.8 GHz, for a total data rate of 3 Gbps. Each subcarrier was root-raised cosine pulse-shaped with an excess bandwidth factor of 0.2. The average RMS EVM across all subcarriers was 14.3% with the BER of every subcarrier within the FEC limit of 2×10^{-3}. The spectral efficiency was measured to be 3.33 b/s/Hz. This downsampling architecture shows great promise for many high instantaneous bandwidth applications.

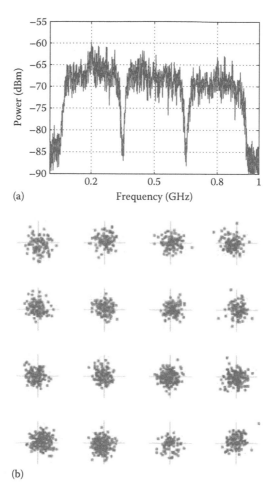

(a)

(b)

FIGURE 9.17 Single subcarrier electrical spectrum (a) and constellation (b) of three 16-QAM subcarriers after downsampling.

9.7 SYSTEM EXAMPLES

This section summarizes the design and measurement results of two experimental millimeter-wave photonic communications systems, one operating in the V-band and the other operating in the W-band, each demonstrating some of the photonic techniques described earlier.

9.7.1 V-Band Photonic Communications System*

The development of photonic technologies to support wideband personal area networking (WPAN) applications, operating in the frequency band 54–64 GHz,

---

* Portions of this section are reproduced with permission from [18] ibid. ©The Johns Hopkins University Applied Physics Laboratory.

has been an active area of research in recent years. This section summarizes the experimental results of one V-band millimeter-wave photonic communications system. The experimental 60 GHz communication system designed using an SBS dual-wavelength laser as the millimeter-wave carrier generator, consisting of a millimeter-wave photonic transmitter and a separate millimeter-wave receiver, is shown in Figure 9.18. For this system, binary-phase-shift keying (BPSK) was used as the data format to achieve data rates up to 3 Gbps. Digital data were generated by a pulse pattern generator and modulated onto one of the demultiplexed optical sidebands using a phase modulator, encoding the binary data values onto its optical phase. The optical signals were then combined and transported on optical fiber over distances up to ~1 km in various configurations. At the photodiode, the baseband data signal was upconverted onto the 60 GHz carrier through photomixing. The millimeter-wave signal was amplified with a power amplifier with 22 dB of gain and transmitted to the receiver via a horn antenna with a gain of 24 dBi. The receiver consisted of an identical horn antenna, followed by a 19 dB gain low-noise amplifier. The 60 GHz signal was first downconverted to an intermediate frequency (IF) of ~6 GHz by mixing the received signal with a 56 GHz RF local oscillator generated by fourfold frequency multiplication of a 14 GHz reference oscillator. The IF signal was then mixed with the 6 GHz output of a dielectric resonator oscillator (DRO) to downconvert the data to baseband. The phase error between the IF and the DRO was monitored with a proportional–integral–differential servo controller (PIC), the output of which was used to close the phase-locked loop on the frequency of the 14 GHz

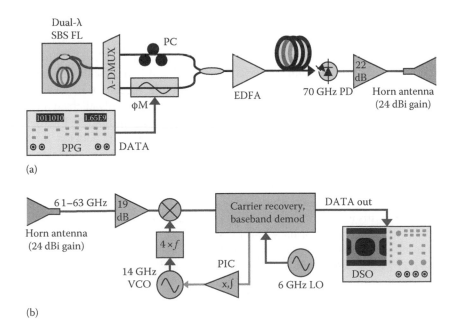

FIGURE 9.18   (a) System block diagram of the 60 GHz transmitter. (b) System block diagram of the 60 GHz receiver.

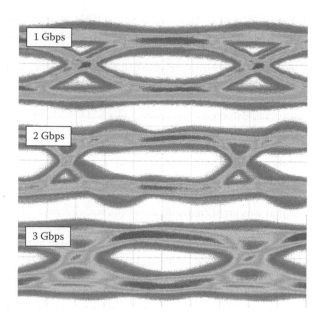

FIGURE 9.19   Eye diagrams of baseband BPSK data with data rates of 1, 2, and 3 Gbps after free-space transmission at 60 GHz over 1.65 m and demodulation.

reference oscillator. The baseband data were recorded using a telecommunications bit error detector and an oscilloscope.

Initial experiments were conducted at short distances to verify system functionality and to explore bandwidth limits. The transmitter and receiver horns were placed 1.65 m apart, and the baseband data were recorded on a LeCroy oscilloscope. Figure 9.19 shows eye diagrams of the demodulated BSPK data with data rates of 1, 2, and 3 Gbps. Because of bandwidth limitations in the receiver millimeter-wave hardware, the data rate was limited to 3 Gbps or less; a photonically enabled receiver would have the potential to increase the data rate beyond this limitation. Figure 9.19 shows eye diagrams for the received BPSK signals at the different data rates, where the clear openings in the eye diagrams indicate low error rates for each data rate. The BER was measured as a function of received power by varying the amplification of the photodiode signal, and in Figure 9.20 the log of the BER is plotted against the received power. At the highest power, the BER was error-free, while at the lowest power level in this test, the BER was 4.9×10^{-5}, within standard FEC limits.

To characterize the performance over distances relevant to WPNA applications, the transmitter and receiver were separated by distances up to 30 m in an indoor hallway. For these tests, the transmitter used a 16 dBi gain circular horn antenna with a beamwidth of approximately 30° while the receiver used a 46 dBi reflector antenna with a 0.9° beamwidth. While the transmit antenna was fairly small in this experiment, a higher gain LNA would alleviate the need for the larger receiver antenna, which in this case was on the order of 12 in. in diameter. The transmitter photodiode, HPA, and antenna were mounted to a tripod, and the dual-wavelength optical signal was remoted to the photodiode from a laboratory workbench over

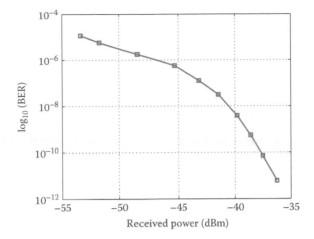

FIGURE 9.20   Measured BER for the 1.65 m link with a data rate of 1.65 Gbps.

100 ft of optical fiber. The transmitter antenna was moved from a range of 15 m to the receiver, to a range of 30 m, transmitting a constant power of approximately 0 dBm. Figure 9.21 shows the receiver setup with the parabolic reflector antenna. The measured BER is shown in Figure 9.22, showing the expected decrease in errors as the distance decreases (note that the *x*-axis plots the decreasing range), which results in increased received power.

FIGURE 9.21   60 GHz receiver setup for hallway testing.

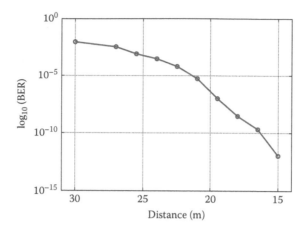

FIGURE 9.22 Measured BER for the 60 GHz hallway test with a data rate of 1.65 Gbps.

9.7.2 W-Band Photonic Communications System

This section discusses an experimental demonstration of a photonic mm-wave wireless system operating in the W-band for point-to-point high-capacity communications [31]. The demonstration used multiple 16-QAM subcarriers for a total of 10 Gbps data transmission over a link distance of 520 m. In this work, the transmitter was as described earlier and is shown in Figure 9.16. For the outdoor 520 m link, the output of the photodiode was amplified with an HPA with 35 dB gain and approximate noise figure of 5 dB to achieve a transmit power of −10.2 dBm, radiated from a Cassegrain antenna with a gain of 50 dBi and a half-power beamwidth of 0.6° at 80 GHz. Figure 9.23 shows the transmit antenna hardware and the 520 m roof-to-roof link on the campus of The Johns Hopkins University Applied Physics Laboratory used for testing. The measured transmit SNR was 19 dB, limited by the measurement instrument.

The receiver for this link utilized an electronic W-band mixer-based design with a photonically generated LO signal; the receiver design is shown in Figure 9.24. A Cassegrain antenna, located 520 m from the transmit antenna, with a gain of 50 dBi at 80 GHz received the −35.3 dBm signal into a low-noise W-band amplifier with a gain of 21 dB and specified noise figure of 5.0 dB. The amplifier output was directed to a balanced Schottky diode mixer for downconversion after mixing with a 65 GHz photonically generated local oscillator. A distributed fiber laser (DFB) and an MZM biased at a null generated two prominent sidebands spaced by 65 GHz, similar to the technique for carrier generation at the transmitter. A photodiode with a 3 dB bandwidth greater than 50 GHz converted the optical signal to a 65 GHz single frequency sinusoid, amplified to a power greater than 12 dBm, to drive the LO port of the mixer. For full fiber remoting, an amplified Ku-band intensity-modulated direct-detection (IMDD) link transported the IF signal from the mixer to the digitizer. A 50 GS/s oscilloscope with 20 GHz bandwidth digitized the IF signal for

(a)

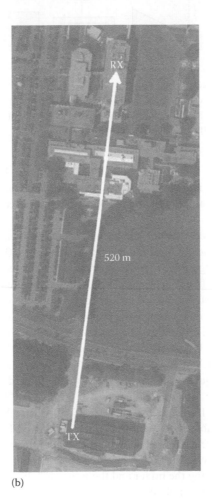

(b)

FIGURE 9.23 (a) Photograph of the transmit antenna and hardware. (Inset) Close-up photograph of the photodiode and the amplifier. (b) Roof-to-roof 520 m free-space path.

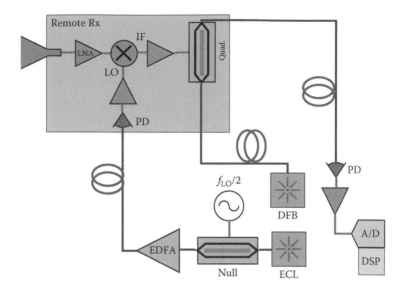

FIGURE 9.24 Photonics-assisted receiver configuration. LNA, low-noise amplifier; LO, local oscillator; IF, intermediate frequency; ADC, analog-to-digital converter; DSP, digital signal processing; MZM, Mach–Zehnder modulator; EDFA, erbium-doped fiber amplifier; PD, photodiode; DFB, distributed feedback laser.

digital signal processing (DSP). Offline DSP performed matched filtering, frequency offset correction, phase recovery, and used a linear, decision-directed, feedforward equalizer to correct for amplitude and phase variation across the bandwidth of each subcarrier.

The photonics-assisted receiver shown in Figure 9.24 was used to demonstrate high data rate communications with a transmit power of −10.2 dBm and a received power of −35.3 dBm. Figure 9.25 shows the received RF spectrum for a transmitted data rate of 10 Gbps on five subcarriers, each with a data rate of 2 Gbps. Figure 9.26 shows the received constellations for an all-electronic receiver, using a W-band synthesizer for the local oscillator, and using the photonics-assisted receiver. For the all-electronic receiver, the EVM of the subcarriers at the transmitted frequencies of 76.1, 76.7, 77.3, 77.9, and 78.5 GHz were measured to be 8.3%, 8.9%, 7.5%, 7.7%, and 7.1%, with the differences in EVM attributed to amplifier ripple. For the photonics-assisted receiver, the lowest frequency subcarrier was not able to be demodulated. The EVM of the subcarriers at the transmit frequencies of 76.7, 77.3, 77.9, and 78.5 GHz were measured to be 14.1%, 13.7%, 13.5%, and 11.5%, respectively, with the differences in EVM primarily attributed to amplifier ripple. The overall reduced sensitivity of the photonics-assisted receiver was attributed to a combination of the quality of the mm-wave amplifier required for the photonically generated local oscillator and the waveguide network driving the mixer and the noise and dynamic range mismatch of the IF photonic link to the digitizer.

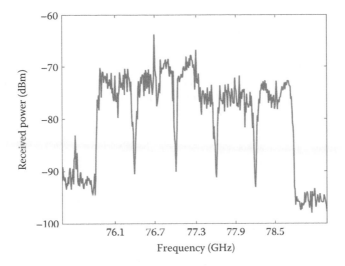

FIGURE 9.25 Received signal spectrum measured after the antenna with a spectrum analyzer using peak detection with a 100 kHz resolution bandwidth.

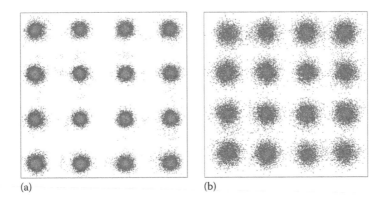

(a)   (b)

FIGURE 9.26 (a) 10 Gbps 16-QAM data transmission with the all-electronic receiver. (b) 8 Gbps 16-QAM data transmission with the photonics-assisted receiver. Symbols are aggregated over all subcarriers.

9.8 SUMMARY

By combining millimeter-wave and photonic technologies in high data rate communications systems, the relative benefits of each technology can be leveraged to develop systems more capable than either technology can support alone. Along with the wide signal bandwidths achievable at millimeter-wave carrier frequencies, the favorable propagation characteristics of some millimeter-wave radiation bands and compact physical size of millimeter-wave components and antennas permits the use of compact, high data rate front-end systems. Along with these millimeter-wave benefits, the ability of photonics to lift bandwidth restrictions on the upconversion and downconversion stages, as well as to generate spectrally pure signals, allows the use of high-modulation format signals, further increasing the potential data rate. Future millimeter-wave communications systems are anticipated to achieve data rates up to and exceeding 100 Gbps over distances of tens of kilometers or more, and while achieving this performance, photonic technologies can significantly reduce the burden on the millimeter-wave technology of producing spectrally efficient waveforms in a small space.

The millimeter-wave front end requires power gain and aperture gain to achieve communication over appreciable distances. Current millimeter-wave amplifier technology leveraging Gallium Arsenide can achieve output powers of up to 2.5 W through the W-band, and this number is likely to increase with improved amplifier technology and power combining techniques. Used with reflecting antenna systems such as Cassegrain or Gregorian reflectors, such amplifier technology allows data rates of multiple Gbps be easily transmitted over tens of kilometers. Ongoing research in millimeter-wave phased arrays shows significant promise for achieving the same power aperture using numerous elements in a significantly smaller aperture footprint, which will allow the use of millimeter-wave comms over great distances on small platforms.

While the bandwidth of photonic technologies remains significantly greater than that which can be achieved with electronics, current research and development in photonics is aimed at increasing the maximum operating frequency of photonic components. Photodiodes and electro-optic modulators operating at frequencies up to 40 GHz are widely available, and those operating up to 100 GHz are becoming commercially available, while electro-optical components operating beyond 100 GHz and beyond are in the realm of emerging technology. Nevertheless, future developments seem primed to enable photodiode and modulator technology operating at frequencies extending to multiple hundreds of gigahertz. Further improvements in efficiency will enable millimeter-wave photonic communications at frequencies well beyond 110 GHz. Even with these developments in component technology, cost and integration are likely to remain significant challenges for millimeter-wave photonic systems. Integrated photonics, where multiple photonic technologies are combined on a single substrate, represents an ideal path toward alleviating these challenges and enabling systems where the electronic–photonic interface is much less pronounced than in current systems. Combined with integrated millimeter-wave advancements, future millimeter-wave photonic wireless systems may be made significantly smaller than the discrete component-based systems common today.

REFERENCES

1. N.C. Currie and C.E. Brown, *Principles and Applications of Millimeter-Wave Radar*. Boston, MA: Artech House, 1987.
2. A.R. Thompson, J.M. Moran, and G.W. Swenson, Jr, *Interferometry and Synthesis in Radio Astronomy*, New York: John Wiley & Sons, 2001.

3. J.A. Nanzer, *Microwave and Millimeter-Wave Remote Sensing for Security Applications*. Norwood, MA: Artech House, 2012.
4. M.C. Gross, P.T. Callahan, T.R. Clark, D. Novak, R.B. Waterhouse, and M.L. Dennis, Tunable millimeter-wave frequency synthesis up to 100 GHz by dual-wavelength Brillouin fiber laser, *Optics Express*, 18, 13321–13330, 2010.
5. J.A. Nanzer, P.T. Callahan, M.L. Dennis, T.R. Clark, D. Novak, and R.B. Waterhouse, Millimeter-wave wireless communication using dual-wavelength photonic signal generation and photonic upconversion, *IEEE Transactions on Microwave Theory and Techniques*, 59(12), 3522–3530, 2011.
6. M. Weiss, A. Stohr, F. Lecoche, and B. Charbonnier, 27 Gbit/s photonic wireless 60 GHz transmission system using 16-QAM OFDM, in *International Topical Meeting on Microwave Photonics, 2009. MWP '09*, Valencia, Spain, 2009, pp. 1–3.
7. F.M. Kuo, C.B. Huang, J.W. Shi, C. Nan-Wei, H.P. Chuang, J.E. Bowers, and P. Ci-Ling, Remotely up-converted 20-Gbit/s error-free wireless on–off-keying data transmission at W-band using an ultra-wideband photonic transmitter-mixer, *IEEE Photonics Journal*, 3(2), 209–219, 2011.
8. A. Stohr, S. Babiel, P.J. Cannard, B. Charbonnier, F. Van-Dijk, S. Fedderwitz, D. Moodie et al., Millimeter-wave photonic components for broadband wireless systems, *IEEE Transactions on Microwave Theory and Techniques*, 58(11), 3071–3082, 2010.
9. T.R. Clark, S.R. O'Connor, and M.L. Dennis, A phase-modulation I/Q demodulation microwave-to-digital photonic link, *IEEE Transactions on Microwave Theory and Techniques*, 58(11), 3039–3058, 2010.
10. J. Kim, M.J. Park, M.H. Perrott, and F.X. Kärtner, Photonic subsampling analog-to-digital conversion of microwave signals at 40-GHz with higher than 7-ENOB resolution, *Optics Express*, 16(21), 16509–16515, October 2008.
11. S.J. Strutz and K.J. Williams, An 8–18-GHz all-optical microwave downconverter with channelization, *IEEE Transactions on Microwave Theory and Techniques*, 49(10), 1992–1995, 2001.
12. C. Middleton and R. DeSalvo, Balanced coherent heterodyne detection with double sideband suppressed carrier modulation for high performance microwave photonic links, in *Avionics, Fiber-Optics and Phototonics Technology Conference, 2009. AVFOP '09. IEEE*, San Antonio, TX, 2009, pp. 15–16.
13. T.P. McKenna, J.A. Nanzer, M.L. Dennis, and T.R. Clark, Fully fiber-remoted 80 GHz wireless communication with multi-subcarrier 16-QAM, *IEEE Photonics Conference*, San Francisco, CA, 2012.
14. P.G. Huggard, B.N. Ellison, P. Shen, N.J. Gomes, P.A. Davies, W.P. Shillue, A. Vaccari, and J.M. Payne, Efficient generation of guided millimeter-wave power by photomixing, *IEEE Photonics Technology Letters*, 14(2), 197–199, 2002.
15. B.S. Williams, Terahertz quantum-cascade lasers, *Nature Photonics*, 1(9), 517–525, 2007.
16. M. Tonouchi, Cutting-edge terahertz technology, *Nature Photonics*, 1(2), 97–105, 2007.
17. M.C. Gross, T.R. Clark, and M.L. Dennis, Narrow-linewidth microwave frequency generation by dual-wavelength Brillouin fiber laser, in *21st Annual Meeting of the IEEE Lasers and Electro-Optics Society, 2008. LEOS 2008*, Newport Beach, CA, 2008, pp. 151–152.
18. J.A. Nanzer, P.T. Callahan, M.L. Dennis, and T.R. Clark, Photonic signal generation for millimeter-wave communications, *Johns Hopkins APL Technical Digest*, 30(4), 299–308, 2012.
19. H. Takahashi, T. Kosugi, A. Hirata, and K. Murata, Supporting fast and clear video, *IEEE Microwave Magazine*, 13(6), 54–64, 2012.
20. C. Lim, A. Nirmalathas, M. Bakaul, K.-L. Lee, D. Novak, and R. Waterhouse, Mitigation strategy for transmission impairments in millimeter-wave radio-over-fiber networks [Invited], *Journal of Optical Networking*, 8(2), 201–214, February 2009.

21. E.L. Wooten, K.M. Kissa, A. Yi-Yan, E.J. Murphy, D.A. Lafaw, P.F. Hallemeier, D. Maack et al., A review of lithium niobate modulators for fiber-optic communications systems, *IEEE Journal of Selected Topics in Quantum Electronics*, 6(1), 69–82, 2000.
22. A. Karim and J. Devenport, High dynamic range microwave photonic links for RF signal transport and RF-IF conversion, *Journal of Lightwave Technology*, 26(15), 2718–2724, 2008.
23. P.W. Juodawlkis, J.J. Hargreaves, R.D. Younger, G.W. Titi, and J.C. Twichell, Optical down-sampling of wide-band microwave signals, *Journal of Lightwave Technology*, 21(12), 3116–3124, 2003.
24. M.B. Airola, S.R. O'Connor, M.L. Dennis, and T.R. Clark, Experimental demonstration of a photonic analog-to-digital converter architecture with pseudorandom sampling, *IEEE Photonics Technology Letters*, 20(24), 2171–2173, 2008.
25. A. Khilo, S.J. Spector, M.E. Grein, A.H. Nejadmalayeri, C.W. Holzwarth, M.Y. Sander, M.S. Dahlem et al., Photonic ADC: Overcoming the bottleneck of electronic jitter, *Optics Express*, 20(4), 4454–4469, February 2012.
26. T.P. McKenna, J.A. Nanzer, and T.R. Clark, Photonic downconverting receiver using optical phase modulation, in *Microwave Symposium Digest (MTT), 2014 IEEE MTT-S International*, Tampa Bay, FL, 2014, pp. 1–3.
27. J.G. Proakis, *Digital Communications*, 4th edn., McGraw-Hill, New York, 2001.
28. T.R. Clark, T.F. Carruthers, P.J. Matthews, and I.N. Duling, III, Phase noise measurements of ultrastable 10 GHz harmonically modelocked fibre laser, *Electronics Letters*, 35(9), 720–721, 1999.
29. J. Macario, P. Yao, S. Shi, A. Zablocki, C. Harrity, R.D. Martin, C.A. Schuetz, and D.W. Prather, Full spectrum millimeter-wave modulation, *Optics Express*, 20(21), 23623–23629, October 2012.
30. T.P. McKenna, J.A. Nanzer, M.L. Dennis, and T.R. Clark, Mixerless 40 GHz wireless link using a photonic upconverting transmitter and a photonic downsampling receiver, in *IEEE MTT-S International Microwave Symposium Digest (IMS)*, Seattle, WA, 2013, pp. 1–3.
31. T.P. Mckenna, J.A. Nanzer, and T.R. Clark, Experimental demonstration of photonic millimeter-wave system for high capacity point-to-point wireless communications, *IEEE Journal of Lightwave Technology*, 32(20), 3588–3594, 2014.

10 CMOS Millimeter-Wave PA Design

Kai Kang, Dong Chen, and Kaizhe Guo

CONTENTS

10.1 INTRODUCTION

With the acquaintance of the characteristics of the electromagnetic wave at the millimeter wave (mm-wave) band and the development of fabrication technology, the applications at the mm-wave band attract people to develop standards and circuits. The 7 GHz ISM band around 60 GHz makes the multi gigabit-per-second (Gbps) communication possible. The high attenuation of 60 GHz electromagnetic waves in oxygen enables the cooperation of more links in the same area with minimal interference, which greatly enhances the utilization rates of the frequency band. And this characteristic also makes the communication channels more secure [1–3]. At the same time, the 77–79 GHz is a proper band to realize highly accurate automobile radar. These characteristics drive people to develop high-performance mm-wave front end to realize these interactive applications. Among all, mm-wave power amplifiers in the front end have great influence on the performance of the system because the power amplifiers usually consume most of the power and occupy a large quantity of chip area. In the past years, power amplifiers were usually made with an individual chip with III–V compounds technology. But with the development of the CMOS technology, the power amplifiers can be made in one chip together with the other parts of the transceiver, which makes the system on chip (SOC) possible.

Many designers have already given some design examples of power amplifiers working in the mm-wave band [4–33]. This chapter discusses the design challenges of mm-wave power amplifiers in CMOS technology and provides a review of circuit and architecture techniques that try to solve these problems. Section 10.2 of this chapter describes the challenges in the design of power amplifiers using CMOS technology. Section 10.3 will introduce some optimizations about the model and layout of active and passive devices. Section 10.4 will present some power amplifier topologies with relatively high performance.

10.2 CHALLENGE

10.2.1 INSUFFICIENT SPEED OF TRANSISTORS

Though the feature size of CMOS technology is scaled down fast, which makes the design of mm-wave circuits easier, the characteristic frequency (f_T) of the transistor is still not enough. The f_T of 90 nm generations reaches 110 GHz [34], which is already beyond 60 GHz, but the transistors still cannot give a significant power gain at the mm-wave frequencies band. However, multistage may be needed to provide sufficient power gain. Many perfect theories, such as the switch mode power amplifier and the harmonic waves control power amplifier, which have shown pretty good performance at the RF band, still cannot achieve a good result at the mm-wave band.

10.2.2 INACCURACY OF MODELS

Most of the transistor modeling work is based on the measurement of fabricated devices. However, due to errors from inaccurate de-embedding becoming quite large at the mm-wave band, it is very hard to obtain an accurate measurement [2]. Especially, for the power amplifier, not only the small signal transistor model but also the large signal model must be accurate.

10.2.3 LOW BREAKDOWN VOLTAGE

Low breakdown voltage is one of the obstacles in designing the power amplifier. The drain-source voltage is usually not more than 1.2 V for 90 nm CMOS transistors, and for 65 nm, it is not more than 1 V. It extremely limits the output power of the amplifier. Large current and transistor size are needed to get a high output power, which results in a higher junction temperature [21]. It increases the useless power consumed, degrades the efficiency of the power amplifier, and even damages the transistors. The problem is exacerbated as the CMOS transistor's minimum feature size is scaled down.

10.2.4 LOW-RESISTIVITY SUBSTRATE

A low-resistivity substrate, with the resistivity of ~10 Ω cm in modern standard silicon progress [3], seriously influences the quality factor of the passive components [9] and the efficiency of the power amplifier. Because of the low-resistivity characteristic, the signals coupling to the substrate cause a significant loss, especially at the mm-wave band [3]. The quality factor of spiral inductors is usually not more than 30 [2].

10.3 DEVICE OPTIMIZATION

10.3.1 Transistor Modeling and Layout Optimization

As depicted in Section 10.2, the accuracy of transistor modeling is usually important. Many designers try to find an accuracy transistor model used for the design of power amplifiers. Some commercial transistor models, such as the BSIM model, EKV model, and IBIC model, have good accuracy of the case of dc character and small signal performance, but the accuracy of large signal behavior still needs to be improved.

The dc power consumed by the power amplifier is usually very high, so the temperature of the amplifier circuits is often far beyond the room temperature. The performance of transistors usually changes with the temperature, so the model of the transistors should be different.

A temperature-dependent scalable transistor model [4] is shown in Figure 10.1. Every parasitic or model parameter is expressed as a function of temperature (T) and device width (W) in the form

$$F(T, W) = a + bT + cW + dT^2 + eW^2 + fTW \qquad (10.1)$$

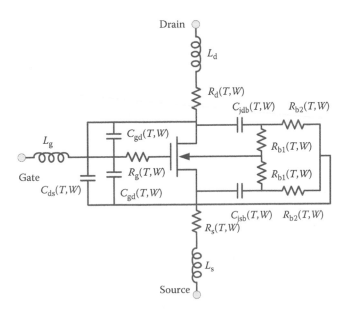

FIGURE 10.1 A temperature-dependent scalable transistor model. (From Dawn, D. et al., 60 GHz CMOS power amplifier with 20-dB-gain and 12 dBm P$_{sat}$, in: *IEEE International Microwave Symposium Digest*, Boston, MA, pp. 537–540, June 2009.)

The dc I–V characteristics and S-parameters up to 65 GHz across a temperature range of –25°C to +80°C for various device sizes from 20 to 80 μm gate width are measured. The de-embedding error is kept below 2% to achieve high accuracy, and significant numbers of chips are measured to account for chip-to-chip variation. The power amplifier designed using the temperature-dependent model shows a good performance over the entire temperature range between –10°C and +80°C.

For the practical use of the power amplifier, the assurance of reliability is very important when the power amplifiers (PA) work in large signal mode [5]. In microwave frequencies, hot carrier injection (HCI) is a dominant mechanism of the power performance degradation [35]. To evaluate the performance degradation, substrate current needs to be known [36]. From the dc test and RF load test, lifetime can be calculated [5]. First, from the dc reliability tests, we can get an expression relating lifetime (T_{50}), where the drain currents in 50% of the transistors decrease by 10%, and substrate current under dc operation. Second, through the RF load circle, the substrate current can be calculated and accumulated. Finally, from the comparison of the average substrate current and that under dc operation, we can calculate the lifetime.

Accurate measurement in the mm-wave frequency band is relatively difficult, which will directly influence the accuracy of models. A major source of inaccuracy is the de-embedding procedure. The inaccuracies are higher when the frequency is higher.

A model-based de-embedding approach dubbed *recursive modeling* is designed to resolve this problem [37]. The modeling is done in a recursive way of multiple steps. At the beginning step of this method, a ground probing pad based on an equivalent circuit is modeled. Then the transmission line is modeled based on the pad models. A transmission line model including the complex characteristic impedance and the propagation constant is used in the transmission line model. Finally, the two models, of the pad and the transmission line, are used for the measurement of the device under test (DUT). The result obtained in each step is used in the next step, so it is called *recursive modeling*. This step improves the accuracy of de-embedding the procedure and makes the result of the other DUTs more dependable.

As the operating frequency gets closer to the transistor cut-off frequency, device parasitics extremely reduce the power gain. Layout optimization is very important to minimize the parasitics of the transistor cell, especially the gate resistance (rg) and gate-to-drain capacitance (Cgd) [37]. Besides, interconnects around the transistor introduce parasitic resistances and inductances, which are not scaled with technology, and degenerate the transistor performance because of the use of large transistors in PA [7].

A new structure using *round-table* connection [37] for the device is also proposed as shown in Figure 10.2. The transistor cells are connected in a matrix or circular fashion. This structure uses external double-contacts (between cells) and multipath connections between sources and drain of the subcells in order to decrease the finger resistance of the device. Devices of several dimensions were fabricated in the 90 nm CMOS process. Measurements were carried out up to 65 GHz.

Both the speed and the desired gain of these devices as compared to regular RF transistors with the same number of fingers are found to be improved from the measurement result. The H21, Mason gain and the maximum stable gain (MSG) of

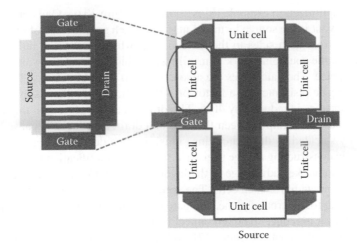

FIGURE 10.2 Conceptual picture of a round-table device. (From Heydari, B. et al., *IEEE J. Solid-State Circ.*, 42(12), 2893, December 2007.)

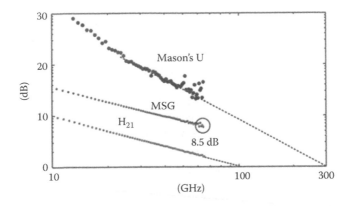

FIGURE 10.3 H_{21}, U, and MSG for a 40 μm round-table device. (From Heydari, B. et al., *IEEE J. Solid-State Circ.*, 42(12), 2893, December 2007.)

a 40 μm transistor using the *round-table* structure in 90 nm CMOS are measured and are shown in Figure 10.3. The f_{max} is improved by about two times, with the f_T remaining almost constant (~100 GHz). The gate resistance is reduced to half of that of a regular transistor. But the source resistance and the capacitance between the gate and the source are increased.

However, for typical power amplifier, the size of transistor is usually very large to provide enough ac power, which will expand the center space of this type of layout. The long interconnects between the transistors will degrade the performance of the transistors. And additional signal loss will occur because the signal delay in the gate and drain path is difficult to balance [6].

To achieve good transistor performance in high frequency, well grounding the source and bulk terminal of the transistor is very important [8].

A layout optimization example [6] is shown in Figure 10.6a. One core optimization of this layout is the source terminal network. Each source node of the transistor is connected from both sides to the ground ring, which is also the bulk ring. The impedance in the source network is minimized, which reduces the voltage drop in the ground plane. The gate nodes are also connected from both sides. This reduces the effective gate resistance by a factor of 4 theoretically [37], and simulation predicts the resistance is reduced by 40%. The drain lines are placed on the top of the transistor, and the overlap between the gate and drain is reduced because the lines are partially shielded by the source line. This, in some degree, degrades the Miller effect. Easily constructing a large transistor for power amplification with minimized parasitic is another key advantage of the layout shown in Figure 10.4a. Figure 10.4b shows the layout with five such cells. The source networks of each unit cell are connected together to the ground plane, and gate and terminals are connected through a thick metal layer to reduce the interconnect resistance.

An *interleaved* layout technique [7] optimized for a differential power amplifier using a creative idea is shown in Figure 10.5. This layout technique also aims to reduce the resistance of the source terminal. The layout is optimized to convert the source degeneration network seen by each transistor to the common-mode impedance.

Figure 10.5a shows the basic principle: by overlapping the source areas of the differential transistors, each source terminal carries the sum of two differential currents, and the interconnect resistance is converted as a common-mode degeneration network. A unit cell consisting of two gate fingers of each transistor is drawn and then large transistors with several unit cells are formed. Figure 10.5b depicts the transformation. Only common-mode currents flow through the source terminals, and the differential operation is not affected. The structure is also easy to be duplicated in the vertical dimension and will not introduce much interconnects.

10.3.2 Passive Component Optimization

For the circuits operating in the mm-wave band, passive components, such as inductor, transformer, and transmission line, are necessary. But the passive components in CMOS technology usually don't have a very good quality factor due to the high loss of the substrate [9]. Some methods have been researched to reduce the influences of the substrate.

Transmission line is one kind of commonly used passive components in mm-wave circuit design, which can be used to implement resonant tanks, impedance matching networks, signal splitters, couplers, balun, and transformers [11]. A transmission line is easier to use in modeling transpiral inductors because it substantially confines the electric and magnetic fields [10]. However, the transmission lines used in the circuits usually have a long length.

Using a microstrip transmission line is one way to isolate the influence of the substrate. A folded microstrip geometry [10] to alleviate layout difficulties is shown in Figure 10.6. In this structure, signal lines are realized using the thick metal, and the bottom metal is used as ground plane. The two ends of the line are near each other, which simplifies the layout.

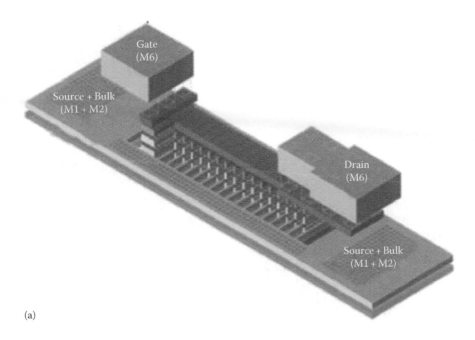

(a)

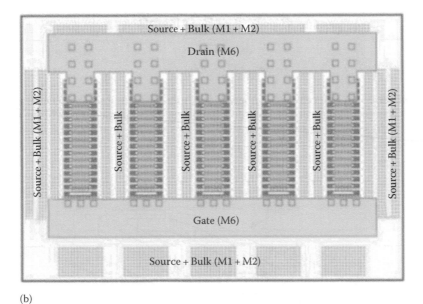

(b)

FIGURE 10.4 Optimized transistor layout [6] for power amplifiers. (a) The layout of a single transistor. (b) The layout of five transistors.

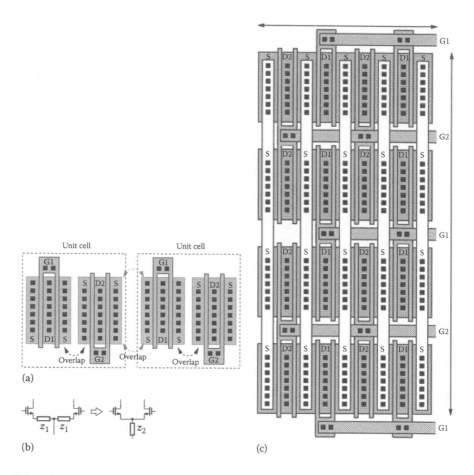

FIGURE 10.5 (a) Unit cells in the proposed structure, (b) transformation of the degeneration impedances, and (c) combined cells in a complete differential pair. (From Liang, C. and Razavi, B., A layout technique for millimeter-wave PA transistors, in: *Proceedings of RFIC Symposium*, Montréal, Québec, Canada, p. 1–4, June 2011.)

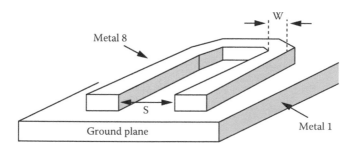

FIGURE 10.6 The 3D view of the folded microstrip. (From B. Razavi, *IEEE J. Solid-State Circ.*, 41(1), January 2006.)

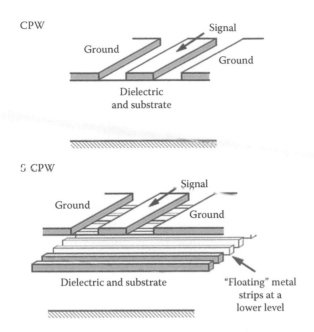

FIGURE 10.7   On-chip transmission line. (From Cheung, T.S.D. et al., On-chip interconnect for mm-wave applications using an all-copper technology and wavelength reduction, in: *IEEE ISSCC*, San Francisco, CA, pp. 396–501, February 2003.)

High characteristic impedance (Z_0) transmission lines are needed to realize low loss matching networks. Keeping large space between signal path and ground is one way to form a low loss and high Z_0 transmission line. But it is hardly possible to realize using microstrip line in most technologies, where the space between different metal layers is 3–10 μm. Though narrowing the signal conductor can raise Z_0, it will also increase the attenuation [11]. The coplanar waveguide (CPW, Figure 10.7) consisting of a center signal and grounds in the plane is one way to solve this problem. Wider conductors can be used to reduce the loss, but electromagnetic waves will be coupled to the substrate, which will increase the attenuation [12]. The slow-wave coplanar waveguide (S-CPW in Figure 10.7) has an advantage of overcoming these limitations. An S-CPW consists of typical CPW conductors, and floating metal shields are placed under the signal conductors vertically, which will minimize the energy coupled to the substrate. At the same time, the wavelength in the S-CPW is lower than the other configuration, which will cost less chip area. This is verified by measurement [11].

Another kind of substrate-shielded coplanar waveguide [12] is shown in Figure 10.8. This is another version of the slow-wave coplanar structure [11]. Figure 10.8c shows the substrate-shielded coplanar structure that is a combination of the microstrip structure and the CPW structure. The bottom plate with many slots forces the return current to be mostly concentrated in the coplanar ground lines and insulates the coupling between the signal line and the substrate.

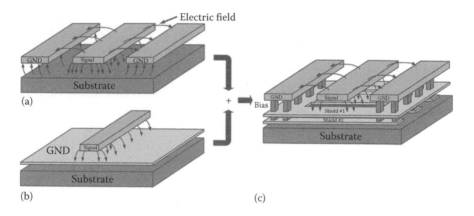

FIGURE 10.8 Combination of (a) CPW and (b) microstrip structures to realize (c) substrate-shielded CPW structure. (From Hasegawa, H. et al., *IEEE JMTT*, MTT-19(11), 869, November 1971.)

10.4 POWER AMPLIFIER TOPOLOGIES

10.4.1 CURRENT COMBINING POWER AMPLIFIER

According to the Federal Communications Commission (FCC) regulations, the radiation power for 60 GHz systems can be as much as 40 dBm [15]. However, due to device limitations such as transistor speed and low breakdown voltage, it is difficult to provide very high power for the power amplifier using the CMOS process. Power combining is necessary, and many different methods can be applied. Wilkinson combiner is a simple way to combine the output power of two power amplifiers together. There is an example of a power amplifier [16] using Wilkinson dividers and combiners, the block diagram of which is shown in Figure 10.9.

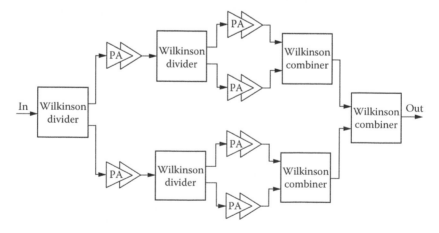

FIGURE 10.9 Block diagram of the 100 mW 60 GHz CMOS power amplifier. (From Law, C.Y. and Pham, A.-V., A high-gain 60 GHz power amplifier with 20 dBm output power in 90 nm CMOS, in: *IEEE ISSCC*, San Francisco, CA, pp. 426–427, 2010.)

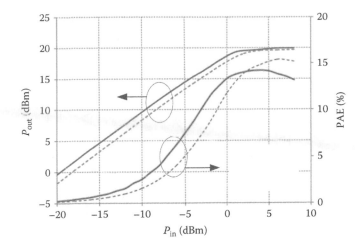

FIGURE 10.10 Measured and simulated output power and power-added efficiency of the PA. (From Law, C.Y. and Pham, A.-V., A high-gain 60 GHz power amplifier with 20 dBm output power in 90 nm CMOS, in: *IEEE ISSCC*, San Francisco, CA, pp. 426–427, 2010.)

This power amplifier achieves +18 dBm of $P_{1\ dB}$ and 20 dBm of P_{sat} at 60 GHz and the power-added efficiency (PAE) of the PA peaks at 14.2% when P_{in} is around 4 dBm. The output power and PAE versus P_{in} is depicted in Figure 10.10.

To obtain high output power, combining the output of a large number of power amplifiers together is the common method used. However, for the typical Wilkinson power combiner, with the number of power amplifiers getting larger, the loss of the power combiner will greatly influence the performance of the whole circuit. The power combiner will also cost a large area. In most of the conditions, the combiner has the same degree inputs. So the quarter-wavelength transmission lines in the Wilkinson combiner, which give a good insulation between ports, are not necessary. If we use shorter lines instead of quarter-wavelength lines, just to realize the function of power combiner, we can save much of the chip area and reduce the loss of the combiner and achieve impedance transformation [17].

Figure 10.11 shows a 16-way zero-degree combiner [17] that achieves low insertion loss and wideband impedance transformation. The combiner combining the power of 16 on-chip PAs achieves an output power of 0.7 W with a PAE of 10% at 42 GHz and a −3 dB bandwidth of 9 GHz. The block diagram of the 16-way power-combined PA is shown in Figure 10.11, and the schematic of each power amplifier is shown in Figure 10.12.

10.4.2 Transformer-Based Power Combining

Transformers have the advantage of being very compact. A transformer power combiner can be consisted of series power combining, as shown in Figure 10.13. The transformer usually includes N primaries and one secondary coil. The voltage on the secondary of the transformer power combiner is the sum of the voltages on all the primaries. On the other hand, current in the secondary is identical with that

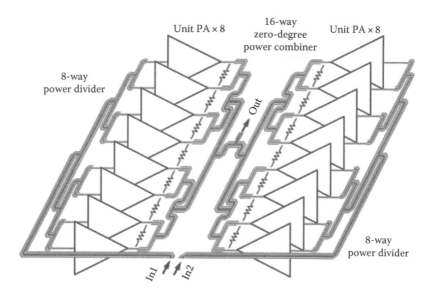

FIGURE 10.11   Block diagram of the high-power PA. (From Tai, W. et al., A 0.7 W fully integrated 42 GHz power amplifier with 10% PAE in 0.13 μm SiGe BiCMOS, in: *IEEE ISSCC*, San Francisco, CA, pp. 142–143, 2013.)

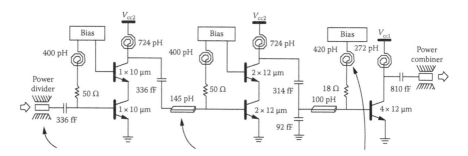

FIGURE 10.12   Schematic of each unit PA. (From Tai, W. et al., A 0.7 W fully integrated 42 GHz power amplifier with 10% PAE in 0.13 μm SiGe BiCMOS, in: *IEEE ISSCC*, San Francisco, CA, pp. 142–143, 2013.)

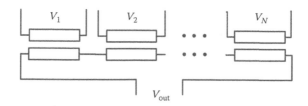

FIGURE 10.13   Transformer combiner with series power combining.

in the primaries if all primaries carry the same current. Thus, a transformer power combiner can connect several PA cells together in series. When N PA cells are connected in series using a transformer power combiner with N primaries, the output voltage swing will be added together and increases by N times. To achieve greater voltage and power enhancement ratio, more input is needed.

An example of a transformer combiner [18] that combines two ways of the power amplifier and enhances the output power of the whole circuit is shown in Figure 10.14. With 1 V supply voltage, the measured 1 dB gain compressed output power ($P_{1\,dB}$) is 15 dBm and the saturated output power (P_{sat}) is 18.6 dBm. The measured peak PAE is 15.1%, and the peak drain efficiency is 16.4%. At the saturated output power level, the amplifier still has 11 dB of power gain.

A power amplifier [19] in Figure 10.15 gives a more complex example, which uses two transformer-based combiners and a transmission-line combiner to combine eight unit power amplifier cells together and gets more output power and higher performance. The power amplifier uses a balun at the output port to convert the differential signal to the single end signal. The 1 dB gain compression output power ($P_{1\,dB}$) is 16.4 dBm and the saturated output power (P_{sat}) is 19.3 dBm with the peak PAE of 19.2% at 79 GHz. The PA provides P_{sat} of more than 19 dBm and $P_{1\,dB}$ of 16 dBm from 77 to 81 GHz.

As we can see, the differential to single end output transformer is not a fully symmetric structure for the single end coil. So the load of the differential transistors is

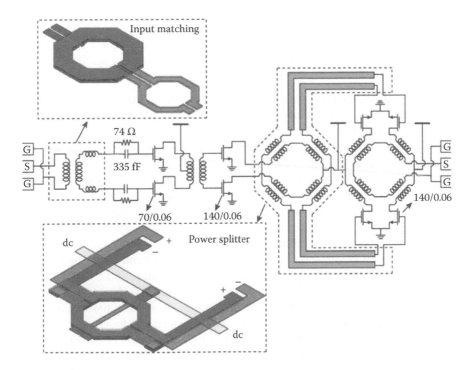

FIGURE 10.14 Schematic of the transformer power combiner PA. (From Chen, J. and Niknejad, A.M., A compact 1V 18.6 dBm 60 GHz power amplifier in 65 nm CMOS, in: *IEEE ISSCC*, San Francisco, CA, pp. 432–433, 2010.)

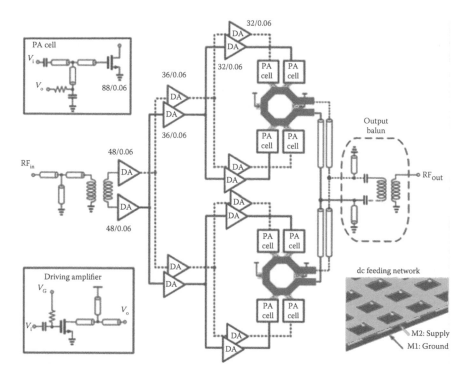

FIGURE 10.15 PA based on two transformers. (From Wang, K.-Y. et al., A 1 V 19.3 dBm 79 GHz power amplifier in 65 nm CMOS, in: *IEEE ISSCC*, San Francisco, CA, pp. 260–261, 2012.)

not the same. This often leads to the drop of the performance of the power amplifier, because the common-mode current is generated. A fully symmetric differential to single end balun [20] is introduced in Figure 10.16.

The output pad is creatively put inside the balun. The secondary coil is split into two windings and connected to the pad in a centrosymmetric way. This minimizes the unbalance seen from each port. This is an ideal solution to implement a fully symmetric balun, but in practical application, we must connect the output pad outward, for example, the GSG probe or the bonding wire. It is difficult to put the whole GSG pad inside the transformer, and the bonding wire will also introduce unbalance unless the flip-chip packing technique is applied.

The rectangular distributed active transformer (DAT) with push–pull amplifiers is an evolution of transformer power combiner as shown in Figure 10.17. The DAT combines several push–pull amplifiers in a more efficient way [21]. The dc and the ac in the DAT are depicted in Figure 10.18. The dc supplies are added at the common node of the transformer, and ac goes around the circle of the transformer. A transformer power combiner with four primaries can deliver four times higher output power.

A 2 W 2.4 GHz single-stage fully integrated DAT switching power amplifier in class E/F has been fabricated and measured using 0.35 μm CMOS transistors in a BiCMOS technology [21]. Driving a balanced load, an output power of 1.9 W at 2.4 GHz is obtained with 8.7 dB gain, using a 2 V power supply. The corresponding

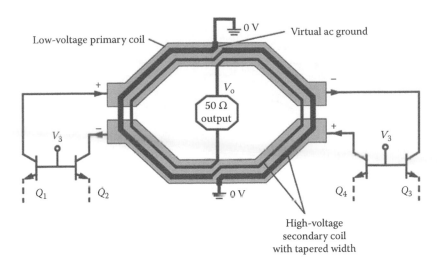

FIGURE 10.16 A symmetric balun. (From Cheung, T.S.D. and Long, J.R., *IEEE J. Solid-State Circ.*, 40(12), 2583, December 2005.)

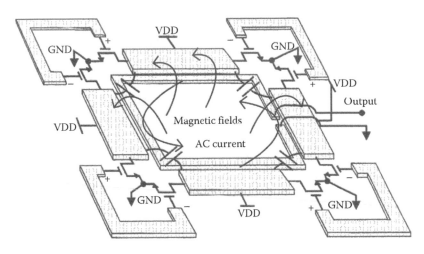

FIGURE 10.17 Distributed active-transformer amplifier with four push–pull blocks. (From Aoki, I. et al., *IEEE J. Solid-State Circ.*, 37(3), 371, March 2002.)

PAE is 41% and drain efficiency is 48%. The amplifier can also drive a single-ended load, achieving a PAE of 31% with an output power of 2.2 W (33.4 dBm), gain of 8.5 dB, and drain efficiency of 36%. It can also produce 450 mW using a 1 V supply with a PAE of 27%.

Another example working at 60 GHz of the DAT is shown in Figure 10.19 [22], and the schematic of the power amplifier is shown in Figure 10.20. When 1 V VDD is applied, 17.9 dBm P_{sat} and 15.4 dBm OP_{1d} are achieved. Including all DAs and UAs, the peak PAE at 1 V VDD is 11.7%.

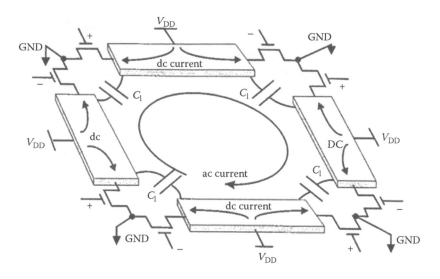

FIGURE 10.18 dc and ac in a DAT. (From Aoki, I. et al., *IEEE J. Solid-State Circ.*, 37(3), 371, March 2002.)

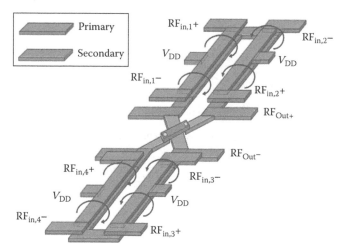

FIGURE 10.19 Another DAT example. (From Lai, J.-W. and Valdes-Garcia, A. A 1 V 17.9 dBm 60 GHz power amplifier in standard 65 nm CMOS, in: *IEEE ISSCC*, San Francisco, USA, pp. 424–425, 2010.)

A stack-up transformer [23] with ground shields is used in the example depicted in Figure 10.21 to reduce the loss of the DAT and achieve a good coupling factor. The transformer has a *sandwich-like* structure. The primary inductor is stacked vertically above the secondary inductor. A ground shield is located under the transformer, which achieves a coupling factor of $k = 0.8$.

Figure 10.22 shows a 3D view of the sandwich-like transformer structure. The top metal is used as the primary coil of the transformer and the other top metal is used as the secondary coil. The dc supply of the amplifiers is from the

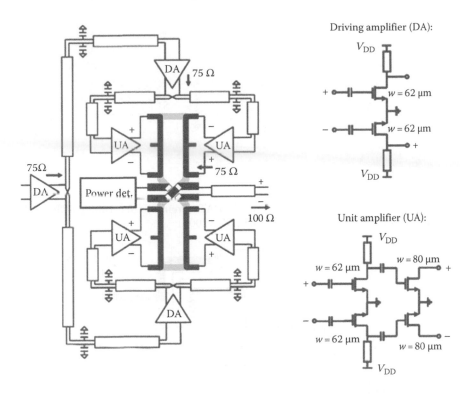

FIGURE 10.20 Schematic of the PA. (From Lai, J.-W. and Valdes-Garcia, A. A 1 V 17.9 dBm 60 GHz power amplifier in standard 65 nm CMOS, in: *IEEE ISSCC*, San Francisco, CA, pp. 424–425, 2010.)

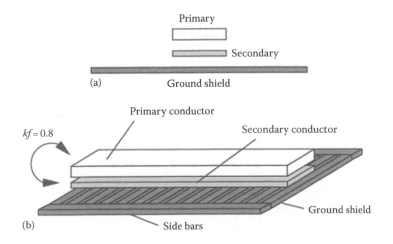

FIGURE 10.21 Stack-up transformer. (a) side view and (b) 3D view. (From Pfeiffer, U.R. and Goren, D., *IEEE Trans. Microw. Theory Tech.*, 55(5), 857, May 2007.)

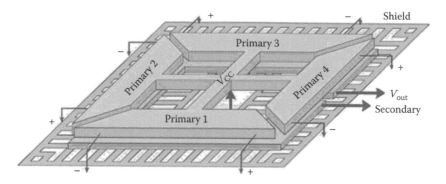

FIGURE 10.22 3D view of the *sandwich-like* DAT. (From Pfeiffer, U.R. and Goren, D., *IEEE Trans. Microw. Theory Tech.*, 55(5), 857, May 2007.)

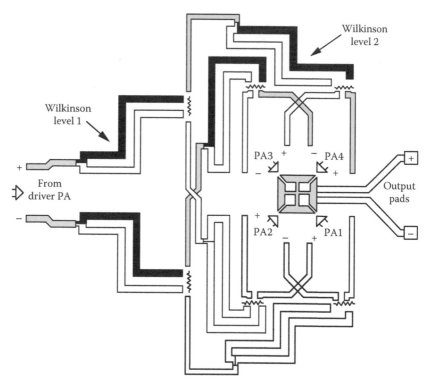

FIGURE 10.23 Schematic of the power amplifier using the *sandwich-like* DAT. (From Pfeiffer, U.R. and Goren, D., *IEEE Trans. Microw. Theory Tech.*, 55(5), 857, May 2007.)

center of the transformer. The schematic of the power amplifier is shown in Figure 10.23. The PA was designed using a 0.13 μm SiGe BiCMOS technology with cutoff frequencies f_{max}/f_T = 240/200 GHz. The output power and PAE versus input power are shown in Figure 10.24.

These power amplifiers give some good examples of the usage of a DAT. But two problems still exist. First, it is very difficult to combine a lot of unit power amplifiers

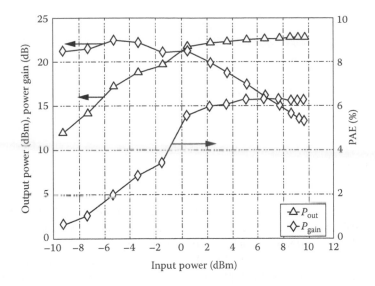

FIGURE 10.24 Output power and PAE versus input power of the proposed. (From Pfeiffer, U.R. and Goren, D., *IEEE Trans. Microw. Theory Tech.*, 55(5), 1054, May 2007.)

together using a rectangular transformer. Figure 10.26 gives a simple example of extending the quantity of the unit power amplifiers. But the adjacent primary windings carry currents in opposite directions [24], which will greatly influence the performance of the transformer. Second, the input impedance of each primary wing is unequal, which generates the common-mode signal and limits the maximum output power [21]. The structure of the traditional transformer power combiner is shown in Figure 10.25. The voltage on the secondary changes from V to 0 from the differential

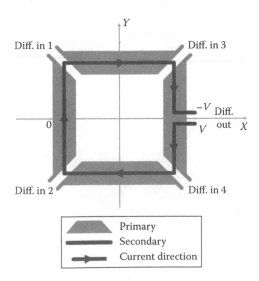

FIGURE 10.25 Traditional DAT structure.

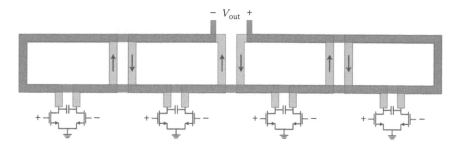

FIGURE 10.26 A simple transformer able to combine a lot of unit PAs. (From Haldi, P. et al., *IEEE J. Solid-State Circ.*, 43(5), May 2008.)

output terminals to the virtual ac ground as shown in Figure 10.25. Therefore, the voltage distribution on the secondary is not symmetrical at the Y-axis. Asymmetric voltage leads to asymmetric coupling between primary and secondary. As a result, differential input impedance at the four differential input ports is asymmetric at the Y-axis. It means that differential input impedances at ports 1 and 2 are different from differential input impedances at ports 3 and 4. Both of the shortcomings aggravate when the circuits are working at the mm-wave band.

A figure 8 power combiner architecture [24] is proposed to solve the first problem. Figure 10.27 shows a single ring of the proposed transformer, and Figure 10.28 shows a simple power-combined PA schematic using the proposed transformer. The primary coils are driven by push–pull amplifiers, and the secondary coil is implemented in alternating orientation, just like "8." This layout minimizes the cancellation effect caused by the adjacent windings because the currents in the primaries of

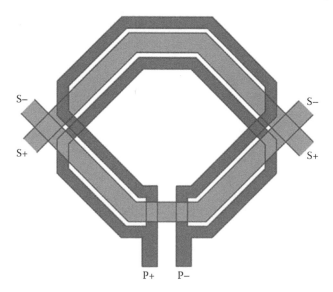

FIGURE 10.27 A single ring of proposed *figure 8* power combiner. (From Haldi, P. et al., *IEEE J. Solid-State Circ.*, 43(5), May 2008.)

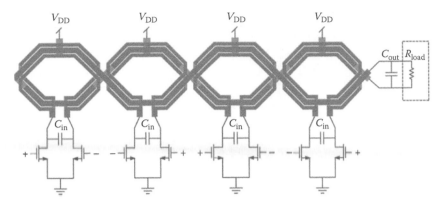

FIGURE 10.28 Simplified schematic of the PA with the implemented *figure 8* power combining network. (From Haldi, P. ct al., *IEEE J. Solid-State Circ.*, 43(5), May 2008.)

the two adjacent stages flow in the same direction. This significantly improves the performance and also reduces the common-mode coupling. The use of the two parallel coils for each primary winding on either side of each secondary winding mitigates the current crowding effect. The current in the secondary windings is spread more uniformly and the loss is therefore reduced.

A way to cope with the asymmetric nature of a DAT is to insert auxiliary components and implement different transistor dimensions for cach combining stage [25]. This is a simple way to solve the problem.

From the idea of the symmetrical balun in Figure 10.28, fully symmetrical power combiner can also be designed. Figure 10.41 shows the structure of an eight-way fully symmetrical DAT. The secondary consists of two parallel windings with an identical structure but opposite direction. The current on two parallel windings flows in the same direction. As the voltage on secondary of the traditional transformer power combiner, the voltage on each wind of the secondary in this transformer power combiner is also unevenly distributed. However, the uneven voltage distributions on two identical windings balance out. Thus, voltage coupling between primaries and secondary is symmetrical to the *Y*-axis. The two parallel secondary windings connect together at the center of the transformer power combiner. Combined power goes from this center connection point to the output port through a differential transmission line. To make the transformer symmetric to the *X*-axis, a floating dummy differential transmission line is designed as shown in Figure 10.29. Symmetries to both the *X*-axis and *Y*-axis lead to balance of the differential input impedance at the four input ports of the primaries.

As shown in Figure 10.30, the primaries of the transformer are designed in the top metal layer with ultrathick metal and the secondary is allocated one layer lower. dc supplies are applied at the virtual ac ground points at the center of each of the four primaries. Simulation results demonstrate that the 100 Ω differential load is transformed to the optimal load impedance of the last power stage at 60 GHz, and the uniformities of the real part and imaginary parts of the four differential input impedances are better than 5% and 7%, respectively. The simulated loss of the transformer is 1 dB at 60 GHz.

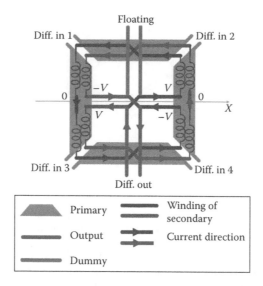

FIGURE 10.29 Diagrammatic sketch of a fully symmetrical eight-way transformer combiner.

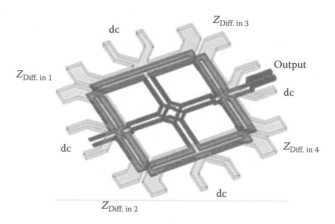

FIGURE 10.30 Implementation of a fully symmetrical eight-way transformer combiner.

A four-stage 1.2 V class A power amplifier is designed using the fully symmetrical eight-way transformer power combiner in 90 nm CMOS. The schematic of the power amplifier is shown in Figure 10.31. Drain voltage is 1.2 V to ensure transistor reliability. Drain current density is designed to be 0.3 mA/μm to maximize f_{max}. Four stages are selected to provide enough gain at 60 GHz. Ai ($i = 1, 2, 3, 4$) stands for each stage of the designed four-stage PA. The transformer power combiner provides optimal load to the last PA stage so that no extra capacitance is needed at the input ports of the transformer. The last stage includes four differential common source (CS) amplifiers. In each differential CS amplifier, three power amplifier cells with a gate width of 76 μm are used for positive signal and three cells with the same gate width are used for negative signal. Load impedance of every power stage is

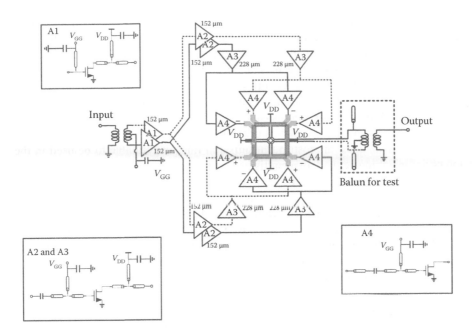

FIGURE 10.31 Diagram of the power amplifier using a fully symmetrical DAT.

selected based on output power consideration. The output port of PA is differential and a balun is used to transform a 100 Ω differential load to a 50 Ω single-ended load to perform a test. Interstage matching networks between A4, A3, and A2 are implemented by using a transmission line. A single stub tuner is adopted to provide conjugate matching between adjacent stages. A balun is used to transform single-ended signal to differential signal at the input port. Meanwhile, it provides impedance matching for the first stage of the PA.

Each CS amplifier stage in the PA consists of a parallel power amplifier cell. Selection of gate width per finger and finger number of each power amplifier cell is important in determining gain of each stage. As the gain of each stage decreases, the total gate width of the power amplifier cell increases; the total gate width of a power amplifier cell should be as small as possible under the precondition that stability is maintained. For certain value of the total gate width, different gate width per finger and finger number can be selected. But too many fingers or too large finger width can increase the total equivalent gate resistance. When power amplifier cells are parallel connected with each other, the large transistor width due to large finger numbers can cause signal imbalance in the input and output port of the power amplifier stage. As a result, less finger number is preferred when choosing a combination of different finger numbers and gate width per finger.

As shown in Figure 10.31, the interstage matching network also has the function of power dividing and the four differential PA cells in the last stage of the amplifier are located at the four corners of the transformer power combiner. It is challenging to connect these four PA cells with the previous stage and ensure that in-phase input signal is able to be delivered to the four differential PA cells. Microstrip transmission line is implemented as the interstage matching network in the proposed PA because it offers great flexibilities in the layout and moderate loss. By controlling the characteristic impedance Z_0, the length of microstrip transmission lines in the matching network can be increased or decreased. Microstrip is also easy to turn a corner, which is inevitable in matching network with power dividing function. Thus, a complicate power dividing and impedance matching network with compact layout can be easily designed. In addition, microstrip transmission lines can provide well-defined current return definite path, which is critical to achieve a first-pass silicon success. With these advantages, microstrip transmission lines are suitable to be used in the interstage matching network of a multistage power amplifier with an eight-way transformer power combining network.

The on-chip balun used for the test is designed separately and measured. The structure of the transformer balun is shown in Figure 10.32. As loss of the balun can be deembedded from the circuit measurement result, good port balance and compact layout constitute the main aim in the design of this balun. The balun comprises a transformer and parallel terminal open microstrip. The transformer transforms single-ended terminal to two balanced terminals, and the terminal open microstrip resonates with input inductive susceptance. To reduce the layout area, the width of the balun is not very important because in most cases the width of a chip is not determined by the width of the balun but is determined by the width of the core circuit. In this condition, the parallel microstrip stub does not increase the chip area and attention should be paid to reduce the distance between the single-ended port and the balanced port of this balun.

The die photo of the power amplifier is shown in Figure 10.33. Figure 10.34a shows comparison between measured results and simulated results. Measured S_{21} is 16.8 dB. S_{11} and S_{22} are better than −13 dB from 55 to 65 GHz. As shown in the figure, 3 dB bandwidth is extended from 54 to 65 GHz. At 65 GHz, S_{21} is 16 dB, so 3 dB bandwidth of the power amplifier is larger than 11 GHz. Figure 10.34b shows the measured output power, gain, and PAE versus input power at 60 GHz. At 60 GHz, 1 dB compression point output power is 17.3 dBm, peak PAE is 13%, and saturated output power is 20 dBm. The power amplifier occupies an area of 0.675 mm^2, excluding the pads and output balun.

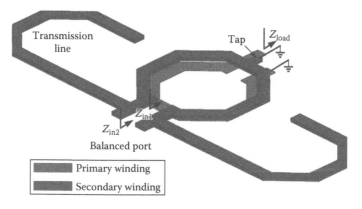

FIGURE 10.32 Structure of the transformer balun for measurement.

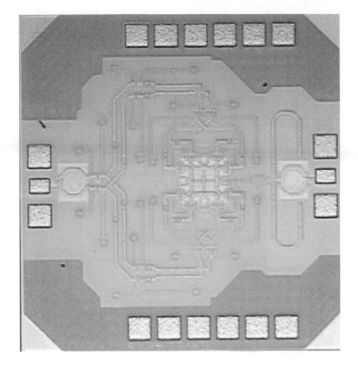

FIGURE 10.33 Die photo of the power amplifier using a fully symmetric DAT.

10.4.3 STACKED POWER AMPLIFIERS

An important method to overcome the low breakdown voltage and low power gain of the CMOS transistors is using the stacked transistors. If the max voltage on a single transistor is VDD, the voltage on n transistors can reach n*VDD, which allows higher output power. At the same time, the output impedance will increase because it is the cascode of the output impedance of each transistor. Figure 10.35 depicts two kinds of schematics of the stacked power amplifier with different input connection.

For the cascode amplifiers, the gates of the common-gate transistors are connected with ac ground. But they are often connected to proper impedance in a stacked amplifier, and this will reduce the gain of the amplifier [26]. This makes the stacked power amplifier more suitable for the large signal condition. Higher output power and drain efficiency can be achieved. Extra components are often added at the intermediate node to optimal complex intermediate node impedance. Figure 10.36 shows three examples of the extra components at the intermediate node.

A simple example [27] of a quadruple-stacked power amplifier working at 18 GHz using 130 nm gate-length mHEMTs is given, the schematic of which is shown in Figure 10.37. The transistors are just stacked together with no extra components. This will not maximize the performance of the amplifier but will save the area of the chip. The measured output and PAE versus input are depicted in Figure 10.38.

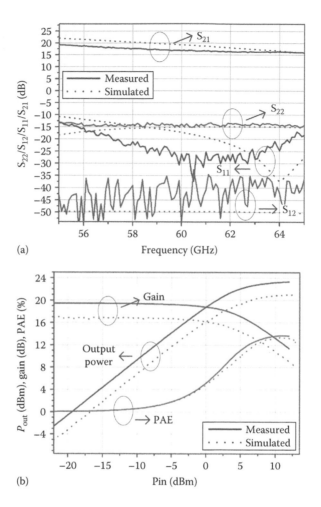

(a)

(b)

FIGURE 10.34 (a) The measured S parameter of the power amplifier, (b) output power, gain, and PAE versus input at 60 GHz.

A 90 GHz power amplifier [28] implemented with three series-connected (stacked) FETs in 45 nm SOI CMOS is shown in Figure 10.39. Shunt elements are used at the intermediate node. The stacked power amplifier has been fabricated in the 45 nm SOI CMOS process, and the performance of the power amplifier is shown in Figure 10.40. It delivers saturated output power of 17.3 dBm in the 88–90 GHz range with peak PAE of 9%.

10.4.4 HIGH PAE PA

The PAE is the key index of a power amplifier. Power amplifiers usually consume most of the dc power among all of the modules. It determines how long the batteries of our movable equipment last and how much energy the base stations will waste.

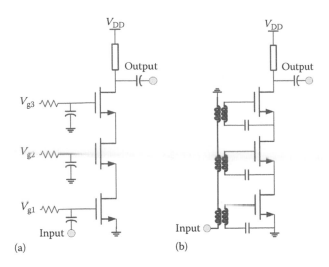

FIGURE 10.35 (a) Series input stacked power amplifier and (b) parallel input stacked power amplifier.

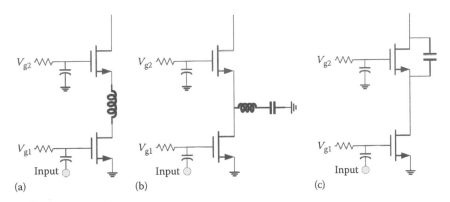

FIGURE 10.36 Stacked power amplifier with extra components: (a) series inductor, (b) parallel inductor, and (c) parallel capacitor.

Limited by the speed of the transistor and the loss of the silicon substrate, the power amplifiers in the mm-wave band often work at class A state. The PAE is usually between 5% and 15%.

With the development of the technology, the cut-off frequency of the transistors is beyond 300 GHz, which makes it possible to design a high PAE power amplifier.

A mm-wave class E tuned power amplifier [29] realized in 0.13 μm SiGe BiCMOS technology is shown in Figure 10.41. The f_{max} and f_T of the transistor are 240 and 200 GHz. The input impedance transformation network provides a low real source impedance to realize the switching-mode operation instead of optimum power match at 60 GHz. At 58 GHz, it achieves a peak PAE of 20.9%, peak power gain of 4.2 dB, and saturated output power of 11.7 dBm. The performance of the power amplifier is shown in Figure 10.42.

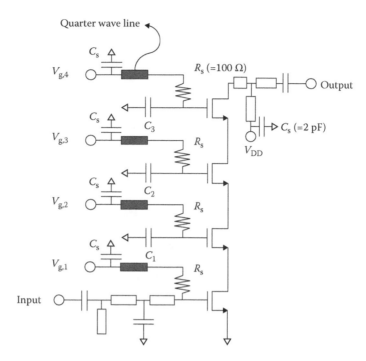

FIGURE 10.37 Schematic of a quadruple-stacked field effect transistor (FET) PA. (From Lee, C. et al., *IEEE Microw. Wireless Compon. Lett.*, 19(12), 828, December 2009.)

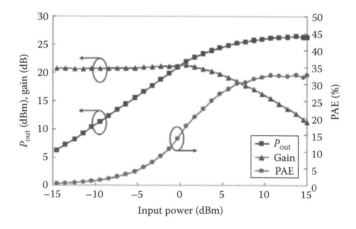

FIGURE 10.38 Measured output and PAE versus input power. (From Lee, C. et al., *IEEE Microw. Wireless Compon. Lett.*, 19(12), 828, December 2009.)

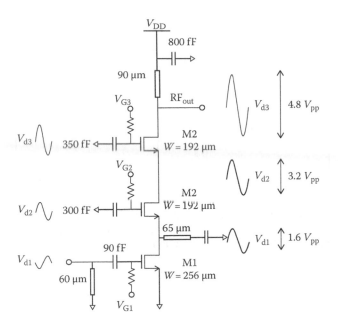

FIGURE 10.39 Schematic of the three-stage power amplifier. (From Jayamon, J. et al., A W-band stacked FET power amplifier with 17 dBm Psat in 45-nm SOI MOS, in: *Radio and Wireless Symposium (RWS), 2013 IEEE*, Austin, TX, pp. 256–258, 2013.)

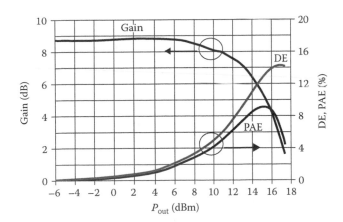

FIGURE 10.40 Performance of the three-stage power amplifier. (From Jayamon, J. et al., A W-band stacked FET power amplifier with 17 dBm Psat in 45-nm SOI MOS, in: *Radio and Wireless Symposium (RWS), 2013 IEEE*, Austin, TX, pp. 256–258, 2013.)

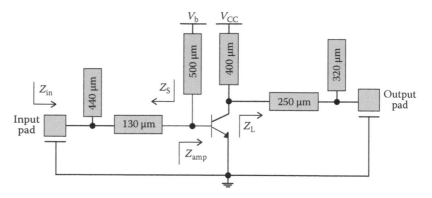

FIGURE 10.41 Measurement result of the class E PA. (From Valdes-Garcia, A. et al., A 60 GHz class-E power amplifier in SiGe, in: *IEEE Asian Solid-State Circuits Conference*, Hangzhou, China, pp. 199–202, November 2006.)

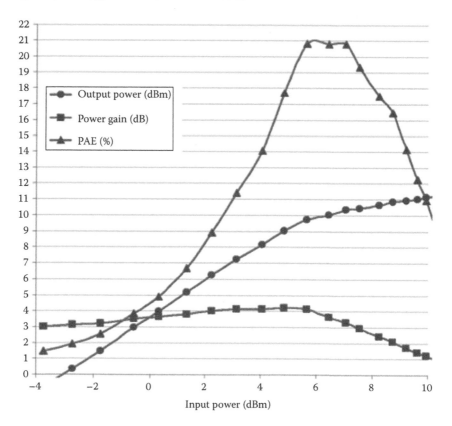

FIGURE 10.42 Schematic of the class E PA. (From Valdes-Garcia, A. et al., A 60 GHz class-E power amplifier in SiGe, in: *IEEE Asian Solid-State Circuits Conference*, Hangzhou, China, pp. 199–202, November 2006.)

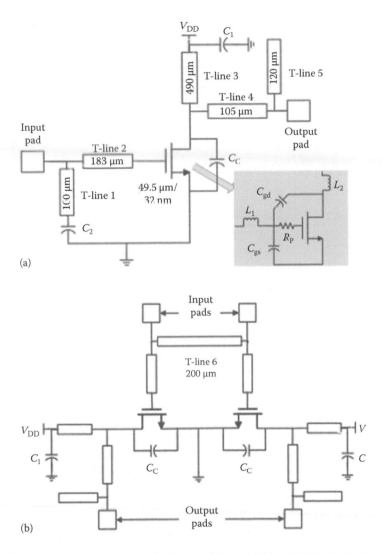

FIGURE 10.43 Schematic of the single end differential class E PA. (a) single end and (b) differential. (From Ogunnika, O.T. and Valdes-Garcia, A., A 60 GHz class-E tuned power amplifier with PAE >25% in 32 nm SOI CMOS, in: *IEEE RFIC*, Montréal, Québec, Canada, pp. 65–68, 2012.)

Another class E tuned power amplifier [30] is depicted in Figure 10.43. The power amplifier is implemented in 32 nm SOI CMOS. Both the single end and the differential amplifiers are designed.

The large signal simulation is shown in Figure 10.44. It can be seen that the drain voltage reaches its maximum value when the current is very low. The peak PAE of the power amplifier is >21%.

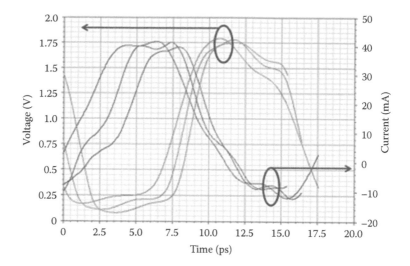

FIGURE 10.44 The large signal simulation of the class E PA. (From Ogunnika, O.T. and Valdes-Garcia, A., A 60 GHz class-E tuned power amplifier with PAE >25% in 32 nm SOI CMOS, in: *IEEE RFIC*, Montréal, Québec, Canada, pp. 65–68, 2012.)

Because of the high performance of the stacked power amplifiers, it is widely used to realize the high PAE power amplifiers in the mm-wave band. A schematic of a high PAE power amplifier [31] using a stacked transistor is shown in Figure 10.45. In order to achieve the best performance of the power amplifier, both M1 and M2 should have the optimized load impedance. Shunt transmission lines are used to make it satisfy the condition.

The PA achieves 34.4% peak PAE at 18.2 dBm output power and P_{sat} of 18.6 dBm with $P_{1\,dB}$ of 17.5 dBm measured at 42.5 GHz. The measured output power, gain, and PAE as a function of input power at 42.5 GHz are shown in Figure 10.46.

Modern modulation technique makes the wave of a signal to have peak-to-average power ratios (PAPRs), but most of the power amplifiers have high PAE only at a high input power state [33]. The Doherty power amplifier is a solution for this problem. The examples of high PAE power amplifiers make it possible to realize the good performance of the Doherty power amplifier.

A Doherty power amplifier example [32] using the stacked power amplifiers is fabricated using 45 nm SOI CMOS technology. The main amplifier and the auxiliary are both stacked power amplifiers with a shunt transmission line at the intermediate nodes. This amplifier achieves peak PAE of 23% and at 6 dB back-off power it reaches 17%. The PAE and drain efficiency of the Doherty power amplifier are depicted in Figure 10.47.

A modified Doherty power amplifier [33] is introduced using an active phase-shift preamplifier to replace one-quarter wave transmission line. The block diagram of the proposed Doherty power amplifier is shown in Figure 10.48. A slow wave transmission line (Figure 10.7) is also used to reduce the loss and length. The measured PAE and drain efficiency are shown in Figure 10.49.

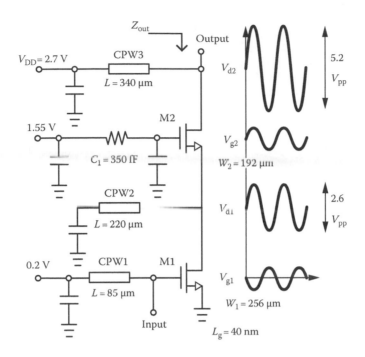

FIGURE 10.45 Schematic of the stacked PA with parallel inductor. (From Agah, A. et al., A 34% PAE, 18.6 dBm 42–45 GHz stacked power amplifier in 45 nm SOI, in: *IEEE RFIC*, Montréal, Québec, Canada, pp. 57–60, 2012.)

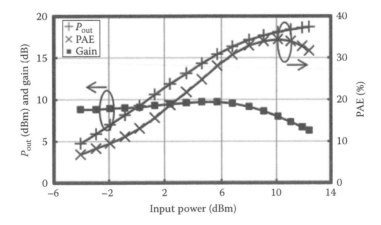

FIGURE 10.46 Measured P_{out}, gain, and PAE versus P_{in} of the stacked PA. (From Agah, A. et al., A 34% PAE, 18.6 dBm 42–45 GHz stacked power amplifier in 45 nm SOI, in: *IEEE RFIC*, Montréal, Québec, Canada, pp. 57–60, 2012.)

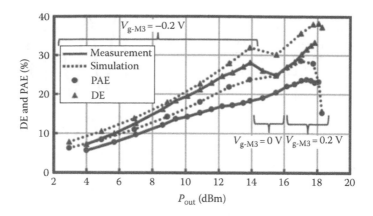

FIGURE 10.47   PAE and drain efficiency (DE) of the Doherty PA. (From Agah, A. et al., A 45 GHz Doherty power amplifier with 23% PAE and 18 dBm, in 45 nm SOI CMOS, in: *IEEE IMS*, Montréal, Québec, Canada, pp. 1–3, June 2012.)

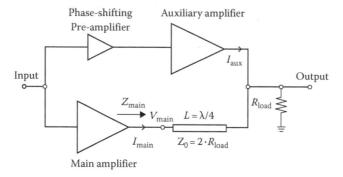

FIGURE 10.48   Active phase-shift Doherty amplifier. (From Agah, A. et al., *IEEE J. Solid-State Circ.*, 48(10), 2338, October 2013.)

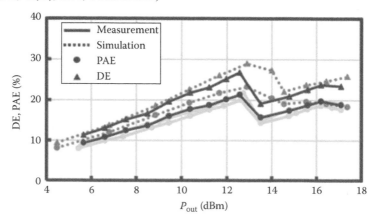

FIGURE 10.49   PAE and DE of the active phase-shift Doherty amplifier. (From Agah, A. et al., *IEEE J. Solid-State Circ.*, 48(10), 2338, October 2013.)

10.5   CONCLUSION

Millimeter electromagnetic wave has many advantages that attract people to develop applications at this band. The challenges in designing a power amplifier using CMOS technology are summarized and the literature on solving these problems is reviewed. When the first circuit was designed for the mm-wave band, only basic function could be realized. But with the development of the technique of manufacturing and circuits, circuits with higher performance are designed and more applications can be implemented. It does not take a genius to figure out that in the foreseeable future applications in the mm-wave band with amazing performance will be all around us and make our lives more convenient.

REFERENCES

1. K. Kang, F. Lin, D.-D. Pham *et al.*, A 60-GHz OOK receiver with an on-chip antenna in 90 nm CMOS, *IEEE J. Solid-State Circ.*, 45(9), 71720–1731, September 2010.
2. B. Razavi, Design of millimeter-wave CMOS radios: A tutorial, *IEEE Trans. Circ. Syst. I*, 56(1), 4–16, January 2009.
3. C.H. Doan, S. Emami, A.M. Niknejad *et al.*, Millimeter-wave CMOS design, *IEEE J. Solid-State Circ.*, 40(1), 144–155, January 2005.
4. D. Dawn, S. Sarkar, P. Sen *et al.*, 60 GHz CMOS power amplifier with 20-dB-Gain and 12 dBm P_{sat}, *IEEE International Microwave Symposium Digest*, pp. 537–540, June 2009.
5. M. Tanomura, Y. Hamada, S. Kishimoto et al., TX and RX front ends for the 60-GHz band in 90 nm standard bulk CMOS, in *ISSCC Digest of Technical Papers*, pp. 558–559, February 2008.
6. D. Zhao and P. Reynaert, A 60-GHz dual-mode class AB power amplifier in 40-nm CMOS, *IEEE J. Solid-State Circ.*, 48(10), 2323–2337, October 2013.
7. C. Liang and B. Razavi, A layout technique for millimeter-wave PA transistors, in *Proceedings of RFIC Symposium*, June 2011.
8. U. Gogineni, H. Li, J.A. del Alamo et al., Effect of substrate contact shape and placement of RF characteristics of 45 nm low power CMOS devices, *IEEE J. Solid-State Circ.*, 45(5), 998–1005, May 2010.
9. S.P. Voinigescu, D. Marchesan, J.L. Showell *et al.*, Process- and geometry-scalable bipolar transistor and transmission line models for Si and SiGe MMIC's in the 5–22 GHz range, in *International Electron Devices Meeting—Technical Digest*, pp. 11.7.1–11.7.4, December 1998.
10. B. Razavi, A 60-GHz CMOS receiver front-end, *IEEE J. Solid-State Circ.*, 41(1), 17–22, January 2006.
11. T.S.D. Cheung, J.R. Long, K. Vaed et al., On-chip interconnect for mm-wave applications using an all-copper technology and wavelength reduction, in *IEEE ISSCC*, pp. 396–501, February 2003.
12. H. Hasegawa, M. Furukawa, and H. Yanai, Properties of microstrip line on Si–SiO2 system, IEEE JMTT, MTT-19(11), 869–881, November 1971.
13. A. Komijani, A. Natarajan, and A. Hajimiri, A 24-GHz, +14.5-dBm fully integrated power amplifier in 0.18-μm CMOS, *IEEE J. Solid-State Circ.*, 40(9), 1901–1908, September 2005.
14. B. Kleveland, C.H. Diaz, D. Vooket *et al.*, Exploiting CMOS reverse interconnect scaling in multigiga-hertz amplifier and oscillator design, *IEEE J. Solid-State Circ.*, 36(10), 1480–1488, October 2001.
15. A.M. Niknejad and H. Hashemi, Mm-Wave Silicon Technology 60 GHz and Beyond, Springer Science, New York, 2007.

16. C.Y. Law and A.-V. Pham, A high-gain 60 GHz power amplifier with 20 dBm output power in 90 nm CMOS, in *IEEE ISSCC*, pp. 426–427, 2010.

17. W. Tai, L. Richard Carley, and D.S. Ricketts, A 0.7 W fully integrated 42 GHz power amplifier with 10% PAE in 0.13 μm SiGe BiCMOS, in *IEEE ISSCC*, pp. 142–143, 2013.

18. J. Chen and A.M. Niknejad, A compact 1 V 18.6 dBm 60 GHz power amplifier in 65 nm CMOS, in *IEEE ISSCC*, pp. 432–433, 2010.

19. K.-Y. Wang, T.-Y. Chang, and C.-K. Wang, A 1 V 19.3 dBm 79 GHz power amplifier in 65 nm CMOS, in *IEEE ISSCC*, pp. 260–261, 2012.

20. T.S.D. Cheung and J.R. Long, A 21–26-GHz SiGe bipolar power amplifier MMIC, *IEEE J. Solid-State Circ.*, 40(12), 2583–2597, December 2005.

21. I. Aoki, S.D. Kee, D.B. Rutledge et al., Fully integrated CMOS power amplifier design using the distributed active transformer architecture, *IEEE J. Solid-State Circ.*, 37(3), 371–383, March 2002.

22. J.-W. Lai and A. Valdes-Garcia, A 1 V 17.9 dBm 60 GHz power amplifier in standard 65 nm CMOS, in *IEEE ISSCC*, 2010.

23. U.R. Pfeiffer and D. Goren, A 23-dBm 60-GHz distributed active transformer in a silicon process technology, *IEEE Trans. Microw. Theory Tech.*, 55(5), 857–865, May 2007.

24. P. Haldi, D. Chowdhury, P. Reynaert et al., A 5.8 GHz 1 V linear power amplifier using a novel on-chip transformer power combiner in standard 90 nm CMOS, *IEEE J. Solid-State Circ.*, 43(5), 1054–1063, May 2008.

25. J. Essing, R. Mahmoudi, Y. Pei et al., A fully integrated 60 GHz distributed transformer power amplifier in bulky CMOS 45 nm, in *Proceedings of IEEE RFIC Symposium Digest*, pp. 1–4, 2011.

26. H.-T. Dabag, B. Hanafi, F. Golcuk et al., Analysis and design of stacked-FET millimeter-wave power amplifiers, *IEEE Trans. Microw. Theory Tech.*, 61(4), 1543–1556, April 2013.

27. C. Lee, Y. Kim, Y. Koh et al., A 18 GHz broadband stacked FET power amplifier using 130 nm metamorphic HEMTs, *IEEE Microw. Wireless Compon. Lett.*, 19(12), 828–830, December 2009.

28. J. Jayamon, A. Agah, B. Hanafi et al., A W-band stacked FET power amplifier with 17 dBm Psat in 45-nm SOI MOS, in *Radio and Wireless Symposium (RWS), 2013 IEEE*, pp. 256–258, 2013.

29. A. Valdes-Garcia, S. Reynolds, and U.R. Pfeiffer, A 60 GHz class-E power amplifier in SiGe, in *IEEE Asian Solid-State Circuits Conference*, Hangzhou, China, pp. 199–202, November 2006.

30. O.T. Ogunnika and A. Valdes-Garcia, A 60 GHz Class-E tuned power amplifier with PAE >25% in 32 nm SOI CMOS, in *IEEE RFIC*, 2012.

31. A. Agah, H. Dabag, B. Hanafi et al., A 34% PAE, 18.6 dBm 42–45 GHz stacked power amplifier in 45 nm SOI, in *IEEE RFIC*, 2012.

32. A. Agah, B. Hanafi, H. Dabag et al., A 45 GHz Doherty power amplifier with 23% PAE and 18 dBm, in 45 nm SOI CMOS, in *IEEE IMS*, pp. 1–3, June 2012.

33. A. Agah, H. Dabag, B. Hanafi et al., Active millimeter-wave phase-shift Doherty power amplifier in 45-nm SOI CMOS, *IEEE J. Solid-State Circ.*, 48(10), 2338–2350, October 2013.

34. CMOS ITRS Roadmap. [Online]. Available http://www.itrs.net/

35. E. Xiao, Hot carrier effect on CMOS RF amplifiers, in *Proceedings of IEEE 43rd Annual International Reliability Physics Symposium*, pp. 680–681, April 2005.

36. M. Ruberto, T. Maimon, Y. Shemesh et al., Consideration of age degradation in the RF performance of CMOS radio chips for high volume manufacturing, *Proceedings of IEEE RFIC Symposium*, pp. 549–552, June 2005.

37. B. Heydari, M. Bohsali, E. Adabi et al., Millimeter-wave devices and circuit blocks up to 104 GHz in 90 nm CMOS, *IEEE J. Solid-State Circ.*, 42(12), 2893–2902, December 2007.

11 Millimeter-Wave Frequency Multiplier with Performance Enhancement

Chiennan Kuo

CONTENTS

11.1 INTRODUCTION

Recently, complementary metal-oxide-semiconductor (CMOS) technology has been successfully extended to wireless transceiver circuits in the regime at millimeter-wave (mmWave) frequencies. CMOS circuits are feasible for various applications, such as high-speed wireless data transmission in the 60 GHz band and automotive radar detection at 77 GHz. No matter what kind of applications, a signal source is a must in a mmWave system. In typical cases, a local oscillator (LO) signal is required for frequency conversion. In some cases, data can be directly modulated on a source to obtain amplitude, phase, and frequency modulating signals. Signal generation is a process that converts dc energy to radiofrequency (RF) power using the circuit block of an oscillator. Oscillator design faces performance challenge with the fundamental frequency directly at the mmWave frequency and beyond. It is difficult to obtain stable oscillation, and the phase noise is not so good. Besides, the energy conversion process is typically of low efficiency at high frequencies. A circuit normally consumes large dc power.

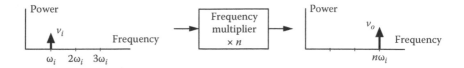

FIGURE 11.1 A frequency multiplier generates an output at the n times of the input frequency.

To alleviate the difficulty of fundamental oscillator design at high frequencies, an effective option is indirect generation by means of frequency multiplication. A frequency multiplier takes a low-frequency input and produces a high-frequency output. This is a popular method to generate a high-frequency signal from the harmonic outputs of a low-frequency fundamental source. Its operation can be illustrated using the frequency spectrum in Figure 11.1. The input sinusoidal signal of $v_i \, cos(\omega_i t)$ enters the multiplier. The output becomes $v_o \, cos(n\omega_i t)$, where n is the frequency multiplication factor. In the case that a very high-frequency signal is in need, several multipliers can be cascaded in series. Alternatively, it is preferred to have a multiplier with a large multiplication factor. In most circuit applications, n is in the range from 2 to 4.

The advantage of frequency multiplication is manifold. It helps quickly undergo a mmWave project using available and proved designs to save the time to market. Usually the designs of low-frequency sources have been well developed with good performance. This was seen in the early development of 60 [1] and 160 GHz [2] wireless communication systems. In these two references, a multiplier with a factor of 3 was proposed to generate the required LO for frequency conversion. Besides, frequency multiplication gives appropriate phase noise performance. As predicted by the Leeson equation [3], phase noise theoretically degrades by the factor of 20 $log(n)$ if the frequency increases by n times. Direct generation of fundamental oscillation usually has a worse result. On the contrary, multipliers have better performance, beneficial from the low-frequency source of high quality. Furthermore, frequency multiplication is possible to offer an output frequency exceeding the maximum oscillation frequency (f_{max}) of active devices. Huang reported a quadrupler at 324 GHz, which is twice of the device f_{max} given by the applied 90 nm CMOS technology [4].

The microwave society has made a spectacular progress in design techniques of frequency multipliers over the past few decades. Several approaches are available. The basic design principle is based on device nonlinearity. For a high multiplication factor, frequency mixing is more effective. If the operation frequency range is not large, we can apply injection locking to lift the output power. Additional design effort is on spur suppression. In this chapter, we will discuss these design techniques in CMOS frequency multipliers.

11.2   CHARACTERIZATION OF FREQUENCY MULTIPLIER

The design of a frequency multiplier is associated with optimization of several parameters. This section is a review of those design parameters typically specified for a frequency multiplier.

11.2.1 Output Power, P_{out}, and Output Efficiency, η_n

Output power is measured at the multiplied frequency, $P_{out}(n\omega_i)$, referring to the load impedance of 50 Ω typically. As a signal source, the output power of a multiplier circuit is better as high as possible. It is therefore important to specify the maximum or the saturation output power. From the viewpoint of circuit operation, large signal is often involved in the design. Thus, it is necessary to characterize circuit efficiency. The conversion efficiency can be assessed by

$$\eta_n = \frac{P_{out}(n\omega_i)}{P_{DC}}. \tag{11.1}$$

11.2.2 Conversion Gain, G_c

A multiplier generates the output corresponding to the input signal. How the signal amplitude transfers from the input to the output is of great interest to multiplier design. The arrangement of different signal frequencies is similar to that in an upconversion mixer. Accordingly, the conversion gain can be defined as

$$G_V = \frac{v_{out}(n\omega_i)}{v_{in}(\omega_i)}, \tag{11.2}$$

if characterized in the voltage domain. It can be specified in the power domain as well.

11.2.3 Harmonic Rejection Ratio

Signal purity at the output can be inspected by harmonic rejection ratio (HRR). Ideally only one desired signal of interest appears at the output. Many other frequency components, however, are also visible in reality. For example, the multiplication factor is two in a frequency doubler. It is not uncommon to find other output components at the third and high-order harmonics of the input frequency. They are considered as spurious noise. Depending on the applications, these spurs are specified not to exceed a maximum level relative to that of the desired frequency, defined as HRR,

$$HRR_m = \frac{P_{out}(n\omega_i)}{P_{out}(m\omega_i)}, \tag{11.3}$$

where
   n refers to the desired output
   m stands for the order of the spurs

In most designs, a filter is usually placed at the output to suppress spurs and improve the rejection ratio. If the voltage ratio is specified, the impedance at different frequencies needs to be taken into account.

11.3   FREQUENCY MULTIPLICATION BY NONLINEARITY

As far as the frequency arrangement is concerned, a frequency multiplier is essentially a nonlinear circuit. The basic realization of frequency multiplication is by means of device current/voltage nonlinear characteristic. If driven by a large sinusoidal input signal, a nonlinear circuit generates a harmonic-rich waveform. This is different from the design of a linear amplifier, which is aimed at prevention of waveform distortion. The notorious outcome of harmonics generation, however, becomes favorable to a frequency multiplier. Nevertheless, the downside is the presence of many harmonics. The only desired one is the harmonic with the order of the frequency multiplication factor. Filtering is generally necessary to suppress all other harmonics.

Several factors determine device nonlinearity and harmonic generation. The major effectiveness comes from the bias condition, which dominates transconductance non linearity and low-frequency multiplication. The second-order nonlinear effect arises from voltage-dependent parasitic capacitances. Camargo gives tutorial discussions on the design steps using a quasi-linearization procedure to resolve the frequency components [5]. Consequently, the equivalent circuits at each harmonic are derived. Note that the analysis shall include the impedance conditions at the source and load, which affect harmonic generation and in turn impact on output power.

The first thought of devices with strong nonlinearity leads to diodes. Diode multi-pliers have been developed early in the microwave society. The capability of current rectification utilizing the exponential characteristic is very advantageous to doubler design. The half-wave rectified waveform has abundant harmonics. A filter at the output helps select the desired second-order harmonic. For better efficiency, we can choose the configuration of the balanced antiparallel connection due to full-wave rectification [6]. The fundamental currents are 180° out of phase in the antiparallel diodes, while the second-order harmonics are in phase. Consequently, the circuit gives an output of the second-order harmonic without the fundamental feedthrough. Among various types of diodes, the Schottky-barrier diode can carry out a very high maximum frequency [7]. Here, we focus the discussion on the implementation of CMOS transistor multipliers.

A CMOS transistor has nonlinear output current in response to the input gate volt-age. Although a CMOS transistor can operate like a diode if the drain and gate nodes are connected together, the design would be more attractive to utilize it as a three-terminal device. It is important to bias the gate voltage properly to maximize the output harmonic currents. As a transconductor, the transistor bias can be categorized into different classes traditionally according to the conduction time percentage or the conduction angle. The transistor of class A turns on all the time and only provides linear responses. Unless the transistor is overdriven by a large input swing to result in a clipped waveform, it is of no use to a multiplier. Operated in class B, the transistor turns on in a half of the duty cycle and behaves like a current rectifier. Biased in class C, the transistor shows waveform distortion very severely. To demonstrate the design principle, we will review the bias condition.

Figure 11.2 shows a typical time response of the input-to-output transfer in a common-source transistor. Without loss of generality, the plot shows the bias condi-tion in class AB. A piecewise linear model is applied to simplify the nonlinear MOS

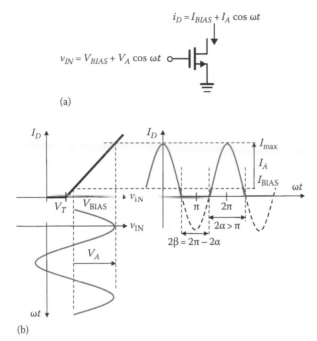

(a)

(b)

FIGURE 11.2 (a) A common-source transistor as a frequency multiplier and (b) the time waveforms of a class AB transistor with the conduction angle of 2α.

transistor characteristic. Let the transistor turn on if the input voltage v_{IN} is larger than a threshold voltage V_T. Then the governing equation can be written as

$$I_D = \begin{cases} 0 & v_{IN} < V_T \\ g_m(v_{IN} - V_T) & v_{IN} \geq V_T \end{cases}. \tag{11.4}$$

The gate port of the transistor is biased at V_{BIAS}, which is higher than V_T. Assume the input is a sinusoidal voltage of amplitude V_A at the frequency ω, expressed as

$$v_{IN} = V_{BIAS} + V_A \cos \omega t. \tag{11.5}$$

If operated in class A, the transistor is fully on. The output drain current would be of the amplitude, $I_A = g_m V_A$. In class AB operation, the transistor is turn on with the conduction angle $2\alpha < 360°$, but larger than $180°$. The following formulation, nevertheless, is applicable to other class types as well. The time response of the drain current waveform can be written as

$$i_D(\omega t) = \begin{cases} I_{BIAS} + I_A \cos(\omega t) & -\alpha \leq \omega t \leq \alpha \\ 0 & \text{otherwise} \end{cases}, \tag{11.6}$$

where the bias current I_{BIAS} corresponds to the bias voltage V_{BIAS}. Given the condition of zero drain current at $\omega t = \alpha$, the parameter α can be derived as

$$I_A \cos \alpha = -I_{BIAS}. \tag{11.7}$$

Consequently, we can rewrite the drain current waveform as

$$i_D(\omega t) = I_{\max} \frac{\cos(\omega t) - \cos\alpha}{1 - \cos\alpha}, \tag{11.8}$$

where $I_{\max} = I_{BIAS} + I_A = I_A(1 - \cos\alpha)$.

How effective can we apply the transistor as a frequency multiplier? The time response in (11.8) contains all the harmonics. We can do the spectral analysis of the drain current. The Fourier series of the drain current in (11.8) can be expanded as

$$i_D(\omega t) = I_0 + \sum_{n=1}^{\infty} [I_{an} \sin n\omega t + I_{bn} \cos n\omega t], \tag{11.9}$$

where I_{an} and I_{bn} are the coefficients of the nth-order harmonic in the Fourier series. It can be found that all $I_{an} = 0$. The expression of I_{bn} can be generally derived using the integral

$$I_{bn} = \frac{1}{\pi} \int_{-\alpha}^{\alpha} i_D(\omega t)\cos(n\omega t)d(\omega t) = \frac{I_{\max}}{\pi(1 - \cos\alpha)} \left[\frac{\sin(n-1)\alpha}{n(n-1)} - \frac{\sin(n+1)\alpha}{n(n+1)} \right]. \tag{11.10}$$

The equation is valid if $n > 1$. This is good enough for our interest to understand the application as a multiplier. For instance, I_{b2}, I_{b3}, and I_{b4} are associated with a doubler, a tripler, and a quadrupler, respectively. I_0 stands for the dc current. It can be calculated from the integral

$$I_0 = \frac{1}{2\pi} \int_{-\alpha}^{\alpha} i_D(\omega t)d(\omega t) = \frac{I_{\max}}{\pi} \frac{\sin\alpha - \alpha\cos\alpha}{1 - \cos\alpha}, \tag{11.11}$$

which gives an estimate of the power consumption of the circuit. It is obvious the output currents depend on the bias. Figure 11.3 plots the magnitudes of harmonic currents normalized to $I_{\max}$ with the parameter of the transistor conduction angle

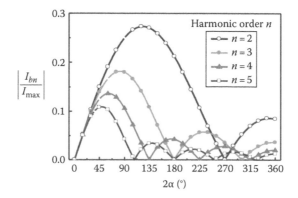

FIGURE 11.3 The normalized magnitude of the nth-order harmonic current with the parameter of the transistor conduction angle 2α.

according to (11.10). The optimal design occurs at the condition of maximum output currents, or at the conduction angle such that $d(I_{bn})/d\alpha = 0$. Take a frequency doubler as an example. The optimal conduction angle is determined by

$$\frac{dI_{b2}}{d\alpha} = \frac{I_{max}}{\pi} \frac{d}{d\alpha}\left[\frac{3\sin\alpha - \sin(3\alpha)}{6(1-\cos\alpha)}\right] = 0. \tag{11.12}$$

Thus, it can be found that the conduction angle, 2α, is $120°$. Similarly, the optimal angle is $79.7°$ for a tripler, $59.7°$ for a quadrupler, and so on. These bias conditions are actually in the class C operation.

Although the aforementioned results give the maximum output current, it is worth extending the concern to spurious noise as well. Three cases of different transistor conduction angles are summarized in Table 11.1 for comparison. Case 1 gives the maximum second-order harmonic output current, about 11.2 dB below I_{max}. The HRRs to the third and the fourth harmonic currents are 6 and 20 dB, respectively. If the conduction angle is chosen as $146.4°$ instead of $120°$, the output current decreases by 0.4 dB but the third harmonic spur is reduced by extra 4.8 dB. If the conduction angle is chosen as $180°$, the third harmonic spur can then be removed at the expense of extra 2.3 dB reduction of the output current. The tradeoff really depends on the application. A designer shall determine the bias condition for proper performance.

Since dc energy is converted to RF power, it is interesting to know the output efficiency. Assume the impedance remains the same at all harmonic frequencies. We simply calculate the current ratio as the efficiency,

$$\eta_n = \frac{I_{bn}}{I_0} = \left[\frac{\sin(n-1)\alpha}{n(n-1)} - \frac{\sin(n+1)\alpha}{n(n+1)}\right]\frac{1}{\sin\alpha - \alpha\cos\alpha}. \tag{11.13}$$

Generally speaking, the analysis is valid to all nonlinear devices. Given the transfer function of the device characteristic, we can obtain the appropriate bias condition. The next step of the design is to incorporate the supply voltage bias circuitry. We need to pay some attention to reduce the loading effect of the bias circuit. A high impedance inductor is usually used as an RF choke. At mmWave frequencies, a parallel LC resonator at the signal frequency is more operative. The microwave approach is through quarter-wavelength interconnect lines, which work as an impedance transformer. The ac ground at the supply voltage node is converted to high impedance at the signal line. Along with the impedance concern, quarter-wavelength

TABLE 11.1
Three Cases of Different Bias Conduction for Doubler Design

Parameter	Case 1	Case 2	Case 3
Conduction angle	$120°$	$146.4°$	$180°$
Normalized output current ($|I_{b2}/I_{max}|$)	−11.2 dB	−11.6 dB	−13.5 dB
HRR_3 ($|I_{b2}/I_{b3}|$)	6 dB	10.8 dB	∞
HRR_4 ($|I_{b2}/I_{b4}|$)	20 dB	20 dB	13.9 dB

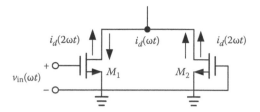

FIGURE 11.4   The push–push configuration in a frequency doubler.

open stubs are often used to enhance the output power and suppress spurious noise. In spite of impedance conjugate matching at the signal frequencies, the additional out-of-band impedance conditions at harmonic frequencies have an effect as well. In Boudiaf's tripler circuit [8], two open stubs are placed at the input gate port at the frequency of 2ω and 3ω, and two other open stubs at the output drain port at ω and 2ω. Those open stubs provide preferred low impedance and improve the efficiency.

The remaining issues always exist how to boost the output signal and how to suppress spurious noise furthermore. As mentioned, proper transistor bias can suppress harmonic spurs. The fundamental leakage usually appears at the output with a significant level. If considered from the viewpoint of circuit architecture, the issue can be alleviated effectively. An interesting example is the push–push circuit as shown in Figure 11.4, consisting of a differential transistor pair $M_{1,2}$. The circuit is commonly used as a doubler if differentially driven by $v_{in}(\omega t)$. The second-order harmonic currents $i_d(2\omega t)$ are in phase and sum up at the output. The fundamental currents $i_d(\omega t)$ appears out of phase and cancel out each other after combination. The circuit configuration gives no odd-order harmonic current at the output, no matter what bias condition is chosen. As a matter of fact, $M_{1,2}$ can be a part of oscillators. The circuit then becomes two coupled oscillators with differential drain currents. A transmission-line resonator of half wavelength at the fundamental frequency ensures differential coupling at oscillation [9]. Another example is an LC cross-coupled oscillator [10]. In a word, the push–push configuration is likely the most popular for a doubler design.

11.4   FREQUENCY MULTIPLICATION BY FREQUENCY MIXING

Alternative approach of multiplier design is based on the concept of a frequency upconversion mixer. An upconversion mixer is typically applied to a wireless system to modulate the radio carrier. It takes two inputs of an LO and an intermediate frequency (IF), and makes use of the second-order nonlinearity to generate the RF output at the mixed frequency that is the sum of the two input frequencies. Frequency multiplication can utilize the operation. If the two inputs are of the same frequency, the summed frequency is therefore a double. The difference between a mixer and a multiplier is not much. In a mixer, the LO level is usually very large in order to switch the circuit on and off for high conversion gain. The circuit still requires to be linear as far as the IF signal is concerned. In a frequency multiplier, there is only one input such that the signal levels are obviously the same. The operation requires boosting nonlinearity and enhancing frequency mixing. There is no linearity concern. Nevertheless, the same circuit schematic can be applied to both.

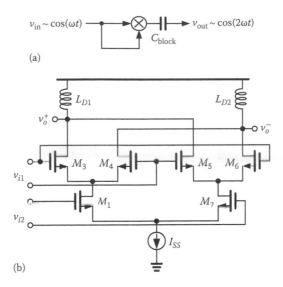

$v_{in} \sim \cos(\omega t)$ $v_{out} \sim \cos(2\omega t)$

C_{block}

(a)

(b)

FIGURE 11.5 (a) A frequency doubler using frequency mixing and (b) a double-balanced Gilbert mixer circuit.

The block diagram of a frequency doubler as shown in Figure 11.5a can illustrate the frequency-mixing concept. The input is at the frequency of ω. The final output becomes $\cos(2\omega t)$ after removing dc by the blocking capacitor. It is quite popular to apply the well-known Gilbert multiplier circuit to realize the active mixer design, as shown in Figure 11.5b. It shows a double-balanced circuit driven by two fully differential inputs of v_{i1} and v_{i2}. In the circuit, transistors M_{1-6} provide four current paths, controlled by the input voltages to allow currents flowing through in different phases. These four current combines at the output nodes, yielding to a different output voltage. The total current is constrained by the tail current source of I_{SS}. It is current commutation to cause frequency conversion. The commutative currents result in the output frequency at 2ω, which is selected by a simple LC filter of the inductive load and the output parasitic capacitance. We have to be cautious to the phase difference between v_{i1} and v_{i2}. To a simple mixer, v_{i1} and v_{i2} have different frequencies such that the phase difference does not matter. To a multiplier, keeping v_{i1} and v_{i2} in phase ensures maximum conversion gain. Careful layout work is therefore necessary to avoid introducing undesired phase shift.

Frequency mixing is advantageous to implement a high multiplication factor easily. Many other outputs are generated in the operation. The output frequencies can be written as the linear combination of the two inputs as $\omega_{out} = m\omega_{i1} + n\omega_{i2} = (m + n)\omega_i$. That is, it is possible to have a large multiplication factor out of the mixer circuit with appropriate filtering to pick up the correct high-order harmonic. Usually the high-order output power is less than the second order. Similar to the nonlinearity approach in Section 11.3, bias optimization can enhance the output power to some extent. Yet changing the configuration appears to be more effective. A cascaded architecture for a frequency tripler is depicted in Figure 11.6a, consisting of two mixer stages. The first stage is essentially a doubler and generates the frequency of 2ω. The second stage, taking this signal and mixing with the input fundamental, gives the output of $\cos(3\omega t)$.

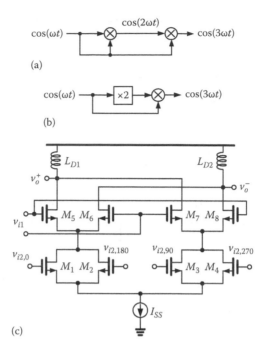

FIGURE 11.6 A frequency tripler using (a) cascaded mixers, (b) the equivalent architecture of a doubler and a mixer, and (c) a double-balanced subharmonic mixer.

Of course, the doubler can be realized by a different approach, instead of frequency mixing. An equivalent architecture is illustrated in Figure 11.6b. Consequently, circuit implementation can utilize a type of mixer circuits, called the subharmonic mixer (SHM) [11], which combines the doubler and mixer in the cascode configuration as shown in Figure 11.6c. This current-reuse topology is advantageous in low power consumption. Two differential input signals, v_{i1} and v_{i2}, of the same magnitude are fed into the transistor pairs. The double-balanced circuit schematic looks very similar to the Gilbert circuit in Figure 11.5b, except those two transistor pairs of $M_{1,2}$ and $M_{3,4}$. Readers might have identified that those transistor pairs are efficient frequency doublers in the push–push configuration. It generates the output current at 2ω to drive the top source-coupled differential pairs $M_{5,6}$ and $M_{7,8}$, which work as the mixer switching stage. This mixer produces the third-order harmonic by frequency upconversion of current commutation. Conversion gain relies on the second-order nonlinearity in transistor pairs. To maximize the efficiency of frequency conversion, each transistor should be biased at the gate voltage for the maximum voltage derivative of device transconductance (g'_m). It can be seen that the entire multiplier circuit is realized in single circuit. The cost is the requirement of quadrature phases, which bring differential second harmonic output currents from M_{1-4}.

Circuit design can be further extended to a quadrature tripler, including in phase and quadrature (*I/Q*) signal paths. This is useful for LO generation in a practical system. An interesting design of a K-band tripler in a 0.18 μm CMOS technology has quadrature signal generation using coupled *I/Q* signals [12]. The design is motivated

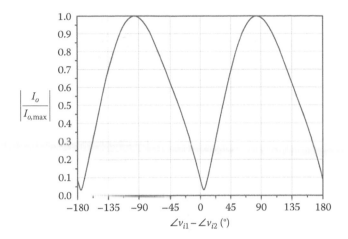

FIGURE 11.7 The normalized third harmonic output current in a single-balanced SHM achieves the maximum if the two inputs are in quadrature phase.

by evaluating the SHM output current. Analysis of a single-balanced SHM reveals that the output current magnitude is strongly dependent on the phase difference between the two inputs but insensitive to the amplitude imbalance. The mixer generates small output current if the signals are in phase, and maximum if the inputs are around ±90° out of phase. As the operation frequency increases, the optimal phase difference moves away due to nonideal parasitics. Figure 11.7 shows the simulation result of the normalized third-order harmonic output current by sweeping the input phase difference of the signals at 7 GHz. As can be seen, quadrature input phases can still achieve the optimal condition of the maximum conversion gain. Therefore, the tripler circuit calls for a pair of I/Q input signals. As such, we can take two SHMs to form a quadrature tripler. The block diagram is as shown in Figure 11.8a, consisting of two single-balanced SHMs as shown in Figure 11.8b. This tripler features quadrature signals at both the input and the output. The fundamental quadrature I/Q input signals at the frequency ω are first used to generate the second-order harmonics at 2ω by frequency doublers. Then the doubler outputs are mixed with the fundamentals to produce the quadrature third-order harmonic outputs at the frequency 3ω. The cross-coupling between I and Q paths not only keeps the quadrature characteristic but also allows maximum frequency conversion efficiency. At the mixer output, inductive load is preferred. Inductors $L_{D1,2}$ nullify the parasitic capacitance at the output nodes and form resonators at 21 GHz. The resonators provide bandpass responses and suppress undesired spurs. Besides, the impedance of the resonators is low at the frequencies ω and 2ω, beneficial to boost the conversion efficiency [8].

The measured power transfer curve of the quadrature tripler is plotted in Figure 11.9a. The circuit is fully differential, but only one single-ended output node is probed. The input power is calibrated to the input of an on-chip balun. The output power increases as the input signal level increases, but eventually saturates at P_{in} of 14 dBm. The frequency response of the conversion gain with two input power levels is plotted in Figure 11.9b. The maximum conversion gain of the core quadrature

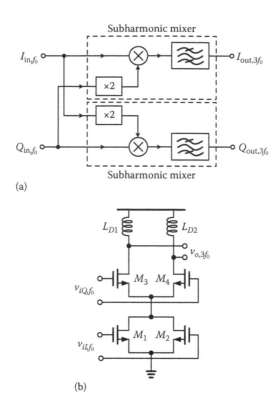

(a)

(b)

FIGURE 11.8 (a) The block diagram shows a quadrature tripler using coupled *I/Q* signals, including (b) two of the single-balanced SHM circuits.

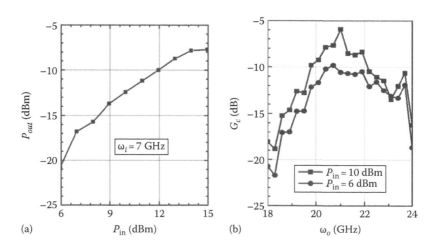

(a)

(b)

FIGURE 11.9 (a) The transfer curve of the output power to the input power with the input signal frequency at 7 GHz and (b) the frequency response of the quadrature tripler.

TABLE 11.2

Performance Summary of the Designed Quadrature Tripler

Technology	CMOS 0.18 μm
Fundamental frequency range	6–8 GHz (42.8%)
Conversion gain	−5.7 dB (at 7 GHz)
HRR_1	22.4 dBc
HRR_2	16.1 dBc
HRR_4	20.2 dBc
Power consumption[a]	7.5 mW

[a] Power consumption includes I/Q paths with the input power of 10 dBm at the balun input.

tripler achieves −5.7 dB, after subtracting the simulated loss of the balun. The HRRs of the fundamental, second-order, and fourth-order harmonics with the input signal at 7 GHz are 22.4 dBc, 16.1 dBc, and 20.2 dBc, respectively. Since the circuit is fully differential, the second-order leakage is not critical in practical applications. It is worth noting the measured phase noise of the input and output signals. The theoretical degradation is 20 log(3) = 9.5 dB in a tripler. The phase noise measured at 1 MHz offset of the 7 GHz input signal source and the third-order output are −140.5 and −130.5 dBc/Hz, respectively. This difference around 10 dB is in well agreement with the theoretical value. Table 11.2 summarizes the performance of the designed quadrature tripler.

Frequency mixing can be put in use with other techniques to further improve the circuit performance. Figure 11.10a is the simplified circuit of a frequency quadrupler in Huang's design [4], consisting of two differential transistor pairs with tail current sources, $I_{SS1,2}$. The circuit is driven by four quadrature input signals. Each transistor conducts in a half of the duty cycle and produces a rectified drain current. Four drain currents are of the same amplitude and 90° phase difference among one another. The design makes use of linear superposition to combine the four rectified drain currents at the output node and generate the fourth harmonic such that the conversion gain achieves the maximum of −15.4 dB in theory. A distinct feature exists in the configuration. Because of circuit symmetry, ideally the fundamental, second, and third harmonics are all cancelled without filtering. In the original design, the tail current sources simply provide the required current biasing to the differential pairs. We can add frequency mixing into the circuit to further enhance the conversion gain without extra dc power consumption. The idea is as shown in the revised circuit in Figure 11.10b. Transistor pairs, $M_{5,6}$ and $M_{7,8}$, acting as frequency doublers replace the two current sources. The two doublers result in even-order harmonic currents, which in turn inject to the rectifier stage and bring a larger output of the fourth harmonic. Note that quadrature phase difference is necessary between the upper and lower transistor pairs to allow maximum conversion gain. The conversion gain increases by 5 dB with the help of frequency mixing.

Same as the approach by device nonlinearity, a critical issue of frequency mixing is spurious noise. Although filtering is a common method to pass the desired

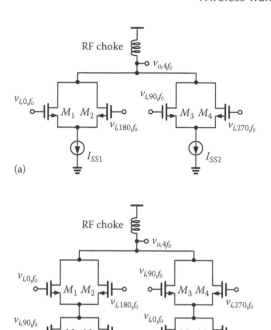

FIGURE 11.10 (a) The quadrupler circuit using linear superposition [4] and (b) the revised circuit with frequency mixing.

signal and suppress noise, on-chip implementation is usually limited by the quality factor of inductive passive elements. On the other hand, a good strategy is to utilize cancellation. Two signals of the same magnitude cancel each other if they are 180° out of phase. If we design two circuit paths in the multiplier to carry out the relative responses of sufficient amplitude and phase matching, the spurious noise will be cancelled. It is well known that the *IC* process features good device matching. Spur rejection can easily achieve 20–30 dB without much design effort. If calibration is introduced, rejection can be even better than 40 dB. The push–push doubler circuit discussed in the previous section essentially utilizes cancellation to remove the fundamental. As to a tripler, the fundamental spur normally goes with the third harmonic at a significant level. In the following, we would like to discuss the design for fundamental cancellation in a frequency tripler of a SHM.

Jackson proposed to apply a feedforward path to cancel the fundamental, as shown in Figure 11.11a [13]. This feedforward path allows the input signal to pass through directly to the output node with appropriate amplitude and phase adjustment of $Ae^{j\varphi}$. Phase shifting is realized using a simple R–C network with a varactor for capacitance tuning. The network introduces a phase shift of $\varphi = -\tan^{-1}(\omega CR)$. A following inverting amplifier tunes for gain matching and also compensate for its loss. This feedforward signal cancels the fundamental spur after the subtractor and improves signal purity. The reported rejection achieves 30 dB. If the tunable range

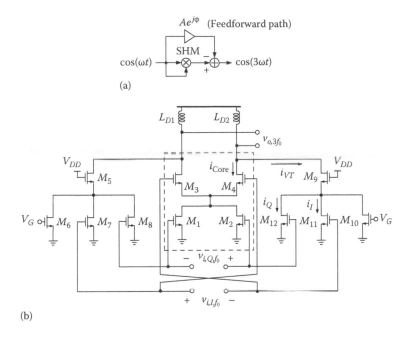

FIGURE 11.11   (a) A feedforward path is applied to cancel the fundamental spur in a frequency tripler of a subharmonic mixer and (b) the circuit implementation using an *I/Q* combiner in [14].

is concerned, the *R–C* network allows a quite limited phase range. Another thought is to apply an *I/Q* vector combiner for a larger range [14]. The combiner circuit is depicted along with a SHM tripler in Figure 11.11b. Transistors M_{5-12} are two sets of the combiner circuit. $M_{7,8}$ and $M_{11,12}$ provides vector combination of the input *I/Q* signals such that the output current i_{VT} is tuned to the correct amplitude and phase, and match with the tripler output current, i_{core}. The cascode configuration using the common-gate transistors $M_{5,9}$ avoids loading effect on the tripler output. The bias condition is assisted by $M_{6,10}$ to sustain the bias current. This combiner design operates in current domain and consumes less dc power. The measured fundamental rejection achieves 35 dB.

11.5   FREQUENCY MULTIPLICATION BY INJECTION LOCKING

Injection locking is a fascinating phenomenon in physics. It occurs to oscillators in circuit systems. Upon injection of an external signal, the oscillation frequency is changed and locked to that of the signal. The mathematical formulation of this non linear behavior was pioneered by Adler [15] and recently reworded by Razavi [16]. Numerous authors have studied in the past. Readers can follow their tutorial discussions on the fundamental theory and find more references in details. The behavior has been found useful in RF integrated circuit design. In this section, we will discuss the application to a frequency multiplier.

The general requirement to sustain injection locking is that the difference between the free-running frequency ω_0 and the injecting frequency ω_{inj} must be within the locking range ω_L, derived as

$$\omega_L = \frac{\omega_0}{2Q} \cdot \frac{I_{inj}}{I_{osc}} \cdot \frac{1}{\sqrt{1-(I_{inj}/I_{osc})^2}}, \qquad (11.14)$$

where

Q stands for the resonator quality factor of the oscillator
I_{inj} is the injecting current
I_{osc} is the oscillator current

It can be seen that the locking range is larger if Q is smaller and if I_{inj} is higher. The beauty of injection locking is also observed that I_{inj} could be less than I_{osc}. That means a small signal can control the system of a large signal, similar to the switching control by a low-power signal in a relay circuit. The application of injection locking to frequency multipliers can utilize this property. Consider a multiplier for high-frequency signal generation. It is preferred the multiplier carries out a large multiplication factor without cascading several stages. Although the conventional design using a nonlinear device or a hard-limiter brings in high-order harmonics, the conversion efficiency tends to be low. The output power of the specified harmonic might require further amplification for practical usage. On the other hand, a fundamental oscillator is power-efficient but with poor phase noise. If we *purify* the fundamental oscillator output through the low-level high-order harmonic injection, the advantage of this composite method becomes obvious. That is, we inject the harmonic to lock and improve the phase noise of the fundamental oscillator. The oscillator can be considered as an efficient amplifier to boost the signal level at the high frequency.

Let's apply injection locking to design a tripler. We begin with the oscillator. The *LC* cross-coupled oscillator is commonly used at mmWave frequencies, as shown in Figure 11.12a. Together with parasitic capacitance at the transistor drain ports, the inductors L_{S1} and L_{S2} form resonators at a fixed frequency in this circuit. The cross-coupled transistor pair, $M_{1,2}$, results in negative input resistance to compensate for energy loss in the resonator. Given sufficient tail current I_{SS}, we can obtain stable oscillation. As a tripler, the free-running frequency of the oscillator needs to be close to three times of the input frequency. The injection point of the external current I_{inj} is chosen at the transistor drain ports, which are also the oscillator outputs nodes. Normally I_{inj} is smaller than I_{osc}. If it meets the locking condition, the oscillation frequency of V_{out} will follow that of I_{inj}.

The circuit schematic of the harmonic generator is illustrated in Figure 11.12b. The differential injection current comes from the third harmonic current of the differential pair, $M_{3,4}$. Operated in strong nonlinearity, $M_{3,4}$ have the output impedances causing significant loading effect on the oscillator. To reduce the effect, we add common-gate transistors $M_{5,6}$ as current buffers so that the cascode configuration offers better isolation. Although the fundamental and second harmonic currents also go injecting to the oscillator, only the third harmonic can lock the oscillator and

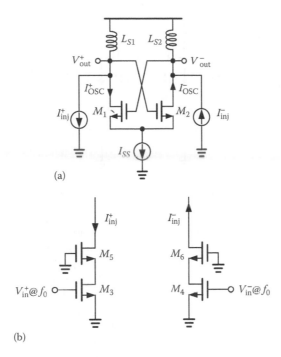

FIGURE 11.12   (a) An LC cross-coupled oscillator with differential injecting currents at the drain ports. (b) The differential injecting current comes from a harmonic generator using device nonlinearity.

turn it into a tripler. Due to the property of injection locking, the operation frequency range is typically smaller than that by other approaches. We need to pay attention to two things. The locking range is determined by the magnitude of third harmonic injection current and the Q-value of the LC tank. Consequently, we optimize the bias condition of $M_{3,4}$ to maximize the third-order harmonic output current, and we choose a low Q-value to enlarge the locking frequency range, namely, the operation range. If driven by V_{in} of sufficient amplitude, the circuit in Figure 11.12 becomes a frequency tripler.

A critical issue of the injection-locked tripler still exists. It is the spectral purity. To an LO signal, spurious noise is required as low as possible to avoid any undesired frequency conversion of interferers. Although oscillator locking is limited to the third-order harmonic frequency range, the fundamental and the second harmonic still directly feed through and appear at the output. Suppression is insufficient by the low-Q oscillator tank. It is typical to apply additional external filtering. We will discuss a harmonic generator with spur suppression.

Instead of using an external filter at the oscillator output, spurious noise is minimized in the harmonic generator before injecting into the oscillator. The revised harmonic generator in [17] is as shown in Figure 11.13. The circuit is verified by a 60 GHz tripler in 0.13 μm RF CMOS technology. Several design techniques are utilized in the circuit, including a harmonic generator $M_{3,4}$ with inductive load $L_{S1,2}$ for the third-order harmonic enhancement, a capacitive cross-coupled pair

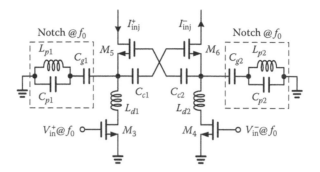

FIGURE 11.13 Improved harmonic generator design, which suppresses the fundamental and the second harmonic output currents.

$M_{5,6}$ for the even-order harmonic rejection, and a notch filter for the fundamental suppression. $M_{3,4}$ generate all harmonic currents of the fundamental signal at the frequency f_0. Inductors $L_{d1,2}$ are placed to result in the loading impedance as low, low, and high at the frequency of f_0, $2f_0$, and $3f_0$, respectively, to enhance the third-order output current. Also, the gate bias voltage, V_{GS}, affects the third-order nonlinearity. The optimal bias voltage, $V_{GS3,4}$, and the load inductance, $L_{d1,2}$, are determined as V_{GS} of 0.6 V and L_d of 200 pH. The third-order harmonic output current achieves 612.5 µA. The cascode configuration of $M_{5,6}$ buffers the third harmonic current and reduces the loading effect on the oscillator tank. Furthermore, the capacitive cross-coupling can effectively reject all common-mode outputs of even-order harmonic currents from $M_{3,4}$ as in [18]. Connected in shunt to the signal path, the notch filter bypasses the fundamental spur with low input impedance at f_0, and sustains the third-order harmonic injection current with high impedance at $3f_0$. A high-order filter is applied, consisting of L_p, C_p, and C_g. Its input impedance is derived as

$$Z_N = \frac{1+s^2 L_p(C_p+C_g)}{s(1+s^2 L_p C_p)C_g},$$
(11.15)

which carries out a zero frequency at

$$\omega_z = \frac{1}{\sqrt{L_p(C_p+C_g)}},$$
(11.16)

and a pole frequency at

$$\omega_p = \frac{1}{\sqrt{L_p C_p}}.$$
(11.17)

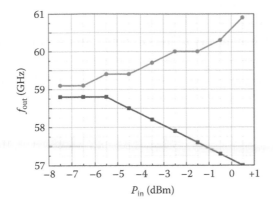

FIGURE 11.14   The measured locking range of the LC oscillator.

The values of L_p, C_p, and C_g are selected to result in f_z at f_0 and f_p at $3f_0$. By doing so, we ensure the fundamental is suppressed significantly.

The differential tripler circuit is characterized single-ended at one output while the other output is terminated into 50 Ω load. The measured free-running frequency of the oscillator is at 58.9 GHz. The locking range increases as the input power increases. It achieves 3.9 GHz at the input power of 0.5 dBm as shown in Figure 11.14. The measured harmonic output power after calibration of the cable loss is shown in Figure 11.15a, over different input power levels and the frequency range of interest. The suppression of the undesired fundamental and the second-order harmonic is quite consistent over the entire frequency band of interest. The maximum conversion gain of −10 dB occurs at the input frequency of 19.7 GHz. The output 3 dB bandwidth is about 2.18 GHz. The HRRs of the fundamental and the second order with the input signal at 19.7 GHz are 31.3 dBc and 45.8 dBc, respectively. The measured phase noise using an input signal at 19.65 GHz shows

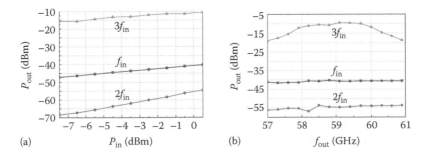

FIGURE 11.15   The measured harmonic output power over (a) the input power level and (b) the frequency.

TABLE 11.3

Performance Summary of the Designed Injection-Locked Tripler

Technology	CMOS 0.13 μm
Free-running frequency	58.92 GHz
Locking range	3.9 GHz
f_{in}	19.7 GHz
P_{in}/P_{out} (dBm)	0.5/−9.5
HRR_1	31.3 dBc
HRR_2	45.8 dBc
Power consumption	9.96 mW

an increase around 10 dB at the injection-locked output, in good agreement with the theoretical value of 9.5 dB. The circuit performance of this tripler is summarized in Table 11.3.

REFERENCES

1. S.K. Reynolds, B.A. Floyd, U.R. Pfeiffer, T. Beukema, J. Grzyb, C. Haymes, B. Gaucher, and M. Soyuer, A silicon 60-GHz receiver and transmitter chipset for broadband communications, *IEEE J. Solid-State Circ.*, 41(12), 2820–2831, December 2006.
2. U.R. Pfeiffer, E. Ojefors, and Y. Zhao, A SiGe quadrature transmitter and receiver chipset for emerging high-frequency applications at 160 GHz, in *IEEE International Solid-State Circuits Conference, Digest of Papers*, San Francisco, CA, pp. 416–417, February 2010.
3. D.B. Leeson, A simple model of feedback oscillator noise spectrum, *Proc. IEEE*, 54, 329–330, February 1966.
4. D. Huang, T.R. LaRocca, M.-C.F. Chang, L. Samoska, A. Fung, R.L. Campbell, and M. Andrews, Terahertz CMOS frequency generator using linear superposition technique, *IEEE J. Solid-State Circ.*, 43(12), 2730–2738, December 2008.
5. E. Camargo, *Design of FET Frequency Multipliers and Harmonics Oscillators*, Artech House, Norwood, MA, 1998.
6. M. Cohn, J.E. Degenford, and B.A. Newman, Harmonic mixing with an antiparallel diode pair, *IEEE Trans. Microw. Theory Tech.*, 23(8), 667–673, August 1975.
7. J. Ward, E. Schlecht, G. Chattopadhyay, A. Maestrini, J. Gill, F. Maiwald, H. Javadi, and I. Mehdi, Capability of THz sources based on schottky diode frequency multiplier chains, in *IEEE International Microwave Symposium, Digest of Papers*, Fort Worth, TX, pp. 1587–1590, June 2004.
8. A. Boudiaf, D. Bachelet, and C. Rumelhard, A high-efficiency and low-phase-noise 38-GHz pHEMT MMIC Tripler, *IEEE Trans. Microw. Theory Tech.*, 48(12), 2546–2553, December 2000.
9. L. Dussopt and G.M. Rebeiz, A low phase noise silicon 18-GHz push–push VCO, *IEEE Microw. Wireless Compon. Lett.*, 13(1), 4–6, January 2003.
10. P.-C. Huang, R.-C. Liu, H.-Y. Chang, C.-S. Lin, M.-F. Lei, H. Wang, C.-Y. Su, and C.-L. Chang, A 131 GHz push–push VCO in 90-nm CMOS technology, in *IEEE RFIC Symposium, Digest of Papers*, Long Beach, CA, pp. 613–616, June 2005.

11. Z. Zhang, Z. Chen, L. Tsui, and J. Lau, A 930 MHz CMOS DC-offset free direct-conversion 4-FSK receiver, in *IEEE International Solid-State Circuits Conference, Digest of Papers*, San Francisco, CA, pp. 290–291, February 2001.
12. C.-N. Kuo, H.-S. Chen, and T.-C. Yan, A K-band CMOS quadrature frequency tripler using sub-harmonic mixer, *IEEE Microw. Wireless Compon. Lett.*, 19(12), 822–824, December 2009.
13. B.R. Jackson, F. Mazzilli, and C.E. Saavedra, A frequency tripler using a subharmonic mixer and fundamental cancellation, *IEEE Trans. Microw. Theory Tech.*, 57(5), 1083–1090, May 2009.
14. C.-C. Tsai, D. Chang, H.-S. Chen, and C.-N. Kuo, A 11-mW quadrature frequency tripler with fundamental cancellation, in *IEEE Topical Meeting on Silicon Monolithic Integrated Circuits in RF Systems, Digest of Papers*, New Orleans, LA, pp. 100–103, January 2010.
15. R. Adler, A study of locking phenomena in oscillators, *Proc. IEEE*, 61, 1380–1385, October 1973.
16. B. Razavi, A study of injection locking and pulling in oscillator, *IEEE J. Solid-State Circ.*, 39, 1415–1424, September 2004.
17. C.-N. Kuo and T.-C. Yan, A 60 GHz injection-locked frequency tripler with spur suppression, *IEEE Microw. Wireless Compon. Lett.*, 20(10), 560–562, October 2010.
18. M.-C. Kuo, S.-W. Kao, C.-H. Chen, T.-S. Hung, Y.-S. Shih, T.-Y. Yang, and C.-N. Kuo, A 1.2 V 114 mW dual-band direct-conversion DVB-H tuner in 0.13-μm CMOS, *IEEE J. Solid-State Circ.*, 44, 950–961, March 2009.

Section III

Biomedical and Short-Range Radios

12 CMOS UWB Transceivers for Short-Range Microwave Medical Imaging

Andrea Bevilacqua

CONTENTS

Medical imaging techniques are commonly used for clinical purposes to create images of the human body and to reveal, diagnose, or examine diseases. Conventionally, waves at both extrema of the frequency spectrum are used, such as ultrasound waves and x-rays. However, recent research efforts have pointed out how the microwave radiation can constitute a valuable medium for imaging the human body, as the electromagnetic (EM) properties of various human tissues differ from each other. Microwaves are sensitive to mismatches in the EM parameters of the materials since their propagation is altered by them. As a consequence, by monitoring the reflection (or the transmission) of the waves, one can detect the EM properties and create an image of the volume that is being illuminated.

As a matter of fact, the availability of custom hardware is a necessary condition to perform microwave imaging in an effective way. Although microwave imaging systems based on off-the-shelf components are in principle possible, dedicated integrated circuits tailored to cover the specific requirements of diagnostic imaging can improve the performance and reduce the size and cost of the system. In perspective, application-specific integrated circuits (ASICs) realized in mainstream technologies, such as complementary metal-oxide-semiconductor (CMOS), have the potential of enabling a wider and more diffused screening of the population with respect to several diseases, increasing the possibility of an early diagnosis and improving the chances of success of the cures.

12.1 OVERVIEW OF MICROWAVE MEDICAL IMAGING

In the last few years, a significant growth of the research involving the use of microwaves to image the human body has been taking place. There are several possible approaches to perform short-range microwave imaging, spanning from tomographic approaches [4,20,34,55,66] to radar-based ones [14,18,19,23,25,26,33, 37,42,48,49,63]. Microwave imaging has been explored as a complementary technique for the early diagnosis of breast cancer [9,14,23,26,29,45,51], as well as for the monitoring of brain stroke [27], analysis of joint tissues [52], cardiac tissues [56], soft tissues [54], and bones [58].

In particular, among the many examples of ongoing research, the use of microwaves for breast cancer diagnostic imaging has seen an increase of interest [9,23,29,39,45,51]. As reported in [1], and illustrated in Figure 12.1, breast cancer is one of the most incident tumors among female population. Since 95% cure rates are possible if the tumor is detected in its early stages, early diagnosis is a key factor in delivering long-term survival of breast cancer patients [1]. The mammography, consisting of x-ray imaging of the compressed breast, is the most commonly used diagnostic technique to detect nonpalpable breast tumors [57]. However, ionizing radiations cause health hazards, while breast compression induces considerable discomfort in patients. Moreover, circumstances like the presence of dense glandular tissue around the tumor, the absence of microcalcifications in the early stages of the disease, and tumors located close to the chest wall or underarm result in a 10%–30% rate of false negatives. The fraction of positive mammograms ending up in the diagnosis of an actual malignancy is <10% [32,57].

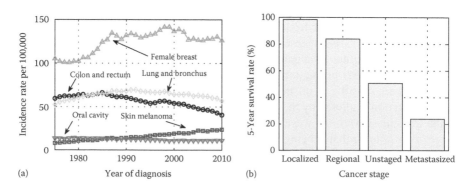

FIGURE 12.1   (a) Age-adjusted tumor incidence rates per 100,000 people grouped by cancer site and (b) 5-year survival rate for breast cancer. (Modified from Bassi, M. et al., *IEEE Trans. Circ. Syst. I: Regular Pap.*, 59(6), 1228, June 2012.)

The limitations of x-ray mammography have stimulated the development of complementary imaging tools, such as those based on microwaves [24,51]. The ultra wideband (UWB) radar imaging technology [9,14,18,19,23,25,26,29,33,37,39,42,48, 49,63] seems particularly promising. This technology leverages the contrast between the dielectric properties of benign and neoplastic tissues at microwave frequencies [44] to identify the presence and location of significant scatterers. The general approach is to illuminate the breast with a UWB pulse from a number of antenna locations. Due to the difference in the dielectric properties, the waves scatter at each dielectric interface. Consequently, the retrieved waveforms contain information about the scattering object, such as the distance and size. By collecting and postprocessing the backscattered signals, a high-resolution dielectric map of the breast can be generated, enabling the detection of tumors, even in their early stage of development.

The first clinical trials employing UWB radar imaging have already been performed [23,39], leading to encouraging results. In these tests, a common feature is the use of microwave laboratory equipment to implement the radar transceiver. As an example, the system presented in [39] requires an electromechanical switching system to interface its 60-element antenna array to a vector network analyzer (VNA), used as the radar transceiver. The switching system introduces losses and limits the number of simultaneous measurements, thus increasing the acquisition time. Overall, the system is quite bulky, and it requires to be mounted on a hydraulic trolley for transportation. In order to improve the performance and reduce the size and cost of the system, the development of a dedicated integrated circuit, tailored to cover the specific wide bandwidth required by medical imaging, while achieving very large dynamic range, is seen as a critical need [51]. The miniaturization involved by the system integration leads to envisioning an antenna array made of modules in which each antenna of the array is directly assembled together with the radar transceiver ASIC. A switching system is therefore avoided along with any high-frequency interconnects. Only signals at low frequencies are to be distributed to the array elements. At the same time, having a transceiver for each antenna removes any limitation on the number of simultaneous measurements that can be performed. A more compact, higher performance, and lower-cost system results.

Enabled by the advances in microelectronic technologies, a clear trend toward the integration of radar imaging systems for medical as well as for security (such as detection of concealed weapons) and industrial applications (such as nondestructive material testing) has emerged in recent years [3,9,35,47,59,60]. As the scaling of the device features improves the performance of standard CMOS technologies, the development of integrated solutions for the radar front-ends in such mainstream technologies has also become a reality [16,35,47,59], with possible benefits in terms of reduced costs and larger diffusion of the radar systems.

In the following, the development of an ASIC for microwave medical radar imaging is described. All the challenges faced during the design, both at the system and at the circuit level, are discussed in detail. As a case study, the implementation of a complete radar transceiver in a 65 nm CMOS technology [16] is reported.

12.2   RADAR IMAGING TECHNIQUE

Conceptually, the radar operation is based on the transmission of a short EM pulse, and the subsequent reception of its echo originated by a reflective target. Assuming, for simplicity, a point scatterer, from the time of flight τ of the EM pulse and the wave velocity v, the distance R of the target from the radar can be retrieved as

$$R = \frac{v \cdot \tau}{2}. \tag{12.1}$$

Clearly, a single radar measurement with an omnidirectional antenna is not sufficient to univocally determine the target location in the space as the target can lay anywhere on a spherical surface of radius R. Traditionally, highly directive antennas are used to solve this issue. Antenna arrays are another possible approach. By repeating the radar measurement using antennas in different (known) physical locations, and combining the results, univocal localization of the target is possible, at the intersection of the various spherical surfaces associated to the antennas, as sketched in Figure 12.2.

Since the radar pulse has a finite duration, as opposed to an infinitesimal one, the measurement of R is performed with an uncertainty, related to the pulse duration itself. In other words, the spherical surface associated to each radar measurement is not an infinitesimal surface, but rather it has a finite thickness due to the pulse duration, as illustrated in Figure 12.2. The shorter the pulse duration, or, correspondingly, the larger the pulse bandwidth B, the more precise the ranging measurement.

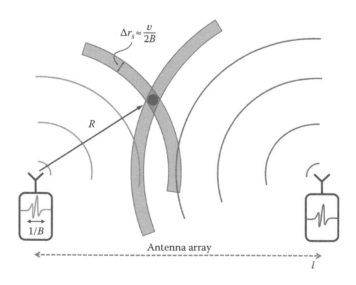

FIGURE 12.2   Antenna array radar concept.

The achievable spatial resolution in the slant range, Δr_s, that is, in the direction of the pulse propagation, is directly related to B as [62]

$$\Delta r_s \approx \frac{v}{2B}. \tag{12.2}$$

Geometrical parameters, such as the number of antennas in the array and the antenna-to-antenna distance, influence the resolution in the cross range, that is, on the plane orthogonal to the direction of the pulse propagation. The resolution in the cross range, Δr_c, is [30,31]

$$\Delta r_c \approx \frac{R}{l}\frac{v}{B}, \tag{12.3}$$

where l is the equivalent synthetic aperture length of the antenna array, that is, the size of the antenna array.

Instead of using the radar array operation to determine the position of a single scatterer, one can also use a beam-forming algorithm to determine whether the volume that has been illuminated contains targets at all. There are many possible algorithms to achieve this result. The delay-and-sum algorithm (see, e.g. [48]) is the simplest kind of processing approach. The idea is to combine the signals detected by the antennas in the array to calculate the intensity of the radiation backscattered by a point of coordinates (x, y, z) belonging to the illuminated volume. Repeating the procedure for a set of different pixels, an image is created. The N waveforms received at the array antennas be

$$I_i(t), \quad i = 1,\ldots,N. \tag{12.4}$$

The image is built as follows. First, the round-trip time $\tau_i(x, y, z)$ from the ith antenna of coordinates (x_i, y_i, z_i) to the pixel of coordinates (x, y, z) is derived:

$$\tau_i(x, y, z) = \frac{2\sqrt{(x - x_i)^2 + (y - y_i)^2 + (z - z_i)^2}}{v}. \tag{12.5}$$

Then, the time-domain signals $I_i(t)$ are time-shifted by an amount equal to the calculated round-trip time $\tau_i(x, y, z)$. In this way, the information on the considered pixel embedded in the various time-domain signals is aligned in time to the point $t = 0$. Finally, the intensity $\mathcal{I}(x, y, z)$ of the pixel of coordinates (x, y, z) is calculated by coherently summing the contributions of all the antennas

$$\mathcal{I}(x, y, z) = \left[\sum_{i=1}^{N} I_i(\tau_i(x, y, z))\right]^2. \tag{12.6}$$

The set of intensities $\mathcal{I}(x,y,z)$ for all the points (x, y, z) belonging to the volume/ plane of interest is nothing but a map of the EM properties of the illuminated object, that is, it is a radar image.

12.2.1 PULSED RADAR VERSUS STEPPED-FREQUENCY CONTINUOUS WAVE RADAR

The spatial resolution of the radar imaging technique is directly related to the band width of the transmitted and received EM pulses, as indicated by (12.2) and (12.3). In order to achieve sub-cm resolution, bandwidths in excess to 5 GHz are needed, even taking into account that the EM wave velocity into the human body is at least three times slower than in air.

Dealing with pulses with such large bandwidths can be extremely difficult, especially if large dynamic ranges are concurrently required. Incidentally, this is exactly the case of the medical imaging applications. Think, for example, to the requirements of the analog-to-digital converter (ADC) placed at the end of the radar receiving chain. Guaranteeing at the same time a wide analog bandwidth (>5 GHz) and a high resolution (≥16 bit) is almost impossible in any available microelectronic technology, unless at the price of an extremely high power consumption.

A way out of this dead end is provided by radar operation performed in the frequency domain as opposed to the pulsed time-domain approach discussed so far. The idea is to leverage the stepped-frequency continuous wave (SFCW) approach [9,16,62]. A single-frequency tone is transmitted, backscattered by the radar target and then received, as sketched in Figure 12.3. The procedure is repeated stepping the tone frequency $k\Delta f$ over a broad span, that is, for $k = 1, ..., M$. From the collected spectral data, $S_i(k\Delta f)$, a synthetic time-domain pulse is retrieved by means of the inverse Fourier transform:

$$S_i(k\Delta f) \xrightarrow{\mathcal{F}^{-1}} I_i(k\Delta t) . \tag{12.7}$$

The synthetic pulse $I_i(k\Delta t)$, with $\Delta t = 1/(N_{\text{IFFT}}\Delta f)$ and $N_{\text{IFFT}} = 2M + 1$, carries the same information of the time-domain pulse defined in (12.4), allowing for further image processing to be performed in the same fashion as in (12.5) and (12.6). The SFCW transceiver operates similarly to a VNA, taking advantage of narrow instantaneous noise bandwidths yielding large dynamic ranges, while preserving the overall wide system bandwidth.

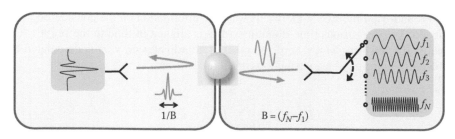

FIGURE 12.3   Pulsed versus SFCW radar approach.

The frequency step Δf sets, in the SFCW approach, the maximum unambiguous range, R_{max}, that is, the maximum distance a target can be in order to be reliably detected without spatial aliasing:

$$R_{max} = \frac{v}{2\Delta f}. \tag{12.8}$$

This is quite different as compared to the pulsed approach, where, in the common case that more than one pulse is transmitted and received for each antenna in the array, R_{max} is set by the maximum time of flight allowed to each pulse, that is, the time interval between two consecutive pulses, T_{PRI}:

$$T_{PRI} = \frac{2R_{max}}{v}. \tag{12.9}$$

In the pulsed approach, R_{max} thus influences the overall time needed for a measurement, while in the SFCW radar the time the system spends transmitting and receiving a tone at each frequency $k\Delta f$ is set by the system noise bandwidth, and not by Δf. Note that, typically, SFCW measurements start from a lower bound frequency that is different from Δf, but rather is equal to $k_{min} \Delta f$, with $k_{min} > 1$. The inverse Fourier transform (12.7) is thus performed on the samples $S_i(k\Delta f)$ setting $S_i(k\Delta f) = 0$ for $k = 0, \ldots, k_{min} - 1$. Moreover, in order to guarantee that $I_i(k\Delta t)$ is a real-valued pulse, Hermitian symmetry is ensured by setting

$$S_i((N_{IFFT} - k)\Delta f) = S_i^*(k\Delta f). \tag{12.10}$$

One interesting feature of the SFCW approach is that the inverse Fourier transform operation leads to a processing gain, that is, an improvement in the signal-to-noise ratio (SNR), roughly equal to M. To understand this, consider that at the radar receiver end, the desired signal samples, $S_i(k\Delta f)$, are corrupted by noise samples, $\Xi_i(k\Delta f)$. The variance of the noise samples, σ_Ξ^2, is set by the receiver noise figure, F, and the noise bandwidth, B_{noise}. Assuming, for simplicity, that $k_{min} = 1$, and that all the signal samples have the same amplitude, $S_i(k\Delta f) = H_0$, the SNR for each single-tone measurement is

$$SNR_{CW} = \frac{H_0^2}{\sigma_\Xi^2}. \tag{12.11}$$

After the inverse Fourier transform, the frequency-domain noise samples are converted to the time domain as

$$\xi(k\Delta t) = \Delta f \sum_{m=0}^{N_{IFFT}-1} \Xi(m\Delta f)e^{j2\pi mk/N_{IFFT}}. \tag{12.12}$$

Taking into account that the noise samples $\Xi_i(k\Delta f)$ are pair-wise correlated because of (12.10), and otherwise uncorrelated, the variance of $\xi(k\Delta t)$ is calculated as

$$\sigma_\xi^2 = \frac{4M\sigma_\Xi^2}{(2M+1)^2(\Delta t)^2} \approx \frac{\sigma_\Xi^2}{M(\Delta t)^2}. \tag{12.13}$$

The time domain desired signal is, with the assumptions previously made, a discrete-time Dirac pulse:

$$I_i(k\Delta t) = \frac{H_0}{\Delta t}\delta_{kp}, \tag{12.14}$$

where
   δ_{kp} is the Kronecker delta function
   p is the index of the time bin where the synthetic pulse is located

Hence, the SNR calculated after the inverse Fourier transform is

$$\mathrm{SNR}_{\mathrm{IFFT}} = \frac{H_0^2/(\Delta t)^2}{\sigma_\xi^2} = M\frac{H_0^2}{\sigma_\Xi^2}, \tag{12.15}$$

that is M-fold larger as compared to (12.11).

The overall measurement time required for a complete SFCW frequency sweep, $t_{m,\,\mathrm{SFCW}}$, is set by the number of frequency samples to be acquired, M, and by the receiver noise bandwidth, B_{noise}. Assuming the time needed for a single-tone measurement is roughly $T_{\mathrm{SFCW}} = 1/B_{\mathrm{noise}}$, we have

$$t_{m,\mathrm{SFCW}} = M \cdot T_{\mathrm{SFCW}} = \frac{M \cdot L}{B}, \tag{12.16}$$

where $L = B/B_{\mathrm{noise}}$.

As compared to (12.16), a pulsed approach would require a longer time. To make a fair comparison, we assume that the single received pulse is processed by the radar receiver analog front-end and subsequently sampled to become a discrete-time Dirac delta of amplitude H_0. Further, we assume that the variance of the associated noise is $L\sigma_\Xi^2$ since the noise bandwidth of the radar receiver is L-times larger in the pulsed approach as compared to the SFCW radar. For the pulsed radar system to feature the same SNR as the SFCW counterpart, that is, (12.15), multiple $(M \cdot L)$ pulses must be transmitted and received and averaging performed. Thus, the overall measurement time required by the pulsed radar is

$$t_{m,\mathrm{Pulsed}} = M \cdot L \cdot T_{\mathrm{PRI}} = \frac{M^2 \cdot L}{B}, \tag{12.17}$$

where the same maximum unambiguous range is assumed for both the SFCW and the pulsed radar, and the simplifying hypothesis $M = B/\Delta f$ is used.

In summary, the SFCW radar has two main advantages over the pulsed radar approach. The first, and more important, is that noise filtering (averaging) is performed ahead of sampling and quantization, hence greatly simplifying the design of the receiver ADC. The second is that, for a given SNR, the overall acquisition time is shorter in the SFCW approach, as shown by (12.16) and (12.17).

12.2.2 MICROWAVE RADAR IMAGING OF THE HUMAN BODY

As discussed, radar imaging is based on the detection of EM waves backscattered off a target. Consequently, the quality and accuracy of the obtainable radar image is related to the capability of an illuminated object to reflect the incoming EM radiation. In other terms, it is related to its reflectivity. A radar imager can distinguish between different objects only if their reflectivity is remarkably different. At a physical level, the EM radiation gets reflected at an interface between two different materials only if they have different EM parameters, that is, they either have different dielectric (or magnetic) constants or electrical conductivities or both.

The tissues of the human body have EM properties that vary depending on various physical or physiological factors such as water content, vascularization/angiogenesis, blood flow rate, and temperature [25,44]. Moreover, they vary depending on the frequency, that is, human tissues are highly dispersive. The characteristics of each tissue can be described by means of the four-term one-pole Cole–Cole parametric dispersion model for complex permittivity [43]:

$$\epsilon_r(\omega) - j\frac{\sigma(\omega)}{\omega\epsilon_0} = \epsilon_\infty + \frac{\Delta\epsilon_c}{1+(j\omega\tau_c)^{(1-\alpha_c)}} + \frac{\sigma_s}{j\omega\epsilon_0} \qquad (12.18)$$

where
  ω is the angular frequency
  $\epsilon_r(\omega)$ is the frequency-dependent dielectric constant
  $\sigma(\omega)$ is the frequency-dependent conductivity

As an example, a large-scale study on the UWB dielectric properties of normal, benignant, and malignant breast tissues from 0.5 to 20 GHz is documented in [44]. The parameters of the Cole–Cole model for the adipose-dominated normal tissue are $\epsilon_\infty = 3.581$, $\Delta\epsilon_c = 3.337$, $\tau_c = 15.21$ ps, $\alpha_c = 0.052$, $\sigma_s = 0.053$ S/m, while for a malignant tumor, they are $\epsilon_\infty = 6.749$, $\Delta\epsilon_c = 50.09$, $\tau_c = 10.50$ ps, $\alpha_c = 0.051$, $\sigma_s = 0.794$ S/m. The corresponding frequency-dependent dielectric constants and conductivities are shown in Figure 12.4.

The data in Figure 12.4 suggest that the dielectric contrast between normal adipose-dominated tissue and malignant tissue is ranging from a minimum of 6:1 to a maximum of 8:1, depending on the frequency of interest. As a consequence, micro wave radar imaging is in this case not only possible, but indeed effective, showing a

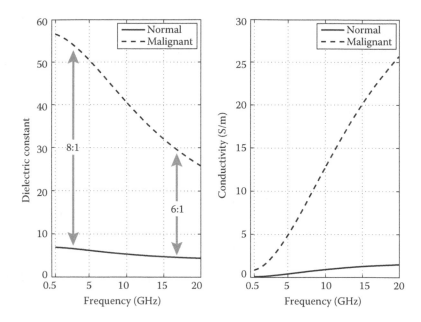

FIGURE 12.4 Comparison between the Cole–Cole curves of the normal (solid) and malignant (dashed) breast tissue. (Reproduced from Bassi, M. et al., *IEEE Trans. Circ. Syst. I: Regular Pap.*, 59(6), 1228, June 2012.)

larger contrast as compared to x-rays [25]. Microwave imaging of other tissues, such as bones or joint tissues, is likewise possible [52,58].

An important issue related to microwave medical radar imaging is that the skin acts as a good reflector itself, such that it is relatively difficult to couple the EM radiation into the body. The use of some coupling media, such as water [51], canola oil [23,52], soybean oil [45,46], alcohol [53], or a matching liquid simulating the EM properties of the body [38,40], has been proposed to reduce the backscattering of the skin and improve the radiation coupling into the body. However, the presence of an air interface between the skin and the coupling medium has experimentally been proven to be a limiting factor [29], such that the case the antennas of the array are in air represents a significant, yet worst-case scenario.

As an example of the magnitude of the skin reflection as compared to the echo of the target to be imaged, the case of diagnostic imaging for breast cancer detection [7] is considered in Figure 12.5. The simulated round-trip attenuation of a signal in the antenna–skin–antenna path ($H_S(f)$) is shown along with the attenuation experienced by the signal in the antenna–tumor–antenna path ($H_T(f)$). The radar antenna is supposed to be in air, at a 1 cm distance from the breast skin. The radar target is a 4 mm diameter tumor placed at 4 cm distance from the antenna. The EM properties of the skin, the healthy tissue, and the cancerous tissue are modeled by means of (12.18) [7]. From the data in Figure 12.5, it is clear that the backscatter from the skin results only attenuated by some 20 dB as the EM waves travel from the radar transmitter to the skin and back to the radar receiver. Moreover, the attenuation is quite constant as a function of frequency. On the other hand, the echo from the actual radar target,

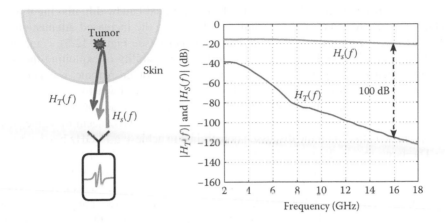

FIGURE 12.5 Attenuation of the EM wave in the antenna–skin–antenna ($H_S(f)$) and antenna–tumor–antenna path ($H_T(f)$) path. (Modified from Bassi, M. et al., *IEEE Trans. Microw. Theory Tech.*, 61(5), 2108, 2013.)

the tumor we want to image, is much weaker. The attenuation of the target backscatter strongly increases with frequency. As a result, the dynamic range the radar transceiver must feature in order to be able to correctly detect the signal due to the tumor backscatter is quite large, and it significantly increases with increasing frequencies. In the scenario of Figure 12.5, the required dynamic range is in excess of 100 dB at 16 GHz. Such a result indicates the great challenge of microwave medical imaging: being able to see objects that are basically laying behind a mirror. As the attenuation the EM wave experiences inside the body at higher frequencies is very high, Figure 12.5 also suggests that the use of mm-waves or THz radiation might not be suitable for the particular application of medical radar imaging.

12.3 TRANSCEIVER ARCHITECTURE AND SYSTEM-LEVEL REQUIREMENTS

The main two tasks the radar transceiver must accomplish are

1. To be able to support an appropriate spatial resolution for medical imaging
2. To feature a high dynamic range to enable the detection of small reflective targets inside the human body

The first requirement is tightly related to the bandwidth of the radar device, as pointed out by (12.2) and (12.3). The second is due to the EM properties of the body and, in particular, to the high attenuation the EM waves undergo within the body at higher frequency, as shown in Figure 12.5.

In order to reach a sub-cm spatial resolution, multi-GHz bandwidths are required. For example, according to [26,29,37,45,50,51,64], to be able to correctly image breast tumors, a 3 mm resolution is adequate. This corresponds to a bandwidth of 14 GHz. Obtaining such a bandwidth at higher center frequencies might seem a good idea.

The system benefits from the reduction of the fractional bandwidth, making it possible to use narrowband design techniques. However, the increased attenuation of the EM waves within the body at higher frequencies (see Figure 12.5), combined with the need of large dynamic ranges, limits the possibility of exploiting the frequency spectrum at the mm-waves and beyond. At the other extreme, it is not possible to leverage lower frequencies and use the spectrum down to dc because of practical (and physical) limitations related to antenna design, ac coupling, etc. As a consequence, frequency ranges such as 1–15 GHz or 2–16 GHz seem to be the best compromise between all the discussed constraints to achieve a 14 GHz bandwidth.

The large required system dynamic range (in excess of 100 dB, as discussed) translates in a requirement of a high-resolution ADC at the end of the radar receiver chain. An ADC with a resolution in excess of 16-bit is necessary. Such an ADC is not compatible with a passband of 14 GHz. Consequently, a pulsed radar approach is not a doable option. On the other hand, a SFCW radar is compatible with a high-resolution ADC, as the instantaneous bandwidth of the system can be set as narrow as desired, while preserving the overall system bandwidth, and hence the radar spatial resolution. In practice, a lower bound to the noise bandwidth is set by constraints on the overall measurement time, as described by (12.16).

The choice of a SFCW radar architecture sets another challenge to the transceiver design. In fact, if, on the one hand, this approach tackles the dynamic range constrains, on the other hand, a new requirement of phase coherence between the transmitted continuous wave signal and the receiver local oscillator (LO) emerges. Since the SFCW radar operates in the frequency domain, the time-of-flight measurements typical of the pulsed radar operation translate into phase measurements. In order to be able to perform such measurements, the phase relationship between the transmitted tone and the receiver LO must be known a priori. In addition, it must be constant over the frequency range spanned by the radar transceiver.

The easiest way to ensure the phase coherence is to use a direct conversion architecture, that is, to leverage the very same signal that is transmitted as the LO for the radar receiver. Incidentally, the use of such an architecture is compatible with the discussed need for a high-resolution ADC, as the desired signal is translated to dc at the end of the receiver path. A super heterodyne scheme with a low intermediate frequency (IF) could in principle also be used. However, the super heterodyne architecture does not allow an easy solution to the transmitter–receiver phase coherence issue. As in a super heterodyne transceiver the transmitted signal and the LO are at different frequencies, multiple frequency synthesizers are required. The latter are required to be all phase locked to a common reference to guarantee the phase coherence. This complexity is likely to result in higher power consumption to achieve the same set of specifications, further emphasizing a preference for the direct conversion solution. Finally, a marginally higher robustness to phase inaccuracies makes the direct conversion perform slightly better in terms of radar image signal-to-clutter ratio, as singled out in [7].

The SFCW radar must behave as a phase coherent system, in which the phase of the transmitted signal relative to the receiver LO is well known, to be able to correctly determine the range-delay τ of the target. All the quadrature-phase errors in the receiver are therefore critical, as well as any variations in the phase relationship

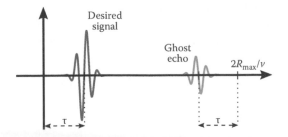

FIGURE 12.6   Impact of a quadrature error in the receiver downconversion: ghost echo.

between the transmitted tone and the LO signal. The latter may be due to the LO phase noise or any circuit mismatches [7].

The presence of a fixed quadrature error in the LO signals used in the receiver downconversion process results in the formation of a ghost target echo in the synthetic time-domain portrait $I_i(k\Delta t)$ obtained by the inverse Fourier transform described by (12.7). A sketch of the signal after the Fourier processing in presence of a large fixed quadrature error is shown in Figure 12.6. The spurious pulse is located at the time instant $2R_{max}/v - \tau$, and it contributes to increase the clutter level in the resulting radar image. The presence of a fixed error in the phase offset between the transmitted tone and the receiver LO results instead in an error in the determination of the pulse range, and consequently in smearing and blurring effects in the derived radar image. In general, the mentioned phase errors are frequency dependent, or even random in nature, as in the case of phase noise. In both cases, they can be modeled as stochastic nonidealities, whose effect is to reduce the amplitude of the desired target response and increase the clutter level. In order to guarantee a large dynamic range, as required by the medical radar imaging applications, the transceiver must feature quadrature errors and cumulative phase noise standard variation well below 1.5° [7].

The required transmitter output power depends on the specific application scenario. However, some general considerations can be carried out. Considering the example of the breast cancer detection, the signal bouncing off the skin and reaching the radar receiver is just attenuated by some 20 dB, as shown in Figure 12.5. Hence, if the transmitter outputs some −15 to −10 dBm, the maximum input signal at the receiver side will be in the neighborhood of −35 to −30 dBm. The receiver 1 dB compression point will be specified consequently. If the transmitted power is increased, the receiver 1 dB compression point will have to increase as well, such that the overall transceiver power consumption will also increase. With a receiver maximum input signal of −35 dBm, a dynamic range in excess of 100 dB can be obtained with a receiver noise figure of 9 dB (or less) and a noise bandwidth of 1 kHz. As a consequence, there is not any particular reason to specify a much larger transmitter output power.

The direct conversion is the preferable architecture in the design of a radar transceiver for medical imaging applications, as discussed. However, using a direct conversion architecture means that the spurious tones (e.g., the reference spurs arising from the frequency generation) may be problematic as intermodulation distortion products may fall on top of the desired signal. Assuming the receiver has an input-referred third-order intercept point (IIP3) of about −20 dBm (a number compatible with the aforementioned −30 dBm 1 dB compression point specification), the maximum relative level of the spurs (S_l) that can be tolerated is

$$S_l \leq -\frac{3P_{RX} - 2\,\text{IIP3} - P_d}{3} = -23\,\text{dBc}, \qquad (12.19)$$

where
  P_{RX} = −35 dBm is the maximum signal we expect to receive
  P_d = −135 dBm is the maximum distortion we can tolerate to have a 100 dB dynamic range

Clearly, a very loose spurious specification is set on the radar transmitter. Conversely, the harmonics of the transmitted signal may limit the performance of the radar system. In fact, while a quadrature downconverter implemented in differential fashion inherently rejects the even harmonics, it is sensitive to the odd ones. The signal at the Nth harmonic has a phase information equal to $N\Delta\varphi$, if $\Delta\varphi$ is the desired information at the fundamental; it can therefore be a source of error in the phase measurement. At first sight, a harmonic rejection of some 100 dBc should be requested for the transmitter. However, some considerations are in order: First, the harmonics fall in the receiver band only for fundamentals in the lower part of the operating frequency band. For example, assuming the radar band is from 2 to 16 GHz, the third harmonic falls in band only for fundamental frequencies up to 5.3 GHz. Second, assuming the mixer is operated in such a way the LO signal can be modeled as a square wave (because of both the native waveform of the LO and the large-signal operation of the mixer devices), the receiver will suppress the odd harmonics as $1/N$, that is, in excess of 10 dBc. Third, as the output signal is tapped at the mixer LO port, it is also a square wave. Hence, the odd harmonics of the transmitted signal are, right from the beginning, at least 10 dBc smaller than the fundamental tone. Fourth, since the attenuation of the target echo increases with the frequency, the dynamic range requirement for the system is more relaxed at lower frequencies. For example, at 5.3 GHz it is lower than 60 dB, in the scenario of Figure 12.5. As a consequence, the harmonic rejection that is to be targeted for the transmitter can be relaxed from 100 dBc to about 40 dBc.

The narrow baseband bandwidth dictated by the SFCW approach and by the high dynamic range requirement make the main drawbacks of a direct conversion architecture, namely second-order distortion and $1/f$ noise, critical impairments and potential show-stoppers. As medical imaging is supposed to be performed in a screened medical environment [2], it is reasonable to assume that no signals other than those generated by the radar system occupy the spectrum, relaxing the required intermodulation performance of the receiver. Still, a large in-band blocker is there, due to the reflection of the skin. The input-referred second-order intercept point (IIP2), calculated in the in-band case, that is, by means of a two-tone test with both the downconverted tones and the intermodulation product within the baseband bandwidth, is the figure of merit used to specify the second-order linearity of the receiver. As counterintuitive as it may sound, a direct conversion receiver for SFCW radar applications appears to be quite immune to second-order nonlinearities. Assuming, as previously done, that the skin backscatter sets the maximum signal level we expect to receive, with a power of −35 dBm, the IIP2 needs to be >20 dBm [7]. At the same time, the $1/f$ noise corner frequency, f_{crn}, must be kept as low as possible not to increase the receiver noise figure and impair the radar dynamic range. The $1/f$ noise contribution is smaller than the thermal noise contribution if the $1/f$ noise corner frequency is $f_{crn} < B_{noise}/[\ln(M)]$.

12.4 DESIGN ISSUES OF THE TRANSCEIVER BUILDING BLOCKS

The discussion carried out so far leads to delineating a block diagram of an integrated SFCW radar transceiver for medical imaging applications, as depicted in Figure 12.7. A direct conversion architecture is preferred over the super heterodyne one, and a wide bandwidth is covered. A reasonable set of specifications for the transceiver follows the guidelines and the derivations previously discussed:

1. Frequency range from 2 to 16 GHz, enabling a 3 mm radar resolution within the body
2. Output power of about −15 dBm, resulting in an expected maximum received signal of about −35 dBm, and thus in a requirement of 1 dB compression point greater than −30 dBm for the receiver
3. Receiver noise figure <9 dB and a receiver noise bandwidth of 1 kHz to achieve a dynamic range in excess of 100 dB
4. IIP2 and IIP3 specified to be >20 and >−20 dBm, respectively
5. Standard deviation of the cumulative phase noise and the quadrature error lower than 1.5° over the entire frequency band of operation
6. Frequency step of $\Delta f = 100$ MHz, as 0.5 m is an adequate maximum unambiguous range to perform short-range imaging of the human body

The impact all these specifications have on the circuit-level design and implementation of the various transceiver building blocks is addressed next.

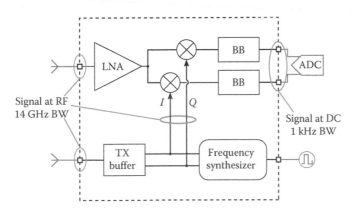

FIGURE 12.7 Conceptual block diagram of a direct conversion transceiver for SFCW medical radar imaging.

12.4.1 Low-Noise Amplifier

The low-noise amplifier (LNA) has the task of providing input matching and gain on the entire frequency band of the radar receiver (2–16 GHz), while featuring a low-noise figure.

Given the SFCW operation, there are several possible options to achieve such a goal. The most straightforward approach is to design a bank of narrow band LNAs to cover sub-bands of the operating frequency range, as illustrated in Figure 12.8a. The amplifiers should be selected depending on the particular frequency tone being transmitted at any given moment during the radar measurement. This approach enables an optimized design for each LNA, depending on the limited frequency range of operation. Each amplifier could be designed as an inductively degenerated single-stage tuned amplifier for minimum noise figure and high gain [41]. An important disadvantage of this approach is the large chip area required to implement the bank of amplifiers, as each one of them requires several integrated inductors. Another big issue is the multiplexing of the received signal to feed the inputs of the LNAs, and the subsequent collection and multiplexing of the LNA outputs. Realizing a low-loss wideband analog multiplexer operating over multiple octaves in the GHz-frequency range is a daunting challenge.

A more compact approach is to implement a single reconfigurable narrow band LNA, as shown in Figure 12.8b. The design philosophy for the amplifier can basically be the same as in the LNA bank, but the reconfigurable LNA requires tunable reactive elements, such as varactors, or banks of switched capacitors. The main burden related to this approach is that the LNA should be automatically tuned and calibrated, as well as it should automatically track the variations in the frequency of the transmitted tone. The overhead of the tuning, calibration, and tracking circuitry greatly increases the complexity of the receiver. Moreover, designing a single reconfigurable LNA able to cover the entire 2–16 GHz band without compromising the performance is still an open issue.

An alternative approach is to design a single wideband LNA, able to operate simultaneously on the entire operation band, without the need of any tuning

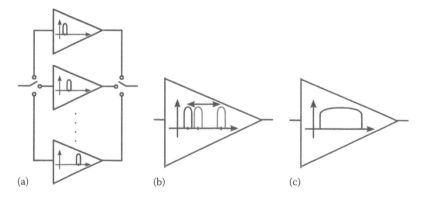

(a)   (b)   (c)

FIGURE 12.8 LNA design for SFCW radar receivers: (a) bank of tuned LNAs, (b) reconfigurable narrow band LNA, and (c) wideband LNA.

or calibration (see Figure 12.8c). Given the specific application of medical imaging, typically carried out in a controlled and screened environment [2], designing a wideband LNA seems the most promising approach. The substantial absence of interferers in a medical environment compensates for the lack of selectivity of the broadband LNA and filtering of the desired input signal in the radio-frequency section of the receiver.

Multiple LNA topologies feature multioctave input matching and broadband amplifier gain. Distributed amplifiers typically require high levels of power consumption and have a large footprint because of the multiple reactive components (inductors and/or transmission lines) they need. Shunt-shunt feedback amplifiers have limited input match at higher frequencies due to the parasitic capacitance of the input devices of the amplifier. Moreover, they also tend to be quite power hungry if a low-noise figure is aimed for. Other topologies leveraging resistive terminations to achieve broadband input matching, such as common-gate amplifiers, or balanced amplifiers, suffer from large noise figures. A multisection reactive input matching network, as proposed in [11,12], is an option to expand the benefits of the inductively degenerated amplifier to wideband operation. However, the combination of the wide fractional bandwidth and the relatively low frequency at the lower edge of the band in medical radar imaging results in an input reactive network featuring many large inductors. As a consequence, this approach is impractical for the considered application. An interesting topology is the noise-cancelling amplifier [13,15], where common-gate and common-source amplifiers are combined. The wideband input matching provided by the common-gate stage is combined to the feedforward cancellation of the noise generated by the common-gate transistor. The noise figure is therefore primarily limited by the common-source stage. Much lower noise figures are achievable as compared to classic common-gate designs.

12.4.2 Receiver Downconverter

In the receiver downconverter, the received signal is mixed with a quadrature-phase LO. The resulting in-phase and quadrature baseband signals make a complex signal whose phase is a measurement and estimate of the phase shift the transmitted tone experiences in the round-trip path to the radar target. In a direct conversion architecture, the SFCW operation dictates that the desired downconverted signal is a purely dc signal, as single tones are transmitted at each step of the radar measurement. This feature, on the one hand, is beneficial in that the bandwidth of the baseband circuitry (i.e., the noise bandwidth) can be as narrow as desired, relaxing the design. On the other hand, dc offsets and $1/f$ noise are absolutely critical impairments. In particular, in CMOS technologies, $1/f$ noise is a problem, as the amount of $1/f$ noise in MOS devices is much larger than, for example, in bipolar transistors.

The dc offsets are due to transistor mismatches and can be calibrated out before the radar measurement is performed. Conversely, $1/f$ noise is by nature unpredictable. The key ingredient to overcome the challenge of $1/f$ noise in CMOS devices is to use passive mixers. This choice results in both good linearity and good noise performance, preventing the flicker noise of the commutating devices to corrupt the downconverted signal [17]. Current-mode mixers have recently received a lot of attention, although voltage-mode operation is also an option.

To reject the $1/f$ noise, dc offsets, and second-order distortion due to the radio-frequency section of the receiver, capacitive ac coupling of the front-end to the passive mixer can be used in a SFCW radar receiver. Typically, transconductor stages are used, in cascade to the LNA, to convert the voltage signal at the LNA output into a signal current that feeds the mixers. A common transconductor stage for the in-phase and quadrature paths can be used if the quadrature LO signals are nonoverlapping 25% duty-cycle clocks. Otherwise, if the LO waveform are more conventional 50% duty-cycle square waves, separate transconductors are required to drive and reciprocally isolate the in-phase and quadrature paths.

The use of passive mixers is not sufficient to address the high flicker noise of MOS transistors. Typically, current-mode mixers are loaded by baseband transimpedance amplifiers (TIAs), based either on common-gate stages or on op-amps with resistive feedback. The flicker noise of the devices of the TIAs is not suppressed, ultimately setting the flicker noise corner of the receiver. To overcome this limitation, chopper stabilization can be used to reduce the flicker corner below 100 Hz. Chopper stabilization is a widespread technique usually applied to voltage amplifiers [5,21,22,65]. The combination of passive current-mode switches and chopper-stabilized TIAs results in a highly linear, low-noise downconversion mixer with a very low flicker noise corner.

12.4.3 FREQUENCY GENERATION

The frequency synthesizer is the most demanding building block in a wideband SFCW radar transceiver. All the complexity saved by avoiding the large instantaneous bandwidth of pulsed operation is transferred to the generation of the frequency tones. The frequency synthesizer is required to generate all the frequencies in the broad range spanned by the SFCW operation. Moreover, it has to produce the quadrature signals needed for the downconversion in the receiver chain with a high accuracy. Achieving these goals is not trivial, when the frequency range of interest is as broad as the one needed for medical imaging, that is, some 14 GHz bandwidth from 1 to 15 GHz or 2 to 16 GHz, as discussed.

A typical approach to generate quadrature signals is leveraging frequency division by two, for example, using a master-slave toggle flip-flop (see Figure 12.9a). However, if the range to convert is, for example, 2–16 GHz, this approach will require a signal source spanning the 16–32 GHz range in order to obtain the other frequencies by means of a cascade of divide-by-two circuits. Designing a frequency synthesizer able to operate in the 16–32 GHz range requires the use of multiple (more than two) voltage-controlled oscillators (VCOs) because of the limited tuning range attainable at such high frequencies. Moreover, the high frequencies involved make this a power-hungry approach.

If a phase-locked loop (PLL) is employed to generate the tones in the higher octave (e.g., 8–16 GHz), a technique different from frequency division is needed to obtain the quadrature signals. Polyphase filters (PPFs) are an option [10,28], as shown in Figure 12.9b. However, in order to achieve a small quadrature error over

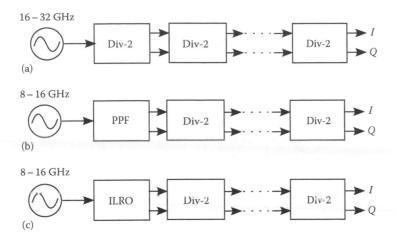

FIGURE 12.9 Frequency and quadrature generation schemes for multioctave SFCW transceivers: (a) PLL at double the higher frequency octave followed by a chain of divide-by-two frequency dividers, (b) PLL at the higher octave followed by a PPF and frequency dividers, (c) PLL at the higher octave followed by an injection-locked ring oscillator (ILRO) and frequency dividers.

an octave frequency range, in a robust fashion with respect to process variations, multisection filters are to be used. Such passive circuits introduce losses that have to be compensated by driving the PPF with suitable active circuits, while regenerating the signals after they have been processed by the filter. Such auxiliary circuits tend to consume a lot of power in excess to the one needed for the PLL.

Having a source capable of producing quadrature signals in the higher octave seems the preferable option as frequency division can be then used to extend the range of generated tones toward lower frequencies. Quadrature VCOs embedded in a PLL can indeed achieve this goal. However, there is a trade-off between the phase noise of quadrature oscillators and the accuracy of the quadrature sequence. Moreover, LC quadrature VCOs need multiple LC tanks, occupying a larger silicon area, while ring oscillators do not achieve the good phase noise performance required by the radar imaging applications. A technique to break all the aforementioned trade-offs is injection locking. A PLL is employed to produce spectrally clean tones, which are then injected into a ring oscillator to which is demanded the task of outputting accurate quadrature phases (see Figure 12.9c). Due to the injection-locking operation, the poor phase noise performance of the ring oscillator is suppressed and replaced by that of the LC VCO-based PLL. This approach has the additional advantage that cascading multiple ring oscillators, that is, by injecting the outputs of one ring oscillator into another ring oscillator, the quadrature error is progressively reduced without impairing the phase noise performance [36].

Another issue related to the frequency generation for a multioctave SFCW radar transceiver is that if the quadrature tones covering each octave are output by different blocks, then some means of analog multiplexing is needed to interface such signals to the receiver mixers and the transmitter output stage, as illustrated in

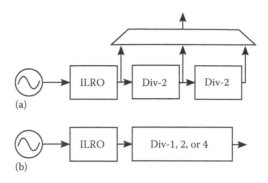

FIGURE 12.10 Multioctave frequency generation: (a) a cascade of frequency dividers needs signal multiplexing; (b) the use of a programmable divider avoids multiplexing.

Figure 12.10a. For example, if the 2–16 GHz range is addressed by using a 8–16 GHz PLL, and ring oscillator injection locking, to cover the higher octave, followed by the cascade of two conventional divide-by-two circuits, then some circuitry is needed to select among the three outputs. In this way, leakages, or changes in the load or the circuit environment, could possibly affect the quadrature accuracy, or introduce sudden phase jumps in the LO between different configurations, all being critical impairments in a radar system.

A preferable strategy is to employ a reconfigurable frequency divider, able to divide by one, two, or four, as proposed in [16] (see Figure 12.10b). In this way, the LO distribution chain is unaltered in any possible configuration, regardless the generated tones are in the higher frequency octave or in a lower one. The key to achieve reconfigurable operation at the high frequencies needed for radar imaging is multiphase injection locking. Starting from a PLL producing all the frequencies from 8 to 16 GHz and then obtaining the quadrature phases by means of injection-locked ring oscillators, multimodulus frequency division is achieved by feeding a particular phase sequence to a single four-stage injection-locked ring oscillator. If a correct phase sequence is injected in multiple points of a ring oscillator, the locking range is widened. Conversely, a wrong phase sequence results in a very narrow locking range, and injection locking is unlikely to take place. Therefore, a ring oscillator injected with signals at a given frequency will select, among different possible modes of operation, the one that matches the provided input phase sequence. In a differential four-stage ring oscillator, the phase difference between the input of each delay cell and the corresponding noninverting output is 45°. Hence, the output nodes of the ring oscillator are in an octet-phase sequence. To force locking at the fundamental frequency (divide-by-one mode), quadrature signals must be injected into alternate delay cells, as shown in Figure 12.11a. In divide-by-two mode, the octet-phase sequence at the output corresponds to a quadrature sequence to be injected into all the delay cells (see Figure 12.11b). Finally, in divide-by-four mode, the octet-phase sequence at the divider output corresponds to a differential sequence at the divider input. Thus, differential phases need to be fed to the ring oscillator delay cells, as illustrated in Figure 12.11c.

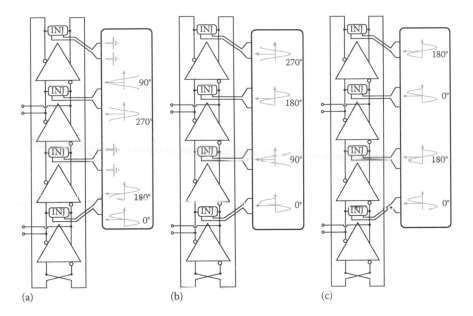

FIGURE 12.11   Multiple-phase injection locking into a ring oscillator: (a) divide-by-one mode, (b) divide-by-two mode, and (c) divide-by-four mode.

12.4.4   Transmitter Output Stage

The transmitter output stage must drive the transmitter antenna and isolate the LO port of the receiver mixer from the transmitter antenna port. Contrary to other systems, in short-range medical imaging SFCW radars, the required power level is quite low (−15 to −10 dBm) such that the transmitter output stage is not the block spending the most power in the system. As a consequence, a proper power amplifier is not strictly needed, nor power efficiency is the first parameter to be optimized. On the other hand, a very broad bandwidth is required, ruling out classic tuned amplifiers (both linear and nonlinear) from the list of possible implementations of this block.

A particular requirement for the transmitter output stage is that it must feature some harmonic rejection. Because of the phase coherence needed by the SFCW operation, the input signal of the transmitter output stage is the same LO signal as fed to the receiver. Typically, for maximum mixer conversion gain, the LO waveform resembles a square wave, that is, it is rich in harmonics. Due to the broad band width of the system, some of these harmonics can fall in band, impairing the system dynamic range, and must be rejected. Harmonic rejection could be in principle implemented in the receiver. However, this approach requires a receiver LO with more phases than simply the quadrature ones. As a consequence, the LO generation, which is already very complex in a UWB SFCW radar system, would be further complicated to an unfeasible level.

Achieving wideband harmonic rejection is not trivial at all. Using low-pass or band-pass filtering requires a high-order structure to get a steep roll-off and a high out-of-band attenuation. Moreover, such a topology must be made programmable and able to automatically track the position of the harmonics in the frequency spectrum to attenuate them while leaving the desired signal pass through. A tunable notch filter, as the one in [61], could be an alternative solution. However, the circuit occupies a large silicon area due to the reactive components. Moreover, it is extremely difficult to make it tunable over a multioctave frequency range. A single notch only solves the issue with one specific harmonic tone; thus multiple notch filters are required. The overhead due to the need of automatic tuning and calibration is also not negligible.

An alternative solution that is simple, robust, wideband, and inductorless is based on PPFs. It leverages the available quadrature LO sequence as follows. It is well known [10,28] that PPFs are capable of discriminating between positive and negative frequency components as they operate on sets of quadrature signals that can be interpreted as complex signals. If the quadrature signals do not have sinusoidal waveforms, and yet they are evenly spaced in time, the fundamental tones make a quadrature sequence that can be seen as a complex tone at a positive frequency $+\omega$. The third harmonics, however, make a reverse quadrature sequence that can be seen as a complex tone at a negative frequency -3ω. A PPF can thus be used to let the fundamental through while notching out the third harmonic [16]. Due to the moderate output power requirement of the SFCW transmitter for medical imaging, the attenuation introduced by the PPF is not a paramount issue. Hence, a multisection PPF design, featuring a large attenuation rejection of the negative frequencies over a very wide bandwidth, can be used.

12.5   A CASE STUDY: A 65 nm CMOS FULLY INTEGRATED RADAR TRANSCEIVER

A SFCW fully integrated radar transceiver, implemented in a 65 nm CMOS technology, has recently been reported in the literature [9,16]. It is the first attempt of an ASIC for the detection of breast cancer in its early development stage by means of microwave radar imaging. The transceiver leverages the direct conversion architecture, and many of the design concepts previously described.

The receiver signal path is made of a wideband LNA, linearized transconductors, passive current-mode mixers, and chopper-stabilized TIAs. The LNA is a three-stage design, based on the noise-cancelling technique [13] and on shunt-peaked loads, to expand the bandwidth. Transconductors linearized by means of local resistive feedback feed passive current-mode mixers through ac-coupling capacitors. This arrangement maximizes the linearity while avoiding the down-converted signal to be corrupted by the $1/f$ noise of the mixer switches. Current-to-voltage conversion of the baseband quadrature signal is performed by TIAs. To cope with the $1/f$ noise of the transistors of the TIAs, the chopper stabilization technique is employed. An integer-N PLL synthesizes all the frequencies from 8 to 16 GHz starting from a 22.6 MHz reference. The PLL relies on two VCOs not to impair the phase noise performance for the tuning range. Each VCO is followed by an injection-locked prescaler by four. A programmable divider, phase-frequency detector (PFD), charge pump, and third-order loop filter close the control loop.

A reconfigurable injection-locked divider by one, two, or four generates the quadrature phases over the entire 2–16 GHz range and drives the mixers. The divider is interfaced to the PLL and to the mixers by means of injection-locked ring oscillators used as regenerative buffers. The signal is tapped at the mixer LO port by a buffer and output to be irradiated. The transmitter buffer is made of a regenerative buffer, similar to the ones used in the LO distribution chain, which drives a PPF, used to suppress the harmonics of the transmitted signal.

The silicon area of the implemented transceiver prototypes is 1.3 by 1 mm², and the overall power consumption is 203 mW from a 1.2 V supply. A breakdown of the power consumption is shown in Figure 12.12. Clearly, the frequency generation chain is the most demanding block in the system.

Table 12.1 shows a summary of the measured transceiver performance. The receiver input matching is lower than −9 dB over the entire 2–16 GHz bandwidth. The receiver conversion gain and average noise figure are 36 and 7 dB, respectively. The 1/f noise corner is 30 Hz, showing the effectiveness of the chopper-stabilized TIA. The input-referred 1 dB compression point is larger than −29 dBm, the IIP3 larger than −13 dBm, and the IIP2 larger than 22 dBm. The I/Q phase mismatch is lower than 1.5° over frequency [8]. The transmitter average output power is −14 dBm, and the harmonic rejection is in excess to 40 dBc. The two VCOs tune from 6.5 to 11.8 GHz and from 11.0 to 18.4 GHz, effectively covering the higher octave of operation. The phase noise at 10 MHz offset from the carrier is lower than −129 dBc/Hz, while the RMS jitter is lower than 680 fs [16].

The SFCW CMOS radar transceiver consistently achieves a dynamic range in excess of 107 dB (calculated combining the 1 dB compression point and the integrated thermal and 1/f receiver noise) over more than 3 octaves, while the 14 GHz bandwidth allows to detect a tumor with 3 mm resolution. This performance enables the radar transceiver to be effectively employed as a breast cancer imaging diagnostic tool to detect small tumors up to a depth of about 6–7 cm [7].

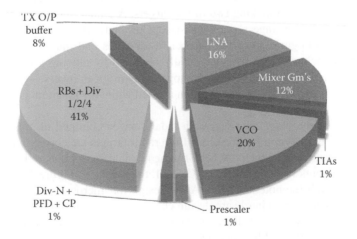

FIGURE 12.12 Breakdown of the power consumption of the radar transceiver.

TABLE 12.1

Radar Transceiver Performance Summary

Receiver Performance

Conversion gain	36 dB
Bandwidth	2–16 GHz
Input match	<–9 dB
Average noise figure	7 dB
$1/f$ noise corner frequency	30 Hz
$P_{1\,dB}$	>–29 dBm
IIP3	>–13 dBm
IIP2	>22 dBm
I/Q phase mismatch	<1.5°

Transmitter Performance

Output power	–14 dBm
Harmonic rejection	>40 dBc
Phase noise at 10 MHz offset	<–129 dBc/Hz
RMS jitter	<680 fs

Source: Reproduced from Bassi, M. et al., *IEEE Trans. Microw. Theory Tech.*, 61(5), 2108, 2013.

An actual imaging experiment carried out using the integrated radar transceiver is reported in [9]. The experiment is performed on a phantom mimicking the EM properties of the human breast over a wide range of frequencies. The material modeling the healthy tissue is a mixture of glycerin, double distilled/de-ionized water, ethylene glycol, polyethylene powder, and agar [6]. The recipe is easy to make, consistent, accurate, and has been extensively tested over more than 2 years [6]. Two enclosures of different diameters (6 and 9 mm) filled with tap water are buried inside the breast phantom, mimicking two tumor targets. A planar antenna array configuration is chosen to arrange a proof-of-concept testbed, allowing an easy and fast mechanical reconfiguration of the number of antennas and mutual distance, such that the array behavior can be synthetically obtained by moving a single imaging element. To further ease the testing, an approach similar to inverse synthetic array radar (ISAR) is used: the antenna position is fixed while the phantom is moved over all the desired positions.

A picture of the phantom along with radar images obtained on three different cross sections is shown in Figure 12.13. Both radar targets are correctly located and clearly identified, despite the presence of some residual clutter due to the implementation impairments of the transceiver. Overall, the results in Figure 12.13 show the feasibility as well the potential of an ASIC design for microwave medical imaging applications. The custom hardware design can effectively contribute to the development of the microwave radar imaging technology, which has the chance of significantly improving the health of people.

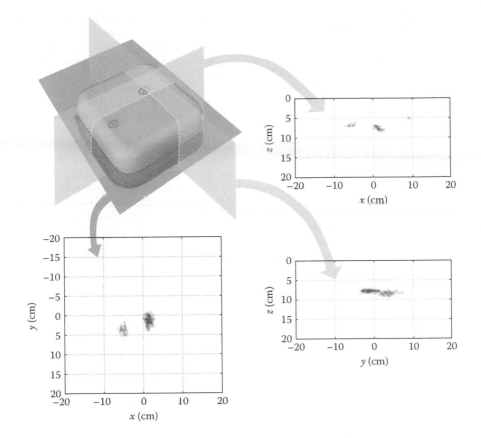

FIGURE 12.13   Radar imaging experiment on a breast phantom using the fully integrated SFCW transceiver. (Modified from Bassi, M. et al., *IEEE Trans. Microw. Theory Tech.*, 61(5), 2108, 2013.)

REFERENCES

1. Surveillance, Epidemiology, and End Results (SEER) Program Research Data (1973–2008), National Cancer Institute, DCCPS, Surveillance Research Program, Cancer Statistics Branch, released April 2011, based on the November 2010 submission.
2. American National Standard Recommended Practice for an on-site, ad hoc test method for estimating radiated electromagnetic immunity of medical devices to specific radio-frequency transmitters. ANSI C63.18-1997, 1997.
3. S.S. Ahmed, A. Schiessl, F. Gumbmann, M. Tiebout, S. Methfessel, and L. Schmidt. Advanced microwave imaging. *IEEE Microwave Magazine*, 13(6):26–43, 2012.
4. M.A. Ali and M. Moghaddam. 3d nonlinear super-resolution microwave inversion technique using time-domain data. *IEEE Transactions on Antennas and Propagation*, 58(7):2327–2336, 2010.
5. A. Bakker, K. Thiele, and J.H. Huijsing. A CMOS nested-chopper instrumentation amplifier with 100-nV offset. *IEEE Journal of Solid-State Circuits*, 35(12):1877–1883, December 2000.

6. Y. Baskharoun, A. Trehan, N.K. Nikolova, and M.D. Noseworthy. Physical phantoms for microwave imaging of the breast. In *IEEE Topical Conference on Biomedical Wireless Technologies, Networks, and Sensing Systems*, Santa Clara, CA, pp. 73–76, January 2012.

7. M. Bassi, A. Bevilacqua, A. Gerosa, and A. Neviani. Integrated SFCW transceivers for UWB breast cancer imaging: Architectures and circuit constraints. *IEEE Transactions on Circuits and Systems I: Regular Papers*, 59(6):1228–1241, June 2012.

8. M. Bassi, M. Caruso, A. Bevilacqua, and A. Neviani. A 65-nm CMOS 1.75–15 GHz stepped frequency radar receiver for early diagnosis of breast cancer. *IEEE Journal of Solid-State Circuits*, 48(7):1741–1750, 2013.

9. M. Bassi, M. Caruso, M.S. Khan, A. Bevilacqua, A. Capobianco, and A. Neviani. An integrated microwave imaging radar with planar antennas for breast cancer detection. *IEEE Transactions on Microwave Theory and Techniques*, 61(5):2108–2118, 2013.

10. F. Behbahani, Y. Kishigami, J. Leete, and A.A. Abidi. CMOS mixers and polyphase filters for large image rejection. *IEEE Journal of Solid-State Circuits*, 36(6):873–887, 2001.

11. A. Bevilacqua, C. Sandner, A. Gerosa, and A. Neviani. A fully integrated differential CMOS LNA for 3–5-GHz ultrawideband wireless receivers. *IEEE Microwave and Wireless Components Letters*, 16(3):134–136, 2006.

12. A. Bevilacqua and A.M. Niknejad. An ultrawideband CMOS low-noise amplifier for 3.1–10.6-GHz wireless receivers. *IEEE Journal of Solid-State Circuits*, 39(12):2259–2268, 2004.

13. S.C. Blaakmeer, E.A.M. Klumperink, D.M.W. Leenaerts, and B. Nauta. Wideband balun-LNA with simultaneous output balancing, noise-canceling and distortion-canceling. *IEEE Journal of Solid-State Circuits*, 43(6):1341–1350, June 2008.

14. E.J. Bond, X. Li, S.C. Hagness, and B.D. Van Veen. Microwave imaging via space-time beamforming for early detection of breast cancer. *IEEE Transactions on Antennas and Propagation*, 51(8):1690–1705, 2003.

15. F. Bruccoleri, E.A.M. Klumperink, and B. Nauta. Wide-band CMOS low-noise amplifier exploiting thermal noise canceling. *IEEE Journal of Solid-State Circuits*, 39(2):275–282, February 2004.

16. M. Caruso, M. Bassi, A. Bevilacqua, and A. Neviani. A 2–16 GHz 204 mW 3 mm-resolution stepped frequency radar for breast cancer diagnostic imaging in 65 nm CMOS. In *IEEE ISSCC Digest of Technical Papers*, San Francisco, CA, pp. 240–241, February 2013.

17. S. Chehrazi, A. Mirzaei, and A.A. Abidi. Noise in current-commutating passive FET mixers. *IEEE Transactions on Circuits and Systems I: Regular Papers*, 57(2):332–344, February 2010.

18. Y. Chen and P. Kosmas. Detection and localization of tissue malignancy using contrast-enhanced microwave imaging: Exploring information theoretic criteria. *IEEE Transactions on Biomedical Engineering*, 59(3):766–776, 2012.

19. S.K. Davis, B.D. Van Veen, S.C. Hagness, and F. Kelcz. Breast tumor characterization based on ultrawideband microwave backscatter. *IEEE Transactions on Biomedical Engineering*, 55(1):237–246, 2008.

20. J. De Zaeytijd, A. Franchois, C. Eyraud, and J.M. Geffrin. Full-wave three-dimensional microwave imaging with a regularized Gauss-Newton method—Theory and experiment. *IEEE Transactions on Antennas and Propagation*, 55(11):3279–3292, 2007.

21. C.C. Enz and G.C. Temes. Circuit techniques for reducing the effects of op-amp imperfections: Autozeroing, correlated double sampling, and chopper stabilization. *Proceedings of the IEEE*, 84(11):1584–1614, November 1996.

22. C.C. Enz, E.A. Vittoz, and F. Krummenacher. A CMOS chopper amplifier. *IEEE Journal of Solid-State Circuits*, 22(3):335–342, June 1987.

23. E.C. Fear, J. Bourqui, C. Curtis, D. Mew, B. Docktor, and C. Romano. Microwave breast imaging with a monostatic radar-based system: A study of application to patients. *IEEE Transactions on Microwave Theory and Techniques*, 61(5):2119–2128, 2013.

24. E.C. Fear, S.C. Hagness, P.M. Meaney, M. Okoniewski, and M.A. Stuchly. Enhancing breast tumor detection with near-field imaging. *IEEE Microwave Magazine*, 3(1):48–56, March 2002.
25. E.C. Fear, X. Li, S.C. Hagness, and M.A. Stuchly. Confocal microwave imaging for breast cancer detection: Localization of tumors in three dimensions. *IEEE Transactions on Biomedical Engineering*, 8(10):812–821, 2002.
26. E.C. Fear, P.M. Meaney, and M.A. Stuchly. Microwaves for breast cancer detection? *IEEE Potentials*, 22(1):12–18, 2003.
27. A. Fhager, Y. Yu, T. McKelvey, and M. Persson. Stroke diagnostics with a microwave helmet. In *2013 Seventh European Conference on Antennas and Propagation (EuCAP)*, Gothenburg, Sweden, pp. 845–846, 2013.
28. M.J. Gingell. The synthesis and application of polyphase filters with sequence asymmetric properties. PhD thesis, University of London, London, U.K., 1975.
29. T. Henriksson, M. Klemm, D. Gibbins, J. Leendertz, T. Horseman, A.W. Preece, R. Benjamin, and I.J. Craddock. Clinical trials of a multistatic UWB radar for breast imaging. In *Loughborough Antennas and Propagation Conference*, pp. 1–4, November 2011.
30. M.G.M. Hussain. Ultra-wideband impulse radar—An overview of the principles. *IEEE Aerospace and Electronics Systems Magazine*, 13(9):9–14, September 1998.
31. M.G.M. Hussain. Principles of space-time array processing for ultrawide-band impulse radar and radio communications. *IEEE Transactions on Vehicular Technology*, 51(3):393–403, May 2002.
32. P.T. Huynh, A.M. Jarolimek, and S. Daye. The false-negative mammogram. *Radiograph*, 18(5):1137–1154, 1998.
33. N. Irishina, M. Moscoso, and O. Dorn. Microwave imaging for early breast cancer detection using a shape-based strategy. *IEEE Transactions on Biomedical Imaging*, 56(4):1143–1153, 2009.
34. J.E. Johnson, T. Takenaka, and T. Tanaka. Two-dimensional time-domain inverse scattering for quantitative analysis of breast composition. *IEEE Transactions on Biomedical Engineering*, 55(8):1941–1945, 2008.
35. T.-Y.J. Kao, A.Y.-K. Chen, Y. Yan, T.-M. Shen, and J. Lin. A flip-chip-packaged and fully integrated 60 GHz CMOS micro-radar sensor for heartbeat and mechanical vibration detections. In *IEEE Radio Frequency Integrated Circuits Symposium*, Montreal, QC, Canada, pp. 443–446, June 2012.
36. P. Kinget, R. Melville, D. Long, and V. Gopinathan. An injection-locking scheme for precision quadrature generation. *IEEE Journal of Solid-State Circuits*, 37(7):845–851, July 2002.
37. M. Klemm, I. Craddock, J. Leendertz, A. Preece, and R. Benjamin. Experimental and clinical results of breast cancer detection using UWB microwave radar. In *IEEE Antennas and Propagation Society International Symposium*, San Diego, CA, pp. 1–4, 2008.
38. M. Klemm, I.J. Craddock, J.A. Leendertz, A. Preece, and R. Benjamin. Radar-based breast cancer detection using a hemispherical antenna array-experimental results. *IEEE Transactions on Antennas and Propagation*, 57:1692–1704, June 2009.
39. M. Klemm, D. Gibbins, J. Leendertz, T. Horseman, A.W. Preece, R. Benjamin, and I.J. Craddock. Development and testing of a 60-element UWB conformal array for breast cancer imaging. In *Proceedings of the Fifth European Conference on Antennas and Propagation*, Rome, Italy, pp. 3077–3079, April 2011.
40. M. Klemm, J.A. Leendertz, D. Gibbins, I.J. Craddock, A. Preece, and R. Benjamin. Microwave radar-based breast cancer detection: Imaging in inhomogeneous breast phantoms. *IEEE Antennas and Wireless Propagation Letters*, 8:1349–1352, 2009.
41. D.K. Shaeffer and T.H. Lee. A 1.5-V, 1.5-GHz CMOS low noise amplifier. *IEEE Journal of Solid-State Circuits*, pp. 745–759, 32, May 1997.

42. D.J. Kurrant, E.C. Fear, and D.T. Westwick. Tumor response estimation in radar-based microwave breast cancer detection. *IEEE Transactions on Biomedical Engineering*, 55(12):2801–2811, 2008.
43. M. Lazebnik, M. Okoniewski, J.H. Booske, and S.C. Hagness. Highly accurate Debye models for normal and malignant breast tissue dielectric properties at microwave frequencies. *IEEE Microwave and Wireless Components Letters*, 17(12):822–824, 2007.
44. M. Lazebnik, D. Popovic, L. McCartney, C.B. Watkins, M.J. Lindstrom, J. Harter, S. Sewall et al., A large-scale study of the ultrawideband microwave dielectric properties of normal, benign and malignant breast tissues obtained from cancer surgeries. *Physics in Medicine and Biology*, 52:6093–6115, 2007.
45. X. Li, E.J. Bond, B.D. Van Veen, and S.C. Hagness. An overview of ultra-wideband microwave imaging via space-time beamforming for early-stage breast-cancer detection. *IEEE Antennas and Propagation Magazine*, 47(1):19–34, February 2005.
46. X. Li, S.K. Davis, S.C. Hagness, D.W. van der Weide, and B.D. Van Veen. Microwave imaging via space-time beamforming: Experimental investigation of tumor detection in multilayer breast phantoms. *IEEE Transactions on Microwave Theory and Techniques*, 52(8):1856–1865, 2004.
47. Y.-A. Li, M.-H. Hung, S.-J. Huang, and J. Lee. A fully integrated 77 GHz FMCW radar system in 65 nm CMOS. In *IEEE ISSCC Digest of Technical Papers*, San Francisco, CA, pp. 216–217, February 2010.
48. H.B. Lim, T.T.N. Nguyen, E. Li, and D.T. Nguyen. Confocal microwave imaging for breast cancer detection: Delay-multiply-and-sum image reconstruction algorithm. *IEEE Transactions on Biomedical Engineering*, 55(6):1697–1704, 2008.
49. P.M. Meaney, M.W. Fanning, T. Zhou, A. Golnabi, S.D. Geimer, and K.D. Paulsen. Clinical microwave breast imaging—2D results and the evolution to 3D. In *International Conference on Electromagnetics in Advanced Applications*, Torino, Italy, pp. 881–884, 2009.
50. J.S. Michaelson, M. Silverstein, J. Wyatt, G. Weber, R. Moore, E. Halpern, D.B. Kopans, and K. Hughes. Predicting the survival of patients with breast carcinoma using tumor size. *Journal of Cancer*, 95(4):713–723, August 2002.
51. N. Nikolova. Microwave imaging for breast cancer. *IEEE Microwave Magazine*, 12(7):78–94, December 2011.
52. S.M. Salvador, E.C. Fear, M. Okoniewski, and J.R. Matyas. Exploring joint tissues with microwave imaging. *IEEE Transactions on Microwave Theory and Techniques*, 58(8):2307–2313, 2010.
53. S.M. Salvador and G. Vecchi. Experimental tests of microwave breast cancer detection on phantoms. *IEEE Transactions on Antennas and Propagation*, 57(6):1705–1712, 2009.
54. S. Semenov, J. Kellam, P. Althausen, T. Williams, A. Abubakar, A. Bulyshev, and Y. Sizov. Microwave tomography for functional imaging of extremity soft tissues: Feasibility assessment. *Physics in Medicine and Biology*, 52(18):5705, 2007.
55. S.Y. Semenov, A.E. Bulyshev, A. Abubakar, V.G. Posukh, Y.E. Sizov, A.E. Souvorov, P.M. van den Berg, and T.C. Williams. Microwave-tomographic imaging of the high dielectric-contrast objects using different image-reconstruction approaches. *IEEE Transactions on Microwave Theory and Techniques*, 53(7):2284–2294, 2005.
56. S.Y. Semenov, V.G. Posukh, A.E. Bulyshev, T. Williams, P. Clark, Y.E. Sizov, A.E. Souvorov, and B.A. Voinov. Development of microwave tomography for functional cardiac imaging. In *IEEE International Symposium on Biomedical Imaging: Nano to Macro*, 2:1351–1353, 2004.
57. I. Craig Henderson, S.J. Nass, and J.C. Lashof. Mammography and beyond: Developing technologies for the early detection of breast cancer. National Academy Press, Washington, DC, 2001.

58. N. Sunaguchi, T. Yuasa, and M. Ando. Iterative reconstruction algorithm for analyzer-based phase-contrast computed tomography of hard and soft tissue. *Applied Physics Letters*, 103(14):143702–143702-4, 2013.

59. A. Tang, G. Virbila, D. Murphy, F. Hsiao, Y.H. Wang, Q.J. Gu, Z. Xu, Y. Wu, M. Zhu, and M.-C.F. Chang. A 144 GHz 0.76 cm-resolution sub-carrier SAR phase radar for 3D imaging in 65 nm CMOS. In *IEEE ISSCC Digest of Technical Papers*, San Francisco, CA, pp. 264–266, February 2012.

60. M. Tiebout, H.D. Wohlmuth, H. Knapp, R. Salerno, M. Druml, M. Rest, J. Kaeferboeck et al. Low power wideband receiver and transmitter chipset for mm-wave imaging in SiGe bipolar technology. *IEEE Journal of Solid-State Circuits*, 47(5):1175–1184, May 2012.

61. A. Vallese, A. Bevilacqua, C. Sandner, M. Tiebout, A. Gerosa, and A. Neviani. Analysis and design of an integrated notch filter for the rejection of interference in UWB systems. *IEEE Journal of Solid-State Circuits*, 44(2):331–343, 2009.

62. D.R. Wehner. *High Resolution Radar*. Artech House Radar Library, Boston, MA, 1994.

63. D.W. Winters, J.D. Shea, P. Kosmas, B.D. Van Veen, and S.C. Hagness. Three-dimensional microwave breast imaging: Dispersive dielectric properties estimation using patient-specific basis functions. *IEEE Transactions on Medical Imaging*, 28(7):969–981, 2009.

64. J.Y. Wo, K. Chen, B.A. Neville, N.U. Lin, and R.S. Punglia. Effect of very small tumor size on cancer-specific mortality in node-positive breast cancer. *Journal of Clinical Oncology*, May 2011.

65. R. Wu, K.A.A. Makinwa, and J.H. Huijsing. A chopper current-feedback instrumentation amplifier with a 1 mHz 1/f noise corner and an ac-coupled ripple reduction loop. *IEEE Journal of Solid-State Circuits*, 44(12):3232–3243, December 2009.

66. C. Yu, M. Yuan, Y. Zhang, J. Stang, R.T. George, G.A. Ybarra, W.T. Joines, and Q.-H. Liu. Microwave imaging in layered media: 3-D image reconstruction from experimental data. *IEEE Transactions on Antennas and Propagation*, 58(2):440–448, 2010.

13 Ultralow-Power RF Systems and Building Blocks

Mahdi Parvizi, Karim Allidina,
and Mourad El-Gamal

CONTENTS

13.1 INTRODUCTION

Wireless sensor networks (WSNs) have become highly sought after in myriad of applications, including health care, environmental monitoring, industrial settings, and agriculture. The nature of these applications imposes severe restrictions on the power consumption of a WSN node. As a result, power-efficient communication schemes and ultralow-power (ULP) radio frequency (RF) front-end circuits are required to maximize battery lifetime and to enable operation using energy harvested from the environment.

Impulse radio ultrawideband (IR-UWB) is a promising technology that has shown great potential for ULP, low-cost, and integrated radio solutions. This form of communication sends information using a stream of very short pulses, with pulse widths on the order of nanoseconds or less. A transceiver can take advantage of these short pulses by cycling its operation. Since power need only be consumed when the system is actively transmitting or receiving, cycled operation can lead to a drastic reduction in power consumption.

Accordingly, IR-UWB is suitable for WSN applications in which power consumption is the main constraint and relatively low data rates are required [1]. The pulsed nature of IR-UWB also allows for simple noncoherent energy detection architectures to be employed. This simplifies the receiver architecture and reduces the cost and power consumption of the transceiver.

The design of power-efficient transceivers also requires careful optimization at the circuit level. As the feature size in standard CMOS technologies is shrunk, the maximum allowed supply voltage is reduced as well. While operation from a low-supply voltage is desirable in systems powered by energy harvesting to minimize conversion losses and power consumption, it also leads to restrictions on the usable circuit topologies, intrinsic gain, and the speed transistors can operate. Consequently, circuits operating from very-low-supply voltages have become very important and are under active research [2–6].

In this chapter, we will first discuss design challenges inherent in ULP and ultralow-voltage (ULV) circuits and techniques to overcome them. These techniques will be demonstrated in the design of two ULP and ULV low-noise amplifiers (LNAs) suitable for UWB systems. Finally, a compact, low-power IR-UWB transceiver for WSNs with a low-complexity synchronization scheme will be presented.

13.2 ULTRALOW-VOLTAGE CIRCUIT DESIGN CHALLENGES

Supply voltage reduction is desirable in applications such as WSNs and systems operating on the energy scavenged from the environment to lower the power consumption and the conversion losses in the dc–dc converters. However, ultralow-supply voltages ($V_{DD} < 0.6$) that force the V_{DS} of the transistors to be close to $V_{DS,sat}$ have severe impacts on the characteristics of a MOS transistor, for example, intrinsic gain, transit frequency (f_T), noise figure (NF), and linearity. Moreover, low-supply voltage restricts the overdrive voltages available for the transistors and the circuit topologies that can be used. Here, we will address the major impacts of ULV on the performance of a MOS transistor [7].

13.2.1 TRANSCONDUCTANCE EFFICIENCY

The ratio of transconductance to dc drain current g_m/I_D is a conventional approach for designing low-power analog CMOS circuits [8]. It is required to study the impact of V_{DS} on the current efficiency. As V_{DS} is reduced, the achievable I_D and correspondingly the g_m of the device decrease due to channel length modulation. Interestingly, the I_D and g_m are reduced by the same factor; hence, the g_m/I_D stays almost constant for different V_{DS} values. Figure 13.1 illustrates the simulation results for g_m/I_D curve for two V_{DS} values with respect to the inversion coefficient (IC) for a transistor in a 90 nm CMOS technology with $W = 40$ μm and $L = 100$ nm, and using a BSIM$_4$ model. The IC is defined by Ref. [9]

$$IC = \frac{I_D}{I_{D0}} = \ln^2\left(1 + e^{\frac{V_{GS}-V_{TH}}{2nU_T}}\right), \tag{13.1}$$

where
I_{D0} is the technology current as defined by $I_{D0} = 2\,n\mu_0 C_{OX} U_T^2(W/L)$
W is the effective channel width
L is the effective channel length
n is the substrate factor whose value depends on process and varies from 1 to 2
μ_0 is the carrier mobility
U_I defined as $U_T = kT/q$ is the thermal voltage
C_{OX} is the gate-oxide capacitance per unit area
V_{GS} is the gate-source voltage
V_{TH} is the threshold voltage

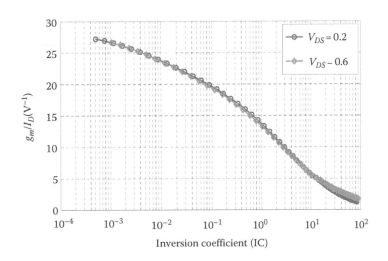

FIGURE 13.1 Simulated g_m/I_D characteristics for an NMOS transistor in a 90 nm CMOS technology.

In general, weak inversion (WI) corresponds to $IC < 0.1$. If $0.1 < IC < 10$ then the transistor is in moderate inversion (MI) and if $IC > 10$ the transistor is in strong inversion (SI). As can be seen in Figure 13.1, the transconductance efficiency has a maximum in the deep WI region. The efficiency reaches 0.5 of the maximum at the center of the MI region and decreases in the SI region.

13.2.2 INTRINSIC VOLTAGE GAIN

The intrinsic voltage gain of a MOS transistor is the small signal low-frequency gain of a common-source MOSFET with an ideal current source as the load. This is simply the ratio of g_m/g_{ds}. Drain-source voltage reduction causes g_{ds} to increase and g_m to decrease. The g_{ds} rise can be explained through channel length modulation (CLM). According to CLM, g_{ds} decreases directly by increasing the gate length (up to a certain point) and excess drain-source voltage above $V_{DS, sat}$ [10]. Hence, reducing the drain-source voltage causes the intrinsic voltage gain of the transistor (g_m/g_{ds}) to be reduced. The simulation results for the intrinsic voltage gain of an NMOS transistor in a 90 nm CMOS technology with $W = 40$ μm and $L = 100$ nm in $BSIM_4$ model for different values of V_{DS} are shown in Figure 13.2. As can be seen in the figure, there is about 45% reduction in the achievable intrinsic voltage gain in the MI region.

13.2.3 TRANSIT FREQUENCY

Another important characteristic of a MOSFET that should be studied is the transit frequency, f_T. As the V_{DS} of the device is reduced, the f_T also gets reduced. This is because f_T is proportional to g_m, and g_m decreases as V_{DS} is reduced. Figure 13.3 illustrates the simulation results for the transit frequency of an NMOS transistor in a

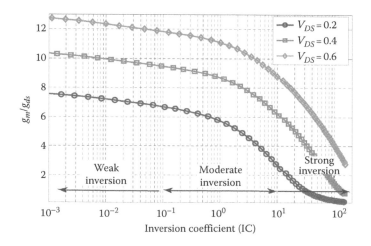

FIGURE 13.2 Simulated intrinsic voltage gain of an NMOS transistor in a 90 nm CMOS for three different V_{DS} values.

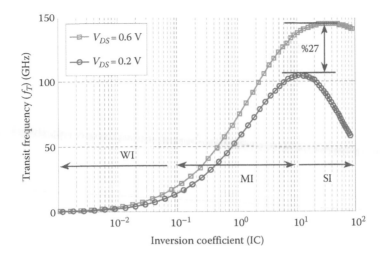

FIGURE 13.3 Simulated transit frequency of an NMOS transistor in a 90 nm CMOS technology for two different V_{DS} values.

90 nm CMOS technology with $W = 40$ μm and $L = 100$ nm for two different values of V_{DS}. As demonstrated in the figure, the achieved f_T for a V_{DS} of 0.2 V decreases by 27% compared with f_T for a V_{DS} of 0.6 V. It should be noted that, when the V_{DS} is at 0.2 V, the peak of the f_T happens at lower ICs since the transistor enters the triode region sooner. This f_T reduction deteriorates the performance, namely, the bandwidth, NF, and gain of the RF front-end circuits.

13.2.4 NOISE FIGURE

The noise characteristics of a MOSFET are highly important for LNA design. The minimum noise figure, NF_{min}, of a MOS transistor is the noise figure at the optimum source resistance. The NF_{min} also varies with respect to V_{DS}. The NF_{min} is inversely proportional to the square root of g_m and hence increases as the V_{DS} decreases. The simulated NF_{min} of an NMOS in a 90 nm CMOS technology with $W = 40$ μm and $L = 100$ nm for two values of V_{DS} is shown in Figure 13.4. It is interesting to note that the IC in which the minimum happens is almost unchanged.

13.2.5 LINEARITY

The linearity of the LNA will decrease as the supply voltage is lowered. To further investigate the effects of supply voltage on the nonlinear behavior of a MOSFET as a weakly nonlinear system, the nonlinear drain current (i_{ds}) of a transistor can be expressed in terms of v_{gs} and v_{ds} by a 2D Taylor series

$$I_{ds}\left(v_{gs}, v_{ds}\right) = g_m v_{gs} + g_{ds} v_{ds} + \frac{g'_m}{2} v_{gs}^2 + \frac{g'_{ds}}{2} v_{ds}^2 + \frac{g''_m}{6} v_{gs}^3 + \frac{g''_{ds}}{6} v_{ds}^3, \quad (13.2)$$

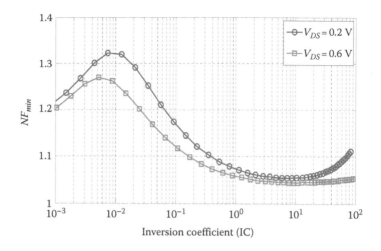

FIGURE 13.4 Simulated minimum NF of an NMOS transistor in a 90 nm CMOS technology for two different V_{DS} values.

where the Taylor coefficients can be derived from

$$g_m^k = \frac{\partial^k I_{ds}}{\partial V_{gs}^k}, \quad g_{ds}^k = \frac{\partial^k I_{ds}}{\partial V_{ds}^k}. \tag{13.3}$$

The cross terms have been ignored for simplicity. It is known that g_m is the strongest contributor to third-order distortion in circuits. However, it will be shown that in deep submicron technologies and specifically at low V_{DS} values, g_{ds} deteriorates the linearity of the circuit as well. Figure 13.5a shows the g_m'' variation with respect to the

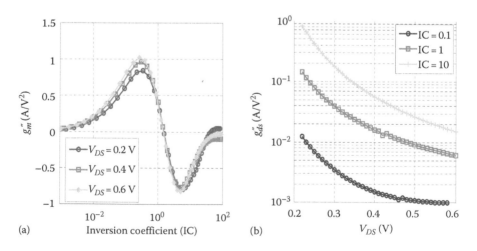

FIGURE 13.5 (a) Simulated third-order distortion of a MOSFET due to g_m for three different V_{DS} values and (b) the third-order distortion due to g_{ds} for three IC values.

TABLE 13.1

Ultralow-Voltage Design Challenges Summary

Design Parameter	Ultralow-Voltage Supply Impact
Transconductance efficiency (g_m/I_D)	Unchanged
Intrinsic gain (g_m/g_{ds})	45% reduction
Transit frequency (f_T)	27% reduction
Noise factor	Increases slightly
Nonlinearity due to g_m	Unchanged (up to the triode region)
Nonlinearity due to g_{ds}	20 times larger (at $IC = 1$)

IC for multiple values of V_{DS}. As can be seen, as long as the transistor is in saturation, g_m'' has the same characteristics for different V_{DS} values. It should be noted that the zero crossing of g_m'' is not dependent on the V_{DS} and happens at IC of 1.2. Despite the variation against process and temperature, this sweet spot was used for circuit linearization [11]. Moreover, Figure 13.5b illustrates g_{ds}'' versus V_{DS} for multiple values of the IC. It can be seen that g_{ds}'' increases by increasing the IC, and this is due to higher current and lower output impedance at higher ICs. It is also interesting to note the variation of g_{ds}'' with V_{DS}. It is clear that g_{ds}'' decreases significantly by V_{DS}. For example, at an IC of 1, g_{ds}'' is 20 times higher at $V_{DS} = 0.2$ compared to $V_{DS} = 0.6$. More importantly, at this point g_{ds}'' is comparable to g_m'' (almost half of it). Hence, the nonlinearity due to g_{ds} must be considered at low-supply voltages.

13.2.6 SUMMARY

Table 13.1 summarizes all the ULV impacts that were discussed in this section on the characteristics of an NMOS transistor in a 90 nm CMOS technology by reducing V_{DS} from 0.6 to 0.2 V.

13.3 ULTRALOW-POWER, ULTRALOW-VOLTAGE, AND WIDEBAND DESIGN TECHNIQUES

Now that some of the issues with ULV circuit design have been presented, methods of overcoming these challenges will be discussed while keeping in mind the goal of low-power consumption with ultralow-supply voltages. At first, general techniques to realize ULP circuits that are suitable for application in ULV environments are discussed. Following this, bandwidth enhancement schemes and wideband matching techniques that are suitable for ULP and ULV circuits will be presented.

13.3.1 ULTRALOW-POWER DESIGN TECHNIQUES

13.3.1.1 Ultralow-Power Biasing Scheme

In this section, we will address the optimum biasing point for a single transistor. The goal is to find a biasing region where the transistor has acceptable transit frequency (bandwidth), gain, and noise performance while consuming very little power from a low-supply voltage. For a common-source MOSFET with a current source at the drain, the low-frequency gain is g_m/g_{ds} and the noise factor can be found by $F = 1 + \left(\dfrac{\gamma}{\alpha}\right)\left(\dfrac{f}{f_T}\right)^2 g_m R_S$ [12].

As stated before, the transconductance efficiency is defined by g_m/I_D. This has been incorporated in various biasing schemes like figures of merit (FOMs) for low-power circuit design, such as [13] $\left(FOM = \left(^{g_m}\!/_{I_D}\right) f_T\right)$. However, as discussed earlier, V_{DS} variation has severe impacts on the intrinsic gain (specifically on the g_{ds} of a transistor). Hence, the effect of V_{DS} on the intrinsic gain must be included to reflect the impact of low-supply voltages. As such, the modified FOM used here is defined as the product of the transconductance efficiency, intrinsic gain, and transit frequency as given in the following [7]:

$$FOM = \frac{g_m}{I_D} \cdot \frac{g_m}{g_{ds}} \cdot f_T. \tag{13.4}$$

This FOM represents the optimum bias point to achieve low-power operation with reasonable noise figure, gain, and bandwidth. However, the bandwidth and NF can be traded with higher voltage gain and lower power consumption as we move to the lower inversion coefficients, and vice versa. Figure 13.6 highlights the proposed FOM for two values of V_{DS}. As can be seen, the optimum point for both curves occurs in the MI region; nevertheless, the overall performance of the transistor is degraded by 60% at ULV supplies. Also, it is interesting to note the dependency of the FOM on V_{DS}. As can be seen in the figure, the peak of the FOM is shifted toward the weak inversion region for $V_{DS} = 0.2$.

13.3.1.2 Current Reuse

Current reuse is a very important technique in the implementation of ULP circuits. In this scheme, the dc current is shared between two or more transistors while

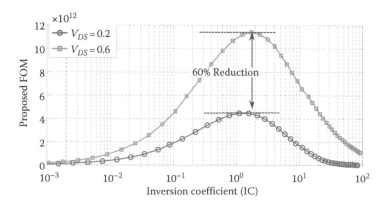

FIGURE 13.6 Proposed FOM for ULP and ULV circuit design for two V_{DS} values.

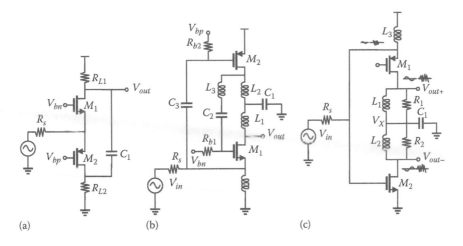

FIGURE 13.7 Complementary current reuse LNA (a) with second-order distortion cancellation (From Wei-Hung, C. et al., *IEEE J. Solid-State Circ.*, 43, 1164, 2008), (b) g_m-boosting (From Khurram, M. and Rezaul Hasan, S.M., *IEEE Trans. Very Large Scale Integr. Syst.*, 20, 400, 2012), and (c) with noise cancellation. (From Parvizi, M. et al., A 0.4 V ultra low-power UWB CMOS LNA employing noise cancellation, in *IEEE International Symposium on Circuits and Systems*, Beijing, China, 2013, pp. 2369–2372.)

each transistor contributes to the total gain. This technique has been widely used in the literature [14–16] to reduce the overall power consumption of the LNAs. Moreover, complementary current reuse [17–20] is a technique that employs both NMOS and PMOS transistors to take advantage of their complementary characteristics, which has led to ULP designs and some additional advantages like distortion and noise cancellation. Figure 13.7 highlights three complementary current reuse architectures that provide additional performance benefits to an LNA. Figure 13.7a shows a complementary current reuse scheme that provides second-order distortion cancellation through the complementary characteristics of the NMOS and PMOS transistors [21]. A g_m-boosted current reuse LNA for UWB applications is highlighted in Figure 13.7b, where the PMOS stage boosts the g_m of the input common-gate stage to reduce the power required for wideband input matching [19]. Figure 13.7c shows a noise-cancelling complementary current reuse scheme for UWB receivers [20]. In this architecture, the PMOS transistor is the input common-gate stage, while the NMOS is the common-source amplifier amplifying the input signal and the noise of M_1. When the signals from these two paths are combined at the output node, the input signal adds constructively while the noise from M_1 is cancelled out.

13.3.1.3 Forward Body Biasing

Low-supply voltages restrict the overdrive and drain-source voltages that can be used for the MOS transistors. Considering the fact that the threshold voltage of the devices has not been reduced noticeably by technology scaling due to leakage currents,

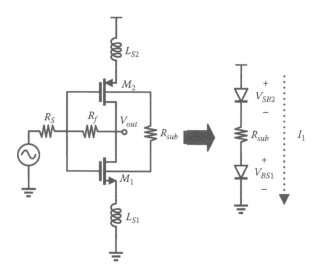

FIGURE 13.8 Schematic of the self-forward biasing scheme. (From Chen-Ming, L., et al., *IEEE Microw. Wireless Compon. Lett.*, 20, 100, 2010.)

forward body biasing can be an important technique for realization of ULP and ULV circuits. The threshold voltage of a MOS transistor can be found by

$$V_{TH} = V_{TH0} + \alpha\left(\sqrt{2\varphi_F - V_{BS}} - \sqrt{2\varphi_F}\right), \tag{13.5}$$

where
   α is the process-dependent parameter
   φ_F is semiconductor parameter with typical values of 0.3–0.4

Hence, by increasing body-source voltage the threshold voltage will decrease. A direct impact of threshold voltage reduction is the lower gate-source voltage to bias the transistor in the desired inversion region.

Figure 13.8 outlines a self-forward-body-biasing scheme. Using this technique eliminates the need for separate dc voltages by connecting the bodies of M_1 and M_2 together through R_{sub} and creating an ULP self-bias loop, including the source-body diodes of M_1 and M_2 [22]. The loop has the voltage equation of $V_{DD} = V_{SB2} + I_1 R_{sub} + V_{BS1}$.

13.3.2 WIDEBAND MATCHING AND BANDWIDTH ENHANCEMENT TECHNIQUES

13.3.2.1 Shunt Feedback

One of the main challenges in the design of ULP and ULV LNAs is to provide wideband input matching with the available power budget. The use of cascaded resonant circuits to provide wideband matching with very little additional noise is possible, but this can consume significant chip area [23]. A common-gate topology can also provide a wideband input match; however, a current of at least 1.7 mA

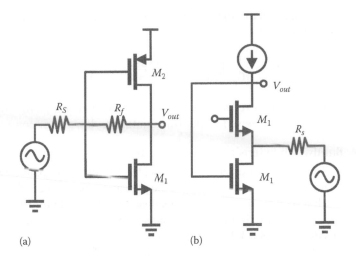

(a)   (b)

FIGURE 13.9 Shunt feedback in a (a) common-source amplifier and in a (b) common-gate stage. (From Allidina, K. and El-Gamal, M.N., A 1 V CMOS LNA for low power ultra-wideband systems, in *IEEE International Conference on Electronics, Circuits and Systems,* 2008, pp. 165–168.)

is required to provide 50 Ω matching in the chosen 130 nm CMOS technology if the amplifier is in moderate inversion (with a g_m/I_D of 12 V^{-1} in this region). This criterion leads to the overall power consumption being greater than 1.7 mW with $V_{DD} = 1$ V and 0.85 mW with $V_{DD} = 0.5$ V. Hence, design techniques are required to enable low-power input matching, preferably without increasing the noise figure.

Shunt feedback has shown great potential for providing wideband low-power input matching [15,24,25]. Figure 13.9 illustrates a possible shunt feedback implementation for both common-source and common-gate gain stages. In Figure 13.9a, the conventional common-source resistive shunt feedback amplifier is shown. The input resistance of this architecture can be found by

$$R_{in} = \frac{R_O + R_f}{1 + g_m R_O},$$ (13.6)

where
R_O is the output resistance of the NMOS in parallel with PMOS
$g_m = g_{m1} + g_{m2}$

Since the feedback is resistive, this results in a wideband and low-power solution. Figure 13.9b outlines another viable option for low-power wideband matching in which shunt feedback around a common-gate amplifier is used. The effect of feedback on the input resistance of the amplifier can be found by

$$R_{in} = \frac{1}{g_m (1 + A)},$$ (13.7)

where A is the feedback factor. In the circuit shown in Figure 13.9b, the input resistance can be found by

$$Z_{in} = \frac{1}{sC_{gs}} \| \frac{R_L(g_{o1}+g_{o2})+1}{(g_{m1}+g_{m2}+g_{o1}+g_{o2})(1+A)}. \tag{13.8}$$

Since the input resistance is reduced by the gain of the feedback path, a wideband 50 Ω input match can be achieved while still maintaining low-power consumption.

13.3.2.2 Bandwidth Extension and g_m-Boosting

A conventional technique to extend the bandwidth without additional power consumption is to use inductors to resonate with the parasitic capacitances of the transistors. Figure 13.10a shows a resistive shunt feedback LNA, and Figure 13.10b and c shows modifications to this circuit with the inductor placed at the input of the LNA [26,27] and inside the feedback loop [28], respectively.

To compare all three topologies in Figure 13.10, the effective g_m is calculated for each. The gate–drain capacitance, C_{ds}, is ignored in this analysis for simplicity. The effective g_m for the circuit shown in Figure 13.10a is given by

$$g_{m,eff} = \frac{g_m}{1 + j\omega C_{gs}R_s + R_s(R_f - (R_L \| R_f)(g_mR_f-1))/R_f^2}. \tag{13.9}$$

At high frequencies, the equation simplifies to

$$g_{m,eff} = \frac{\omega_T}{\omega} \cdot \frac{1}{jR_s}. \tag{13.10}$$

The effective transconductance for the circuit shown in Figure 13.10b with the inductor at the input can be found by

$$g_{m,eff} = \frac{g_m}{1 - \omega^2 L_1 C_{gs} + j\omega C_{gs}R_s + \left((R_s + j\omega L_1)(R_f - (R_L \| R_f)(g_mR_f-1))/R_f^2\right)}. \tag{13.11}$$

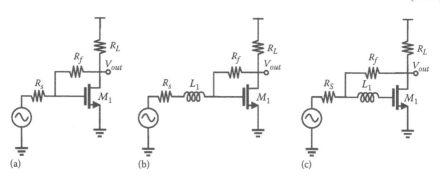

(a)   (b)   (c)

FIGURE 13.10 Resistive shunt feedback LNA (a) without bandwidth extension technique, (b) with inductive series peaking at the input, and (c) with inductive series peaking in the feedback path.

The effective transconductance for the LNA with inductive series peaking in the feedback path can be found by

$$g_{m,eff} = \frac{g_m}{\left(1 - \omega^2 L_1 C_{gs}\right)\left(1 + \dfrac{R_s}{R_f} + \dfrac{R_s}{R_f^2}\cdot\left(R_L \parallel R_f\right)\cdot\left(g_m R_f + \omega^2 L_1 C_{gs} - 1\right) + \dfrac{j\omega C_{gs} R_s}{\left(1 - \omega^2 L_1 C_{gs}\right)}\right)}.$$

(13.12)

As can be seen, by adding the inductor inside the feedback loop, the denominator ends up having only the term $j\omega C_{gs} R_s$ at the resonance frequency and as a result, the g_m will be boosted. However, for the case where the inductor is placed at the input, Equation 13.11, the last term in the denominator exists at the resonant frequency and dampens the response so no g_m-boosting occurs. Figure 13.11 shows MATLAB® simulation results for the effective g_m for the three circuits. The bandwidth enhancement for series peaking in the feedback loop is obviously higher than the inductive peaking at the input.

From pole-zero perspective, this technique pushes the dominant poles of the circuit to higher frequencies. Figure 13.12 illustrates the effect of inductor values on the position of dominant poles of the circuit shown in Figure 13.10c. As can be seen, the conjugate poles are pushed into the left side by increasing the value of inductors. However, after a certain limit, the magnitude of the dominant poles from the origin gets reduced, leading to lower bandwidth.

Input matching of the amplifier is also affected by the inductive peaking in the feedback and hence it should be taken into consideration while choosing inductor values. Staggering the resonance frequencies of the two feedback loops further broadens the bandwidth and limits the amount of peaking in the response of the amplifier.

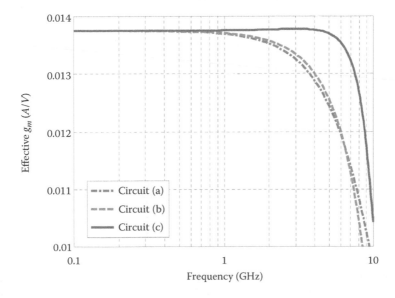

FIGURE 13.11 Effective transconductance of the circuits shown in Figure 13.10.

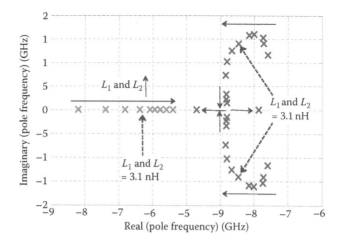

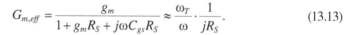

FIGURE 13.12 Effect of inductive series peaking in the feedback loop on the dominant poles of the circuit.

Another technique to enhance the bandwidth in common-gate transistors can be realized by adding an inductor at the gate [20]. This technique boosts the g_m of the transistor at high frequencies without any extra power consumption. To verify this scheme, the effective transconductance ($G_{m,\,eff}$) of the common-gate transistor is found with and without an inductor at its gate as shown Figure 13.13. The $G_{m,\,eff}$ of a common-gate transistor can be found by finding the output current of a device versus the input voltage and by taking into account the source resistance and parasitic capacitances. The exact $G_{m,\,eff}$ of a common-gate transistor and the high-frequency approximation can be found by

$$G_{m,eff} = \frac{g_m}{1 + g_m R_S + j\omega C_{gs} R_S} \approx \frac{\omega_T}{\omega} \cdot \frac{1}{jR_S}. \tag{13.13}$$

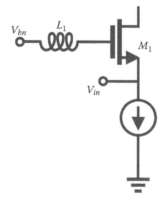

FIGURE 13.13 A common-gate stage with inductive g_m-boosting at the gate. (From Parvizi, M. et al., A 0.4 V ultra low-power UWB CMOS LNA employing noise cancellation, in *IEEE International Symposium on Circuits and Systems*, Beijing, China, 2013, pp. 2369–2372.)

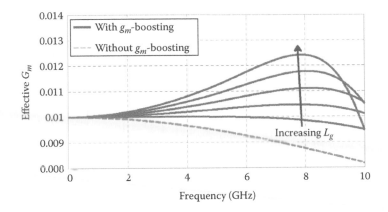

FIGURE 13.14 $G_{m,\,eff}$ of a common-gate transistor with and without g_m-boosting.

From (13.13), it can be seen that $G_{m,\,eff}$ decreases with frequency. However, by adding an inductor to the gate of CG transistor, the $G_{m,\,eff}$ becomes

$$G_{m,eff} = \frac{g_m}{1 + g_m R_S + j\omega C_{gs} R_S - \omega^2 L_g C_{gs}},$$
(13.14)

and at high frequencies, this simplifies to

$$G_{m,eff} \approx \frac{\omega_T}{\omega} \cdot \frac{1}{jR_S - \omega L_g}.$$
(13.15)

Consequently, $G_{m,\,eff}$ is boosted at the desired frequency. These simplified models are plotted in MATLAB and shown in Figure 13.14. For a common-gate transistor with g_m-boosting, $G_{m,\,eff}$ is plotted for multiple values of L_g. As can be seen, $G_{m,\,eff}$ decreases with frequency for a conventional CG transistor. However, the proposed g_m-boosting technique increases the $G_{m,\,eff}$ until the resonance frequency of L_g with C_{gs}.

13.4 ULTRALOW-POWER, ULTRALOW-VOLTAGE, WIDEBAND LNAs

13.4.1 ULP UWB LNA Employing Complementary Current-Reuse Noise Cancellation

Using the aforementioned design techniques, two ULV and ULP LNAs are designed and implemented in a 90 nm CMOS technology. In this section, we will briefly discuss the performance of these LNAs.

Figure 13.15 illustrates a current reuse noise-cancelling UWB LNA with supply of 0.4 V and power consumption of 0.41 mW. The proposed LNA utilizes a complementary current reuse noise cancellation scheme, through M_1 and M_2, to lower the power consumption and NF. Inductive g_m-boosting is employed by placing L_4 and L_5 at the gate of M_1 and M_2, respectively, to improve the bandwidth, input matching

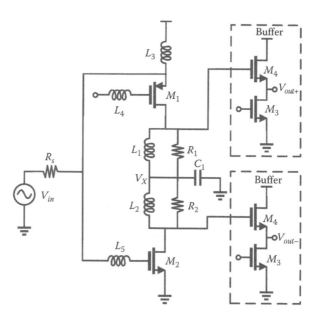

FIGURE 13.15 Circuit schematic of the proposed ULV and ULP current reuse LNA with g_m-boosting.

and gain of the LNA without additional power consumption. The optimal bias point for these transistors was obtained by considering the modified FOM shown in Equation 13.4. Noise cancellation of the input common-gate stage, M_1, is realized through M_2 in the common-source configuration. The output balancing and noise cancellation criteria are determined using

$$g_{m1} \cdot Z_1 = -g_{m2} \cdot Z_2, \tag{13.16}$$

where the left side of (13.16) is the gain of the CG stage and the right side is the gain of common-source stage. Z_1 represents the total impedance at the drain of the common-gate transistor, and Z_2 is the total impedance at the drain of the CS transistor.

The noise factor of this LNA is calculated assuming that the transistors have infinite output impedance and that the gate resistance is negligible for simplicity. The noise factor of this LNA is determined using the following formula:

$$F = 1 + \frac{\left(\dfrac{\gamma}{\alpha}\right) g_{m1} \left(Z_1 - g_{m2} Z_2 R_S\right)^2}{R_S A_V^2} + \frac{\left(\dfrac{\gamma}{\alpha}\right) g_{m2} Z_2 \left(1 + g_{m1} R_S\right)^2}{R_S A_V^2} + \frac{\left(R_1 + R_2\right)\left(1 + g_{m1} R_S\right)^2}{R_S A_V^2},$$

$$\tag{13.17}$$

where
$A_V = (g_{m1} Z_1 + g_{m2} Z_2)$
γ is the MOSFET noise parameter
$\alpha = g_m/g_{d0}$

The second term is the contribution of PMOS transistor, the third term is due to the NMOS transistor, and the last term comes from the load resistors.

The proposed g_m-boosting technique also improves the input matching condition, especially at high frequencies. The input impedance of the proposed circuit is given by

$$Z_{in} \cong \frac{\left(1 + g_{ol}Z_1\right)\left(1 - \omega^2 L_{g1}C_{gs1}\right)}{\left(g_{m1} + j\omega C_{gs1}\left(1 + g_{ol}Z_1\right)\right) + g_{ol}\left(1 - \omega^2 L_{g1}C_{gs}\right)} \left\| \left(\frac{1 - \omega^2 L_{g2}C_{gs2}}{j\omega C_{gs2}}\right), \quad (13.18)$$

where g_{Ol} is the output conductance of transistor M_1. This equation shows that the effect of parasitic capacitances at the input will be reduced by adding L_4 and L_5.

The LNA is implemented in a 90 nm TSMC CMOS technology and is simulated using $BSIM_4$ models in SpectreRF. Figure 13.16 highlights the input matching (S11), reverse isolation (S12), output matching (S22), and S21 of the LNA. The maximum gain of the LNA is 15 dB and its 3 dB bandwidth is between 3.2 and 10 GHz. The S11 is well below −10 dB in this band thanks to resonance at the input and inductive g_m-boosting. The S22 is below −10 dB as well, and the S12 is less than −35 dB. The noise figure of the proposed LNA is shown in Figure 13.17, and it varies between 4.5 and 5.3 dB across the bandwidth.

It is important to examine the performance of the noise-cancelling LNA at different process corners. Figure 13.18 illustrates the contribution of the drain noise current of the common-gate and common-source transistors in the bandwidth of operation. As can be seen, the noise contribution of common-gate transistor is very small (less than 1.5%), between 3 and 5 GHz. However, it gradually increases at high frequencies. This is due to a phase imbalance between the two outputs, which increases at high frequencies and reduces the drain current noise cancellation. A comparison of the performance of the LNA with the state-of-the-art works in the literature will be provided at the end of this section.

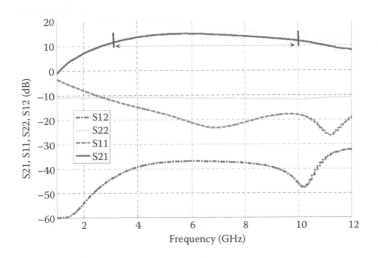

FIGURE 13.16 Simulated S21, S11, S12, and S22 of the proposed LNA.

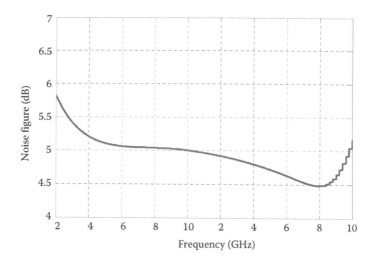

FIGURE 13.17 Simulated noise figure of the proposed LNA.

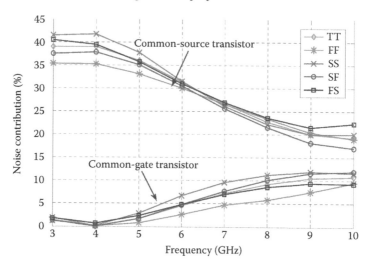

FIGURE 13.18 Noise contribution of the common-source and common-gate transistors indicating the effectiveness of noise cancellation at different process corners.

13.4.2 SUB-mW WIDEBAND LNA WITH RESISTIVE SHUNT-FEEDBACK AND INDUCTIVE SERIES PEACKING

The next LNA that will be covered in this section is a 0.5 V 750 μW resistive shunt feedback LNA with inductive series peaking [7]. The schematic of the LNA is shown in Figure 13.19. As with the previous LNA, the design techniques illustrated earlier in this chapter have been employed to realize the low-power LNA. Complementary current reuse technique is implemented through M_1 and M_2 to reduce the power consumption and simultaneously cancel the second-order distortion. Moreover, as

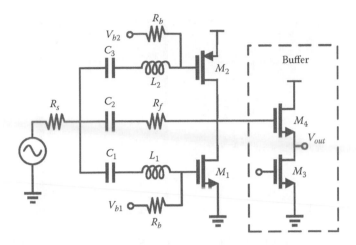

FIGURE 13.19 Complete schematic of the proposed ULP and ULV LNA with buffer.

discussed previously, inductive series peaking in the feedback path allows for bandwidth extension, NF reduction, and input matching improvement. Furthermore, the transistors are biased according to the ULP and ULV biasing methodology introduced earlier.

To better understand the operation of the LNA, the design equations are provided later. At low frequencies, the input impedance of the LNA shown in Figure 13.19 can be found by Equation 13.6. However, at high frequencies, the parasitic capacitances degrade the input matching, but this is compensated for by the addition of the inductors as described in Section 13.3.2.2. The overall input impedance of the proposed LNA can be found by

$$Z_{in} = Z_1 \| Z_2 \| \left(R_f + r_{o1} \| r_{o2} \right) \Big/ \left(\frac{g_{m1} \left(r_{o1} \| r_{o2} \right)}{j\omega C_{gs1} Z_1} + \frac{g_{m2} \left(r_{o1} \| r_{o2} \right)}{j\omega C_{gs2} Z_2} \right), \quad (13.19)$$

where Z_1 and Z_2 are $(j\omega L_1 + 1/j\omega Cgs_1)$ and $(j\omega L_2 + 1/j\omega Cgs_2)$, respectively. The voltage gain of the LNA is boosted by the inductive series peaking in the feedback path at high frequencies. The overall voltage gain of the LNA is given by

$$A_V = \frac{\left(r_{o1} \| r_{o2} \right) \left(1 - \dfrac{g_{m1} R_f}{j\omega C_{gs1} Z_1} - \dfrac{g_{m2} R_f}{j\omega C_{gs2} Z_2} \right)}{\left(R_f + \left(r_{o1} \| r_{o2} \right) \right) \left(1 + \dfrac{R_s}{Z_1 \| Z_2} \right)}. \quad (13.20)$$

The main noise sources in this LNA are the channel noises of M_1 and M_2 and of the feedback resistor, R_f. The noise factor of the LNA can be expanded as

$$NF \approx 1 + \frac{R_f}{R_s} \left(\frac{1 + g_{mt} R_s}{1 - g_{mt} R_f} \right)^2 + \frac{\gamma g_{mt}}{\alpha R_s} \left(\frac{\left(R_s + R_f \right)}{\left(1 - g_m R_f \right)} \right)^2. \quad (13.21)$$

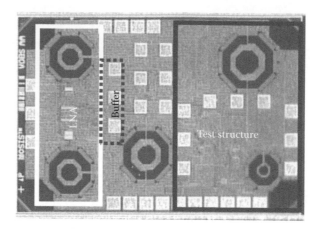

FIGURE 13.20 Die micrograph of the ULP and ULV resistive shunt feedback LNA.

The effective g_m of the transistors gets boosted due to inductive series peaking in the feedback path, giving rise to lower NF at the resonance frequency.

The die photograph of the LNA is shown in Figure 13.20, and the active area is 0.27 mm² × 0.88 mm². The circuit was tested using a probe station, and an on-chip buffer was used to drive the 50 Ω impedance of the measurement equipment. The buffer was measured separately and de-embedded from the S21 and NF results presented here. The dc pads were bonded and the dc voltages were supplied using a custom printed circuit board (PCB). The length of the bond wires was minimized to lower the parasitic inductors due to bond wires. Figure 13.21 shows the measurement results along with simulation results for S21 and S11.

The fabricated LNA reaches a maximum S21 of 12.6 dB at 4.5 GHz with a 3 dB cut-off frequency of 7 GHz and achieves an S11 of less than −10 dB along the whole bandwidth. The gain roll-off seen in the measured voltage gain can be explained by the change in the resonant frequency of the inductors due to increased inductances

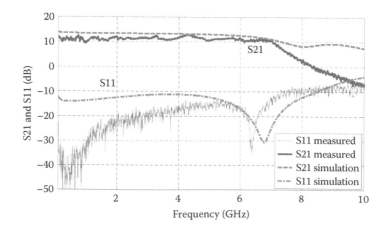

FIGURE 13.21 Measurement and simulation results for S21 and input matching.

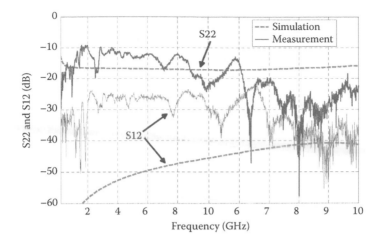

FIGURE 13.22 Reverse isolation and output matching of the LNA.

and/or parasitic capacitances in the fabricated LNA. Figure 13.22 shows the S22 and S12 of the LNA. The S22 is less than −10 dB in the whole bandwidth, and S12 is better than −23 dB in the band of interest. The parasitic coupling of the LNA and the buffer have deteriorated the S12 at low frequencies compared with the simulations.

The NF of the LNA is shown in Figure 13.23. The minimum measured NF is 5.5 dB while the average NF in the whole band is 6 dB. The measured NF is higher than the simulation results. This is mainly due to inaccurate transistor noise modeling in moderate inversion and to a reduced voltage gain compared with the simulation results. The bond wires used in this test setup also contribute to noise injection, since there are insufficient on-chip decoupling capacitances at the DC biasing nodes to create a strong small-signal ground. The power consumption of the LNA is only 0.75 mW from a 0.5 V supply voltage.

Table 13.2 presents a comparison of the presented LNAs with other LNAs in the literature. A figure of merit expressed in (13.22) [29] is employed to compare

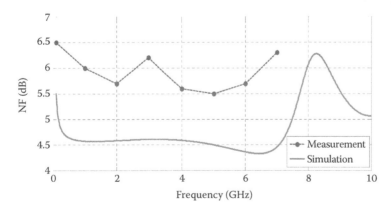

FIGURE 13.23 Measurement and simulation results for NF.

TABLE 13.2
Performance Summary and Comparison with the State-of-the-Art LNAs

Parameter	This Work (NC)	This Work (Feedback)	[30]	[31]	[32]	[33]	[16]	[29]
3 dB BW (GHz)	3.2–10	0.1 ~ 7	0.8 ~ 8.4	1.3 ~ 12.3	2.6 ~ 10.5	3.1 ~ 10.6	2.6 ~ 10.2	0 ~ 6
Power (mW)	0.41	0.75	2.6	4.5	0.99	9	7.2	9.2
Supply (V)	0.4	0.5	1.3	1.8	1.1	1.2	1.2	1.2
S11 (dB)	<−10	<−10	<−9	<−6	<−10	<−9.9	<−9	<−10
S21 (dB)	15	12.6	12.6	8.2	7.9	16.5	12.5	16.5
NF (dB)	4.5 ~ 5.3	5.5 ~ 6.5	3.3 ~ 5.5	4.6 ~ 5.5	5.5 ~ 6.5	2.1 ~ 2.9	3 ~ 7	2.5
IIP3 (dBm)	−2 ~ −7	−6 ~ −9	3.9 ~ 8.5	8	NA	−5.5 ~ −8.5	NA	−10
IIP2 (dBm)	NA	+5	1.8 ~ 13.9	NA	NA	NA	NA	NA
Technology	90 nm	90 nm	0.13 μm	0.18 μm	0.18 μm	0.13 μm	90 nm	90 nm
Area (mm²)	NA	0.23	0.58	1	0.73	0.87	0.64	0.0017
FOM	31.5	20.89	15.55	7.76	15.93	15.67	4.76	12.1

the overall performance of the LNAs. It can be seen that LNAs consume much less power and require a lower supply voltage compared with other works, while simultaneously achieving a higher FOM. However, it should be noticed that a small variation in performance will be noted when the LNA is fabricated and measured:

$$FOM = 20\log_{10}\left(\frac{Gain_{average}[lin] \times BW[GHz]}{P_{dc}[mW] \times (F_{average}[lin]-1)}\right).$$
(13.22)

13.5 LOW-POWER, LOW-COMPLEXITY IR-UWB RECEIVER DESIGN

The previous sections of this chapter focused on low power and low-voltage design techniques for individual amplifiers in RF receivers. This section will focus on implementing a receiver using some of these techniques, as well as system-level choices that allow for ultralow-power operation. A low-power transmitter will also be presented.

As mentioned earlier, the inherent duty-cycled operation of pulse-based transceivers allows UWB radios to achieve energy-efficient communications for low data rate systems, such as those used in many WSN applications. This work focuses on one such transceiver for use in the 100–960 MHz band to increase energy efficiency compared to the operation above 3.1 GHz. While the antenna size for this frequency band renders this system unattractive for mobile devices, applications such as structural health monitoring of urban infrastructure (e.g., the structural integrity of bridges) do not have this size constraint.

Efficient low data rate receivers based on energy detection and correlation have previously been implemented in Refs. [1,34–39]. Due to their windowed nature of operation, they employ relatively complex pulse synchronization schemes involving delay chains for timing, multiple switching integrators (or correlators), and digital backend processing. This leads to complex and/or lengthy synchronization procedures, and this problem is aggravated at low data rates where pulse widths are small compared to the pulse repetition rate.

This same type of situation exists in burst mode transceivers, which have a high instantaneous data rate but implement duty cycling at the packet level. In this case, efficient and fast packet-level synchronization is required to create an optimal solution. Additionally, since the UWB pulses in burst mode communications are condensed into a smaller time period, each pulse must have less energy in order to meet FCC regulations, in which the power spectral density of the pulse train is measured over a 1 ms time span. For this reason, this work focuses on non-burst-mode UWB communications.

Synchronization time and complexity can be decreased by increasing the window of integration in energy detection receivers, but this decreases the performance of the communications link due to the collection of additional noise and interference. This work focuses on peak detection instead of energy detection to create a single path receiver that does not suffer from cumulative noise and interference integrated within the detection window. The synchronization scheme is based on a clock and data recovery (CDR) system, which effectively transfers the pulse synchronization process to the analog domain. This reduces the amount of digital circuitry and its associated leakage current and allows for a high-efficiency, low-complexity implementation with a simple synchronization scheme.

The transceiver implemented here uses on–off keying modulation with a nominal data rate of 1 Mbps. The transceiver achieves a receiver efficiency of 292 pJ/pulse and a transmitter efficiency of 115 pJ/pulse, making it attractive for sensor nodes that require increased battery longevity, and also provides for the possibility of powering sensor nodes using energy harvesting techniques. Section 13.5.1 will describe the receiver design, and the transmitter design will be presented in Section 13.5.2. Measurement results will be presented in Section 13.5.3, along with a comparison with other recent works.

13.5.1 RECEIVER DESIGN

Figure 13.24 shows a simplified block diagram of the receiver. The RF front-end is responsible for amplifying a received UWB pulse to digital levels, and the CDR circuitry synchronizes the receiver and outputs the baseband data and clock signals.

13.5.1.1 RF Front-End

An expanded view of the RF front-end is shown in Figure 13.25. The signal is first amplified by an LNA and gain stages and is then compared with a low-pass version of itself to remove any slow-varying offsets. A large input signal will trigger the comparator, which indicates that a pulse was received.

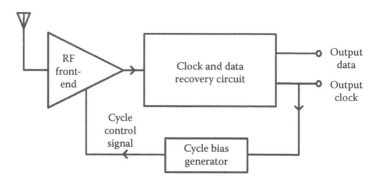

FIGURE 13.24 Block diagram of the UWB receiver.

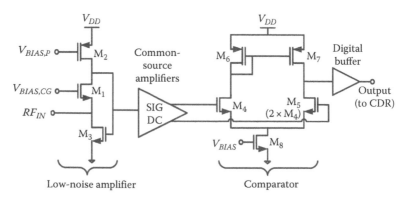

FIGURE 13.25 Expanded view of the RF front-end.

The LNA is based on a common-gate amplifier (M_1) with feedback provided by M_3 to reduce the current needed to achieve a 50 Ω input match. As discussed earlier, the use of a single current path allows for low-power operation by allowing the bias current to be reused. Additional discussion of the circuit can be found in Ref. [15]. The LNA is followed by two common-source gain stages with resistive feedback to decrease their sensitivities to bias voltage variations. The signal (SIG) and low-pass (dc) outputs are then compared using a differential pair in which M_5 is twice the size of M_4 to avoid false detections in the presence of noise and process variations as determined from Monte Carlo simulations. The output pulse is then converted to digital levels using a digital buffer, before being synchronized to the local clock by the CDR circuitry described in the next subsection. All front-end stages include switches to enable duty-cycled operation. To ensure that noise from this switching operation (amplified by the RF front-end) does not incorrectly trigger an output pulse, the enabled signal for the digital buffer is delayed relative to the amplifier and comparator enable signals.

It should be noted that alternative comparators based on positive feedback [40] or rectification to detect both positive and negative peaks [41] can also be used with this receiver architecture.

13.5.1.2 Clock and Data Recovery System

The schematic of the CDR system is shown in Figure 13.26. It consists of a phase-frequency detector (PFD), two charge pumps, a loop filter, and a voltage-controlled oscillator (VCO). As mentioned earlier, the CDR system is responsible for synchronizing the receiver's clock to the incoming data.

Standard phase detectors (PDs) for use with random data, such as the Hogge and Alexander PDs, are based on sampling the data with the clock [42]—an approach that does not work with the low duty cycle pulses in UWB communications. A PD architecture that provides a linear output from pulsed data inputs was devised here specifically for this application and is shown in Figure 13.27a. The data pulse is used to trigger two flip-flops, which are reset by the rising edges of the clock and its inverse. Subtracting the output of the XOR gate from the *phase up* signal yields a signal whose average varies linearly with the delay between the rising edges of the data and clock signals. Using the data signal to trigger the flip-flops ensures that no output is produced for a "0". Additionally, when the clock and data signals are synchronized, the *phase up* signal also represents the output data.

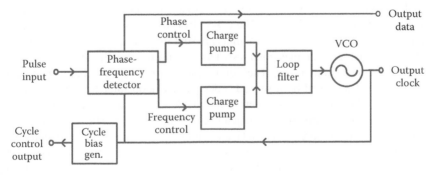

FIGURE 13.26 Block diagram of the clock and data recovery system.

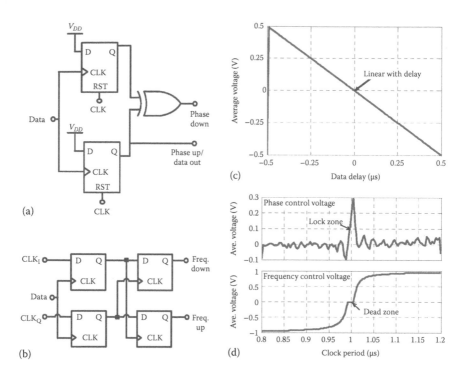

FIGURE 13.27 Schematics of the (a) phase and (b) frequency detectors, as well as simulation results for the average output voltage versus (c) data delay, and (d) clock period.

A frequency detector (FD) based on sampling the in-phase and quadrature clocks is included to increase the lock range of the CDR and is shown in Figure 13.27b [42]. Subtracting the *freq down* signal from the *freq up* signal results in bang-bang frequency detection with a narrow dead zone around the region where the clock and data periods are equal.

Simulation results for both the phase and frequency detectors are shown in Figure 13.27c and d. The PD presents a linear output voltage versus delay (Figure 13.27c); however, there is only a narrow region over which the PD exhibits a linear average output voltage versus the clock period (Figure 13.27d, top). The FD is seen to have the desired average output voltage versus the clock period, except during the dead zone centered around the data period of 1 μs. The PD controls the synchronization in this region.

The phase and frequency control signals are sent to separate charge pumps that use dummy branches and complementary switches to minimize charge sharing, charge injection, and feedthrough [43]. The output currents of the two charge pumps are summed up and averaged by a second-order passive loop filter to minimize power consumption. This output voltage controls the VCO, which is based on a two-stage differential ring oscillator structure, with quadrature output clocks. The block diagram of the VCO is shown in Figure 13.28a and b and Figure 13.28c shows the block and circuit level schematics of a single VCO delay stage, respectively. A positive feedback latch is used in each stage to create an additional delay and allow for oscillations with only two stages. Tuning of the VCO frequency is achieved through current-starved inverters

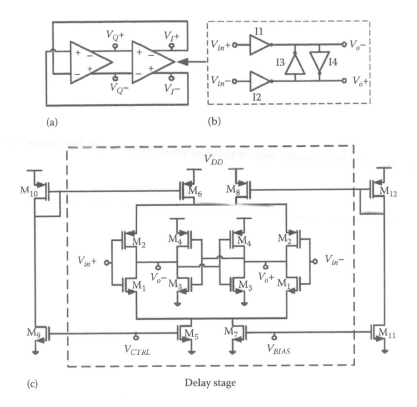

(a)

(b)

(c)   Delay stage

FIGURE 13.28 Schematics of (a) the VCO architecture and a (b) block level and (c) circuit level view of a single delay stage.

(I1 and I2 in Figure 13.28b). A fixed current is injected into these inverters to reduce the VCO gain and account for shifts in frequency due to process variations. The signals V_{CTRL} and V_{BIAS} (in Figure 13.28c) are used to implement these functions. The inverse quadrature clock is used to derive the signal for the cycled biasing of the RF front-end.

The loop bandwidth is an important parameter in determining the dynamics of the receiver. A large loop bandwidth will result in a quick synchronization time and the ability to better track jitter present in the transmitted data. This will allow the receiver to perform more aggressive duty cycling since there will be a greater certainty as to when the next UWB pulse will arrive. However, if the loop bandwidth is too large, the receiver will not be able to tolerate long runs of consecutive 0s and may become unstable. The loop was designed to have a loop bandwidth of 40 kHz and a phase margin of 50° at a nominal data rate of 1 Mbps. The system can accommodate data rates from 100 kbps to 1.8 Mbps as long as the loop bandwidth is modified to ensure a stable system. For instance, a lower data rate would require a smaller loop bandwidth. This would increase the time in seconds required to synchronize the receiver to the transmitter; however, the number of bit periods required for synchronization would remain the same. The receiver would also need to be on for a longer time period to account for the increased jitter with respect to the transmitted signal, but the level of duty cycling will remain the same since the data rate has also been reduced.

For this work, the total loop filter capacitance is 415 pF, and the loop filter resistance is 16 kΩ.

13.5.2 TRANSMITTER DESIGN

One of the advantages of UWB communications is the ability to design energy-efficient digital transmitters. The transmitter designed in this work is shown in Figure 13.29. It uses a current-starved inverter (I1) and an AND gate (I2) to create a short pulse from the rising edge of the data input, which is then filtered by a cascade of three current-starved inverters (I3). A large inverter (I4) is used to drive the 50 Ω load of an antenna (or measurement equipment). Tuning the pulse width can be accomplished by varying the delay through I1, and the bandwidth can be tuned by varying the speed at which the inverters in I3 charge and discharge the capacitive input of I4. The bandwidth of the transmitter varies by 50% across process variations, but can be tuned to the nominal value using the aforementioned tuning mechanisms.

13.5.3 MEASUREMENT RESULTS

The transceiver was designed and fabricated in an ST Microelectronics 90 nm CMOS process, and the chip micrograph is shown in Figure 13.30a. As the top level metal fill covers up the chip features, the layout is shown in Figure 13.30b. The loop filter was not integrated on-chip to allow for more flexible testing, but the typical sizes of its components are small enough to allow for complete integration in the 0.94 mm² chip area once the test structure in Figure 13.30b is removed.

The measured time and frequency domain responses of the transmitted pulse are shown in Figure 13.31a and b, respectively. The transmitted pulse spectrum meets the most stringent interpretation of the FCC regulations [44].

Figure 13.32 shows measurement results when the transmitter is connected to the receiver through a 33 dB attenuator. The pulse output of the comparator has a width of ~20 ns, a design choice that ensures detection by the PFD and prevents quick multipath echoes from triggering multiple pulses. The front-end enable signal rises prior to the clock edge to allow for the turn-on time of the RF front-end. The quadrature clock is used to provide optimal sampling of the output data.

Figure 13.33 shows synchronization times for a 1 Mbps data stream composed of 50% of 1s for various initial values of the VCO control voltage. As expected, setting the initial VCO frequency close to that of the anticipated data rate reduces the synchronization time. The receiver also synchronizes faster when the VCO is initially operating at a frequency higher than that of the data stream, owing to mismatches

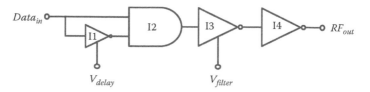

FIGURE 13.29 Schematic of the UWB transmitter.

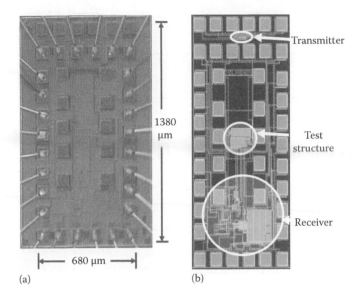

(a) (b)

FIGURE 13.30 (a) Chip micrograph (covered by metal fill), along with (b) the layout showing chip features.

in the charge pumps. Figure 13.34 shows the pk–pk clock jitter for runs of 0s. This defines the worst-case uncertainty when the next UWB pulse will be, and thus the level of duty cycling that can be achieved. Even with runs of 14 consecutive 0s, the *on* period of the receiver can be less than 2.5% at 1 Mbps.

Table 13.3 provides a summary of the receiver's performance, along with comparisons to similar recently published works. The receiver sensitivity was measured

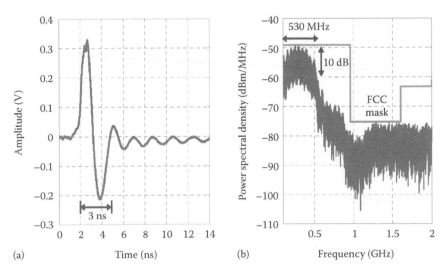

(a) Time (ns) (b) Frequency (GHz)

FIGURE 13.31 Measurement results of the transmitter in (a) the time and (b) the frequency domain.

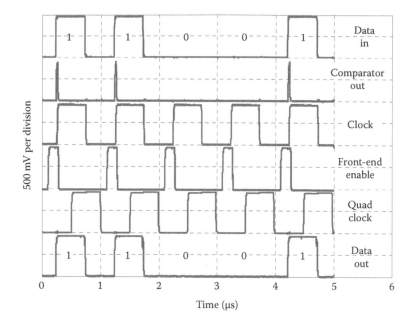

FIGURE 13.32 Measurement results when the transmitter is connected to the receiver through a 33 dB attenuator.

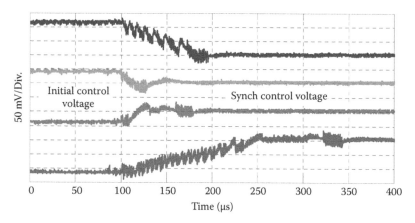

FIGURE 13.33 Measured control voltage synchronization results for various initial control voltages.

by using an FPGA as a bit error rate tester (BERT). A 23-bit pseudo random bit sequence was transmitted and the errors in the received bitstream were evaluated at different levels of attenuation. As in other works, correct receiver operation was defined as having a BER of less than 10^{-3}. The receiver sensitivity of −65 dBm in this implementation can be improved by adding additional gain stages after the LNA at the cost of higher power consumption. Specifically, each additional gain stage would increase the sensitivity by ~11 dB (until the noise floor is reached) at a cost of

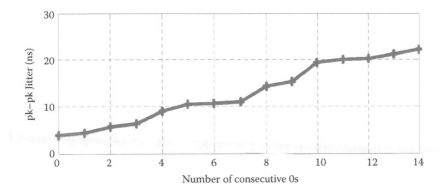

FIGURE 13.34   Measured pk–pk clock jitter versus the number of consecutive 0s.

230 µW (34 µW during 15% cycled operation). With the two gain stages used here and the proposed CDR-based synchronization scheme, the energy efficiency when receiving a signal is better than in Refs. [1,37,45]. While Ref. [39] appears to be more efficient, it is important to note that the number quoted is energy/pulse, not energy/ bit. Multiple pulses/bit are used in Ref. [39] to obtain a processing gain, therefore the pulse repetition rate must be higher than the data rate. However, due to the limits imposed by regulators on the transmitted pulse power in UWB communications, a faster pulse rate means that less energy can be transmitted per pulse, which negates this processing gain. (It should be noted that this assumes the peak power of the pulses do not surpass the FCC regulation.) The efficiency in terms of energy/bit at the data rate of 1.3 Mbps reported in Ref. [39] is 3.3 nJ/bit.

Comparing the power consumption in the digital back-ends shows the benefits of the analog CDR-based synchronization scheme proposed here. The overall power consumption in this work is comparable to that in Ref. [45]; however, the data rate here is higher by more than an order of magnitude.

By setting the initial VCO frequency close to that of the anticipated data rate (which would be fixed in advance), the number of pulse periods needed for pulse-level synchronization is 25. While this is higher than the simple counter scheme reported in Ref. [45], the receiver implemented here only requires a clock frequency equal to the data rate. The receiver in Ref. [45] requires a stable clock that is orders of magnitude higher than the data rate to provide sufficient resolution for accurate duty cycling. Furthermore, the scheme presented here allows for real-time correction of clock drift between the transmitter and receiver with no additional power consumption compared to the digital tracking algorithm implemented in Ref. [39].

The receiver's performance in the presence of in-band interference was tested by injecting a tone at 300 MHz, which is at the center of the pulse spectrum. A BER of less than 10^{-3} was observed for signal-to-interference (SIR) ratios of −14 dB and higher at a data rate of 1 Mbps. This performance is ~10 times better than the −15 dB SIR achieved at the lower data rate of 100 kbps with the energy detection receiver in Ref. [35], which collects energy over a (relatively long) 30 ns window.

Table 13.4 provides a summary of the transmitter performance, along with a comparison to recently published transmitters in the 0–960 MHz band. The better energy

TABLE 13.3
Comparison with Recent IR-UWB Receivers

Receiver Information	This Work	[1]	[37]	[39]	[45]
Technology	90 nm CMOS	90 nm CMOS	0.13 wm CMOS	0.13 μm CMOS	0.15 μm FD-SOI CMOS
Supply	1 V	1 V	1.2 V	1.2 V	1 V
Die size	0.94 mm²	5.46 mm²	8 mm²	4.52 mm²	0.35 mm²
Data rate	100 kbps to 1.8 Mbps	Up to 16 Mbps	Up to 31 Mbps	Up to 40 Mbps	25 kbps
Modulation	OOK	OOK, PPM	OOK, PPM	BPSK	BPSK
Measured Receiver Results					
Receiver sensitivity	−65 dBm at 1 Mbps	−78 dBm at 16 Mbps	−78 dBm at 975 kbps	−69 dBm at 1.3 Mbps	Not reported
Front-end noise figure	<9 dB[a]	9 dB	Not reported	23.5 dB	Not reported
Front-end bandwidth	80 MHz to 1.1 GHz[a]	3.25–4.75 GHz	4–5 GHz	500 MHz in the 0–960 MHz band	140–350 MHz
S11	<−12 dB	<−7.8 dB	<−15 dB	Not reported	<−10 dB
Front-end gain	Tunable up to 40 dB	6–45 dB	Tunable up to 80 dB	20–50 dB	Tunable up to 50 dB
Pulse synchronization time	<25 pulse periods	Not reported	Not reported	190 pulse periods[b]	2 pulse periods
Front-end power consumption (not cycled)	1.1 mW	5.9–20.2 mW	7 mW	1.8–2.81 mW	1.67 mW
Digital/back-end power consumption	127 μW (at 1 Mbps)	2.48 mW	26.3 mW	1.5–2 mW	30 μW
Energy efficiency	292 pJ/pulse (15% cycled at 1 Mbps)	0.5 nJ/pulse to 1.4 nJ/pulse	34.2 nJ/pulse (at 975 kbps)	108 pJ/pulse (at 40 Mpulse/s)[c]	12 nJ/pulse (cycled operation at 25 kbps)
Average power at highest efficiency	292 μW at 1 Mbps	8.38 mW at 16 Mbps	33.3 mW at 31 Mbps	4.2 mW at 1.3 Mbps	299 mW at 25 kbps

a Simulation results (output unavailable on-chip).
b At a pulse repetition rate of 40 MHz with Ns = 5.
c Unlike the other implementations, each bit is made up of multiple pulses. Please refer to the text for more details.

TABLE 13.4
Comparison with Recent IR-UWB Transmitters

Measured Transmitter Results	This Work	[45]	[46]	[47]
Technology	90 nm CMOS	0.15 μm FD-SOI CMOS	0.18 μm CMOS	90 nm CMOS
Center frequency	275 MHz	250 MHz	250 MHz	250 MHz
10 dB bandwidth	530 MHz	300 MHz	300 MHz	400 MHz
FCC spectral efficiency	55%	31%	31%	42%
Power consumption	115 μW (at 1 Mbps)	1 μW (at 25 kbps)	234 μW (at 100 Mbps)	220 μW (at 100 Mbps)
Energy efficiency	115 pJ/pulse (at 1 Mbps)	40 pJ/pulse (at 25 kbps)	2.3 pJ/pulse (at 100 Mbps)	2.2 pJ/pulse (at 100 Mbps)

efficiency in Refs. [46, 47] is due solely to the higher data rate since the output power spectral densities of UWB transmitters are limited by FCC regulations. The work in this chapter offers the highest FCC spectral efficiency (pulse bandwidth divided by FCC allowed bandwidth) while still meeting the mask requirements, allowing for the transmission of more energy and enabling communications over a larger distance.

13.5.4 CONCLUSION

In this chapter, ULP and ULV CMOS circuit design challenges were introduced and analyzed. Design techniques to overcome these challenges in wideband LNAs were discussed, including a new scheme to find the optimum bias point for transistors in ULP and ULV circuits, current reuse techniques to reduce the power consumption and distortion, forward body-biasing to account for reduced voltage headroom, passive and active shunt feedback to provide low power and wideband input matching, and bandwidth extension and g_m-boosting techniques to achieve high-frequency wideband operation. Two ULP and ULV LNAs that achieve sub-mW power consumption with high FOMs were provided as design examples.

Additionally, a low-power IR-UWB transceiver was discussed. The transceiver presents a low complexity solution for IR-UWB applications. By transferring the pulse synchronization to the analog domain in the form of a clock and data recovery loop, a reduction in the size and power consumption of the digital circuitry is obtained. Using this technique, the implemented receiver has a nominal data rate of 1 Mbps using on–off keying, a receiver efficiency of 292 pJ/pulse, and a synchronization time of 25 pulse periods. A transmitter capable of using 59% of the allotted FCC spectral mask was also designed with an efficiency of 115 pJ/pulse at a data rate of 1 Mbps. The entire transceiver consumes an average power of 407 μW and occupies a chip area of 0.94 mm². This level of power consumption and efficiency is well suited to WSN nodes that require increased battery lifetime or that rely on energy harvesting techniques for power.

REFERENCES

1. D.C. Daly, P. Mercier, M. Bhardwaj, A.L. Stone, Z.N. Aldworth, T.L. Daniel et al., A pulsed UWB receiver SoC for insect motion control, *IEEE J. Solid-State Circ.*, 45, 153–166, 2010.
2. N. Stanic, P. Kinget, and Y. Tsividis, A 0.5 V 900 MHz CMOS receiver front end, in *IEEE Digest of Technical Papers VLSI Circuits*, Honolulu, HI, 2006, pp. 228–229.
3. M. Brandolini, M. Sosio, and F. Svelto, A 750 mV fully integrated direct conversion receiver front-end for GSM in 90-nm CMOS, *IEEE J. Solid-State Circ.*, 42, 1310–1317, 2007.
4. C. Hsiao-Chin, W. Tao, C. Hung-Wei, K. Tze-Huei, and L. Shey-Shi, 0.5-V 5.6-GHz CMOS receiver subsystem, *IEEE Trans. Microw. Theor. Tech.*, 57, 329–335, 2009.
5. A. Balankutty, Y. Shih-An, F. Yiping, and P.R. Kinget, A 0.6-V zero-IF/low-IF receiver with integrated fractional-N synthesizer for 2.4-GHz ISM-band applications, *IEEE J. Solid-State Circ.*, 45, 538–553, 2010.
6. A. Balankutty and P.R. Kinget, An ultra-low voltage, low-noise, high linearity 900-MHz receiver with digitally calibrated in-band feed-forward interferer cancellation in 65-nm CMOS, *IEEE J. Solid-State Circ.*, 46, 2268–2283, 2011.
7. M. Parvizi, K. Allidina, and M. El-Gamal, A sub-mw, ultra-low voltage, wideband low-noise amplifier design technique, Accepted for publication, *IEEE Trans. Very Large Scale Integ. Syst.*, 2014.
8. F. Silveira, D. Flandre, and P.G.A. Jespers, A $g_m I_D$ based methodology for the design of CMOS analog circuits and its application to the synthesis of a silicon-on-insulator micropower OTA, *IEEE J. Solid-State Circ.*, 31, 1314–1319, 1996.
9. C.C. Enz and E.A. Vittoz, *Charge-Based MOS Transistor Modeling: The EKV Model for Low-Power and RF IC Design*. John Wiley & Sons, New York, NY, 2006.
10. D.M. Binkley, *Tradeoffs and Optimization in Analog CMOS Design*. John Wiley & Sons, New York, NY, 2008.
11. V. Aparin, G. Brown, and L.E. Larson, Linearization of CMOS LNA's via optimum gate biasing, in *IEEE International Circuits and Systems Symposium*, Vancouver, Canada, 2004, pp. IV-748–751.
12. D.K. Shaeffer and T.H. Lee, A 1.5-V, 1.5-GHz CMOS low noise amplifier, *IEEE J. Solid-State Circ.*, 32, 745–759, 1997.
13. A. Shameli and P. Heydari, A novel power optimization technique for ultra-low power RFICs, in *Proceedings of the International Symposium on Low Power Electronics and Design*, Tegernsee, Germany, 2006, pp. 274–279.
14. J.S. Walling, S. Shekhar, and D.J. Allstot, A gm-boosted current-reuse LNA in 0.18 μm CMOS, in *IEEE Radio Frequency Integrated Circuits Symposium*, Honolulu, HI, 2007, pp. 613–616.
15. K. Allidina and M.N. El-Gamal, A 1 V CMOS LNA for low power ultra-wideband systems, in *IEEE International Conference on Electronics, Circuits and Systems*, St. Julien's, Malta, 2008, pp. 165–168.
16. G. Sapone and G. Palmisano, A 3–10-GHz low-power CMOS low-noise amplifier for ultra-wideband communication, *IEEE Trans. Microw. Theor. Tech.*, 59, 678–686, 2011.
17. H. Hsieh-Hung and L. Liang-Hung, Design of ultra-low-voltage RF frontends with complementary current-reused architectures, *IEEE Trans. Microw. Theor. Tech.*, 55, 1445–1458, 2007.
18. C. Hyouk-Kyu, M.K. Raja, Y. Xiaojun, and J. Minkyu, A CMOS medradio receiver RF front-end with a complementary current-reuse LNA, *IEEE Trans. Microw. Theor. Tech.*, 59, 1846–1854, 2011.
19. M. Khurram and S.M. Rezaul Hasan, A 3–5 GHz current-reuse gm-boosted CG LNA for ultrawideband in 130 nm CMOS, *IEEE Trans. Very Large Scale Integr. Syst.*, 20, 400–409, 2012.

20. M. Parvizi, K. Allidina, F. Nabki, and M. El-Gamal, A 0.4 V ultra low-power UWB CMOS LNA employing noise cancellation, in *IEEE International Symposium on Circuits and Systems*, Beijing, China, 2013, pp. 2369–2372.

21. C. Wei-Hung, L. Gang, B. Zdravko, and A.M. Niknejad, A highly linear broadband CMOS LNA employing noise and distortion cancellation, *IEEE J. Solid-State Circ.*, 43, 1164–1176, 2008.

22. L. Chen-Ming, L. Ming-Tsung, H. Kuang-Chi, and T. Jenn-Hwan, A low-power self-forward-body-bias CMOS LNA for 3–6.5-GHz UWB receivers, *IEEE Microw. Wireless Compon. Lett.*, 20, 100–102, 2010.

23. A. Bevilacqua and A.M. Niknejad, An ultrawideband CMOS low-noise amplifier for 3.1–10.6-GHz wireless receivers, *IEEE J. Solid-State Circ.*, 39, 2259–2268, 2004.

24. W. Sanghyun, K. Woonyun, L. Chang-Ho, L. Kyutae, and J. Laskar, A 3.6 mW differential common gate CMOS LNA with positive-negative feedback, in *ISSCC Digest of Technical Papers*, San Francisco, CA, 2009, pp. 218–219, 219a.

25. F. Belmas, F. Hameau, and J. Fournier, A low power inductorless LNA with double gm enhancement in 130 nm CMOS, *IEEE J. Solid-State Circ.*, 47, 1094–1103, 2012.

26. K. Chang-Wan, K. Min-Suk, A. Phan Tuan, K. Hoon-Tae, and L. Sang-Gug, An ultrawideband CMOS low noise amplifier for 3–5-GHz UWB system, *IEEE J. Solid-State Circ.*, 40, 544–547, 2005.

27. C. Hsien-Ku, C. Da-Chiang, J. Ying-Zong, and L. Shey-Shi, A compact wideband CMOS low-noise amplifier using shunt resistive-feedback and series inductive-peaking techniques, *IEEE Microw. Wireless Compon. Lett.*, 17, 616–618, 2007.

28. C. Tienyu, C. Jinghong, L.A. Rigge, and J. Lin, ESD-protected wideband CMOS LNAs using modified resistive feedback techniques with chip-on-board packaging, *IEEE Trans. Microw. Theor. Tech.*, 56, 1817–1826, 2008.

29. J. Borremans, P. Wambacq, C. Soens, Y. Rolain, and M. Kuijk, Low-area active-feedback low-noise amplifier design in scaled digital CMOS, *IEEE J. Solid-State Circ.*, 43, 2422–2433, 2008.

30. Z. Heng, F. Xiaohua, and E.S. Sinencio, A low-power, linearized, ultra-wideband LNA design technique, *IEEE J. Solid-State Circ.*, 44, 320–330, 2009.

31. S. Shekhar, J.S. Walling, and D.J. Allstot, Bandwidth extension techniques for CMOS amplifiers, *IEEE J. Solid-State Circ.*, 41, 2424–2439, 2006.

32. J.F. Chang and Y.S. Lin, 0.99 mW 3–10 GHz common-gate CMOS UWB LNA using T-match input network and self-body-bias technique, *Electron. Lett.*, 47, 658–659, 2011.

33. M.T. Reiha and J.R. Long, A 1.2 V reactive-feedback 3.1–10.6 GHz low-noise amplifier in 0.13 um CMOS, *IEEE J. Solid-State Circ.*, 42, 1023–1033, 2007.

34. T. Terada, S. Yoshizumi, M. Muqsith, Y. Sanada, and T. Kuroda, A CMOS ultrawideband impulse radio transceiver for 1-Mb/s data communications and 2.5-cm range finding, *IEEE J. Solid-State Circ.*, 41, 891–898, 2006.

35. F.S. Lee and A.P. Chandrakasan, A 2.5 nJ/bit 0.65 V pulsed UWB receiver in 90 nm CMOS, *IEEE J. Solid-State Circ.*, 42, 2851–2859, 2007.

36. A. Gerosa, S. Solda, A. Bevilacqua, D. Vogrig, A. Neviani, An energy-detector for noncoherent impulse-radio UWB receivers, *IEEE Trans. Circ. Syst. I Reg. Papers*, 56, 1030–1040, 2009.

37. D. Lachartre, B. Denis, D. Morche, L. Ouvry, M. Pezzin, B. Piaget et al., A 1.1 nJ/b 802.15.4a-compliant fully integrated UWB transceiver in 0.13 μm CMOS, in *IEEE International Solid-State Circuits Conference–Digest of Technical Papers*, San Francisco, CA, 2009, pp. 312–313, 313a.

38. L. Lechang, T. Sakurai, and M. Takamiya, A 1.28 mW 100 Mb/s impulse UWB receiver with charge-domain correlator and embedded sliding scheme for data synchronization, in *IEEE Digest of Technical Papers VLSI Circuits*, Kyoto, Japan, 2009, pp. 146–147.

39. N. Van Helleputte, M. Verhelst, W. Dehaene, and G. Gielen, A reconfigurable, 130 nm CMOS 108 pJ/pulse, fully integrated IR-UWB receiver for communication and precise ranging, *IEEE J. Solid-State Circ.*, 45, 69–83, 2010.
40. R. Dokania, X. Wang, S. Tallur, C. Dorta-Quinones, and A. Apsel, An ultralow-power dual-band UWB impulse radio, *IEEE Trans. Circ. Syst. II Exp. Briefs*, 57, 541–545, 2010.
41. M. Crepaldi, L. Chen, K. Dronson, J. Fernandes, and P. Kinget, An ultra-low-power interference-robust IR-UWB transceiver chipset using self-synchronizing OOK modulation, in *IEEE International Solid-State Circuits Conference Digest of Technical Papers*, San Francisco, CA, 2010, pp. 226–227.
42. B. Razavi, *Design of Integrated Circuits for Optical Communications*. New York: McGraw-Hill Companies, Inc., 2003.
43. J. Craninckx and M.S.J. Steyaert, A fully integrated CMOS DCS-1800 frequency synthesizer, *IEEE J. Solid-State Circ.*, 33, 2054–2065, 1998.
44. FCC, Part 15—Radio frequency devices. Available at: http://wireless.fcc.gov/, 2010. Accessed January 2014.
45. A. Tamtrakarn, H. Ishikuro, K. Ishida, M. Takamiya, and T. Sakurai, A 1-V 299 μW flashing UWB transceiver based on double thresholding scheme, in *IEEE Digest of Technical Papers VLSI Circuits*, Honolulu, HI, 2006, pp. 202–203.
46. Z. Xiaodong, S. Ghosh, and M. Bayoumi, A low power CMOS UWB pulse generator, in *IEEE Midwest Symposium on Circuits and Systems*, Covington, KY, 2005, pp. 1410–1413, Vol. 2.
47. L. Lechang, Y. Miyamoto, Z. Zhiwei, K. Sakaida, J. Ryu, K. Ishida et al., A 100 Mbps, 0.41 mW, DC-960 MHz band impulse UWB transceiver in 90 nm CMOS, in *IEEE Digest of Technical Papers VLSI Circuits*, Honolulu, HI, 2008, pp. 118–119.

14 Energy-Efficient High Data Rate Transmitter for Biomedical Applications

Chun-Huat Heng and Yuan Gao

CONTENTS

14.1 INTRODUCTION

Recent development on wireless medical applications, such as wireless endoscopy and multichannel neural recording IC, has spurred the need for energy-efficient high data rate transmitter. For example, wireless endoscopy with image resolution of 640×480, 6 fps and 8-bit color depth requires a few tens of Mbps if on-chip image compression is not available. Similar requirement applies to 256-channel neural recording IC with 10-bit resolution and sampling rate of 10 kS/s. Such applications are often characterized by asymmetric data link as shown in Figure 14.1 where high data rate uplink is required to upload critical biomedical data and low data rate downlink is used only for configuring the implanted device. Therefore, complex modulation, which supports high data rate, does not necessarily lead to power-hungry solution if power-efficient transmitter architecture can be realized as the complex receiver function is usually implemented off-body without stringent power constraint. In addition,

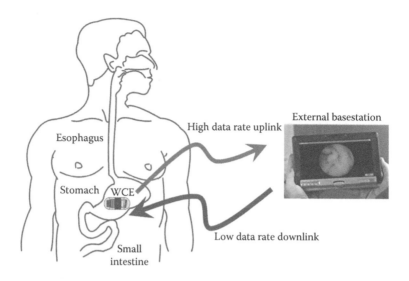

FIGURE 14.1   Asymmetric data link.

it has been found that transmission frequency below 1 GHz is generally preferred for such applications due to smaller body loss (Kim et al. 2010). As this frequency range is often crowded, complex modulation with better spectrum efficiency would thus be more favorable.

Conventional mixer-based or phase-locked loop (PLL)-based approach, which supports phase shift keying (PSK) or quadrature amplitude modulation (QAM), does not lead to low-power solution. They are often limited by the need of power-hungry high-frequency building blocks, such as mixer, radio frequency (RF) combiner, high-frequency divider, and PLL. In this chapter, we will discuss the choice of modulation and examine the various proposed techniques that can eventually lead to an energy and spectral efficient high data rate transmitter.

14.2   MODULATION

Figure 14.2 shows the evolving trends on adopted modulations for transmitters at industrial, scientific, and medical (ISM) band, mm-wave band, and sub-1 GHz band. Due to the improved technology and the demand on higher data rate over limited spectrum resource, designs are moving from simple modulation such as on–off keying (OOK) and frequency shift keying (FSK) toward more complex and spectral efficient modulation such as PSK or QAM. As shown in Figure 14.3, by moving from FSK to 16-QAM, the spectral efficiency can be improved by four times. The relatively high side lobe can be suppressed if pulse shaping is adopted as shown in Figure 14.4 to improve the spectral efficiency further. Nevertheless, there are a few requirements for implementing complex modulation and pulse shaping. First, accurate phase is needed, which demands accurate frequency and phase generation. This is conventionally accomplished with PLL. Second, pulse shaping can be achieved with either mixer-based approach or polar transmitter, which generally leads to

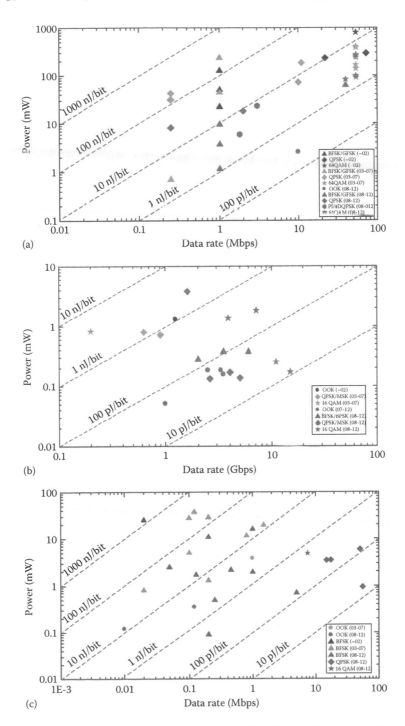

FIGURE 14.2 Transmitter evolution trend for (a) ISM band, (b) mm-wave band, and (c) sub-1 GHz band.

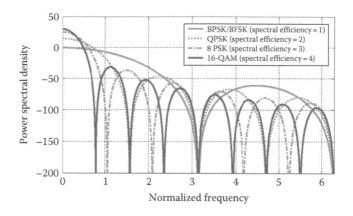

FIGURE 14.3 Spectral efficiency for different modulation.

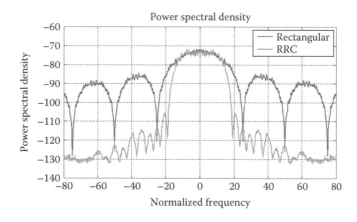

FIGURE 14.4 QPSK modulation with and without band shaping.

complex architecture and higher power consumption as mentioned earlier. In the next section, we will examine a few proposed techniques that can lead to simplified architecture and achieve low-power consumption.

14.3 FREQUENCY AND PHASE GENERATION

PLL is widely adopted for frequency and phase generation. However, due to the need of high-frequency divider, charge pump, and so on, it is not a power-efficient solution. In addition, for LC oscillator, multiphase is obtained by operating at twice larger the desired output frequency and divide down the output through high-frequency divider, which incurs more power penalty due to the higher operation frequency. On the other hand, ring oscillator (RO) can readily provide the desired output phase but suffer from poorer phase noise. Here, we proposed two energy-efficient ways of achieving accurate frequency and phase generation without the aforementioned issues.

14.3.1 LC-Based Injection-Locked Oscillator

Injection-locked oscillator (ILO) has been found to be an energy-efficient way of obtaining accurate frequency reference without the need of additional components, such as frequency divider, phase frequency detector, and charge pump. Hence, it can lead to an energy-efficient solution for frequency carrier generation. Fundamentally, it behaves like a first-order PLL where the free-running frequency of the LC oscillator will lock to the nth harmonic of the injected reference frequency. Here, we proposed an efficient way of obtaining the desired multiphase output without resorting to higher operating frequency and divide down method. As shown in Figure 14.5, depending on the free-running frequency of the LC tank, the output phase exhibits certain relationship with respect to the nth harmonic of the injected reference (Razavi 2004). For example, if the free running frequency of the LC tank is higher than the nth harmonic of the injected reference, the output phase will be lagging the injected reference. On the other hand, if the free-running frequency is lower, the output phase will be leading the injected reference. This observation allows us to manipulate the free-running frequency of the LC oscillator to obtain the desired output phase. As shown, output phase with $\pm45°$ can be obtained through this manner, and the remaining $\pm135°$ can be obtained by swapping the output phase to introduce the $180°$ phase shift (Diao et al. 2012). Finer output phase resolution can be achieved by using switched capacitor bank to obtain fine frequency tuning for the LC oscillator.

From the aforementioned description, it is obvious that the frequency tuning will directly affect the phase accuracy and amplitude variation, which in turn will determine the resulting error vector magnitude (EVM) as follows:

$$\text{EVM}(\%) = \sqrt{\left(\frac{\Delta M}{OI}\right)^2 + \left(\sin\phi\right)^2} \times 100\% \approx \sqrt{\left(\frac{\Delta M}{OI}\right)^2 + \phi^2} \times 100\%, \quad (14.1)$$

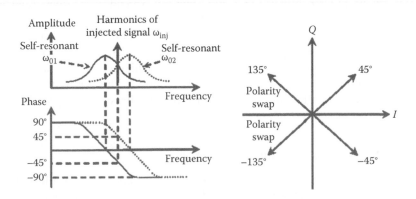

FIGURE 14.5 Phase modulation through modifying self-resonant frequency of LC tank.

where ΔM and ϕ are the resulting magnitude and phase deviations. It has been shown in Diao et al. (2012) that the oscillator output amplitude (V_{osc}) and phase are related as follows:

$$|V_{osc}| = \frac{I_{inj} \times \omega_{inj} L \sqrt{1 + \phi^2}}{\sqrt{\left(1 - \frac{\omega_{inj}^2}{\omega_0^2}\right)^2 + \left(\omega_{inj} L\right)^2 \left(\frac{1}{R_p} - g_m\right)^2}}, \tag{14.2}$$

$$\frac{\omega_0 - \omega_{inj}}{\omega_0} = \frac{1}{2Q} \cdot \frac{K \cdot \sin\left(\pm 45° + \phi\right)}{1 + K \cdot \cos\left(\pm 45° + \phi\right)}, \tag{14.3}$$

where I_{inj} is the injection current amplitude, ω_{inj} is the injected nth harmonic frequency, ω_0 is the free-running frequency of the oscillator, L is the inductance of the LC tank, R_p is the equivalent LC tank parallel resistance, g_m is the transconductance of the cross-coupled pair transistor of the LC oscillator, Q is the quality factor of the LC tank, and K is the injection strength given as

$$K = \frac{I_{inj,Nth}}{I_{osc}} = \frac{4}{N\pi} \frac{I_{inj}}{I_{osc}} = \frac{4\alpha}{N\pi}, \tag{14.4}$$

where I_{osc} is the oscillator current.

Based on (14.1) through (14.4), the resulting EVM for a given phase error can then be obtained and is plotted in Figure 14.6. As shown, for EVM to be smaller than 23%, a phase error as large as 10° can be tolerated. Also, from the studies, the impact of magnitude error on EVM is much smaller compared to phase error as illustrated in Figure 14.7.

To determine the design requirement on capacitor bank, the phase versus frequency characteristic can be obtained based on (14.3) and is shown in Figure 14.8. As illustrated, larger K allows the desired phase to spread over larger frequency range. This will relax the capacitor bank design and is thus desirable. Figure 14.9 shows the design choice on K and the needed smallest unit capacitance per degree phase resolution. To avoid false locking to wrong harmonics, locking range smaller than 100 MHz is desired. On the other hand, to achieve higher data rate (tens of MHz), locking time smaller than 10 ns is required. This limits the range of K to 0.55~0.72, which translate to smallest unit capacitance per degree phase of 3~5 fF/°.

14.3.2 INJECTION-LOCKED RING OSCILLATOR

For sub-1 GHz regime, RO is a good candidate that offers readily available multiphase and compact area. However, it generally suffers from poor phase noise and thus poorer EVM performance. Subharmonically injection-locked ring oscillator (ILRO) offers a compact and energy-efficient way of generating accurate output frequency

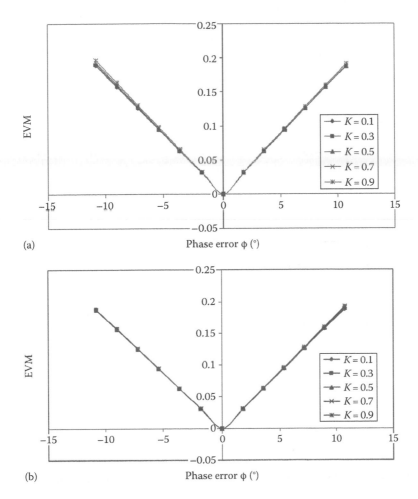

FIGURE 14.6   EVM versus phase error (ϕ) at (a) +45° and (b) −45°.

with multiphase output and good phase noise. As shown in Figure 14.10, the main reason of poor phase noise performance of RO is due to the jitter accumulation along the delay chain. By subharmonically injecting a clean reference signal, the delay jitter is cleaned up during every reference period and eliminate jitter accumulation. This leads to good output phase noise, which generally follows the phase noise of the clean input reference even when a noisy RO is employed. Nevertheless, there are a number of factors that will affect the resulting phase accuracy and thus the EVM performance, which needs to be carefully examined.

As shown in Figure 14.10, the output phase edge is only cleaned up at the beginning of each reference period. For the remaining interval, the oscillator will oscillate at its own free-running frequency. Hence, a small phase error appears at the end of each oscillator output period. This is equivalent to having a phase modulation where the phase error is accumulating over the entire interval and get corrected

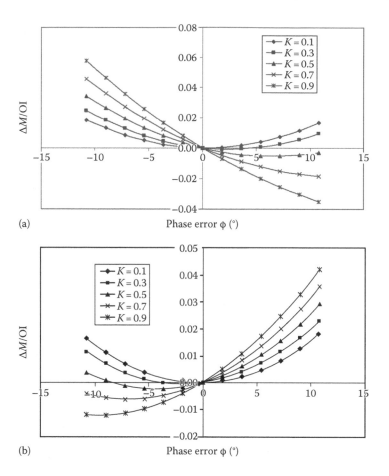

(a)

(b)

FIGURE 14.7 Relative magnitude error versus phase error (ϕ) at (a) +45° and (b) −45°.

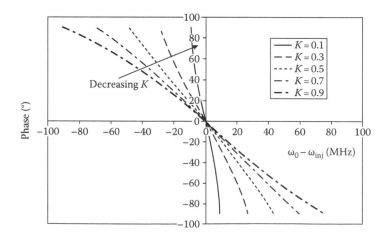

FIGURE 14.8 Phase characteristics for $\omega_0 - \omega_{inj}$ for IL-VCO.

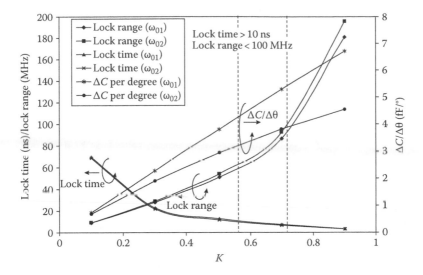

FIGURE 14.9   Choice of K.

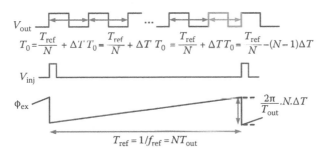

FIGURE 14.10   Systematic phase error accumulation.

at the beginning of next reference period (Izad and Heng 2012). By modeling this systematic phase error as phase modulation using a sawtooth waveform, we obtain

$$\theta_{error,rms} = \sqrt{\frac{1}{T_{ref}} \int_{T_{ref}} \left(\phi_{ex}(t) \right)^2 dt} = \frac{2\pi}{T_{out}} \cdot \frac{N \cdot \Delta T}{\sqrt{3}} = \frac{2\pi \cdot N \cdot \Delta f}{\sqrt{3} f_0}, \qquad (14.5)$$

where f_0 is the free-running frequency of the RO and Δf is the frequency deviation between the nth harmonic of injected reference and f_0. Using (14.5), the desired tunable frequency resolution can be found for a given EVM. In fact, it is also found (Izad and Heng 2012) that the reference spur level at the output is directly related to the resulting frequency deviation as follows:

$$Spur(dBc) = 20 \log \left(\frac{N\Delta T}{T_{out}} \right) = 20 \log \left(N \frac{|f_{out} - f_0|}{f_0} \right) = 20 \log \left(N \frac{|\Delta f|}{f_0} \right). \qquad (14.6)$$

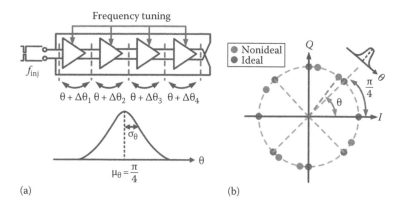

(a)   (b)

FIGURE 14.11 (a) Random mismatch between delay cells in the oscillator. (b) Effect of random phase mismatch on the constellation.

Hence, the measurement of the resulting reference spur level can serve as an indicator for the achieved frequency deviation, and thus the systematic phase error.

The other factor that affects the EVM performance is classified as random phase error, which is mainly caused by device noise and mismatch. Due to the injection-locking mechanism, the RO output phase noise will follow the injected reference closely. Within the locking range, the output phase noise can be modeled as $L_{inj}+20 \log(N)$, where L_{inj} is the phase noise of the injected reference. The total root-mean-square (RMS) phase noise can be found by taking the root of the total integrated phase noise across the entire frequency band. To minimize its impact, low-noise-injected reference and smaller N should be adopted in the design.

Due to mismatch between delay cells, the delay introduced by one stage can differ from another stage. For four-stage RO, each stage provides a mean phase step of $\pi/4$ (delay of $T/8$) with standard deviation of σ_θ. This mismatch manifests itself as a distortion of the constellation. In next section, we will look at ways to minimize this mismatch. The effect of these phase errors on constellation is shown in Figure 14.11.

14.4 BAND SHAPING AND SIDE-BAND SUPPRESSION

The QAM modulation can be described with the following equation:

$$s(t) = A(t)\cos\big(\omega t + \varphi(t)\big) = A_I(t)\cos(\omega t) + A_Q(t)\sin(\omega t). \tag{14.7}$$

As illustrated in (14.7), the middle expression is the basis for polar architecture where fractional-N PLL is adopted to provide the carrier with arbitrary phase output, and the supply modulated power amplifier (PA) is adopted for amplitude modulation. On the other hand, the right expression is the basis for mixer-based approach where the in-phase and quadrature-phase components ($A_I(t)$ and $A_Q(t)$) are generated at baseband, upconverted through mixer, and summed through RF combiner and PA. To avoid the issues faced by these two architectures, we adopt direct quadrature modulation at PA, which can be considered as hybrid approach of the former two

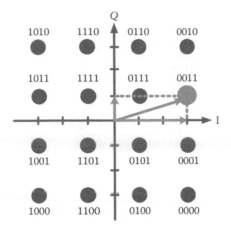

FIGURE 14.12　Constellation of 16-QAM.

architectures. First, the in-phase and quadrature-phase RF components ($A_I(t)\cos\omega t$ and $A_Q(t)\sin\omega t$) are generated by driving PA with variable current strength from the oscillator multiphase output. The variable current strength is obtained by activating different number of transistors that get connected to a specific output phase. This eliminates the need of complex phase generation architecture and upconversion mixers. The summing of two components is then performed at current domain, which can accommodate high-frequency operation. The idea is best illustrated in Figure 14.12. For example, to achieve the constellation point corresponding to (0011), we can combine φ_0 with 3× amplitude and φ_{90} with 1× amplitude. The concept can be easily extended to enable band shaping by providing multiple amplitude levels for the four-phase outputs. This will enable the fine phase and amplitude tuning required for band shaping as illustrated. From system simulation, it is determined that 5-bit amplitude control per phase is needed to achieve more than 38 dB side lobe suppression.

14.5　ENERGY-EFFICIENT TRANSMITTERS

In this section, we examine three energy-efficient architectures that have incorporated the proposed techniques mentioned earlier. The design insights associated with the circuit implementation will also be discussed.

14.5.1　QPSK AND O-QPSK TRANSMITTER BASED ON ILO

The energy-efficient quadrature phase-shift keying (QPSK)/offset quadrature phase-shift keying (O-QPSK) transmitter based on ILO is shown in Figure 14.13. The proposed architecture consists of only an ILO, a polarity swap circuit, a buffer, and a mapping circuitry, which transforms the input I and Q signals to the corresponding output phases. By eliminating the multiphase PLL and operating the VCO at the desired output frequency, the power consumption can thus be further reduced. As explained earlier, by controlling the free-running frequency of the LC tank to be

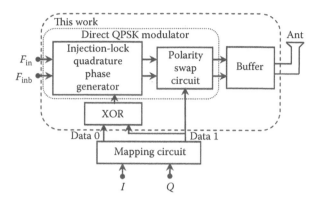

FIGURE 14.13 Block diagram of proposed QPSK/O-QPSK transmitter.

lower than the harmonic frequency of the injected signal (solid lines), the locked LC tank signal leads the harmonic of the injected signal by 45°. On the other hand, by making the free-running frequency higher than the harmonic of the injected signal (dotted lines), the locked LC tank signal lags behind the harmonic of the injected signal by 45°. Therefore, by changing the free-running frequency of the LC tank, we can create a phase difference of 90° in the output signal. The self-resonant frequency of the LC tank can be easily modified through capacitor bank switching. To generate all the four phases required for QPSK/O-QPSK modulation, additional polarity swap circuit is employed to introduce 180° phase shift to the output signal. By employing both the capacitor bank and the polarity swap circuit, we could thus realize +45°, −45°, −135°, and +135° phase shifts required for QPSK/O-QPSK modulation.

The detailed circuit is shown in Figure 14.14. The ILO consists of a symmetrical NMOS cross-coupled pair (NM$_2$, NM$_3$), an LC tank incorporating a center-tapped

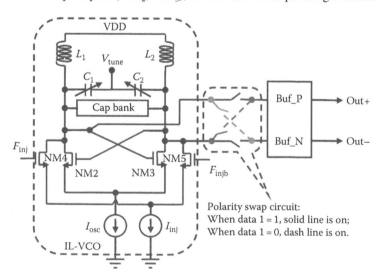

FIGURE 14.14 Schematic of the proposed QPSK/O-QPSK transmitter.

differential inductor $(L_1 + L_2)$ and a capacitor bank, a differential pair transistor (NM_4, NM_5) for signal injection, and tail currents $(I_{inj}$ and $I_{osc})$. F_{inj} and F_{injb} are the differential-injected signals. The free-running LC VCO has a free-running frequency (ω_0) centered around the targeted carrier frequency (ω_c). The designated harmonic of the injected signal (ω_{inj}) is chosen to be the same as ω_c. The free-running frequency can be changed by switching the capacitor bank to generate the desired phase shift as explained earlier. In this design, I_{inj} and I_{osc} are chosen based on the desired K value mentioned earlier. The differential inductance is chosen to maximize its quality factor and parallel tank resistance (R_p) at around 915 MHz. Large R_p will reduce the tank current required for larger voltage swing and lead to lower power consumption. Based on momentum simulation, a differential inductor with a value of 20.8 nH and a Q factor of 5 is designed.

The schematic of capacitor bank is shown in Figure 14.15a. To cover full ±90° phase range with the chosen 10° phase resolution, a 6-bit binary capacitor bank is required. The resonant frequency can be tuned to ω_{01} and ω_{02} by setting the control words $B_{-\omega01}[5:0]$ and $B_{-\omega02}[5:0]$, respectively. To avoid loading the tank, the Q factor of each switch-capacitor unit within the capacitor bank is designed to be around 45. This ensures that the total tank Q is limited by inductor.

The switch-capacitor unit is shown in Figure 14.15b. It consists of MIM capacitors C_1/C_2, shunt switches NM_1/NM_2, and a series switch NM_3, all controlled by B_N. NM_1/NM_2 will set the DC bias to ground when activated and can be kept small. The Q is mainly determined by series switch NM_3. This configuration halves the series

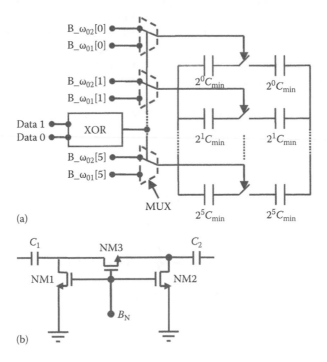

FIGURE 14.15   (a) Switched capacitor bank; (b) switched capacitor unit.

resistance and allows smaller NM_3 to be used. This leads to smaller parasitic capacitance and larger C_{max} to C_{min} ratio. It has been shown in Diao et al. (2012) that getting absolute output phases of $\pm45°$ is not necessary in ensuring QPSK/O-QPSK with good EVM. This is because the resulting magnitude errors due to the phase deviations from $\pm45°$ are usually less than 2% and do not impact the EVM significantly. Hence, free-running frequencies correspond to output phase of ~+60° and ~−30° can be chosen as long as they exhibit a phase difference close to 90°. This helps relax the capacitor array design as larger unit capacitance is allowed, which results in larger frequency step size in self-resonant frequency tuning.

Gray coding is adopted for the phase modulation. Phases of 45°, −45°, 135°, and −135° correspond to Data0 and Data1 of "00," "10," "01," and "11," respectively. It is observed that 180° phase shift occurs whenever Data1 changes as shown in Figure 14.14. Therefore, the polarity swap circuitry is controlled by Data1. On the other hand, "00" and "11" will generate phase that corresponds to self-resonant frequency $\omega_{01,}$ whereas "01" and "10" will generate phase that corresponds to self-resonant frequency ω_{02}. Therefore, XOR gate can be employed to switch between ω_{01} and ω_{02} as illustrated in Figure 14.15a. With this arrangement, QPSK modulation can be easily achieved. For O-QPSK modulation, Data1 is shifted by half symbol period with respect to Data0 before sending to the whole circuit.

An inverter-type output buffer is adopted in this design as shown in Figure 14.16. Its performance is limited by the QPSK modulation due to its nonconstant envelope nature. For O-QPSK with quasi constant-envelope nature, the linearity requirement is much relaxed. Therefore, nonlinear PA with high-power efficiency can be used (Liu et al. 2009). Fortunately, the targeted output power is limited to −3 dBm and the employed inverter-type PA can meet the linearity requirement with careful design (Sansen 2009). As shown in Figure 14.16, the PA includes a trans-impedance stage and an inverter stage. R is around 50 kΩ, which settles the DC bias to around half of the supply voltage. From (Feigin 2003), the linearity requirement of the PA can be worked out as follows:

$$P_{OIP3} = \frac{relative\_sideband + 3}{2} + P_{out}, \tag{14.8}$$

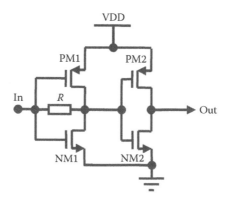

FIGURE 14.16   Inverter-type PA.

where P_{OIP3} is the desired output third-order intermodulation point in dBm, P_{out} is the transmitter output power in dBm, and relative sideband is the relative t first side lobe suppression in dB. With the targeted output power of −3 dBm and the first side lobe suppression of 12 dB (Ideal QPSK), the desired P_{OIP3} is 4.5 dBm. This leads to output 1 dB compression point ($P_{out, 1 dB}$) requirement of −5.5 dBm.

In this design, the aspect ratios of PM_2 and NM_2 are chosen to be 25 µm/0.18 µm and 10 µm/0.18 µm, respectively. The PA output is connected to a 50 Ω load via a 3.5 pF AC coupling capacitor. It achieves $P_{out, 1 dB}$ of −4.2 dBm with 20% power efficiency and the output power saturates at 1.13 dBm.

The measured EVMs are 5.97%/3.96% for QPSK/O QPSK at data rate of 50 Mbps/25 Mbps, respectively, as shown in Figure 14.17. Consuming 5.88 mW under 1.4 V supply, it can achieve energy efficiency of 118 pJ/bit while delivering output power of −3 dBm. Its performance is summarized in Table 14.1.

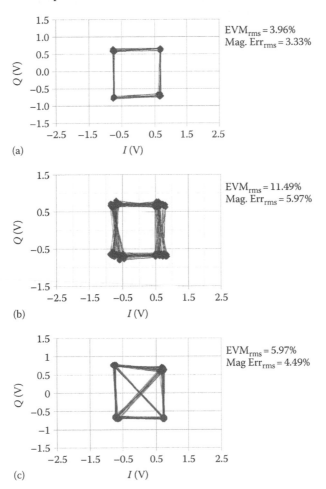

(a)

(b)

(c)

FIGURE 14.17 Measured EVM with 305 MHz injection signal: (a) O-QPSK at 25 Mbps; (b) O-QPSK at 50 Mbps; and (c) QPSK at 50 Mbps.

TABLE 14.1
Performance Summary

Injection Signal	1 V at 101.67 MHz	1 V at 305 MHz
Process	CMOS 0.18 μm	
Center frequency	915 MHz	
Supply voltage	1.4 V	
Die area	0.4 × 0.7 mm (active core)	
Modulation	QPSK/O-QPSK	
Maximum data rate	50 Mbps	100 Mbps
Power consumption (P_{DC})	5.88 mW	5.6 mW
	(PA: 2.24 mW	(PA: 2.24 mW
	ILO: 3.5 mW)	ILO: 3.22 mW)
Phase noise at 1 MHz	−119.4 dBc/Hz	−125 dBc/Hz
EVM	6.41% (QPSK)	5.97%(QPSK)/3.96%(O-QPSK)
Output power (Pout)	−3.3 dBm	−3 dBm

For injection-locked LC oscillator, there is a trade-off between data rate and resulting spurs. To support higher data rate, larger locking range is needed to speed up the setting time. However, this results in larger spurious tones due to limited harmonic suppression. Given the low-quality factor of on-chip inductor, the resulting spurious tones are in the range of −20~−30 dBc and exceed the transmitter spectral requirement of <−40 dBc (Kavousian et al. 2008).

In Izad and Heng (2012), it has been demonstrated that pulse shaping technique can be employed to suppress adjacent harmonics and result in spurious tones suppression of more than 22 dB. As the desired injected harmonic does not change, the resulting pulse shaping circuit is much simpler than Izad and Heng (2012b), and the idea is shown in Figure 14.18. The resulting spurious tones suppression with and without the proposed pulse shaping circuit is shown in Figure 14.19, as illustrated, more than 18 dB suppression has been achieved and the transmitter requirement can thus be met.

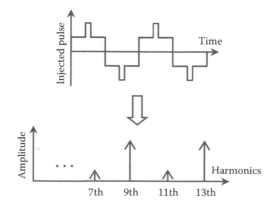

FIGURE 14.18 Pulse shaping with adjacent tones suppression.

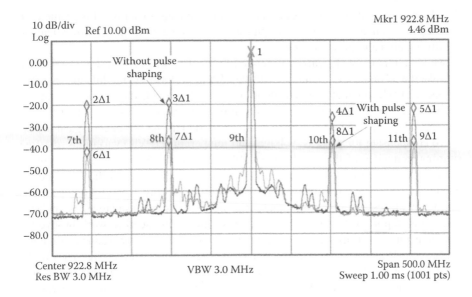

FIGURE 14.19   Spur reduction with and without pulse shaping.

14.5.2   QPSK AND 8 PSK TRANSMITTER BASED ON ILRO

Figure 14.20 shows the block diagram of the proposed architecture. Inherent multi-phases of a four-stage differential RO are directly employed to provide the PSK modulation. Although RO suffers from poor phase noise and frequency instability, these problems are solved by subharmonic injection locking of the oscillator to the 15th harmonic of a 61 MHz *fundamental mode* crystal reference (Micro Crystal CC1F-T1A). Due to the fact that injection-locked oscillator behaves like a first-order PLL,

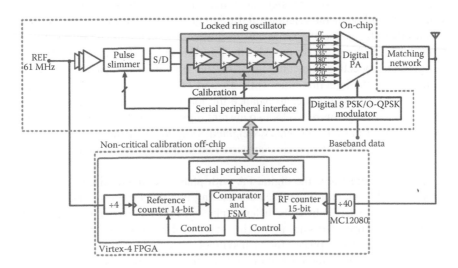

FIGURE 14.20   Proposed QPSK/8 PSK transmitter architecture.

the phase noise and frequency stability are improved significantly. This architecture reduces the complexity of the multiphase carrier generation block, which leads to significant area and power saving.

Multiple phases of the carrier generated by the RO will drive a digital PA. Here, we take the advantage of digitally controlled PA to perform the modulation. A digital 8 PSK/O-QPSK modulator directly controls the PA to provide the modulated output to the antenna. Note that the proposed architecture is amenable to process and supply scaling due to its digital nature and can be easily ported to the newer technological nodes.

To compensate for process, voltage, and temperature (PVT) variations, the free-running frequency of the oscillator is calibrated offline. A digital loop determines the code word required for digitally controlled delay elements in the RO. The frequency of the ring is digitally measured by counting the ring cycles for a fixed period of time. A successive-approximation search algorithm is then used to adjust the delay code. It is important to notice that this transmitter is intended to work in vivo and the temperature variations are small, which abates the need for frequent frequency calibration (Bohorquez et al. 2009, Bae et al. 2011). Thus, all the circuitries in this block can be turned off after calibration and the digital implementation enables digital storage of the control words with only leakage power. In this prototype, the calibration circuit is implemented off-chip using an FPGA for testing flexibility. The calibration engine can be easily ported on chip. After synthesis and place and route, the calibration engine only occupies negligible area of 35×45 µm.

As shown in Figure 14.21a, this prototype uses a four-stage current-starved pseudodifferential RO. The random mismatch between delay cells influences the accuracy of the generated phases. We use mismatch filtering resistors similar to Chen et al. (2010) to improve the phase accuracy. Since each node is coupled to the adjacent nodes by the resistors to average out the transition edges, any random mismatch between delay cells is reduced. According to Monte Carlo simulations, this technique improves the standard deviation of the phase mismatch by 4.7 times without the use of larger transistor for better matching. This leads to better energy efficiency. Furthermore, although the reference pulse is injected only in the first stage of the RO, all other stages use the same cell to maintain matching. The pulse slimmer circuit shown in Figure 14.21b provides the injection pulses. A detailed implementation of each delay cell is shown in Figure 14.21c. The ratio between M_7-M_6 and M_{15}-M_{14} determines the injection ratio while M_8 and M_{13} replicate M_1 to M_4. Moreover, M_9-M_{12} are added as dummies to balance the loading in the differential path. The oscillator is digitally controlled by two current digital-to-analog converters (DACs). Each DAC consists of 10-bit binary weighted array to cover the desired operation frequency range across the process corners. The frequency resolution is determined based on (14.5).

Figure 14.22 illustrates the digital PA with phase modulator circuit topology. It consists of eight unit amplifiers each comprising two transistors in series. The bottom transistors are driven by eight different phases of the RF carrier, tapped from the RO. The cascode transistors act as switches controlled by the digital baseband modulator. They select proper phases for each symbol. Note that each branch only amplifies a constant envelop (and constant phase) carrier while the (phase) modulation is performed by switching between different branches. This allows for higher data rate

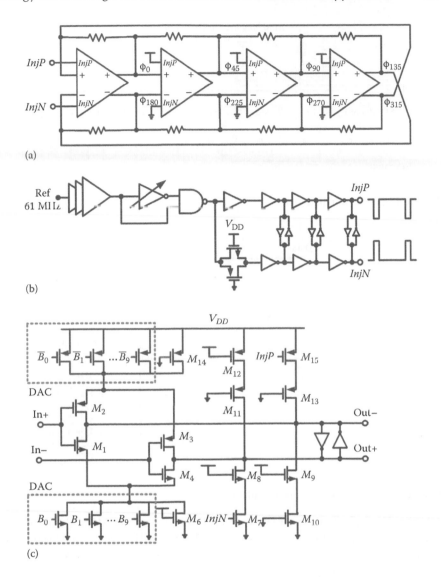

FIGURE 14.21   (a) Implementation of injection-locked RO. (b) Pulse generator. (c) Detailed schematic of the delay cell.

modulation with much lower complexity compared to conventional PLL-based phase modulation and digital PA approach (Mehta et al. 2010). The combined output of all the amplifiers is connected to an off-chip matching network for impedance transformation to drive antenna. One can either use all the unit amplifiers driven by an 8 PSK modulator (Figure 14.23) or only enable four of the unit amplifiers driven by a QPSK/O-QPSK modulator (Figure 14.23) to obtain different modulation scheme.

The 8 PSK/O-QPSK transmitter is fabricated in 65 nm digital CMOS with an active area of 0.038 mm². The RO can be tuned to oscillate from 758 to 950 MHz

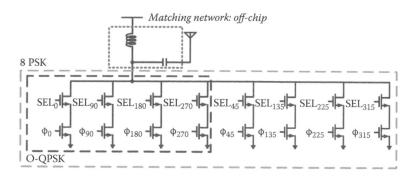

FIGURE 14.22 Power amplifier.

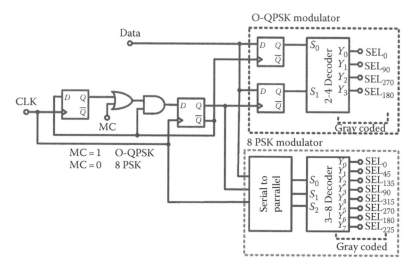

FIGURE 14.23 Digital baseband modulator.

under 0.8 V supply. For the targeted output frequency of 915 MHz, the measured locking range of the oscillator extends from 902 to 927 MHz. Figure 14.24 shows the measured convergence of the successive approximation algorithm, which requires only 10 cycles to converge.

Measured phase noise of the RO in locked and unlocked condition is shown in Figure 14.25a. As illustrated, injection locking significantly improves the phase noise from 1 kHz to 1 MHz offset. Total measured integrated RMS jitter for the locked RO is 4.6 ps (1.52°) while the worst-case reference spurs are less than −47 dBc (Figure 14.25b). The measured settling time of the RO is shorter than 80 ns, which allows aggressive duty-cycled operation. Figure 14.25c shows the output spectrum for 8 PSK mode at 55 Mbps. As illustrated, this prototype meets the FCC spectral mask and thus the interference to other wireless standard would not be a concern.

Figure 14.26 shows the measured constellation for 8 PSK and O-QPSK mode. A similar measurement has been carried out on 10 chips to investigate the effect

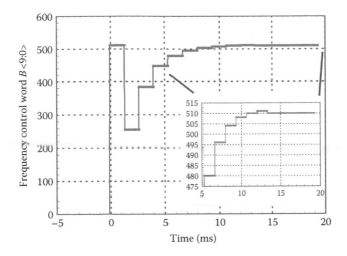

FIGURE 14.24 Measured convergence of the successive approximation algorithm.

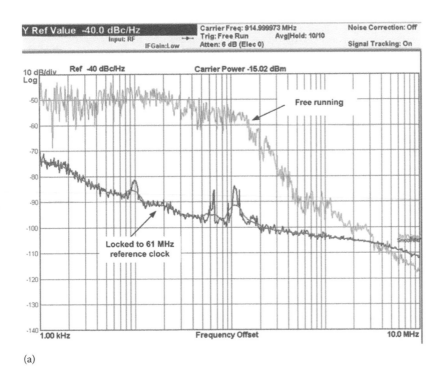

(a)

FIGURE 14.25 (a) Measured phase noise of the RO under locked and unlocked conditions.

(*Continued*)

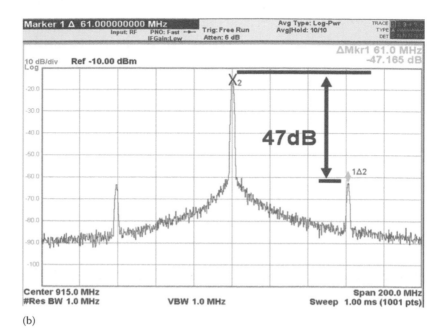

(b)

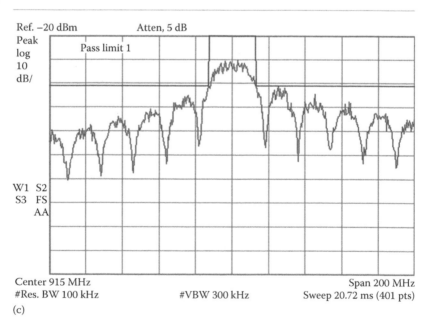

(c)

FIGURE 14.25 (CONTINUED) (b) Measured unmodulated output spectrum of the carrier at −15 dBm output power. (c) Measured output spectrum with output power of −15 dBm at 55 Mbps.

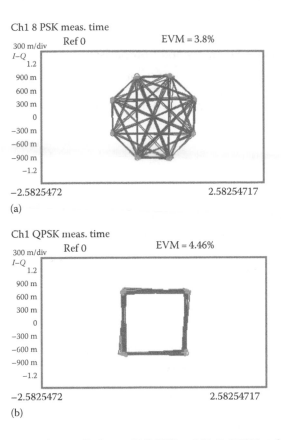

FIGURE 14.26   Measured constellation at (a) 8 PSK and (b) O-QPSK at 55 Mbps.

of mismatch. The worst-case measured EVMs at data rate of 55 Mbps are 3.8% and 4.46%, respectively. A typical 8 PSK/O-QPSK receiver requires EVM of less than 15/23% to achieve bit error rate (BER) better than 10^{-4}. The measured EVM in our transmitter meets this requirement with a good margin.

The RO consumes 538 μW and the PA dissipates 286 μW while delivering −15 dBm output power. The delivered power has included loss of PCB and matching network. The total power consumption (including crystal oscillator) in any mode of operation is less than 938 μW.

Table 14.2 summarizes the performance of the prototype and compares it with the state-of-the-art designs. This work achieves the energy efficiency of 17 pJ/bit, which is 8× smaller than that of the state of the art. It also has the lowest area without requiring any off-chip inductors (other than matching network). This prototype is the first transmitter that provides spectral efficient 8 PSK modulation with power consumption in sub-mW range, confirming the feasibility of achieving higher data rates with low-power consumption.

TABLE 14.2

TX Performance Summary and Comparison

	Daly JSSC Dec07	Vidojkovic ISSCC Feb 11	Liu CICC Sep 08	Bae JSSC Apr 11	This Work
Frequency (MHz)	900	2400	400	920	915
Modulation	OOK	OOK	O-QPSK	FSK	8 PSK or O-QPSK
Data rate (Mbps)	1	10	15	5	55
Output power (dBm)	−11.4	0	−15	−10	−15
Power diss. (mW)	3.8	2.53	3.48	0.7	0.938
Active area (mm²)	0.27	0.882	0.7	1.65[a]	0.038
Energy/bit (nJ/Bit)	3.8	0.253	0.23	0.14	0.017
Supply voltage (V)	0.8–1.4	1	1.2	0.7	0.8
Technology	0.18 μm	90 nm	0.18 μm	0.18 μm	65 nm

[a] Chip area. It uses four off-chip inductors.

14.5.3 QPSK AND 16-QAM TRANSMITTER BASED ON ILRO WITH BAND SHAPING

The proposed transmitter (TX) architecture is shown in Figure 14.27. ILRO forms the core of TX, which provides four-phase output (φ_0, φ_{90}, φ_{180}, φ_{270}) with good phase noise. Direct quadrature modulation at PA is proposed here to provide both phase and amplitude modulations.

To achieve the desired modulation and band shaping, the incoming serial data is first converted into parallel I/Q data depending on desired modulation (2 bits/symbol for QPSK and 4 bits/symbol for 16-QAM). If band shaping is activated, the

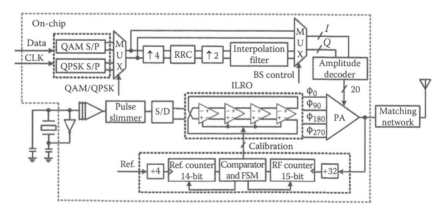

FIGURE 14.27 Proposed QPSK/16-QAM TX architecture.

I/Q data will then be upsampled by four times before passing through root raised cosine (RRC) filter ($\alpha = 0.4$). Finite impulse response (FIR) filters instead of read only memory (ROM) are adopted here for RRC filter implementation to provide flexibility in filter coefficient tuning. Following that, it is further upsampled by two times before going through an interpolation filter. The upsampling will push the unwanted image further away from the targeted output and can thus be suppressed easily by matching network and antenna. If band shaping is deactivated, the I/Q data will be sent to decoder right away, bypassing the intermediate upsampler and RRC filter to save energy.

In this implementation, injected reference of 100 MHz is chosen and the ILRO will lock to the ninth harmonic of the injected reference. Similar to earlier work, a frequency calibration algorithm has been incorporated on chip to fine-tune the RO free-running frequency to match the ninth harmonics of injected reference. Unlike Izad and Heng (2012a), which use off-chip reference source, the fundamental mode 100 MHz crystal oscillator is also built on-chip to better evaluate the phase noise and energy efficiency performance of the TX.

Identical four-stage pseudodifferential RO shown in Figure 14.21 with mismatch filtering technique is employed. This helps improve the energy efficiency of the RO. The digital PA with embedded phase multiplexer and amplitude control is shown in Figure 14.28. It consists of four amplifier cores driven by four output phases (φ_0, φ_{90}, φ_{180}, φ_{270}) from ILRO. The bottom transistors are connected to respective ILRO output phases, whereas the top transistors are used to activate the corresponding phase. To achieve 5-bit amplitude control for each output phase, the transistors within each amplifier core are further segmented into an array of 31 transistor pairs. Direct quadrature modulation is achieved through current summing from two activated phase branches with different current amplitude. The combined output current is then sent to an off-chip impedance matching network before driving the 50 Ω antenna. The 20-bit control (5-bit/phase × 4 phases) for the PA is provided from amplitude decoder.

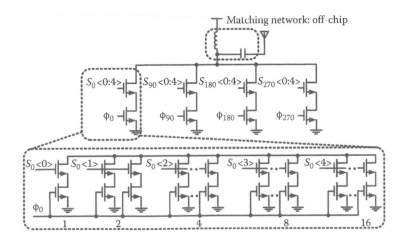

FIGURE 14.28   Digital PA with direct phase and amplitude modulation.

Fabricated in 65 nm CMOS, the TX occupies an active area of 0.08 mm^2. The only off-chip components needed are matching network and 100 MHz crystal. The measured tuning range of RO covers from 0.81 to 1 GHz. The measured locking range of the ILRO is from 885 to 925 MHz.

Figure 14.29 shows the measured phase noise under injection locking. By setting the crystal oscillator power consumption to 115 µW, the ILRO achieves a total integrated RMS jitter of 1.54°. If the crystal oscillator power consumption is allowed to increase to 380 µW, 7 dB improvement in phase noise is noted. The ILRO also shows a fast startup time of less than 88 ns in Figure 14.30, which is critical for

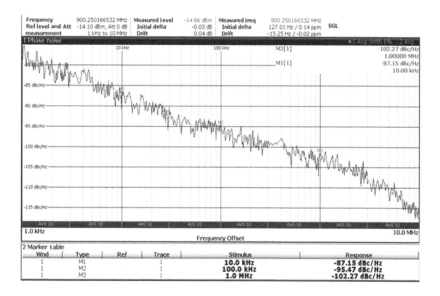

FIGURE 14.29   Measured phase noise under injection locking.

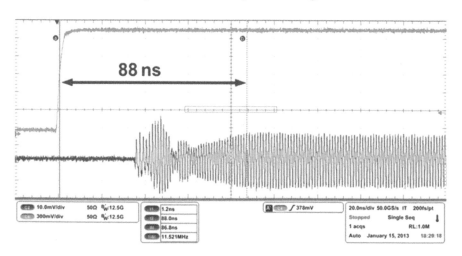

FIGURE 14.30   Measured settling time.

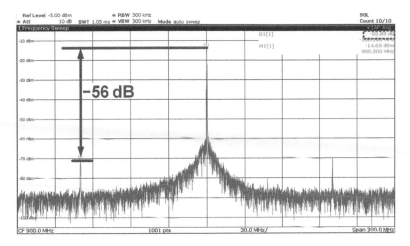

FIGURE 14.31   Measured spurious tones performance of ILRO.

burst-mode operation. With frequency calibration, the reference spur of the ILRO can be lowered to −56 dBc as shown in Figure 14.31.

The measured constellation for QPSK/16-QAM with and without BS is shown in Figure 14.32. EVMs better than 6% are observed for QPSK at 50 Mbps and 16-QAM at 100 Mbps. Figure 14.33 plots the EVM performance at different data rate under

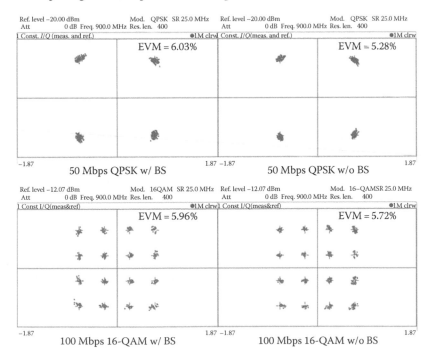

FIGURE 14.32   Measured EVM for QPSK and 16-QAM at 25 MSps with and without band shaping.

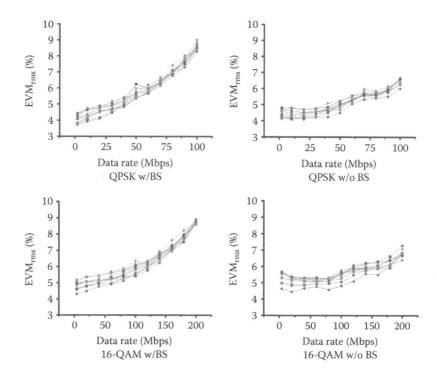

FIGURE 14.33 Measured TX EVM variations versus data rate across 10 chips.

different modulation. Only 1% EVM variation is observed for the collected data over 10 chips, showing the robustness of the proposed TX.

Figure 14.34 presents the output spectrum with fixed data rate of 50 Mbps. As illustrated, 38 dB side lobe suppression is achieved with BS, which is 25 dB more compared to TX without BS. In addition, 16-QAM mode is also twice more spectral efficient than QPSK.

Under 0.77 V supply, the TX consumes 2.6 and 1.3 mW, respectively, with and without BS while transmitting at 25 MSps with −15 dBm output power. The digital portion that provides band shaping consumes 50% of the total power. This power can be further reduced by adopting ROM-based RRC filter.

The TX performance is compared with other similar multi-PSK and 16-QAM TX in Table 14.3. We achieve the highest data rate of 100 Mbps and energy efficiency of 13 pJ/bit (without band shaping) compared with others. Due to the simplicity of our TX architecture, the energy efficiency only worsens to 26 pJ/bit with band shaping.

14.5.4 COMPARISON

As a comparison, we have plotted all low-power transmitters with and without band shaping in Figure 14.35. The x-axis indicates the achievable data rate while the y-axis indicates the power consumption of the corresponding transmitter. The dashed diagonal lines imply a constant energy efficiency boundaries. Transmitter

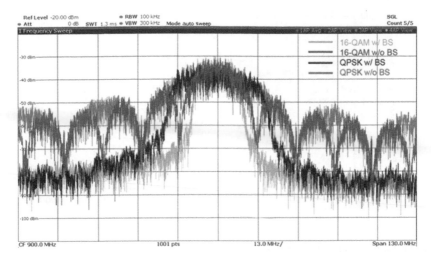

FIGURE 14.34 Comparison of output spectrum with/without band shaping for QPSK and 16-QAM at 50 Mbps.

TABLE 14.3
PSK/QAM TX Performance Summary and Comparison

References	Liu ASSCC'11	Diao ASSCC'10	Izad CICC'12	Zhang RFIC'12	This Work
Frequency (MHz)	2400	900	915	350–578	900
Modulation	HS-OQPSK	O-QPSK/ QPSK	O-QPSK/ 8 PSK	16-QAM[a]	QPSK/ 16-QAM
Data rate (Mbps)	2	50	55	7.5	50/100
Output power (dBm)	−3	−3.3/−15	−15	0.23	−15
Power (mW)	15	5.88/3	0.938	4.9	1.3 (w/o BS) 2.6 (w/BS)
Area (mm²)	0.35	0.28	0.038	0.7	0.08
Band shaping	Yes, >29 dB	No	No	No	Yes, >38 dB
Energy/bit (nJ/bit)	7.5	0.12/0.06	0.017	0.65	0.026/0.013 (w/o BS) 0.052/0.026 (w/BS)
Supply (V)	1.5	1.4	0.8	1.5	0.77
Technology	0.18 μm	0.18 μm	65 nm	0.18 μm	65 nm

[a] Circular constellation-based 16-QAM.

with better energy efficiency will be approaching the bottom-right corner. The energy efficiency of the three fabricated transmitters discussed earlier has been plotted. As illustrated, the three designs reported the lowest energy efficiency for both transmitter with and without band shaping, clearly showing the advantage of the proposed techniques.

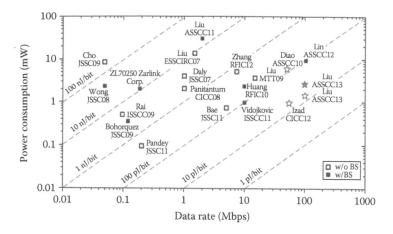

FIGURE 14.35   Energy efficiency comparison of low-power transmitters.

14.6   CONCLUSION

In this chapter, we have reviewed the need of energy efficient high data rate transmitter. We have also discussed the trend of moving toward complex modulation to maximize spectral efficiency. Through the proposed techniques, such as ILO, ILRO, and ILRO with direct quadrature modulation, energy-efficient PSK and QAM transmitter can be achieved, which report one order improvement on energy efficiency, reaching 13 pJ/bit.

ACKNOWLEDGMENT

We thank Mediatek for sponsoring the fabrication. We would also like to thank the Institute of Microelectronics for facilitating the packaging and testing of some of the works. Last but not the least, we would like to thank Shengxi Diao, San Jeow Cheng, Mehran Mohammadi Izad, and Xiayun Liu for the designing and testing of the chips.

REFERENCES

Bae J., L. Yan, and H. J. Yoo. A low energy injection-locked FSK transceiver with frequency-to-amplitude conversion for body sensor applications. *IEEE J. Solid-State Circ.* 46 (4), (2011): 928–937.

Bohorquez J., A. P. Chandrakasan, and J.L. Daeson. A 350 μW CMOS MSK transmitter and 400 μW OOK super-regenerative receiver for medical implant communications. *IEEE J. Solid-State Circ.* 44 (4), (2009): 1248–1259.

Chen M. S. W., D. Su, and S. Mehta. A calibration-free 800 MHz fractional-N digital PLL with embedded TDC. *ISSCC Dig. Tech. Papers* (2010): 472–473.

Diao S., Y. Zheng, Y. Gao et al. A 50-Mb/s CMOS QPSK/O-QPSK transmitter employing injection locking for direct modulation. *IEEE Trans. Microw. Theory Tech.* 60 (1), (2012): 120–130.

Feigin J. Don't let linearity squeeze with accurate models, the range- and data-rate-limiting. *Commun. Systems Design* (2003): 12–16.

Izad M. M. and C. H. Heng. A 17 pJ/bit 915 MHz 8 PSK/O-QPSK transmitter for high data rate biomedical applications. *Proc. CICC* (2012a): 1–4.

Izad M. M. and C. H. Heng. A pulse shaping technique for spur suppression in injection-locked synthesizers. *IEEE J. Solid-State Circ.* 47 (3), (2012b): 652–664.

Kavousian A., D. K. Su, M. Hekmat et al. A digitally modulated polar CMOS power amplifier with a 20-MHz channel bandwidth. *IEEE J. Solid-State Circ.* 43 (10): (2008): 2251–2258.

Kim K., S. Lee, E. Cho, J. Choi, and S. Nam. Design of OOK system for wireless capsule endoscopy. *Proc. ISCAS* (2010): 1205–1208.

Liu Y.H., C.L. Li, and T. H. Lin. A 200-pJ/b MUX-based RF transmitter for implantable multichannel neural recording. *IEEE Trans. Microw. Theory Tech.* 57 (10), (2009): 2533–2541.

Mehta J., R. B. Staszewski, O. Eliezer et al. A 0.8 mm all-digital SAW-less polar transmitter in 65 nm EDGE SoC. *ISSCC Dig. Tech. Papers* (2010): 58–59.

Razavi B. A study of injection locking and pulling in oscillators. *IEEE J. Solid-State Circ.* 39 (9), (2004): 1415–1424.

Sansen W. Low-power, low-voltage design. ISSCC Short Course (2009).

15 Design and Implementation of Ultralow-Power ZigBee/WPAN Receiver

Zhicheng Lin, Pui-In Mak, and Rui P. Martins

CONTENTS

15.1 INTRODUCTION

In recent years, the proliferation of short-range wireless applications for Internet of Things and personal healthcare calls for ultralow-power and ultralow-cost CMOS radios [1]. Ultralow-voltage (ULV) designs have been one of the key directions to approach a better power efficiency [2–5]. Regrettably, an ULV supply will limit the voltage swing, and device's f_T and overdrives, deteriorating the spurious-free dynamic range (SFDR) while necessitating area-hungry inductors (or transformers) to assist the bias and tune out the parasitic capacitances. This chapter describes the design and implementation of a compact, low-power, and high-SFDR receiver suitable for ZigBee or wireless personal area network (WPAN) applications. The research background is outlined as follows.

Four potential ultralow-power receiver architectures are shown in Figure 15.1. The first (Figure 15.1a) employs a single low-noise transconductance amplifier (single LNTA) followed by two passive I/Q mixers and transimpedance amplifiers (TIAs). If a 50% duty-cycle local oscillator (50% LO) is applied, this topology can suffer from image current circulation between the I and Q paths, inducing I/Q crosstalk, unequal high-side and low-side gains, IIP2 and IIP3 [6]. Lowering the LO duty cycle to 25% (Figure 15.1b) can alleviate such issues [7] at the expense of extra sine-to-square LO buffers and logic operation. Another alternative is to add two signal buffers before the mixers (Figure 15.1c), but they must be linear enough (i.e., more power) to withstand the voltage gain of the low-noise amplifier (LNA) [8,9]. The basis of our proposed solution (Figure 15.1d) is to split the LNTA into two, such that a single-ended RF input is maintained, while allowing isolated passive mixing that facilitates the use of a 50% LO for power savings.

This chapter is organized as follows: Section 15.2 overviews the operating principle of the proposed "split-LNTA + 50% LO" receiver. An analytical comparison of it with the existing "single-LNTA + 25% LO" architecture is given in Section 15.3. In Section 15.4, a number of techniques are proposed, including (1) a low-power voltage-mode TIA to enhance the out-channel linearity both at RF and baseband (BB); (2) a mixed-supply (V_{DD}) design approach [10] to alleviate the design trade-offs in RF LNTA (power, gain, and noise) and BB TIA (power, linearity, and signal swing); and (3) a low-power LO generation scheme that consists of a LC voltage-controlled

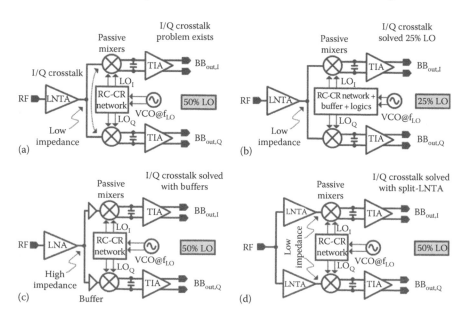

FIGURE 15.1 Four potential receiver architectures: (a) Single-LNTA + 50% LO. (b) Single-LNTA + 25% LO. (c) Single-LNA + 50% LO + signal buffers. (d) Split-LNTA + 50% LO (proposed).

oscillator (VCO) and an input impedance–boosted type II *RC-CR* network. They optimize the VCO's output swing with the LC tank's quality factor, while offering adequate I/Q accuracy at low power. The measurement results are reported in Section 15.5, and the conclusions and future developments are presented in Section 15.6.

15.2   PROPOSED "SPLIT-LNTA + 50% LO" RECEIVER

The split-LNTA (Figure 15.2) is based on two self-biased inverter-based amplifiers (M_1, M_2, and R_F), which have no inner parasitic pole. They also can take the speed advantage of fine linewidth CMOS to lower the device overdrive voltages, featuring a high g_m-to-I_d efficiency at low V_{DD} (V_{DD06} = 0.6 V). Its single-ended RF input avoids the RF balun and its associated insertion loss. In front of the split LNTA, a proper codesign between the RF input capacitance (C_{in}) and bond wire (L_{bw}) facilitates the input impedance matching, while offering a passive pregain (A_v) decisively important to the NF and power efficiency. The two LNTAs convert the RF signal (v_{in}) into two equal currents $i_{out, I}$ and $i_{out, Q}$ for the I and Q channels, respectively. To avoid the parasitic and area impact from AC coupling, $i_{out, I}$ and $i_{out, Q}$ are directly dc-coupled to the passive mixers (M_3 and M_4). As long as the dc passing through M_3 and M_4 is kept small, the 1/f noise induced by the mixers can be minimized [11]. This aim can be achieved by matching the output common-mode level of the LNTA to that of the BB TIA.

The 50% four-phase LO ($LO_{Ip, n}$ and $LO_{Qp, n}$) is generated by a 2.4 GHz LC VCO followed by a new type II RC-CR network, which features a capacitor divider at the input to boost the input impedance. When driving the LO to the mixers (M_3 and M_4), a proper dc level ($V_{LO, b}$) can optimize the switching time. The downconverted

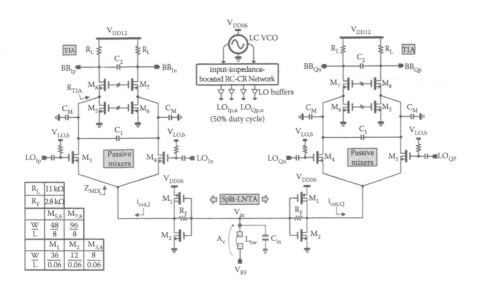

FIGURE 15.2   Schematic of the proposed receiver exploiting passive pregain, split-LNTA, passive mixers, 50% LO, and common-gate TIAs.

low-IF (2 MHz) signal is further amplified by a common-gate TIA (M_{5-8} and R_L), which uses a 1.2 V (V_{DD12}) supply to accommodate more signal swing and enhance linearity. Here, we assume a complex low-IF filter will follow the BB TIA, rendering the 1/f noise and IIP2 not significant and will not be further addressed. Due to the bidirectional transparency of passive mixers [7,8], the BB capacitors (C_1 and C_M) can enhance the selectivity at both RF (the output of the LNTA) and BB, improving the out-band linearity. The grounded C_M also helps suppress the common-mode RF feedthrough, which is limited by the bond wire inductance that appears in series with C_M under common-mode operation.

15.3 COMPARISON OF "SPLIT-LNTA + 50% LO" AND "SINGLE-LNTA + 25% LO" ARCHITECTURES

This section presents an analytical comparison of the two architectures: "split-LNTA + 50% LO" and "single-LNTA + 25% LO." For brevity, "50% LO" and "25% LO" are exploited to represent them, respectively. Figure 15.3a and 15.3b shows their simplified equivalent circuits. For a fair comparison, the two LNTAs in Figure 15.3a are modeled as g_m (transconductance) and $2R_{out}$ (output resistance), whereas the single LNTA in Figure 15.3b is modeled as $2 g_m$ and R_{out}. These models are developed under the same approach described in [12–14], where the harmonic upconversion in passive mixers is modeled as R_{sh}. The impedances looking into the 50% LO and 25% LO mixers are denoted as Z_{MIX1} and Z_{MIX2}, respectively. Each mixer features an on-resistance of R_{sw}. R_{TIA} is the input resistance of the TIA. The single-ended differential mode capacitance is denoted as C_d ($= C_M + 2C_1$).

15.3.1 GAIN

For Figure 15.3a, we summarize in (15.1 through 15.5) the derived expressions of both Z_{MIX1} and the voltage gain (A_{Vx1}) at V_{x1} at the LO + IF frequency ($\omega_{LO} + \omega_{IF}$), the BB output current (I_{BB1}) with respect to v_{in}, the voltage gain (A_{Vy1}) at $V_{ylp, n}$, and finally the voltage gain (A_{Vout1}) at $V_{outlp, n}$:

$$Z_{MIX1}| \text{ at} (\omega_{LO} + \omega_{IF}) \approx R_{sw} + \left(\frac{2Z_{BB}/\pi^2}{R_{sh}} \right) \tag{15.1}$$

where

$$Z_{BB} = \frac{1}{s(2C_1 + C_M)} //R_{TIA}; \quad R_{sh} \approx \frac{2}{3}(2R_{out} + R_{sw}) \tag{15.2}$$

$$A_{Vx1} \text{ at } (\omega_{LO} + \omega_{IF}) \approx g_m(2R_{out} //Z_{MIX1})$$

$$\frac{I_{BB1}}{v_{in}} \text{ at DC} = \frac{I_{BB1p} - I_{BB1n}}{v_{in}} \approx g_m \frac{2R_{out}}{R_{TIA} + 2(2R_{out} + R_{sw})} \frac{4}{\pi} = G_{m1} \tag{15.3}$$

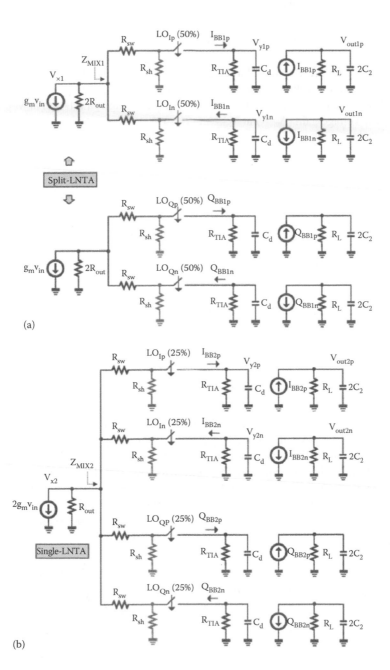

FIGURE 15.3 Small-signal equivalent circuits. (a) Split-LNTA + 50% LO. (b) Single-LNTA + 25% LO.

$$A_{Vy1} \text{ at DC} = A_{Vy1p} - A_{Vy1n} \approx G_{m1} R_{TIA} \tag{15.4}$$

$$A_{Vout1} \text{ at DC} = A_{Vout1p} - A_{Vout1n} \approx G_{m1} R_L \tag{15.5}$$

Similarly, for Figure 15.3b, we have (15.6 through 15.10) the derived expressions of both Z_{MIX2} and the voltage gain (A_{Vx2}) at V_{x2} at the LO + IF frequency ($\omega_{LO} + \omega_{IF}$), the BB output current (I_{BB2}) with respect to v_{in}, the voltage gain (A_{Vy2}) at $V_{y2p, n}$, and finally the voltage gain (A_{Vout2}) at $V_{out2p, n}$:

$$Z_{MIX2} \Big| \text{ at } (\omega_{LO} + \omega_{IF}) \approx R_{sw} + \left(\frac{2Z_{BB}}{\pi^2} / /R_{sh} \right) \tag{15.6}$$

where

$$Z_{BB} = \frac{1}{s(2C_1 + C_M)} / /R_{TIA} ; \quad R_{sh} \approx 4\left(R_{out} + R_{sw} \right)$$
$$A_{Vx2} \text{ at } (\omega_{LO} + \omega_{IF}) \approx 2g_m \left(R_{out} / /Z_{MIX2} \right) \tag{15.7}$$

$$\frac{I_{BB2}}{v_{in}} \text{ at DC} = \frac{I_{BB2p} - I_{BB2n}}{v_{in}} \approx 2g_m \frac{R_{out}}{R_{TIA} + 4(R_{out} + R_{sw})} \frac{4\sqrt{2}}{\pi} = G_{m2} \tag{15.8}$$

$$A_{Vy2} \text{ at DC} = A_{Vy2p} - A_{Vy2n} \approx G_{m2} R_{TIA} \tag{15.9}$$

$$A_{Vout2} \text{ at DC} = A_{Vout2p} - A_{Vout2n} \approx G_{m2} R_L \tag{15.10}$$

Note that the output capacitance of the LNTA was neglected. In fact, the output capacitance of LNTA will induce C_{out} and $2C_{out}$ for the g_m and $2g_m$ LNTA stages, respectively. This will render the output impedance ratio at V_{x1} and V_{x2} slightly larger than 2. Besides, the parasitic capacitor will affect R_{sh} too. The proposed separated g_m stage imposes a smaller C_{out} and thus lowers the degradation of gain and NF than those predicted by Equations 15.11 and 15.12. With proper sizing, one can achieve $R_{sw} \ll R_{out}$ and $R_{sw} \ll R_{TIA}$ and R_L such that the gain difference between 25% LO and 50% LO at different RF and BB nodes can be estimated as

$$\Delta A_{Vx1,2} \text{ at } \omega_{LO} = 20 \log A_{Vx2} - 20 \log A_{Vx1} \approx 20 \log \frac{2\left(R_{out} / / \frac{2R_{TIA}}{\pi^2} / /4R_{out} \right)}{2R_{out} / / \frac{2R_{TIA}}{\pi^2} / / \frac{4R_{out}}{3}} = 6 \, dB$$

$$\Delta A_{Vy1,2} \text{ at DC} = 20 \log A_{Vy2} - 20 \log A_{Vy1} = 20 \log \left(\sqrt{2} \frac{R_{TIA} + 4R_{out} + 2R_{sw}}{R_{TIA} + 4R_{out} + 4R_{sw}} \right) \approx 3 \, dB$$

$$\Delta A_{Vout1,2} \text{ at DC} = 20 \log A_{Vout2} - 20 \log A_{Vout1} = 20 \log \left(\sqrt{2} \frac{R_L + 4R_{out} + 2R_{sw}}{R_L + 4R_{out} + 4R_{sw}} \right) \approx 3 \, dB$$

$$\tag{15.11}$$

From (15.11), the 25% LO should have a higher gain at both RF and BB nodes than the 50% LO. However, as analyzed in Section 15.3.3, a higher gain at RF will penalize the IIP3, while a higher BB gain can be achieved easily by using a larger R_L. For the impact of these gain differences to NF, we analyze them next.

15.3.2   NF

The NF is analyzed according to the equivalent LTI noise model [12–14]. As shown in Figure 15.4a and 15.4b, the four noise sources are the thermal noises from $R_s (V_{n,Rs}^2 = 4kTR_s)$, $LNTA (I_{n,gm}^2 = 4kT\gamma_1 g_m$ or $I_{n,2gm}^2 = 4kT\gamma_1 2g_m)$, $R_{sw} (V_{n,sw}^2 = 4kTR_{sw})$, and the noise from TIA that is $V_{n,TIA}^2 \approx 4kT\gamma_2/g_{m\_TIA} \approx 4kT\gamma_2 R_{TIA}$, given that the output impedance of the mixer is sufficiently large. Here, g_{m_TIA} is the transconductance of the bias transistor for the TIA, while the noise from the CG device is degenerated. An accurate model of the TIA noise can be found elsewhere [11]. The noise of R_F is ignorable and the noise coupling between the I and Q paths under a 50% LO is minor (confirmed by simulations), easing the NF calculation of each path separately. The noise factor (F) can be found by dividing the total output noise by the portion contributed by R_s

$$F = 1 + \frac{\gamma_1}{R_s A_v^2 G_m} + \frac{R_{sw}}{R_s A_v^2 G_m^2 R^2} + \frac{(R + R_{sw})^2}{R_s A_v^2 G_m^2 R^2 \beta \gamma_2 R_{TIA}} + \frac{a \gamma_1}{R_s A_v^2 G_m} + a + \frac{a R_{sw}}{R_s A_v^2 G_m^2 R^2}$$

(15.12)

where $\beta = 2/\pi^2$ is the downconversion scaling factor and a is the harmonic folding factor:

$$a = \left(\frac{\pi^2}{4} - 1\right), G_m = g_m, \text{ and } R = 2R_{out} \quad \text{for Figure 15.4a}$$

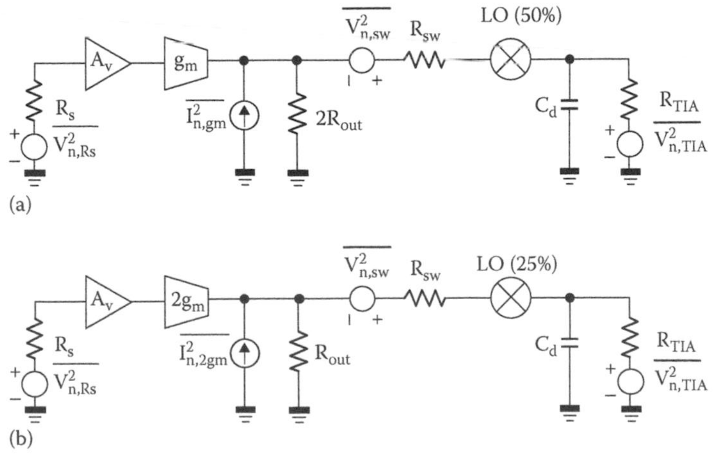

(a)

(b)

FIGURE 15.4   Equivalent LTI noise model with pregain for (a) 50% LO (Figure 15.3a) and (b) 25% LO (Figure 15.3b).

$$a = \left(\frac{\pi^2}{8} - 1\right), G_m = 2g_m, \text{ and } R = R_{out} \quad \text{for Figure 15.4b}$$

In (15.12), the second term is from the LNTA, the third term is from the mixer, and the fourth term is from the TIA. The rest of the terms are the noise folding from the odd harmonics of the LO for LNTA, R_s, and R_{sw}, respectively. The NF calculated from (15.12) for 50% LO is single sideband (SSB). For a double sideband (DSB) NF, it is 3 dB less. Since the harmonic's power of 50% LO is larger than that of 25% LO, the folding terms of 50% LO are also higher. From (15.12), one can plot the DSB NF of 50% LO and 25% LO in Figure 15.5 as a function of A_v, where $\Delta NF = NF_{50\%} - NF_{25\%}$ $R_{sw} = 50\ \Omega$, $\gamma_1 = \gamma_2 = 1$, $g_m = 9$ mS, $R_{out} = 200\ \Omega$, and $R_{TIA} = 2.5$ kΩ. It can be seen that ΔNF is reduced to 0.91 dB (0.51 dB) when A_v is just 2 V/V (3 V/V), which is easily achievable in practice. In fact, a moderated A_v can even eliminate the need of the LNTA (or LNA) [3]. However, when considering also the input matching and LO-to-RF isolation, both pregain and LNTA should be employed concurrently. The simulated LO-to-RF isolation is <–100 dBm. Due to the passive pregain, the IIP3 of the receiver is more demanding than the NF, promoting the use of a 50% LO. Together with its power advantage (i.e., lower VCO frequency and no divider), our proposed topology (i.e., pregain + split-LNTA + 50% LO) should ease the trade-off between NF, IIP3, area, and power.

15.3.3   IIP3

The third-order intermodulation (IM3) distortion is analyzed to assess the linearity. The aim is to find the in-band IIP3 of the receiver under 50% LO and 25% LO in

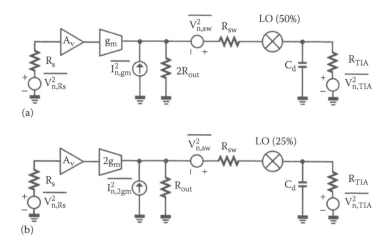

FIGURE 15.5   Simulated NF_{DSB} and ΔNF against A_v for (a) 50% LO and (b) 25% LO.

response to two-tone excitation. Assuming that the nonlinearity of the receiver is dominated by the LNTA, its nonlinearity contributions are considered as

1. Third-order LNTA nonlinearity due to input excitation v_{in} [α_2 (I/V^3)]
2. Third-order LNTA nonlinearity due to output excitation v_x [α_3 (I/V^3)]

Thus, $i_{ds} = \alpha_1 v_{in} + \alpha_2 v_{in}^3 + \alpha_3 v_x^3$. If the coefficients α_1, α_2, and α_3 are assumed to be proportional to the device W/L,

For 50% LO, $\alpha_1 = g_m$, $\alpha_2 = g_{m3}$, $\alpha_3 = g_{03}$
For 25% LO, $\alpha_1 = 2g_m$, $\alpha_2 = 2g_{m3}$, $\alpha_3 = 2g_{03}$

where g_{m3} and g_{03} are the third-order nonlinear transconductance and conductance, respectively. With a two-tone excitation of amplitude A and the first order voltage gain and current gain given in (15.1 through 15.11), the IM3 output voltage for each of the nonlinear coefficients listed earlier can be written as

$$V_{o3\alpha2} = \frac{3}{4} g_{m3} A^3 I_{BB1} R_L; \quad V_{o3\alpha3} = \frac{3}{4} g_{03} A_{Vx1}^3 A^3 I_{BB1} R_L$$

for a 50% LO. Thus,

$$IM_{3\_50\%} = \frac{V_{o3\alpha2} + V_{o3\alpha3}}{V_{01\alpha1}} = \frac{\dfrac{3}{4} g_{m3} A^3 I_{BB1} R_L + \dfrac{3}{4} g_{03} A_{Vx1}^3 A^3 I_{BB1} R_L}{A g_m I_{BB1} R_L}$$

Let

$$IM_{3\_50\%} = 1 \rightarrow IIP_{3\_50\%} = \sqrt{\frac{4 g_m}{3(g_{m3} + g_{03} A_{Vx1}^3)}} \tag{15.13}$$

Following the same procedure, the IIP3 for 25% LO can be derived as

$$IIP_{3\_25\%} = \sqrt{\frac{4 g_m}{3\left(g_{m3} + g_{03} A_{Vx2}^3\right)}} \tag{15.14}$$

Since $A_{Vx2} > A_{Vx1}$, we can find that, from (15.13) to (15.14), the LNTA's third-order nonlinearity term is larger for a 25% LO. Thus, the IIP3 of 50% LO should be better than that of 25% LO, benefiting the SFDR since both architectures will feature a similar NF after adding the pregain.

15.3.4 Current- and Voltage-Mode Operations

Both 25% LO and 50% LO architectures can be intensively designed for current-mode or voltage-mode operation. For a high-performance design like [7,8,12], $R_{TIA} \ll R_{out}$

TABLE 15.1
Proposed Receiver under Current- and Voltage-Mode Operations

Mode	Gain	NF	In-Band IIP3	Power	Suitable for
Current mode (small R_{sw} and R_{TIA})	↗	↘	↗	↗	High performance
Voltage mode (large R_{sw} and R_{TIA})	↘	↗	↘	↘	Ultra low power

and $R_{sw} \ll R_{out}$ are preferred to keep the signals in the deep current mode. As such, (15.3) and (15.8) can be simplified as $G_{m1} = 2g_m/\pi$ and $G_{m2} = 2\sqrt{2}g_m/\pi$, respectively. Both of them are higher when compared with themselves in the voltage-mode operation. In terms of IIP3 and NF, the current mode is also preferable since $A_{Vx1} \approx g_m(R_{sw} + 2/\pi^2 R_{TIA})$ and $A_{Vx2} \approx 2g_m(R_{sw} + 2/\pi^2 R_{TIA})$ will be lower, and the noise due to the folding term and TIA will be also smaller as noted in (15.12).

Nevertheless, the current-mode operation also brings up two sizing constraints being less attractive for *low-power* design: (1) a low R_{sw} entails a large device W/L and a higher overdrive voltage for the mixers, both calling for a larger-power budget in the LO path and (2) a low R_{TIA} implies that the TIA has to draw a large bias current. For example, if a low R_{TIA} of 50 Ω is required from the 1.2-V TIA (a common-gate amplifier), its bias current is as high as $I_{bias} = 2$ mA for a typical overdrive voltage of 200 mV. Thus, for ultralow-power applications like ZigBee/WPAN that has relaxed NF and linearity requirements, higher R_{sw} and R_{TIA} are preferable to operate the receiver more on the voltage mode. A summary of performance differences in current- and voltage-mode operations is given in Table 15.1.

15.4 CIRCUIT TECHNIQUES

15.4.1 IMPEDANCE UPCONVERSION MATCHING

From Section 15.3, we expect a passive pregain A_v of 2–3 V/V. As shown in Figure 15.6a, A_v can be derived under $R_{in} = R_s$,

$$\frac{V_{out}^2}{2R_{out}} = \frac{V_s^2}{8R_s}, \quad V_{out} = V_{in}A_v, V_{in} = 0.5V_s \Rightarrow A_v = \sqrt{\frac{R_{out}}{R_{in}}}$$

Thus, an upconversion matching network is entailed to ensure $A_v > 1$. A convenient way to achieve it is to use L_{bw} to resonant with C_{in}. The schematic is shown in Figure 15.6b. The parallel connection of C_{in} and R_{out} can be transformed into a series connection of C_{ser} and R_{ser}, as shown in Figure 15.6c. At $L_{bw}C_{ser}$ resonance, and with $R_{ser} = R_s$ and $i = V_s/2R_{ser}$, we have

$$V_{out} = V_{R_{ser}} + V_{C_{ser}} = \frac{V_s}{2}\left(1 - j\frac{Q_C}{2}\right)$$

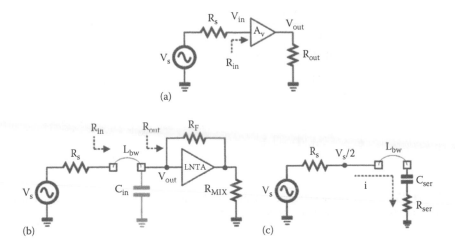

FIGURE 15.6 Input impedance matching: (a) A_v converts R_{out} to R_{in} to match with R_s, (b) L_{bw} C_{in} as an impedance conversion network, and its (c) narrowband equivalent circuit.

where

$$V_{R_{ser}} = -j\frac{Q_C V_s}{2} sC_{ser}R_{ser} = \frac{V_s}{2}$$

$$V_{C_{ser}} = \frac{1}{j\omega_0 C_{ser}} \frac{V_s}{2R_{ser}} = -j\frac{Q_C}{2} V_s,$$

$$\omega_0 = \frac{1}{\sqrt{L_{bw}C_{ser}}} \quad \text{and} \quad Q_C = \frac{\sqrt{L_{bw}/C_{ser}}}{R_{ser}}$$

Interestingly, such a voltage boosting factor $\sqrt{1+Q_c^2/4}$ is larger than the conventional inductively degenerated LNA, which is only $Q_c/2$. In fact, when the capacitance of the PCB trace is accounted, the Q of the matching network will be higher, easing the impedance matching.

15.4.2 MIXER–TIA INTERFACE BIASED FOR IMPEDANCE TRANSFER FILTERING

For the employed single-balanced passive mixers, the RF-to-IF feedthrough has to be addressed. Based on Figure 15.7, we can calculate the currents i_{M7} and i_{M8} with respect to the RF current i_{RF} as given by

$$i_{M7} = \frac{i_{RF}}{2}\left[1 - \text{sign}(\cos\omega_{LO}t)\right] \qquad (15.15)$$

$$i_{M8} = \frac{i_{RF}}{2}\left[1 + \text{sign}(\cos\omega_{LO}t)\right] \qquad (15.16)$$

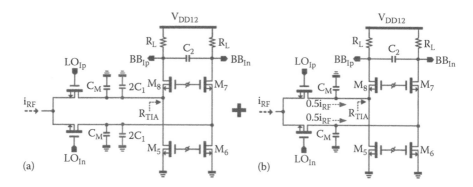

FIGURE 15.7 Equivalent circuits of the mixer–TIA interface for (a) the differential low-IF signal and (b) the common-mode RF feedthrough.

They imply that the currents can be decomposed into the differential mode (Figure 15.7a) with amplitude of $2i_{RF/\pi}$ at BB and into the common mode (Figure 15.7b) with amplitude of $0.5i_{RF}$ at RF. To suppress the latter, C_M was added to create a low-pass pole ($C_M//R_{TIA}$). For the differential IF signal, the pole is located at ($C_M + 2C_1$)$//R_{TIA}$, which suppresses the out-of-channel interferers before they enter the TIA. As such, the TIA can be biased under a very small bias current. The resultant high input impedance of the TIA indeed benefits both BB and RF filtering because of the bidirectional impedance-translation property of the passive mixers [7,8]. Figure 15.8 shows the simulated out-band IIP3, which is subject to the allowed total capacitance of $C_M + 2C_1$. For instance, when $C_M + 2C_1$ is increased from 16 to 42 pF, the out-band IIP3 raises from +2.5 to +4.7 dBm at the expense of the die area. For the on-resistance of the mixer switches (R_{sw}), it involves a trade-off of the LO path's power to the out-band IIP3 and NF. As shown in Figure 15.9, if R_{sw} is increased from 50 to 150 Ω for power savings, the NF and out-band IIP3 will be penalized by ~1 dB.

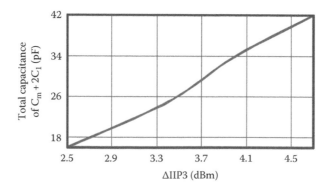

FIGURE 15.8 Out-band IIP3 can be improved by allowing more total capacitance of $C_M + 2C_1$.

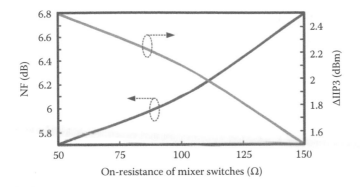

FIGURE 15.9   The on-resistance of the mixer switches represents the trade-off of the LO path's power to the out-band IIP3 and NF.

15.4.3   RC-CR NETWORK AND VCO CODESIGN

The LC VCO (Figure 15.10a) employs a complementary *NMOS-PMOS* (M_{1-4}) negative transconductor. For power savings, M_1 and M_2 are based on ac-coupled gate bias ($V_{vco, b}$) to lower the supply to 0.6 V. Here, we implement a capacitive divider (C_{M1} and C_{M2}) to boost the input impedance of its subsequent two-stage *RC-CR* network (Figure 15.10b). The optimization details are presented next.

    RC-CR network is excellent for low-power and narrowband I/Q generation. With a type II architecture, both phase balancing and insertion loss can be better optimized than its type I counterpart [15]. For instance, the simulated insertion loss

	$M_{1,2}$	$M_{3,4}$	Lp	C_{M1}	C_{M2}	R_{N1}	C_{N1}	R_{N2}	C_{N2}
$\dfrac{W}{L}$	$\dfrac{12}{0.12}$	$\dfrac{24}{0.12}$	4.6 nH	530 fF	280 fF	450 Ω	35 fF	900 Ω	120 fF

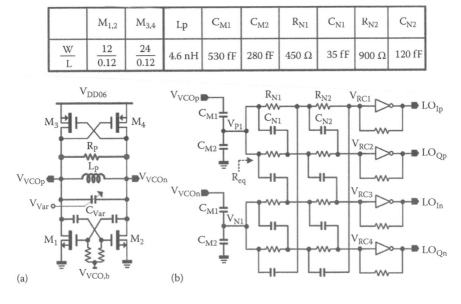

(a)

(b)

FIGURE 15.10   (a) LC VCO and (b) the proposed input impedance–boosted two-stage type II *RC-CR* network for four-phase 50% LO generation.

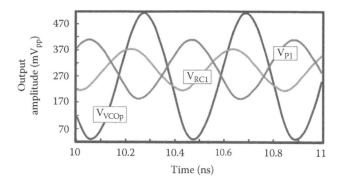

FIGURE 15.11 Simulated time-domain signals at the output of the VCO (V_{vcop}), capacitor divider (V_{p1}), and the *RC-CR* network (V_{RC1}).

of a two-stage type II *RC-CR* network is roughly 2 dB as shown in Figure 15.11, which will be raised to 4–5 dB if a type I topology is applied (not shown). For low-power LO buffering, the amplitude balancing is critical because its imbalance will lead to inconsistent zero-crossing points, resulting in AM to duty-cycle distortion. Figures 15.12 (V_{RC1-4}) and 15.13 ($LO_{Ip, n}$ and $LO_{Qp, n}$) are the simulated transient waveforms, showing the consistent duty cycle and zero-crossing points achieved in the proposed design.

For an *RC-CR* network operated at 2.4 GHz, if we select $R_{N1} = 1$ kΩ, C_{N1} is just 66 fF, which benefits the area, VCO tuning range, and phase noise, but the I/Q accuracy over Process-Voltage-Temperature (PVT) variations should be considered [16]:

$$\frac{\sigma\left(\text{Image out}\right)}{\text{Desired out}} = 0.25\sqrt{\left(\frac{\sigma_R}{R}\right)^2 + \left(\frac{\sigma_C}{C}\right)^2} \tag{15.17}$$

Since ZigBee/WPAN applications call for a low image rejection ratio (IRR) of 20–30 dB [17], according to (15.17), the matching of the resistors (σ_R) and capacitors (σ_C) can be relaxed to 2.93% for a 30 dB IRR (3σ). The sizes of $C_{N1,2}$ and $R_{N1,2}$ are summarized in Figure 15.10. The poles from $C_{N1,2}$ and $R_{N1,2}$ are

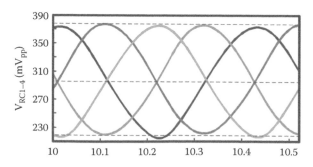

FIGURE 15.12 Simulated time-domain signals at V_{RC1-4}.

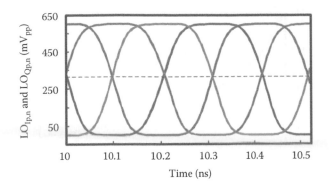

FIGURE 15.13   Simulated time-domain signals at $LO_{Ip, n}$ and LO_{Qp-n}.

distributed around 2.4 GHz to cover the PVT variations. The impact of R_{N1} to the VCO can be analyzed as follows:

When the VCO's inductor is 4 nH with a Q of 20 ($R_P \approx 1.2$ kΩ), we have $R_{tank} \approx 0.5R_P//0.5R_{N1}$. Thus, directly connecting the *RC-CR* network to the VCO will limit the LC tank's Q_{tank} degrading the phase noise [18,19]. To alleviate this, we boost up the equivalent input resistance of the *RC-CR* network (R_{eq}) by adding a capacitive divider (C_{M1} and C_{M2}). For the total tank capacitance C_{tank}, it can be approximated as

$$C_{tank} \approx 2C_{Var} + \frac{(C_{M2} + 2C_{N1})C_{M1}}{C_{M1} + C_{M2} + 2C_{N1}} \tag{15.18}$$

By defining an input impedance boosting factor n,

$$n = \frac{C_{M1}}{C_{M1} + C_{M2} + 2C_{N1}} \tag{15.19}$$

we have

$$V_{P1} \approx nV_{VCOp} \tag{15.20}$$

It means that the signal swing (V_{P1}) delivered to the *RC-CR* network are in trade-off with n. Handily, in our VCO, sweeping $V_{vco, b}$ can track the phase noise with the output swing (Figure 15.14). Given a bias current and a phase noise target, R_{tank} can be set from $V_{VCOp} \approx 2I_{bias}R_{tank}$, and n can be set from (15.21) with a specific R_p and R_{eq},

$$R_{tank} \approx \frac{R_{eq}}{n} \Big|\Big| \frac{R_P}{2} \tag{15.21}$$

In this work, n = 0.6 is selected to balance the output swing with C_{tank} and the total tank resistance (R_{tank}).

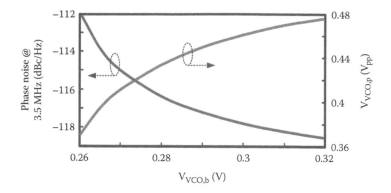

FIGURE 15.14 Trade-off between VCO output amplitude and phase noise with respect to $V_{vco,\,b}$.

15.5 EXPERIMENTAL RESULTS

The receiver (Figure 15.15) fabricated in 65 nm CMOS occupies an active area of 0.14 mm^2 and is encapsulated in a 44-pin CQFP package for PCB-based measurements. The estimated bond wire inductance is ~7 nH for the provided package (13.5 × 13.5 mm^2). Figure 15.16 shows that the measured S_{11} is −8 dB within 2.24–2.46 GHz (for a different package, external inductor or capacitor can be added to optimize S_{11}). The simulation results with and without considering the PCB trace capacitances are also given. The measured voltage gain is 32.8–28.2 dB, and the DSB NF is between 8.6 and 9 dB for an IF spanning from 1 to 3 MHz, as shown in Figure 15.17. We also measured the gain and NF from 2.2 to 2.6 GHz (Figure 15.18).

For a narrowband receiver, the linearity is mainly justified by the out-channel linearity tests. According to the case given in [17,20], two tones are applied at $(f_{LO} + 10\ \text{MHz},\ f_{LO} + 22\ \text{MHz})$ with a power level sweeping from −24 to −32 dBm. Because of the RF and BB filtering associated with the bidirectional property of passive mixers, the out-band IIP3 (Figure 15.19) achieves −7 dBm and the P_{1dB} is −26 dBm.

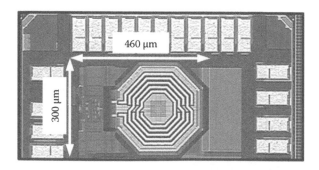

FIGURE 15.15 Chip micrograph of the fabricated receiver.

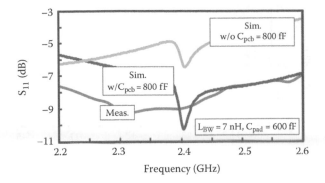

FIGURE 15.16   Measured S_{11} and simulated S_{11} with and without C_{pcb}.

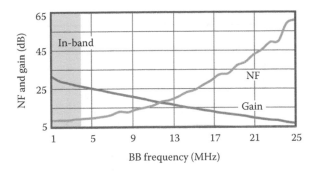

FIGURE 15.17   Measured receiver gain and NF versus BB frequency.

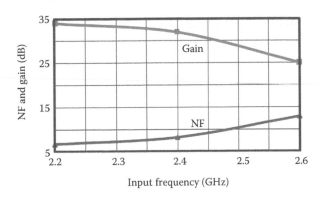

FIGURE 15.18   Measured receiver gain and NF versus input signal frequency.

For the VCO, it measures 21% tuning range from 2.623 to 2.113 GHz, as shown in Figure 15.20. At 3.5 MHz offset, the phase noise (Figure 15.21) is −112.46 dBc/Hz, fulfilling the specification (−102 dBc/Hz [17,20]) with an adequate margin. From frequency 100 kHz to 1 MHz, the result fits the $1/f^3$ slope, and from 1 to 10 MHz, it starts to be saturated, primarily limited by the small output amplitude (−28.31 dBm) of the test buffer.

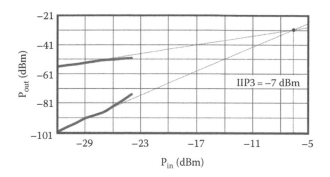

FIGURE 15.19 Measured out-of-band IIP3.

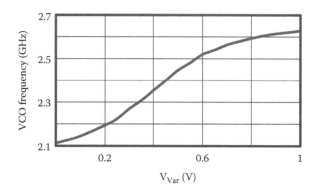

FIGURE 15.20 Measured VCO turning range.

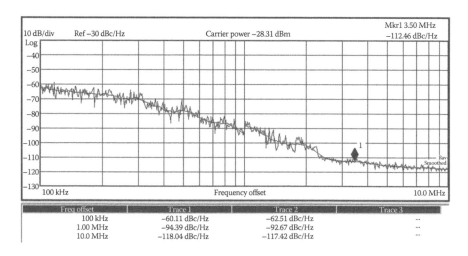

FIGURE 15.21 Measured VCO phase noise at 2.4 GHz.

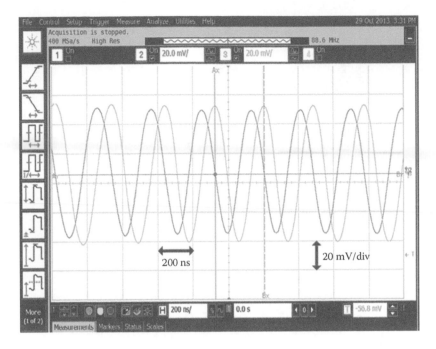

FIGURE 15.22   Measured I/Q BB transient outputs.

Based on transient measurements, the I/Q BB differential outputs (Figure 15.22) have ~0.08 dB gain mismatch and 2° phase match, corresponding to an IRR of ~25 dB.

The performance summary and benchmark are given in Table 15.2 [5,17,21–27]. This work [28] succeeds in achieving the highest power and area efficiencies via proposing a mixed-V_{DD} topology co-optimized with a number of circuit techniques. Only one on-chip inductor is entailed in the VCO. The achieved NF and out-band IIP3 correspond to a competitive SFDR of 59.4 dB according to [17,19]:

$$SFDR = \frac{2(P_{IIP3} + 174dBm - NF - 10logB)}{3} - SNR_{min} \qquad (15.22)$$

where SNR_{min} = 4 dB is the minimum signal-to-noise ratio required by the application and B = 2 MHz is the channel bandwidth. As presented in Figures 15.8 and 15.9, the SFDR can be further optimized by allowing more budgets in area (bigger $C_M + 2C_1$) and/or power (smaller on-resistance of the mixer switches), being a design-friendly architecture easily adaptable to different specifications.

15.6 CONCLUSIONS AND FUTURE DEVELOPMENTS

The design and implementation of a mixed-V_{DD} 2.4-GHz ZigBee/WPAN receiver have been described. It features passive pregain, split-LNTA, high-input-impedance BB TIA, and low-power 50% LO generation. They together lead to improved power

TABLE 15.2

Performance Summary and Benchmark with the State of the Art

Parameters	JSSC'08 [21]	JSSC'10 [22]	JSSC'10 [17]	TCAS-I'10 [24]	TMTT'11 [23]	TMTT'11 [27]	TMTT'06 [28]	ISSCC'13 [5]	ISSCC'13 [25,26]	This Work
Gain (dB)	35	67	75	24.5	51	22.5	30	83	55	32
DSB NF (dB)	7.5	16	9	16.5 (SSB)	3.2	7 (SSB)	7.3 (SSB)	6.1	9	8.8
Out-band IIP3 (dBm)	−10	−10.5	−12.5	−19 (in-band data)	−32 (in-band data)	−21.5 (in-band data)	−8	−21.5	−6	−7
SFDR (dB)	58.3	52.3	55.5	38.3	36.5	51	59.8	51.6	60	59.4
VCO phase noise (dBc/√Hz)	N/A	−127 at 3 MHz	−116 at 3.5 MHz	N/A	N/A	N/A	N/A	−112 at 1 MHz	−115 at 3.5 MHz	−111.4 at 3.5 MHz
Power (mW)	5.4 (w/o VCO)	32.5 (w/ VCO)	3.6 (w/ VCO)	2.52 (w/o VCO)	8.1 (w/o VCO)	1.06 (w/o VCO)	1.8 (w/o VCO)	1.6 (w/ VCO)	2.7 (w/ VCO)	1.4[b] (w/ VCO)
No. of inductor or transformer	2	3	1	3	5	3	2	4	2	1
Die area (mm²)	0.23 (w/o VCO)	2.88 (w/ VCO)	0.35 (w/ VCO)	N/A	1.27 (w/ VCO)	1.1 (w/o VCO)	2.07[a] (w/o VCO)	2.5[a] (w/ VCO)	0.26[a] (w/ VCO)	0.14 (w/ VCO)
Supply (V)	1.35	0.6	1.2	1.8	1.8	1.2	1.8	0.3	0.6/1.2	0.6/1.2
CMOS tech.	90 nm	90 nm	90 nm	0.18 μm	0.18 μm	0.18 μm	0.18 μm	65 nm	65 nm	65 nm

[a] Include more BB gain stages and filters.

[b] The power breakdown is LNTA, 0.4 mW; TIA, 0.18 mW; and VCO + buffer, 0.82 mW.

and area efficiencies, as well as a high SFDR while eliminating the need of an RF balun. These beneficial features render this work a superior receiver candidate for cost and power reduction of ZigBee/WPAN radios in nanoscale CMOS.

New energy-harvesting techniques, such as solar cells, are opening up huge applications for the ZigBee/WPAN radios that must be ultra low power (1–2.5 mW) at very low voltage (50–900 mV) [29]. To address this, a low-cost multi-ISM band (433/860/915/960 MHz) ZigBee receiver [30] has been demonstrated recently, with state-of-the-art power (1.15 mW) and area (0.2 mm²) efficiencies. Unlike the designs in [25,28,31] that entail dual supply voltages, [30] entails only a single 0.5 V supply easing the power management, and being more suitable for energy-harvesting applications.

ACKNOWLEDGMENT

This work was funded by the Macao FDCT and University of Macau—MYRG114 (Y1-L4)-FST13-MPI.

REFERENCES

1. P. Choi, H. Park, I. Nam et al., An experimental coin-sized radio for extremely low-power WPAN (IEEE 802.15.4) application at 2.4 GHz, *IEEE J. Solid-State Circ.*, 38, 2258–2268, 2003.
2. C.-H. Li, Y.-L. Liu, and C.-N. Kuo, A 0.6-V 0.33-mW 5.5-GHz receiver front-end using resonator coupling technique, *IEEE Trans. Microw. Theory Tech.*, 59(6), 1629–1638, 2011.
3. B. W. Cook, A. Berny, A. Molnar et al., Low-power, 2.4-GHz transceiver with passive RX front-end and 400-mV supply, *IEEE J. Solid-State Circ.*, 41, 2767–2775, 2006.
4. A. C. Herberg, T. W. Brown, T. S. Fiez et al., A 250-mV, 352-µW GPS receiver RF front-end in 130-nm CMOS, *IEEE J. Solid-State Circ.*, 46, 938–949, 2011.
5. F. Zhang, K. Wang, J. Koo et al., A 1.6 mW 300 mV-supply 2.4 GHz receiver with –94 dBm sensitivity for energy-harvesting applications, *IEEE ISSCC Dig. Tech. Papers*, 456–457, 2013.
6. A. Mirzaei, H. Darabi, J. C. Leete et al., Analysis and optimization of current-driven passive mixers in narrowband direct-conversion receivers, *IEEE J. Solid-State Circ.*, 44, 2678–2688, 2009.
7. A. Mirzaei, H. Darabi, J. C. Leete et al., Analysis and optimization of direct-conversion receivers with 25% duty-cycle current- driven passive mixers, *IEEE Trans. Circuits Syst. I, Reg. Papers*, 57, 2353–2366, 2010.
8. A. Balankutty and P. R. Kinget, An ultra-low voltage, low-noise, high linearity 900-MHz receiver with digitally calibrated in-band feed-forward interferer cancellation in 65-nm CMOS, *IEEE J. Solid-State Circ.*, 46, 2268–2283, 2011.
9. Y. Feng, G. Takemura, S. Kawaguchi et al., Digitally assisted IIP2 calibration for CMOS direct-conversion receivers, *IEEE J. Solid-State Circ.*, 46, 2253–2267, 2011.
10. P.-I. Mak and R. P. Martins, A 0.46-mm² 4-dB NF unified receiver front-end for full-band mobile TV in 65-nm CMOS, *IEEE J. Solid-State Circ.*, 46, 1970–1984, 2011.
11. N. Poobuapheun, W.-H. Chen, Z. Boos et al., A 1.5-V 0.7–2.5-GHz CMOS quadrature demodulator for multiband direct-conversion receivers, *IEEE J. Solid-State Circ.*, 42, 1669–1677, 2007.
12. C. Andrews and A. C. Molnar, A passive mixer-first receiver with digitally controlled and widely tunable RF interface, *IEEE J. Solid-State Circ.*, 45, 2696–2708, 2010.

13. C. Andrews and A. C. Molnar, Implications of passive mixer transparency for impedance matching and noise figure in passive mixer-first receivers, *IEEE Trans. Circuits Syst. I, Reg. Papers*, 57, 3092–3103, 2010.

14. A. Molnar and C. Andrews, Impedance, filtering and noise in N-phase passive CMOS mixers, *Proc. IEEE CICC*, 1–8, 2012.

15. J. Kaykovuori, K. Stadius, and J. Ryynanen, Analysis and design of passive polyphase filters, *IEEE Trans. Circuits Syst. I, Reg. Papers*, 55, 3023–3037, 2008.

16. F. Behbahani, Y. Kishigami, J. Leete et al., CMOS mixers and polyphase filters for large image rejection, *IEEE J. Solid-State Circ.*, 36, 873–887, 2001.

17. A. Liscidini, M. Tedeschi, and R. Castello, Low-power quadrature receivers for ZigBee (IEEE 802.15.4) applications, *IEEE J. Solid-State Circ.*, 45, 1710–1719, 2010.

18. T. H. Lee, *The Design of CMOS Radio-Frequency Integrated Circuits*, 2nd edn., Cambridge University Press, 2004.

19. B. Razavi, *RF Microelectronics*, 2nd edn., Upper Saddle River, NJ: Prentice-Hall, 2011.

20. W. Kluge, F. Poegel, H. Roller et al., A fully integrated 2.4-GHz IEEE 802.15.4-compliant transceiver for ZigBee™ applications, *IEEE J. Solid-State Circ.*, 41, 2767–2775, 2006.

21. M. Camus, B. Butaye, L. Garcia et al., A 5.4 mW/0.07 mm² 2.4 GHz front-end receiver in 90 nm CMOS for IEEE 802.15.4 WPAN stand, *IEEE J. Solid-State Circ.*, 43, 1372–1383, 2008.

22. A. Balankutty, S. Yn, Y. Feng et al., A 0.6-V zero-IF/low-IF receiver with integrated fractional-N synthesizer for 2.4-GHz ISM-band applications, *IEEE J. Solid-State Circ.*, 45, 538–553, 2010.

23. J.-S. Syu, C. Meng, and C.-L. Wang, 2.4-GHz low-noise direct-conversion receiver with deep N-well vertical-NPN BJT operating near cutoff frequency, *IEEE Trans. Microw. Theory Tech.*, 45, 538–553, 2010.

24. J. Kaykovuori, K. Stadius, and J. Ryynanen, An energy-aware CMOS receiver front end for low-power 2.4-GHz applications, *IEEE Trans. Circuits Syst. I, Reg. Papers*, 57, 2675–2684, 2010.

25. Z. Lin, P.-I. Mak, and R. P. Martins, A 1.7 mW 0.22 mm² 2.4 GHz ZigBee RX exploiting a current-reuse blixer + hybrid filter topology in 65 nm CMOS, *IEEE ISSCC Dig. Tech. Papers*, 448–449, 2013.

26. Z. Lin, P.-I. Mak, and R. P. Martins, A 2.4-GHz ZigBee receiver exploiting an RF-to-BB-current-reuse blixer + hybrid filter topology in 65-nm CMOS, *IEEE J. Solid-State Circ.*, 49, 1333–1344, 2014.

27. J. L. Gonzalez, H. Solar, I. Adin et al., A 16-kV HBM RF ESD protection codesign for a 1-mW CMOS direct conversion receiver operating in the 2.4-GHz ISM band, *IEEE Trans. Microw. Theory Tech.*, 59, 2318–2330, 2011.

28. T.-K. Nguyen, V. Krizhanovskii, J. Lee et al., A low-power RF direct-conversion receiver/transmitter for 2.4-GHz-band IEEE 802.15.4 standard in 0.18 μm CMOS technology, *IEEE Trans. Microw. Theory Tech.*, 54, 4062–4071, 2006.

29. Z. Lin, P.-I. Mak, and R. P. Martins, A 0.14-mm², 1.4-mW, 59.4 dB-SFDR, 2.4-GHz ZigBee/WPAN receiver exploiting a "Split-LNTA + 50% LO" topology in 65-nm CMOS, *IEEE Trans. Microw. Theory Tech.*, 2014.

30. Z. Lin, P.-I. Mak, and R. P. Martins, A 0.5 V 1.15 mW 0.2 mm² sub-GHz ZigBee receiver supporting 433/860/915/960 MHz ISM bands with zero external components, *IEEE ISSCC Dig. Tech. Papers*, 164–165, 2014.

31. S. Bandyopadhyay and Anantha P. Chandrakasan, Platform architecture for solar, thermal, and vibration energy combining with MPPT and single inductor, *IEEE J. Solid-State Circ.*, 47, 2199–2215, 2012.

Section IV

Modulators and Synthesizers

16 All-Digital Phase-Locked Loops for Linear Wideband Phase Modulation

Salvatore Levantino, Giovanni Marzin, and Carlo Samori

CONTENTS

16.1 INTRODUCTION

The scaling of CMOS technology is changing the way analog circuits are designed. While intrinsic analog performances of transistors (e.g. maximum gain, flicker noise, matching) degrade with scaling, powerful digital signal processing assists analog circuits and boosts their performance at lower and lower area and power consumption. Furthermore, the digital assistance of analog subblocks reduces the overall cost of implementation and the time needed for the design as it benefits from the design automation of digital sections.

In this context, all-digital phase-locked loops (ADPLLs) have recently emerged as a valid alternative to traditional analog phase-locked loops (PLLs) in the most scaled technology nodes. ADPLLs not only are suitable as frequency synthesizers for wireless transceivers but also exhibit excellent performance when they are employed as direct phase/frequency modulators of the carrier in RF transmitters. The application of those blocks is not confined to communications based on phase modulation/frequency modulation (FM) of the carrier (such as Gaussian minimum-shift keying [GMSK]). Even in the case of nonconstant envelope modulations (such as those used in the most advanced cellular and wireless local area network [LAN] standards), the adoption of polar or outphasing (OP) transmitter architectures require low-noise, linear, and wideband phase modulators. Such unconventional transmitter architectures are being investigated and implemented in recent years to outperform the power efficiency of the more conventional Cartesian transmitter.

In a polar transmitter, as the one illustrated in the block diagram in Figure 16.1a, the carrier generated by a local oscillator (LO) passes through a phase modulator that provides the correct phase modulation to the carrier and then is fed to a highly efficient saturated power amplifier (PA), whose supply dynamically varies following the signal envelope [9]. Unfortunately, given the nonlinear relationship $\Theta(t) = \tan^{-1}[Q(t)/I(t)]$ between the signal phase and the Cartesian components of the signal, the bandwidth of $\Theta(t)$ is much larger than the original signal bandwidth. Depending on the specific modulation scheme, the bandwidth scales up by a factor between 5 and 10.

A variant of the polar transmitter concept is the original idea of the envelope elimination and restoration (EER) technique as proposed by Kahn [14]. The EER technique consists in extracting the amplitude and phase signals from the modulated carrier and then reapplying them to the carrier in a polar fashion. For this reason, the polar architecture in Figure 16.1a is sometimes referred to as direct polar (DP) [23], since the carrier is modulated directly via its polar components. Recent examples of DP implementations can be found in Refs. [1,3,17,36].

The OP scheme, originally proposed by Chireix [4], is illustrated in the block diagram in Figure 16.1b. The modulated carrier is separated into two carriers with constant envelope and differently modulated phases; thus, this scheme is often referred to as linear amplification with nonlinear components (LINC) [6]. As in the DP case, it relies on the use of saturated or almost saturated PAs that entail good efficiency, and, similarly, the nonlinear relationship between the Cartesian components and the two phases $\Theta_1(t)$ and $\Theta_2(t)$ makes the bandwidth of the phase signals much larger than the original one. Recent examples of high-performance OP transmitters can be found in Refs. [11,26].

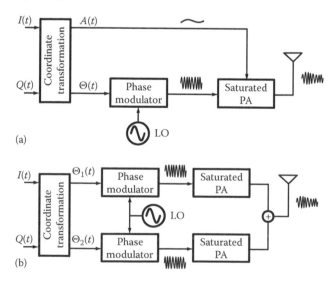

FIGURE 16.1 Simplified diagram of (a) polar and (b) OP radio transmitter.

Among the reasons for the renewed interest in these techniques, the main one is probably the inherently high efficiency of these architectures and the elimination of mixers and filters typically employed in Cartesian transmitters. This is, however, only a part of the story. Another key point is that these architectures seem naturally suited to the digitally intensive approach to circuit design discussed earlier. In both DP and OP systems, the phase modulator is typically driven directly by a digital word and acts effectively as a digital-to-phase converter. The DP approach can be pushed even to a *more digital* implementation by relying on a digital power amplifier (DPA) [5,37] even at watt power level [24].

16.2 PHASE MODULATORS

In both polar and OP transmitters, the phase modulation of the carrier must be performed by means of some sort of wideband phase modulator. The two most typical approaches to the design of a phase modulator are illustrated in Figure 16.2. The first one in Figure 16.2a is the direct phase modulation and is based on the generation of a certain number of phases of the carrier with a constant phase shift among them. The phases may be derived from the output of the frequency synthesizer through frequency division, through polyphase filtering, or by means of a regulated delay line. However, owing to power constrains, only a limited number of phases (typically no more than few tens) can be practically generated in those ways, while a fine phase resolution (in the order of few degrees) is often needed. The typical method to refine phase resolution is either to use an analog phase interpolator [36] or to digitally select one of the phases via a multiplexer (MUX) and dither the phase selection signal by means of a $\Delta\Sigma$ modulator [18], as shown in Figure 16.2a. The quantization noise introduced by the $\Delta\Sigma$ can be cancelled out by employing more advanced structures as proposed in Ref. [29]. Alternatively, the fine phase shift may be achieved by relying on a tuned resonator [35]. The main advantage of the direct phase modulator is

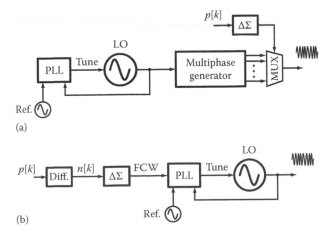

FIGURE 16.2 Typical implementation of the phase modulator: (a) direct and (b) indirect phase modulation.

the high achievable speed of the phase signal. On the other hand, one of the problems is that the generation of several high-frequency phases of the carrier and the subsequent multiplexing operation cost high-power dissipation, especially at high frequency. The linearity of the modulator, which is affected by the mismatches between the time shifts among the phases, is another critical issue [26].

Since the instantaneous frequency of the carrier is given by the first derivative of the carrier phase, an alternative way to implement a phase modulator is shown in the block diagram in Figure 16.2b. The phase signal is differentiated, and then the result is used to modulate the frequency control word of the PLL. This solution is in principle more power-efficient than the previous one, since only one high-frequency signal (i.e., the LO signal) has to be generated at a time. Furthermore, it theoretically entails better linearity thanks to the linearization provided by the PLL closed loop. The challenges of this approach are (1) the limited achievable speed of the modulation and (2) the wide range of the FM signal. As regards the first issue, the bandwidth of a PLL is constrained to be lower than about one-tenth of its reference frequency. Thus, the larger the reference frequency, the wider the bandwidth. Concerning dynamic range, the phase signal in polar and OP transmitters is in general unconstrained and can vary up to $\pm\pi$ [rad]. This variation of the phase may be required in just one clock period, being the clock period equal to the period T_{ref} of the reference signal.* This step increment of the phase signal can be produced by an angular frequency pulse with rectangular shape, whose integrated value is equal to $\pm\pi$. Thus, if $\varphi[kT_{ref}] - \varphi[(k-1)T_{ref}] = \pm\pi$, then $\omega[kT_{ref}] = \pm\pi/T_{ref}$, or equivalently, the frequency pulse must be as wide as $\pm1/(2T_{ref}) = \pm f_{ref}/2$ [Hz], being f_{ref} the reference frequency of the PLL. Hence, the larger the reference frequency, the wider the FM pulse and thus the PLL tracking range. Of course, in order not to degrade the signal-to-noise ratio of the produced modulation, a certain degree of linearity must be guaranteed over the whole FM range. We can conclude that a trade-off exists in the choice of the reference frequency and that the modulation bandwidth of the PLL must be traded with its linear modulation range.

The remainder of this chapter will be devoted to discuss the second approach and, in particular, the main issues related to the design of an all-digital-PLL-based phase modulator.

16.3   ALL-DIGITAL PLLs WITH DIRECT FM

The block diagram of an all-digital (or also known as digital) PLL is shown in Figure 16.3 [13]. The relative time difference between the edges of a reference signal and the frequency-divided output of a digitally controlled oscillator (DCO) is detected and digitized through the use of a time-to-digital converter (TDC). The TDC output is fed to a digital loop filter whose output is in turn used as the digital tuning control of the DCO. As in analog PLLs for frequency synthesis, the modulus control of the frequency divider is dithered by a digital $\Delta\Sigma$ modulator that interpolates the frequency control word (FCW) and realizes a fractional-N division.

---

* Such a large phase variation in one sample period occurs for instance in an unfiltered QPSK modulated carrier.

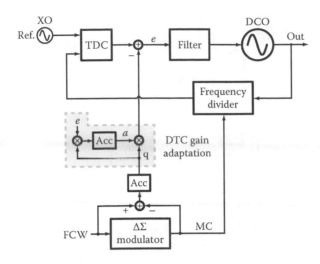

FIGURE 16.3 Diagram of a digital PLL.

Since the phase error induced by $\Delta\Sigma$ quantization is a deterministic signal, it can be cancelled out by subtracting it (after proper scaling) from the output of the TDC. This technique, often referred to as *digiphase technique* [15], was first introduced in analog fractional-N PLLs [10,15], and then applied to digital PLLs [13], where the calibration of the scaling factor (the a signal in Figure 16.3) is easily implemented by a digital loop (the shadowed block in Figure 16.3). At steady state, the a gain tends to the value that nulls the product between e and q and in turn the correlation between phase error and $\Delta\Sigma$ quantization error. This loop can be regarded as a simplified implementation of an least mean squares (LMS) algorithm [27].

Unfortunately, such an implementation of digital PLL requires a TDC with large number of bits, especially at wide PLL bandwidths. In fact, on the one hand, the TDC must accommodate the quantization error introduced by the $\Delta\Sigma$ and convert it linearly, and on the other hand, it must add no significant intrinsic quantization noise. The analyses and simulations reported in Ref. [16] show that the required number of TDC bits is as high as 10 to guarantee a level of residual fractional spurs below −60 dBc, assuming a first-order $\Delta\Sigma$ modulator dithering the divider modulus control. Furthermore, the integral nonlinearity (INL) of the TDC over this wide dynamic range must be as low as one least significant bit (LSB). If a third-order $\Delta\Sigma$ modulator is employed, the dynamic range of the TDC must be as wide as about seven DCO periods and requires about three additional bits. This numerical example leads to 13 equivalent bits required to the TDC. Of course, flash-type TDCs satisfying these specifications would produce excessive power dissipation. For this reason, different types of TDC (such as the oversampling or pipeline) [2,8,12,19,24,28] have been proposed and investigated, as well as several linearization techniques [31,32]. Although considerable effort has been done to improve power efficiency, the TDC still remains one of the main power hungry blocks of the loop.

To solve this issue and improve PLL noise/power trade-off, a new class of digital PLLs has been recently introduced [30,40]. The idea is to relax TDC specifications

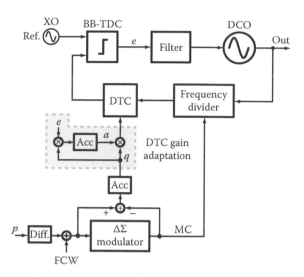

FIGURE 16.4 Diagram of a DTC-based digital PLL.

substantially by adding a digital-to-time converter (DTC) in the PLL feedback branch. The DTC allows to subtract the $\Delta\Sigma$ quantization error and thus reduce its amplitude down to DTC resolution. As a result, the required dynamic range of the subsequent TDC is reduced as well. In Ref. [40], a TDC and a DTC, both with equivalent number of bits equal to four, are employed. In this way, the implementation of the TDC, its linearity requirement, and power consumption are greatly reduced. The DTC is implemented as a delay-locked loop with 16 delay elements whose delay is automatically tuned. A more drastic simplification to TDC design and substantial improvement in the PLL noise/power figure of merit was presented in Ref. [30]. In that case, the resolution of the TDC is reduced to only one bit as shown in the block diagram in Figure 16.4. A single-bit TDC, also known as bang-bang TDC (BB-TDC), detects simply which one of the two input signals leads or lags the other one. In practice, a TDC with coarse mid-rise quantization (whose characteristic is shown in the upper plot in Figure 16.5) is employed to speed up lock transient. However, even in such a case, only two levels of the TDC characteristic are exploited when the PLL is in lock and thus the loop is equivalently controlled by a bang-bang detector. The design of the TDC, which is implemented as a time arbiter, is greatly simplified and power dissipation is substantially reduced. Furthermore, having a single-bit output, the nonlinearity issue is removed, like it happens to the comparator of an analog $\Delta\Sigma$ modulator. Of course, it introduces a hard nonlinearity in the loop, which potentially gives rise to limit cycles and in turn unwanted spurs in the spectrum.

In this context, the presence of random noise has a positive effect. Under proper conditions, it dithers the time delay detected by the BB-TDC avoiding limit cycles in the PLL. More specifically, as demonstrated in Ref. [7], when the random jitter induced by the thermal noise sources is larger than the deterministic error induced by quantization, the low-pass-filtered TDC output (i.e., averaged in time $\langle e \rangle$) as a function of the input phase error φ is given by the integration of the probability

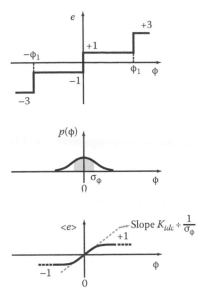

FIGURE 16.5   Bang-bang characteristic and stochastic gain.

density function $p(\varphi)$. This result is schematically shown in Figure 16.5. In the case of Gaussian phase error, the $\langle e \rangle$ curve versus φ is given by an error function, whose slope around zero is inversely proportional to the standard deviation σ_φ of the phase error. Thus, in practice, the mid-rise quantizer in the presence of dominant random noise at its input can be linearized and its linear gain is inversely proportional to the rms value of the input jitter. As discussed earlier, the linearization of the hard-limiting TDC characteristic holds as long as the random component of jitter dominates over the deterministic one. This mode of operation is referred as to *random-noise regime*, and it is opposed to the *limit-cycle regime* in which quantization error dominates over random jitter [41]. For an integer-N synthesized channel, the deterministic component is produced by the limited DCO frequency resolution and in turn by the truncation of the filter output word. Thus, in practice, this condition can be verified by improving DCO resolution [20].

By contrast, for a fractional-N channel, the quantization is dominated by the $\Delta\Sigma$ modulator dithering the divider modulus control. The resulting deterministic quantization error that is as wide as a few multiples of T_{dco} is always much larger than the typical random-noise jitter at the input of the detector. The latter cannot be increased, since it would raise the output jitter as well. Thus, in the digital PLL in Figure 16.4, instead of cancelling the $\Delta\Sigma$ quantization at TDC output, the cancellation is performed at TDC input via the DTC. The DTC allows us to subtract the phase error induced by the $\Delta\Sigma$ modulator, as in the PLL in Figure 16.3. Similarly, the amplitude of the subtracted signal is automatically estimated by means of an LMS-type feedback loop. However, in contrast to that case, since the quantization error at TDC input is cancelled out, there is no need for a multibit TDC. We can thus rely on a BB-TDC and yet get the detector to work in the random-noise regime.

The main advantages of the DTC-based PLL architecture over the conventional one are still the higher power efficiency and the lower design complexity. The BB-TDC is implemented as a single flip-flop and the DTC as a digital buffer stage with switched capacitor load. Thus, if we compare the cascade of the BB-TDC and the DTC against a multibit TDC, we have a single-time arbiter instead of many. Furthermore, the DTC-based topology is also favorable to the implementation of automatic predistortion algorithms [16,40]. This allows to adopt a segmented structure with scaled capacitor banks used as load of the DTC, yet reaching very good linearity.

Finally, the issue of the dependence of the BB-TDC gain on input jitter is solved by adopting the automatic bandwidth control scheme disclosed in Ref. [22]. In this way, the loop bandwidth and the whole frequency response of the PLL is repeatable, regardless of the spread of the analog parameters, including TDC gain. This solution eliminates the dependence of the bandwidth on input jitter.

16.4   WIDEBAND PHASE MODULATOR

To derive the transfer function of the digital PLL in Figure 16.4, we may refer to the linear phase model in the z-domain shown in Figure 16.6. This model is a discrete-time model at reference rate. The phase modulation signal p is differentiated and the result n is added to the FCW and fed to the $\Delta\Sigma$ modulator. For the sake of simplicity, we considered a first-order $\Delta\Sigma$ modulator, although in practice a second- or higher-order $\Delta\Sigma$ is employed. In the model of the $\Delta\Sigma$, the quantizer is linearized and a proper quantization error $q(z)$ is added. The output of the $\Delta\Sigma$ drives the modulus control of the divider. The latter is modeled as a

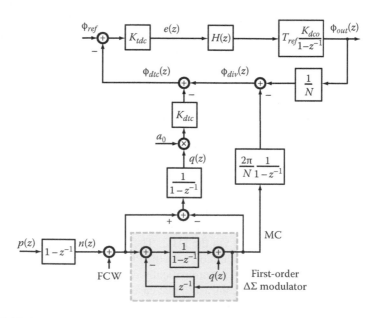

FIGURE 16.6   Equivalent model of the system in Figure 16.3.

discrete-time integrator, whose gain is $2\pi/N$. This follows from the fact that if MC is incremented by one, a time delay equal to the output period T_{DCO} is added to the *div* signal and in turn a shift of $2\pi/N$ is added to the phase. The DTC is simply modeled as a gain K_{dtc} [rad/bit] and an adder that adds/subtracts a certain phase shift in the feedback branch. The TDC is modeled as a detector of the phase error with gain K_{tdc} [bit/rad] and the DCO as an integrator with gain K_{dco} [rad/s/bit]. This means that as the DCO tuning word is incremented by one, the excess phase increases by $T_{ref} \cdot K_{dco}$ [rad] (after T_{ref} [s]). It is easy to verify that after the proper choice of the gain

$$a_0 \cdot K_{dtc} = \frac{2\pi}{N}, \qquad (16.1)$$

this scheme allows to cancel out the effect of the $\Delta\Sigma$ quantization q. The value of a_0 is set automatically in the background by the LMS loop (not shown in this model).

On the basis of the phase model, let us derive the transfer function from the phase modulation signal $p(z)$ to the output phase $\varphi_{out}(z)$. It is given by

$$F(z) = \frac{\phi_{out}}{p} = 2\pi \cdot \frac{G_{loop}(z)}{1 + G_{loop}(z)}, \qquad (16.2)$$

where G_{loop} is the PLL loop gain:

$$G_{loop}(z) = \frac{K_{tdc} K_{dco} T_{ref}}{N} \frac{H(z)}{1 - z^{-1}}. \qquad (16.3)$$

At low frequency, the transfer function $F(z)$ is the desired one, that is, 2π. It means that increasing MC by one produces a time shift at the output equal to one output period T_{dco} (i.e., 2π shift in the phase). The bandwidth of $F(z)$ equals the bandwidth of the PLL, which is constrained to be much lower than f_{ref}. This is a severe limitation, since in both polar and OP transmitters, the bandwidth of the phase modulation signal is typically larger than the signal bandwidth at radiofrequency.

The modulation bandwidth can be enlarged by performing a pre-emphasis of $p(z)$, which compensates for the bandwidth roll-off. This requires a tight matching between analog loop parameters and digital pre-emphasis. The alternative solution is the two-point injection scheme, originally proposed in Ref. [33], and recently adopted in Refs. [21,38,39]. In this topology, the phase modulation signal, after being differentiated, is injected both into the divider (in the feedback branch) and the DCO (in the forward branch). Applying this technique to the DTC-based digital PLL discussed so far, the resulting block scheme is the one sketched in Figure 16.7.

Neglecting the DTC with the *DTC gain adaption* block and the *DCO gain adaptation* block that play no role in the system response, the phase model of the system in the z-domain is shown in Figure 16.8. We will assume that the gain g has reached

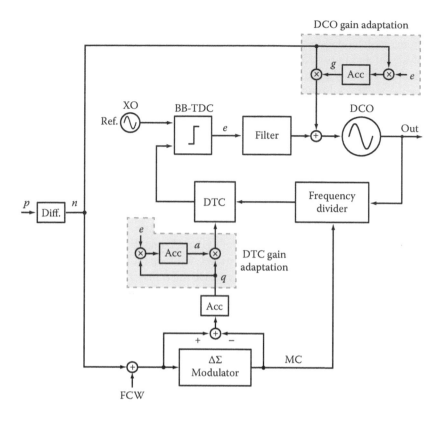

FIGURE 16.7 Two-point injection scheme.

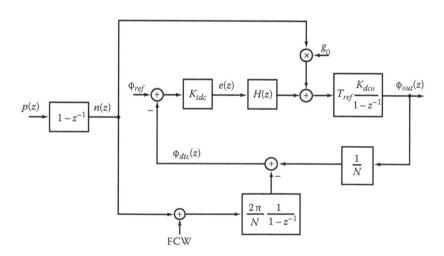

FIGURE 16.8 Equivalent model of system in Figure 16.7.

a constant value g_0. On the basis of this model, we can calculate again the transfer function from $p(z)$ to the output phase $\Phi_{out}(z)$. It becomes

$$T(z) = \frac{\Phi_{out}}{p} = 2\pi \cdot \frac{G_{loop}(z)}{1+G_{loop}(z)} + g_0 K_{dco} T_{ref} \cdot \frac{1}{1+G_{loop}(z)}, \qquad (16.4)$$

where G_{loop} has the same expression (16.3) as in the previous system.

While the first term in (16.4) has a low-pass shape in frequency, the second term, which comes from the DCO injection path, is high pass. Further than that, as long as the following equality holds

$$2\pi = g_0 K_{dco} T_{ref}, \qquad (16.5)$$

$T(z)$ becomes an all-pass transfer function, which allows the direct modulation of wideband signals. The modulation rate is simply limited by Nyquist theorem, since the system is sampled at f_{ref}. Hence, the maximum signal bandwidth (at baseband) is limited to $f_{ref}/2$. The previous equality (16.5) is guaranteed by regulating the gain g_0. Unfortunately, the DCO gain, K_{dco}, is hardly controllable as it varies over process and temperature and over the synthesized channel frequency. Any mismatch in the equality would result in a nonideal zero-pole cancellation and, therefore, in a linear distortion of the signal.

As a result, the automatic regulation of g shown in Figure 16.7 is essentially to guarantee the ideal all-pass shape of the system response. The working principle of the automatic regulation can be understood by noting that the transfer function from $n(z)$ to $e(z)$

$$e(z) = n(z) \cdot \left(2\pi f_{ref} - g_0 K_{dco}\right) \cdot \frac{G_{loop}(z)}{1+G_{loop}(z)} \cdot \frac{1}{K_{dco} H(z)} \qquad (16.6)$$

is proportional to the gain imbalance $(2\pi - g_0 K_{dco} T_{ref})$ between the two injection paths. This means that the correlation between the error signal $e(z)$ and $n(z)$ provides a measure of the gain imbalance. Thus, if we force this correlation to be null, we will get perfect gain balance and all-pass-shaped system response. The DCO gain adaptation in Figure 16.7 implements this concept: the gain g will tend to the g_0 value that nulls the correlation between e and n (i.e., the product between e and n). This automatic gain adaptation is implemented in the digital domain with insignificant resources or design effort.

16.5   LINEARITY ENHANCEMENT OF PHASE MODULATOR

Although the two-point injection and the automatic cancellation of gain mismatches between the two injection paths provide a very wide modulation bandwidth, we still have to discuss the second main challenge of the direct FM modulation of a PLL, that is, the linearity required. As discussed earlier, the largest variation of the output frequency must be equal to $\pm f_{ref}/2$ in order to produce a phase variation up to $\pm\pi$ in one reference clock. On the other hand, as already mentioned, the LSB of the DCO must be sufficiently small to guarantee insignificant truncation error introduced in the loop. In practical systems, this LSB must be in the order of 10 kHz/bit. Therefore, the DCO must cover a linear range of about 12-bit plus margin (if we assume f_{ref} equal to 40 MHz). DCO nonlinearity plays a major role on the modulation accuracy, degrading the error-vector magnitude (EVM) of the constellation.

The main source of nonlinearity in the DCO characteristic is the mismatch among switched capacitors in the resonator. Hence, a thermometric weighting should be preferred. Unfortunately, such a coding scheme would be impractical in a DCO with high number of bits. Therefore, several thermometrically weighted banks of capacitors must be used in the DCO resonator, as shown in the PLL in Figure 16.9, where the DCO resonator contains a varactor controlled by a resistor-string digital-to-analog converter (DAC), a fine and a coarse capacitor bank. The fine bank is implemented employing a single-unit capacitance C, while the coarse with unit capacitance $K \cdot C$ (with $K > 1$). This design reduces the number of capacitors and thus area occupation. As the number of interconnections goes down, the parasitic capacitance decreases and the tuning range of the DCO increases as well. In addition, different types of capacitors can be chosen for different banks, and in this way, the resonator quality factor can be optimized. This solution is somehow similar to the segmentation employed in the design of DACs, although in that case the LSBs are binary weighted. Unfortunately, the segmentation does not assure the DCO characteristic to be monotonic, since the gains of the different banks may be different.

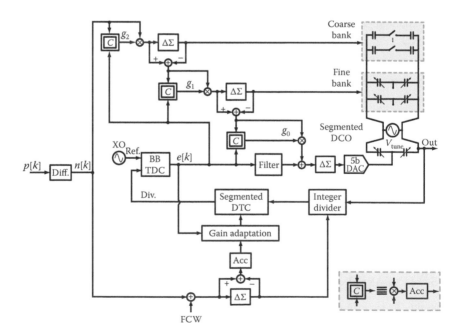

FIGURE 16.9 Schematic diagram of the phase modulator.

The adoption of the digitally intensive design in Ref. [21] proves useful even in this case. As illustrated in Figure 16.9, the FM signal $n(z)$ is multiplied by a proper factor g_2, and then it is quantized via a digital $\Delta\Sigma$ modulator and fed to the coarse capacitor bank. The quantization error of the $\Delta\Sigma$ quantizer is multiplied by a factor g_1 and fed to the fine capacitor bank. Finally, the residual quantization error is multiplied by another factor g_0 and fed to the input of the voltage-mode DAC driving the varactor bias. Let us assume that the gains relative to the three DCO banks from coarsest to finest are K_{dco2}, K_{dco1}, and K_{dco0}, respectively. It is possible to demonstrate that an all-pass-shaped response of the whole modulator systems is assured by the following equalities:

$$2\pi = g_2 K_{dco2} T_{ref} \tag{16.7}$$

$$g_1 \cdot K_{dco1} = K_{dco2} \tag{16.8}$$

$$g_0 \cdot K_{dco0} = K_{dco1}. \tag{16.9}$$

While the first equality expresses the balance between the gain of the two injection paths (divider and DCO), the other two equalities get the finer banks to have the same equivalent gain as the coarsest one. The three conditions may be satisfied by the proper choice of the three gains, g_2, g_1, and g_0, which are automatically regulated in background by the three LMS loop shown in Figure 16.9.

16.6   EXAMPLE OF PRACTICAL IMPLEMENTATION

An example of practical implementation of the phase modulator in Figure 16.9 is reported in Ref. [21]. Among the analog building blocks, the DTC is the most unconventional one and deserves some comment. As described earlier, it is in principle a buffer stage whose input–output delay is controlled by varying its capacitive load. In practice, the required number of bits of the DTC is equal to the required number of TDC bits in the conventional digital PLL architecture in Figure 16.3. As we have already observed, this number may vary between 10 and 13 in high-performance wireless applications. Thus, as for the DCO, a thermometric coding of the load capacitors of the DTC would be impractical. A power-efficient implementation of the DTC at circuit level is illustrated in Figure 16.10. It adopts a segmented architecture. The output of the integer divider (labeled as *div output*) is resampled by a cascade of three latches that are clocked by the two outputs of the differential DCO. Those outputs *dco* and $\overline{dco}$ are 180° out of phase. In this fashion, the time shift between the signals *P0* and *P1* is equal to $T_{dco}/2$, that is, half the DCO period. This value represents the coarser delay generated by the circuit or, in other words, the delay corresponding to the most significant bit (MSB), $a_0[k]$. The analog multiplexer (MUX) controlled by $a_0[k]$ selects one of the *P0/P1* signals and drives the capacitive load. The latter is divided into two banks of thermometrically weighted switched MOS capacitors: a fine and a coarse bank of 64 capacitors each.

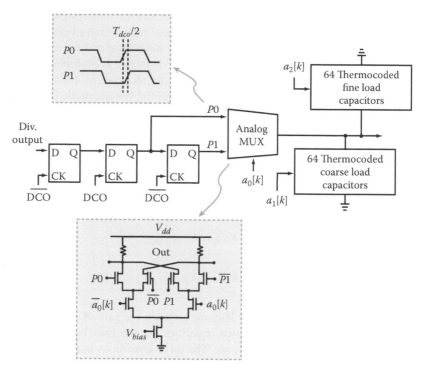

FIGURE 16.10 Circuit schematic of the DTC.

The two banks are controlled by $a_1[k]$ and $a_2[k]$, respectively. To guarantee the overlap between fine and coarse characteristics in the presence of process, voltage, and temperature (PVT) spreads, the delay range of the finest bank covers with margin the resolution of the coarse bank. Similarly, the total delay range of the coarse bank exceeds the delay generated by the MSB (i.e., the $T_{dco}/2$ delay). The digital signals $a_0[k]$, $a_1[k]$, and $a_2[k]$ are automatically regulated in the background by three LMS loops [30]. The differential topology of the MUX provides better immunity to supply bounces compared with a CMOS implementation.

The complete modulator, which is designed to synthesize carriers with frequency tunable between 2.9 and 4.0 GHz from a 40 MHz reference oscillator, embeds three additional blocks (not shown in Figure 16.9): a coarse frequency loop, an automatic bandwidth control (already mentioned), and another digital circuit running in background that corrects for the delay mismatches between the two injection paths [21]. The modulator described in Figure 16.9 has been fabricated in 65 nm CMOS technology. The die photograph is shown in Figure 16.11a. The area of the core circuits (excluding pads used for testing purposes) is slightly larger than 0.5 mm² and includes the pad drivers and the reference oscillator driven by an external crystal (XO). It is interesting to observe that most of the area is occupied by digital circuits, realized with standard cells. In addition to the described digital blocks, this section includes two different baseband signal generators, for GMSK and quadrature phase-shift keying (QPSK) modulation,

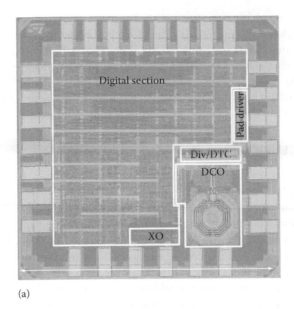

(a)

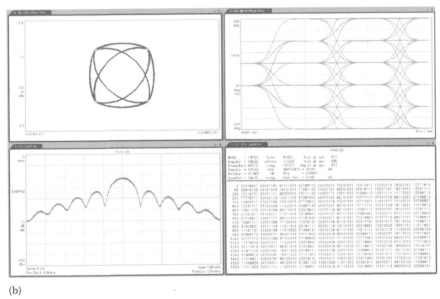

(b)

FIGURE 16.11 Implemented modulator: (a) chip photograph and (b) measured performance for a 20 Mb/s QPSK modulation.

respectively. The overall power consumption (excluding XO reference oscillator and pad drivers) is 5 mW from the 1.2 V supply.

Figure 16.11b shows the measured performance of the phase modulator (signal constellation, phase trellis diagram, spectrum), when the 3.6 GHz carrier is modulated with a 20 Mb/s QPSK signal. At this modulation rate, the measured EVM of the modulation, which accounts mainly for phase noise and nonlinearity of the modulator, is below 1.6% or −36 dB. The efficiency of the modulator that measures the energy required for transmitting one bit is equal to 0.25 nJ/bit. When the carrier is modulated with a continuous-phase modulation, such as a 10 Mb/s GMSK modulation, the EVM is still −36 dB and the energy per bit is 0.5 nJ/bit. In both cases, the efficiency and EVM achieved are much better than the figures obtained in phase modulators based on analog PLLs or phase-switching technique.

16.7 CONCLUSIONS

The scaling of CMOS technologies pushes toward an innovative design approach for analog blocks. The degradation of the intrinsic performance of those blocks induced by device scaling can be overcompensated by adding complex digital calibration algorithms within the building blocks and/or including nonlinear circuits (such as hard limiters) in high-precision RF and analog circuits. In this chapter, we have shown how an all-digital PLL can be designed to realize a linear and wideband phase modulator suitable for wireless transmitters. The DTC-based all-digital PLL exploits a BB-TDC to save power and adopts a two-point injection of the modulation signal to enlarge the modulation bandwidth. Several digital calibration loops operating in the background allow to regulate the gain of the DTC and to match the gains of the injection paths. The very good EVM and energy per bit achieved demonstrate that the digitally intensive design approach for analog circuits is becoming more than just a curiosity, at least for some important RF building blocks.

REFERENCES

1. Z. Boos, A. Menkhoff, F. Kuttner, M. Schimper, J. Moreira, H. Geltinger, T. Gossmann, P. Pfann, A. Belitzer, and T. Bauernfeind. A fully digital multimode polar transmitter employing 17b RF DAC in 3G mode. In *IEEE ISSCC Digest of Technical Papers*, San Francisco, CA, February 2011, pp. 376–377.
2. Y. Cao, P. Leroux, W. De Cock, and M. Steyaert. A 1.7 mW 11b 1-1-1 MASH ΔΣ time-to-digital converter. In *IEEE ISSCC Digest of Technical Papers*, San Francisco, CA, February 2011, pp. 480–482.
3. J. Chen, L. Rong, F. Jonsson, G. Yang, and L.-R. Zheng. The design of all-digital polar transmitter based on ADPLL and phase synchronized ΔΣ modulator. *IEEE J. Solid-State Circ.*, 47(5):1154–1164, 2012.
4. H. Chireix. High power outphasing modulation. *Proc. IRE*, 23(11):1370–1392, 1935.
5. D. Chowdhury, L. Ye, E. Alon, and A. M. Niknejad. An efficient mixed-signal 2.4-GHz polar power amplifier in 65-nm CMOS technology. *IEEE J. Solid-State Circ.*, 46(8):1796–1809, 2011.
6. D.C. Cox. Linear amplification with nonlinear components. *IEEE Trans. Comm.*, 22(12):1942–1945, 1974.
7. N. Da Dalt. Linearized analysis of a digital Bang-Bang PLL and its validity limits applied to jitter transfer and jitter generation. *IEEE Trans. Circ. Syst. I*, 55(11):3663–3675, 2008.
8. A. Elshazly, S. Rao, B. Young, and P. Hanumolu. A 13b 315fsrms 2 mW 500 MS/s 1 MHz bandwidth highly digital time-to-digital converter using switched ring oscillators. In *IEEE ISSCC Digest of Technical Papers*, San Francisco, CA, February 2012, pp. 464–466.

9. J. Groe. Polar transmitters for wireless communications. *IEEE Comm. Mag.*, 45(9): 58–63, 2007.

10. M. Gupta and B.-S. Song. A 1.8-GHz spur-cancelled fractional-N frequency synthesizer with LMS-based DAC gain calibration. *IEEE J. Solid-State Circ.*, 41(12):2842–2851, 2006.

11. M.E. Heidari, M. Lee, and A.A. Abidi, All-digital outphasing modulator for a software defined transmitter. *IEEE J. Solid-State Circ.*, 44(4):1260–1271, 2009.

12. J.-P. Hong, S.-J. Kim, J. Liu, N. Xing, T.-K. Jang, J. Park, J. Kim, T. Kim, and H. Park. A 0.004 mm² 250 µW ΔΣ TDC with time-difference accumulator and a 0.012 mm² 2.5 mW bang-bang digital PLL using PRNG for low-power SoC applications. In *IEEE ISSCC Digest of Technical Papers*, San Francisco, CA, February 2012, pp. 240–242.

13. C.-M. Hsu, M.Z. Straayer, and M.H. Perrott. A low-noise wide-BW 3.6-GHz digital ΔΣ fractional-N frequency synthesizer with a noise-shaping time-to-digital converter and quantization noise cancellation. *IEEE J. Solid-State Circ.*, 43(12):2776–2786, 2008.

14. L.R. Kahn. Single sideband transmission by envelope elimination and restoration. *Proc. IRE*, 40(7):803–806, 1952.

15. A.L. Lacaita, S. Levantino, and C. Samori. *Integrated Frequency Synthesizers for Wireless Systems*. Cambridge University Press, New York, 2007.

16. S. Levantino, G. Marzin, and C. Samori. An adaptive pre-distortion technique to mitigate the DTC non-linearity in digital PLLs. *IEEE J. Solid-State Circ.*, 49(8):1762–1772, 2014.

17. Y.-H. Liu, X. Huang, M. Vidojkovic, K. Imamura, P. Harpe, G. Dolmans, and H. De Groot. A 2.7 nJ/b multi-standard 2.3/2.4 GHz polar transmitter for wireless sensor networks. In *IEEE ISSCC Digest of Technical Papers*, San Francisco, CA, February 2012, pp. 448–449.

18. Y.-H. Liu, C.-L. Li, and T.-H. Lin. A 200-pJ/b MUX-based RF transmitter for implantable multichannel neural recording. *IEEE Trans. Microw. Theor. Tech.*, 57(10):2533–2541, 2009.

19. P. Lu, A. Liscidini, and P. Andreani. A 3.6 mW, 90 nm CMOS gated-Vernier time-to-digital converter with an equivalent resolution of 3.2 ps. *IEEE J. Solid-State Circ.*, 47(7):1626–1635, 2012.

20. G. Marucci, S. Levantino, P. Maffezzoni, and C. Samori. Analysis and design of low-jitter digital bang-bang phase-locked loops. *IEEE Trans. Circ. Syst. I*, 61(1):26–36, 2014.

21. G. Marzin, S. Levantino, C. Samori, and A. Lacaita. A 20 Mb/s phase modulator based on a 3.6 GHz digital PLL with −36 dB EVM at 5 mW power. *IEEE J. Solid-State Circ.*, 47(12):2974–2988, 2012.

22. G. Marzin, S. Levantino, C. Samori, and A. Lacaita. A background calibration technique to control bandwidth in Digital PLLs. In *IEEE ISSCC Digest of Technical Papers*, San Francisco, CA, February 2014, pp. 54–55.

23. E. McCune. Envelope tracking or polar—Which is it? *IEEE Microw. Mag.*, 13(4):34–56, 2012.

24. T. Nakatani, J. Rode, D.F. Kimball, L.E. Larson, and P.M. Ashbeck. Digitally controlled polar transmitter using a watt-class current-mode class-D CMOS power amplifier and Guanella reverse balun for handset applications. *IEEE J. Solid-State Circ.*, 47(5):1104–1112, 2012.

25. S. Pamarti, L. Jansson, and I. Galton. A wideband 2.4-GHz delta-sigma fractional-N PLL with 1-Mb/s in-loop modulation. *IEEE J. Solid-State Circ.*, 39(1):49–62, 2004.

26. A. Ravi, P. Madoglio, H. Xu, K. Chandrashekar, M. Verhelst, S. Pellerano, L. Cuellar et al. A 2.4-GHz 20–40-MHz channel WLAN digital outphasing transmitter utilizing a delay-based wideband phase modulator in 32-nm CMOS. *IEEE J. Solid-State Circ.*, 47(12):3184–3196, 2012.

27. A.H. Sayed. *Adaptive Filters*. Wiley-IEEE Press, Piscataway Township, NJ, 2008.

28. M.Z. Straayer and M.H. Perrott. A multi-path gated ring oscillator TDC with first-order noise shaping. *IEEE J. Solid-State Circ.*, 44(4):1089–1098, 2009.

29. P. Su and S. Pamarti. A 2.4 GHz wideband open-loop GFSK transmitter with phase quantization noise cancellation. *IEEE J. Solid-State Circ.*, 46(3):615–626, 2011.

30. D. Tasca, M. Zanuso, G. Marzin, S. Levantino, C. Samori, and A.L. Lacaita. A 2.9-to-4.0 GHz fractional-N digital PLL with bang-bang phase detector and 560fsrms integrated jitter at 4.5 mW power. *IEEE J. Solid-State Circ.*, 46(12):2745–2758, 2011.

31. E. Temporiti, C. Welti-Wu, D. Baldi, M. Cusmai, and F. Svelto. A 3.5 GHz wideband ADPLL with fractional spur suppression through TDC dithering and feedforward compensation. *IEEE J. Solid-State Circ.*, 45(12):2723–2736, 2010.

32. E. Temporiti, C. Weltin-Wu, D. Baldi, R. Tonietto, and F. Svelto. A 3 GHz fractional all-digital PLL with a 1.8 MHz bandwidth implementing spur reduction techniques. *IEEE J. Solid-State Circ.*, 44(3):824–834, 2009.

33. M.J. Underhill and R.I.H. Scott. Wideband frequency modulation of frequency synthesizers. *IEEE Electron. Lett.*, 15(13):393–394, 1979.

34. L. Vercesi, A. Liscidini, and R. Castello. Two dimensions vernier time-to-digital converter. *IEEE J. Solid-State Circ.*, 45(8):1504–1512, 2010.

35. G. Yahalom and J.L. Dawson. A low-Q resonant tank phase modulator for outphasing transmitters. In *IEEE Symposium on Radio Frequency Integrated Circuits Symposium (RFIC)*, Seattle, WA, June 2013, pp. 221–224.

36. L. Ye, J. Chen, L. Kong, P. Cathelin, E. Alon, and A. Niknejad. A digitally modulated 2.4 GHz WLAN transmitter with integrated phase path and dynamic load modulation in 65 nm CMOS. In *IEEE ISSCC Digest of Technical Papers*, San Francisco, CA, February 2013, pp. 330–331.

37. S.-M. Yoo, J.S. Walling, E.C. Woo, B. Jann, and D.J. Allstot. A switched-capacitor RF power amplifier. *IEEE J. Solid-State Circ.*, 46(12):2977–2987, 2011.

38. M. Youssef, A. Zolfaghari, B. Mohammadi, H. Darabi, and A. Abidi. A low-power GSM/EDGE/WCDMA polar transmitter in 65-nm CMOS. *IEEE J. Solid-State Circ.*, 46(12):3061–3074, 2011.

39. S.-A. Yu and P. Kinget. A 0.65-V 2.5-GHz fractional-N synthesizer with two-point 2- Mb/s GFSK data modulation. *IEEE J. Solid-State Circ.*, 44(9):2411–2425, 2009.

40. M. Zanuso, S. Levantino, C. Samori, and A.L. Lacaita. A wideband 3.6 GHz digital ΔΣ fractional-N PLL with phase interpolation divider and digital spur cancellation. *IEEE J. Solid-State Circ.*, 46(3):627–638, 2011.

41. M. Zanuso, D. Tasca, S. Levantino, A. Donadel, C. Samori, and A.L. Lacaita. Noise analysis and minimization in bang-bang digital PLLs. *IEEE Trans. Circ. Syst. II*, 56(11):835–839, 2009.

17 Hybrid Phase Modulators with Enhanced Linearity

Ni Xu, Woogeun Rhee, and Zhihua Wang

CONTENTS

17.1 INTRODUCTION

As the CMOS technology advances, a digital-intensive design is essential for low-cost multistandard transceiver systems. The $\Delta\Sigma$ phase-locked loop (PLL) enables digital phase modulation without requiring digital-to-analog converters (DACs) and RF up-converters, thus significantly simplifying the overall transmitter architecture. Since the typical PLL bandwidth is not wide enough to accommodate the required modulation symbol rate, a digital compensation method [1–4] or a two-point modulation method is employed [5–15] to overcome the bandwidth limitation. In the digital compensation method, the transfer function of the digital compensation filter needs to be matched well with that of the PLL. However, the loop dynamics of the PLL is highly sensitive to process and temperature variations, making the digital compensation method less attractive for on-chip modulation. On the other hand, the two-point modulation method shown in Figure 17.1 overcomes the bandwidth control problem by having two modulation paths. The high-pass transfer characteristic of the voltage-controlled oscillator (VCO) modulation and the complementary low-pass transfer characteristic of the frequency divider modulation form an all-pass transfer function regardless of the PLL bandwidth. Accordingly, the PLL bandwidth can be

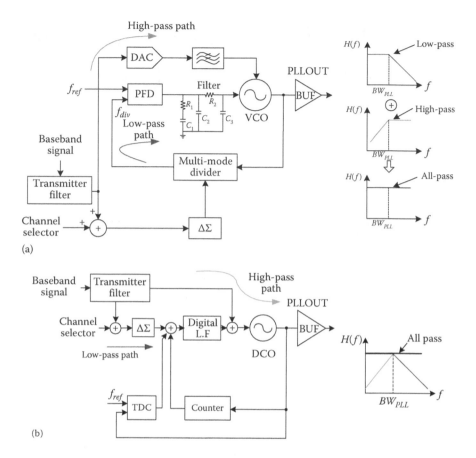

FIGURE 17.1 Two-point modulator: (a) with conventional $\Delta\Sigma$ PLL and (b) with ADPLL.

optimized for other performances. The challenging issues in the two-point modulation method lie in the accurate timing control and gain matching between the two modulation paths.

The all-digital PLL (ADPLL) along with the digitally controlled oscillator (DCO) provides robust two-point modulation with digital-assisted calibration [9–14]. A simplified block diagram of the ADPLL-based two-point modulator is shown in Figure 17.1b. The ADPLL-based modulation is considered a viable solution to realize a robust two-point modulator with the digital calibration of the gain and the delay mismatches. The ADPLL, however, requires a high-performance time-to-digital converter (TDC). To enhance the TDC linearity, a high-resolution delay cell and a high-sensitivity D-type flip-flop (DFF) are needed. Otherwise, the nonlinearity of the TDC becomes another source of the fractional spur and degraded in-band phase noise in the ADPLL. To improve the TDC performance, various methods are proposed in the literature, including a gated ring oscillator TDC with noise shaping [15], a time-amplified TDC [16], a dithered TDC [17], a nonuniform stepping TDC [18], a bang-bang phase detector (BBPD) with sub-gate delay cells [19], and a TDC with an injection-locked ring VCO [20]. In this work, we introduce a hybrid-loop two-point modulator that utilizes a TDC-less semidigital $\Delta\Sigma$ PLL for low-power low-complex design, while offering a digital-intensive architecture for technology scalability and digital-assisted calibration.

17.2 HYBRID-LOOP TWO-POINT MODULATOR

17.2.1 SEMIDIGITAL PLL WITH HYBRID-LOOP CONTROL

Various TDC-less semi-digital PLL architectures have been recently proposed to achieve linear loop dynamics while removing a large integration capacitor [21–27]. Figure 17.2 shows a conceptual diagram of the mixed-mode loop control shown in differential implementation. The digital/voltage-controlled oscillator (D/VCO) is used to accommodate both analog and digital controls. For the proportional-gain path, the conventional analog control with the PFD and the charge pump followed by the second-order passive loop filter is employed. Since the capacitance values for high-order poles are not high, a small area of the loop filter is realized. As for the integral path, digital implementation is done with a BBPD and a finite-state machine (FSM) to compensate for the limited frequency tracking capability of the proportional-gain path, forming a type-II PLL. The $\Delta\Sigma$ modulator is used to provide a fine frequency resolution as done in the ADPLL.

As for the D/VCO gain variation, we consider two paths: a digital control path and an analog control path. The digital control path is the same as the digital input of the conventional DCO, in which the digital gain calibration can be performed. The gain variation of the analog path is more problematic. However, the analog gain nonlinearity of the D/VCO is much more relaxed than that of the conventional VCO. Figure 17.3 illustrates how the analog gain variation is reduced in the D/VCO. Since the digital control path with the $\Delta\Sigma$ modulation provides continuous frequency tracking with very fine resolution, the effective analog control voltage range is significantly reduced. For example, the actual tuning voltage range in the analog

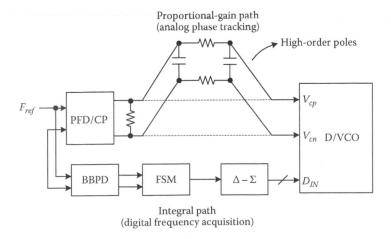

FIGURE 17.2 Hybrid-loop control with D/VCO.

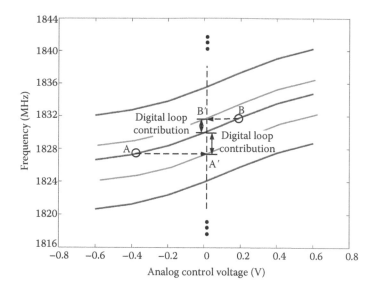

FIGURE 17.3 D/VCO tuning curves with hybrid control.

path can be less than ±50 mV over entire tuning range if the digital control path sets the desired output frequency of the D/VCO with fine resolution. Such a small control voltage range apparently enhances the linearity of the analog gain and also improves the linearity of the charge pump. Since a large capacitor in the integral path is replaced with the digital FSM, an overdamped loop dynamics can be easily designed. As a result, the absolute gain variation of the analog control path can be compensated for by the programmable charge pump current to a certain degree. The settling time can be improved by designing an automatic frequency control (AFC) block in the digital path of the D/VCO, which is similar to the ADPLL.

Offering the direct digital modulation in the digital path and the good linearity in the analog path with the small control voltage range, the semidigital ΔΣ PLL can be a good alternative architecture for the two-point modulation, while avoiding the needs of the high-complex TDC. The semidigital-based modulator for wireless applications is proposed in Ref. [22], but only a simple triangular modulation has been demonstrated. The PLL also suffers from high sensitivity to the supply noise since the hybrid-loop architecture is done with the single-ended charge pump followed by the dc-coupled VCO input, which has a direct coupling path to the supply voltage.

17.2.2 Two-Point Modulation

Figure 17.4 shows a block diagram of the proposed two-point modulator based on the semidigital ΔΣ PLL. Since the high-frequency modulation path is the digital input path of the D/VCO, the DAC is not required. The high-frequency modulation path is considered the same as that of the ADPLL-based two-point modulator. The low-frequency modulation is performed by the ΔΣ modulated frequency divider that

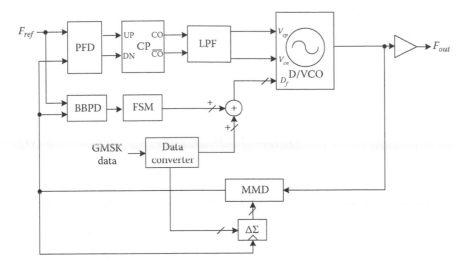

FIGURE 17.4 Proposed two-point modulator.

also takes the digital input directly. Hence, the GMSK-modulated digital data are directly fed to the input of the D/VCO and the input of the $\Delta\Sigma$ modulator without any digital-to-analog conversion. Therefore, the proposed hybrid-loop architecture has nearly the same function as the all-digital architecture as far as the two-point modulation is concerned.

A linear model of the proposed two-point modulator is shown in Figure 17.5, where G_L is the gain parameter in the low-frequency modulation, which sets a normalized frequency deviation for the divider modulation, G_H is the gain parameter in the high-frequency modulation, which sets a compensated gain before the VCO input, τ_L and τ_H stand for the delay control for low- and high-frequency modulation paths, respectively, N is the division ratio, β is the integral-gain coefficient, K_{VCO} is the gain of the analog controlled oscillator, $H_{LPF}(s)$ is the gain of the loop filter, and K_{BBPD} is the gain of the BBPD. Let $H_a(s)$ and $H_d(z)$ be the transfer functions of

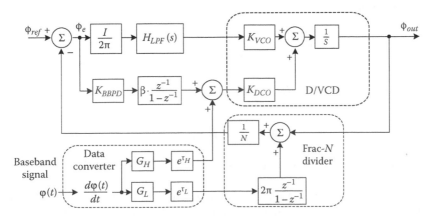

FIGURE 17.5 Linear model of hybrid-loop two-point modulator.

the analog modulation path and the digital modulation path, respectively. Then, the overall transfer function $A(s, z)$ of the semidigital PLL is given by

$$A(s,z) = \frac{H_a(s) \cdot K_{VCO} + H_d(z) \cdot K_{DCO}}{N} \cdot \frac{1}{s}, \tag{17.1}$$

where $H_a(s)$ and $H_d(z)$ are

$$H_a(s) = \frac{I}{2\pi} \cdot H_{LPF}(s), \tag{17.2}$$

and

$$H_d(z) = K_{BBPD} \cdot \frac{\beta \cdot z^{-1}}{1 - z^{-1}}. \tag{17.3}$$

Then, the transfer functions for the high-frequency modulation $H_H(s)$ and the low-frequency modulation $H_L(s, z)$ are given by

$$H_H(s,z) = \frac{G_H \cdot e^{s\tau_H} \cdot K_{DCO}}{1 + A(s,z)} \cdot \frac{1}{s} \cdot s, \tag{17.4}$$

and

$$H_L(s,z) = \frac{G_L \cdot e^{s\tau_L} \cdot A(s)}{1 + A(s,z)} \frac{2\pi \cdot z^{-1}}{1 - z^{-1}} \cdot s. \tag{17.5}$$

By applying the s-domain transform with the reference clock period T as follows [28],

$$\frac{z^{-1}}{1 - z^{-1}} \approx \frac{1}{j2\pi fT} = \frac{1}{sT}, \quad \text{for } f \ll \frac{1}{T}. \tag{17.6}$$

Then, we obtain

$$H_H(s,z) + H_L(s,z) = \frac{G_H \cdot e^{s\tau_H} \cdot K_{DCO}}{1 + A(s,z)} + \frac{G_L \cdot e^{s\tau_L} \cdot A(s)}{1 + A(s,z)} \frac{2\pi}{T}. \tag{17.7}$$

If we assume that $\tau_H = \tau_L$, $G_H = 1/K_{DCO}$, and $G_L = T/2\pi$ with the reference clock period T, then

$$H_H(s,z) + H_L(s,z) = 1, \tag{17.8}$$

proving that the all-pass transfer function is also valid for the hybrid-loop two-point modulator.

Figure 17.6 shows the behavior simulation results of the semi-digital PLL with 270 kb/s GMSK two-point modulation. The PLL bandwidth is set to 40 kHz. As shown in Figure 17.6, the analog differential control voltage goes to 0 V as the digital output

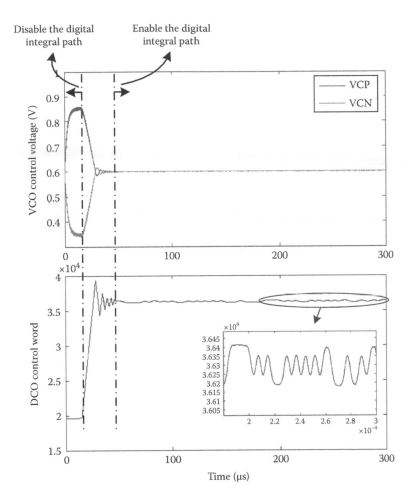

FIGURE 17.6 Transient simulation results of the D/VCO control voltage and digital control bits for the settling behavior.

in the integral path is settled, which supports the previous discussion that the analog control voltage range can be very small, resulting in good linearity of both the D/VCO and the charge pump. The magnified plot for the D/VCO control word shows that the control word clearly reflects the GMSK modulation, verifying that the two-point modulation is successfully performed with the hybrid-loop architecture.

In the two-point modulation, the delay and gain mismatches between the high-frequency modulation path and the low-frequency modulation path, and the gain nonlinearity of the D/VCO degrade the overall performance. With the reference clock frequency of 26 MHz, the delay mismatch is tolerable by implementing a high-frequency DFF to adjust the delay mismatch between the low-frequency modulation path and the high-frequency modulation path. The gain mismatch can be also calibrated by adjusting the gain of the baseband signal. The D/VCO gain nonlinearity is relatively difficult to be calibrated. For the digital-path calibration, the complex calibration method as done in the ADPLL-based modulation is still required [11,12,29,30]. For the analog-path calibration, the analog control is assumed to have good linearity with the small control voltage range, so the calibration of the charge pump mismatch can be mainly considered.

17.3   CIRCUIT DESIGN

17.3.1   270.83 kb/s GMSK Hybrid-Loop Phase Modulator

Figure 17.7 shows a block diagram of the proposed hybrid-loop two-point modulator with 270.83 kb/s GMSK modulation. A 900 MHz output frequency is generated from a 1.8 GHz D/VCO. A reference frequency of 26 MHz is used for a GMSK data converter block that performs proper code mapping for two-point modulation. The digital baseband modulation is performed by a field programmable gate array (FPGA) through a GMSK lookup table, IQ-to-PM signal translation, and differentiation of the PM signal. In the high-pass modulation path, the data are added with the digital integral path for direct D/VCO modulation.

Figure 17.8 illustrates a basic structure of the D/VCO control paths. Since the fully differential charge pump and loop filter are used, a differential input is designed for the analog control path. The digital control path is done with 13 control bits where 7 bits from the LSB are modulated by a second-order ΔΣ modulator to generate an oversampled output. The clock frequency of the modulator is the one-eighth of the D/VCO frequency, or about 225 MHz. The quantization noise of the ΔΣ modulator at the output of the D/VCO can be estimated from Ref. [29]. For the minimum frequency step of 580 Hz and the clock frequency of 225 MHz, the phase noise contribution of the ΔΣ modulator at the D/VCO output would be about −151 dBc/Hz at 400 kHz offset frequency and −183 dBc/Hz at 20 MHz offset frequency, which are much lower than the D/VCO noise.

Figure 17.9 shows the schematic of the D/VCO core in 65 nm design. To achieve good power supply rejection, a current bias and a cross-coupled pair are designed

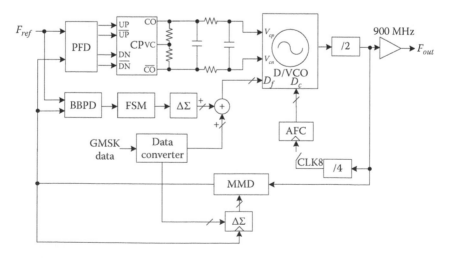

FIGURE 17.7 Proposed two-point modulator with 270 kb/s GMSK modulation.

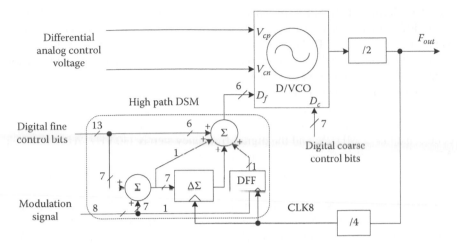

FIGURE 17.8 D/VCO control paths.

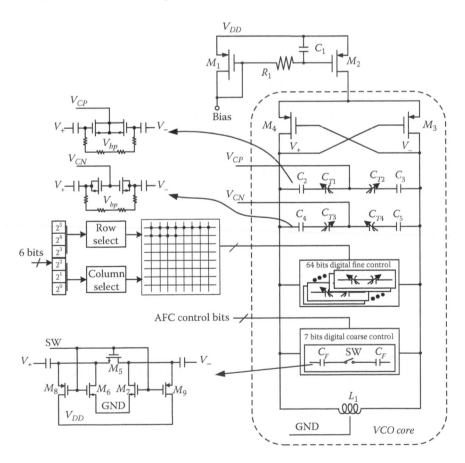

FIGURE 17.9 Schematic of D/VCO core in 65 nm design.

with PMOS transistors. Since the fully differential charge pump and loop filter are used, differential-input varactors C_{T1}–C_{T4} are designed for the analog control path. Metal-insulator-metal (MIM) capacitors C_2–C_5 are used to provide dc bias through V_{CP} and V_{CN} to increase the linear region of the varactors. A 7-bit binary-coded MIM capacitor array is employed as a coarse tuning block to extend the tuning range over process and temperature variations. For fine tuning, a 64-bit thermometer-coded varactor array is used. The thermometer-coded array offers good matching and linearity with guaranteed frequency monotonicity. The sensitivity of the analog control path is 22 MHz/V, and the digital frequency step is 140 kHz. After the 7-bit $\Delta\Sigma$ modulation, the effective frequency resolution of the D/VCO is reduced to be about 580 Hz. With the differential spiral inductor of 2 nH, the tuning range of 1.38–1.90 GHz is achieved.

Figure 17.10 shows the details of the digital integral path. The FSM works at the reference clock frequency of 26 MHz. It contains a 21-bit accumulator and a 1-bit selector, which is controlled by the output of the BBPD. The FSM output increases or decreases by a during each clock period. The value of a is controllable to caliber the gain of the integral path. The bandwidth of the digital path must be much smaller than that of the analog path, so that the digital integral path mostly performs frequency acquisition like a large signal behavior in the overall system, while the analog proportional-gain path performing phase tracking like a small signal behavior in the linear system.

Figure 17.11 shows the schematic of the programmable charge pump followed by a differential loop filter. The differential loop filter is designed to minimize noise coupling and enhance the linearity of the charge pump output. The charge pump contains four switches to control the current outputs from 40 to 160 µA in differential. Current sources with a cascaded output stage improves linearity and matching

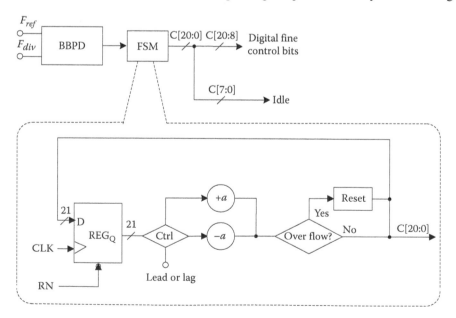

FIGURE 17.10 Detailed diagram of integral-gain path.

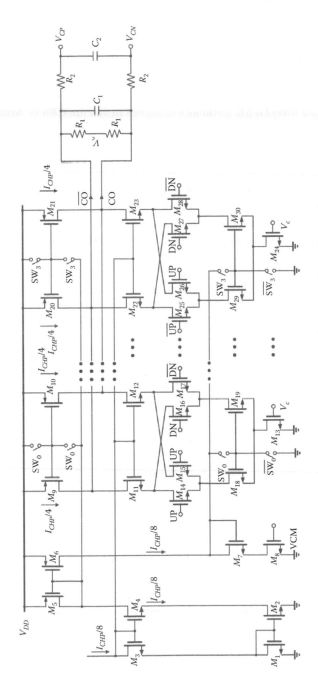

FIGURE 17.11 Schematic of differential programmable charge pump with a differential loop filter.

with high output impedance. NMOS transistors M_8, M_{13}, and M_{24} are in triode region for common-mode feedback to set the common-mode voltage VCM of the differential charge pump.

17.3.2 2.08 Mb/s GFSK HYBRID-LOOP PHASE MODULATOR

Thanks to the TDC-less architecture, a digital-intensive modulator can be designed without using an advanced CMOS technology. Figure 17.12 shows the block diagram of the 2 GHz 2.08 Mb/s GFSK modulator designed in 180 nm CMOS. Similar to the previous design, the integral path is implemented by a BBPD, an FSM, a high-path $\Delta\Sigma$ modulator, and a DAC for frequency tracking. In order to reduce the nonlinear effect from the DAC control path, the modulation path is controlled separately as shown in Figure 17.12.

Compared with 65 nm CMOS, the DCO design in 180 nm CMOS has difficulty in generating the fine frequency resolution with a wide linear tuning range. Therefore, a current-steering DAC-based digital control is implemented as shown in Figure 17.13. The 6-bit DAC has the 3-bit thermometer code in the MSB part and 3-bit binary code in the LSB part with the output range of 0.35–1.25 V. The DAC is designed to work at one-fourth of the VCO frequency. The analog VCO gain controlled by the differential loop filter is about 25 MHz/V. The frequency range controlled by the DAC is about 32 MHz with the gain of 500 kHz/LSB. The differential spiral inductor of 2.4 nH is realized by two asymmetric inductors in series as illustrated in Figure 17.13. With the 5-bit binary-weight MIM capacitor array for the coarse frequency control, the tuning range is about 1.7–2.1 GHz. The fine frequency resolution for the high-pass modulation signal is about 125 kHz/LSB. The tuning range of 4 MHz is wide enough for the 2.08 Mb/s GFSK modulation whose frequency deviation is 665 kHz.

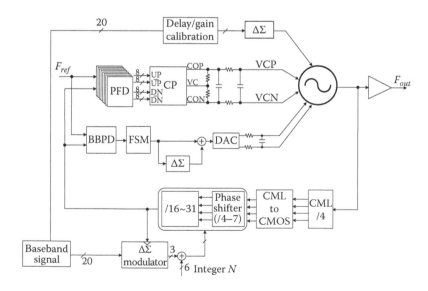

FIGURE 17.12 Proposed two-point modulator with 2 Mb/s GFSK modulation.

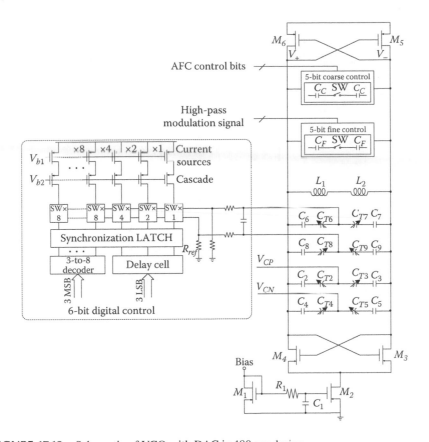

FIGURE 17.13 Schematic of VCO with DAC in 180 nm design.

17.4 EXPERIMENTAL RESULTS

17.4.1 270.8 kb/s GMSK Hybrid-Loop Phase Modulator

A prototype 1.8 GHz $\Delta\Sigma$ PLL with the hybrid-loop control for 270 kb/s GMSK two-point modulation is implemented in 65 nm CMOS. The chip micrograph is shown in Figure 17.14. It occupies an active area of 0.47 mm^2 in which the area of the D/VCO is about 0.35 mm^2 and the area of the analog loop filter is less than 0.018 mm^2. The resistor R_1 for the proportional gain and the capacitor C_1 for the third pole are external to have flexible bandwidth control for the prototype PLL testing. The equivalent on-chip area of both passive devices ($R_1 = 5$ kΩ, $C_1 = 75$ pF) is 0.024 mm^2, which is considered negligible even with fully on-chip implementation. The GMSK modulated data generated by the FPGA are fed to the low-frequency modulation path and the high-frequency modulation path of the $\Delta\Sigma$ PLL.

Figure 17.15 shows the measured output spectrum at 913.2 MHz with and without the digital integral path. When the digital integral path is disabled, the PLL has the same loop dynamic behavior as the type-I PLL, which exhibits a large static phase offset or a large reference spur for center-deviated output frequencies.

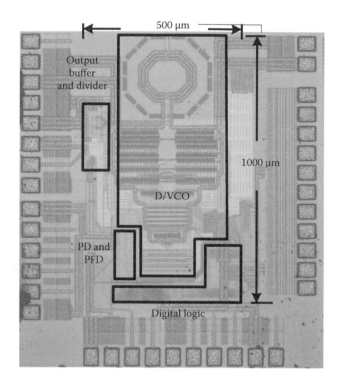

FIGURE 17.14   Chip micrograph of the GMSK modulator.

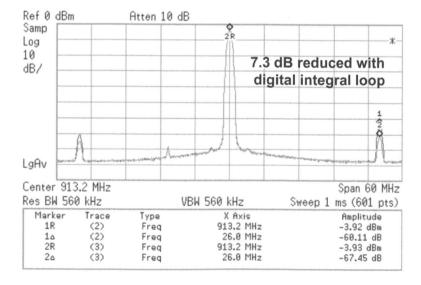

FIGURE 17.15   Measured reference spur with and without digital integral path.

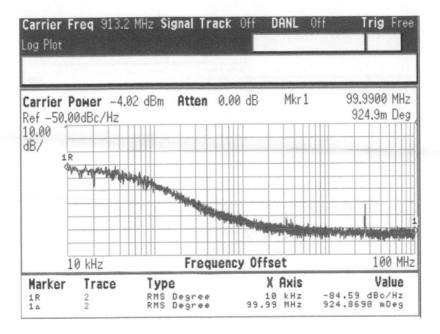

FIGURE 17.16 Measured phase noise performance.

When the digital integral path is enabled, the reference spur is reduced from −60.1 to −67.4 dBc, verifying that the digital integral path carries the frequency information like the conventional type-II PLL.

Figure 17.16 shows the measured closed-loop phase noise at 913.2 MHz output frequency. The measured in-band phase noise is about −85 dBc/Hz with the loop bandwidth of 40 kHz. The out-of-band phase noise is −117 dBc/Hz at 400 kHz offset frequency. When the phase noise is integrated from 10 kHz to 100 MHz, the integrated phase error is $0.92°_{rms}$. To check the capability of the linear bandwidth control with the programmable charge pump, the charge pump current is reduced by half and phase noise is compared to the original one as shown in Figure 17.17. The loop parameters are the same as the previous condition in Figure 17.17 except the high-order poles are slightly reduced to have the optimum phase noise performance for both bandwidths. The upper plot represents the phase noise measured with the charge pump current of 80 μA, and the bottom plot is measured with the charge pump current of 160 μA. The measured loop bandwidths are 20 and 40 kHz, respectively. It shows that the loop bandwidth is linearly changed corresponding to the charge pump variation. Note that noise peaking near the bandwidth is not observed for both cases, showing that the overdamped loop is obtained with the digital integral path.

Figure 17.18 shows an output spectrum with 270.833 kb/s GMSK modulation. The measured spectral density at 200 and 400 kHz offsets are −36 and −63.5 dBc, respectively, with the resolution bandwidth of 30 kHz, showing that the modulated output spectrum satisfies the GSM transmitter requirement with 3.5 dB margin at 400 kHz offset frequency.

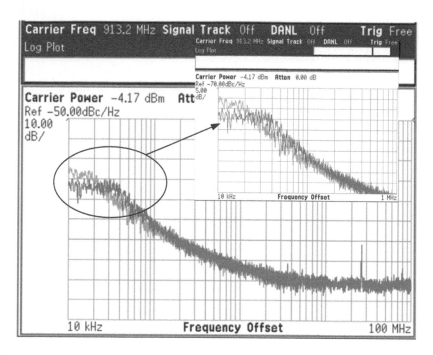

FIGURE 17.17 Measured phase noise with two different bandwidths.

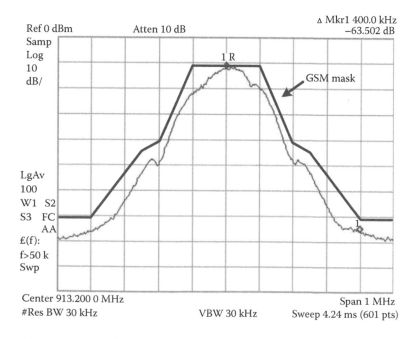

FIGURE 17.18 Measured output spectrum with GMSK modulation.

The proposed modulator consumes 6.9 mW from a 1 V supply where the D/VCO consumes 5.1 mW and the digital blocks consume 0.9 mW. The power and area of the proposed modulator are comparable to those of recent ADPLL-based modulators. It shows that the proposed hybrid-loop architecture offers a good way of realizing digital modulation while avoiding the complicated design effort for the high-performance TDC.

17.4.2   2 Mb/s GFSK Hybrid-Loop Phase Modulator

A 2.08 Mb/s hybrid-loop GFSK modulator is implemented at 2 GHz carrier frequency with 180 nm technology. The chip micrograph is shown in Figure 17.19. The active area of the modulator is 1.63 mm². The 2.08 Mb/s GFSK baseband modulation signal is generated by FPGA with the modulation index of 0.32.

Figure 17.20 shows the phase noise of the loop at 2.005 GHz with 50 MHz reference clock. The in-band noise performance is about −88 dBc/Hz with 100 kHz bandwidth, and the out-band noise at 3 MHz offset is about 131 dBc/Hz. The fractional spur at 1.3 MHz is about −71 dBc with the digital integral path enabled as shown in Figure 17.21. It shows about 13 dB improvement of the fractional spur with the hybrid loop control. Figure 17.22 shows the reference spur performance of −80 dBc with the PLL bandwidth of 100 kHz.

Figures 17.23 and 17.24 show the measured output spectra, the constellations, the EVM values, frequency deviation, and the eye diagrams with 2.08 Mb/s GFSK modulation. The RMS EVM value of 3.31% and the peak EVM value of 9.4% are achieved. The semidigital modulator draws a current of 26 mA from a 1.8 V supply.

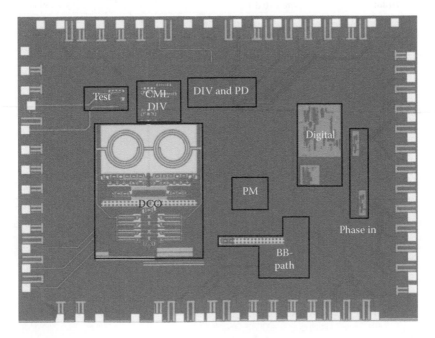

FIGURE 17.19   Chip micrograph of the GFSK modulator.

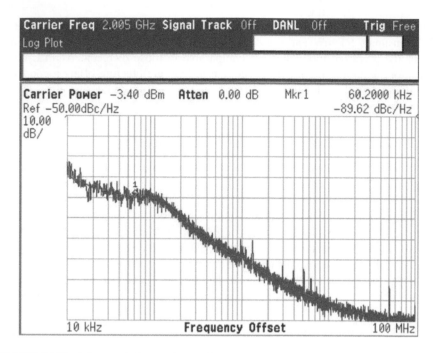

FIGURE 17.20   Measured phase noise performance of the 2 GHz semidigital PLL.

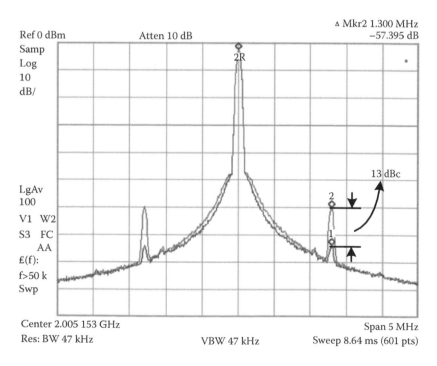

FIGURE 17.21   Measured phase fractional spur improvement of the 2 GHz semidigital PLL.

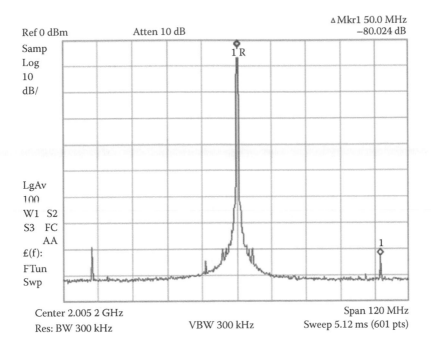

FIGURE 17.22   Measured output reference spur performance for the 2 GHz semidigital PLL.

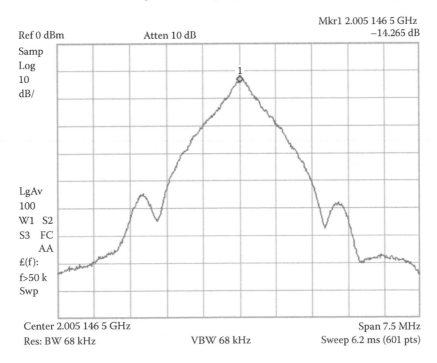

FIGURE 17.23   Measured output spectrum with 2.08 Mb/s GFSK modulation.

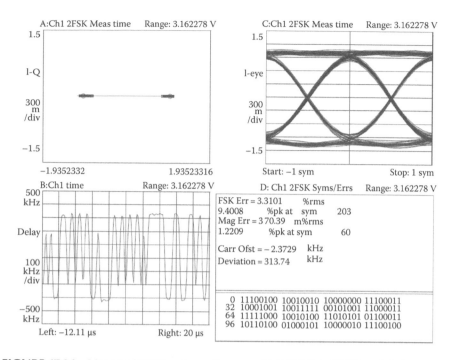

FIGURE 17.24   Measured EVM and eye diagram with 2.08 Mb/s GFSK modulation.

17.5   CONCLUSION

In this chapter, we present a hybrid-loop two-point phase modulator by utilizing a semi digital $\Delta\Sigma$ PLL and avoiding the complex TDC, which can also enable the digital-assisted calibration and offer technology scalability like the ADPLL. The simulation results and analyses show that nearly the same features of the ADPLL-based modulator can be achieved with the hybrid-loop architecture. The TDC-less semi-digital $\Delta\Sigma$ PLL with the fully differential linear gain path is implemented in 65 nm CMOS, consuming a 6.9 mW from a 1 V supply. When the 270.833 kb/s GMSK modulation is applied, the proposed hybrid modulator meets the spectrum mask requirement for GSM standards with 3.5 dB margin at 400 kHz offset frequency. In addition, we present a 2 GHz hybrid loop modulator implemented in 180 nm CMOS. The RMS EVM achieves 3.31% with 2.08 Mb/s GFSK modulation, proving that the proposed TDC-less architecture is also suitable for the low-cost design. The experimental results show that the proposed hybrid-loop architecture offers an alternative way of realizing digital modulation while avoiding the complicated design effort for the high-performance TDC.

REFERENCES

1. M.H. Perrott, T.T. Tewksbury, and C.G. Sodini, A 27-mW CMOS fractional-N synthesizer using digital compensation for 2.5-Mb/s GFSK modulation, *IEEE J. Solid-State Circ.*, 32(12), 2048–2060, 1997.
2. D.R. McMahill and C.G. Sodini, A 2.5-Mbls GFSK 5.0-Mb/s 4-FSK automatically calibrated Σ–Δ frequency synthesizer, *IEEE J. Solid-State Circ.*, 37(1), 18–26, 2002.

3. H. Darabi, A. Zolfaghari, H. Jensen et al., A fully integrated quad-band GPRS/EDGE radio in 0.13 μm CMOS, in *IEEE International Solid-State Circuits Conference (ISSCC) Digest of Technical Papers*, San Francisco, CA, February 2008, pp. 206–207.

4. C.-H. Wang, P.-Y. Wang, L.-W. Ke et al., A direct digital frequency modulation PLL with all digital on-line self-calibration for quad-band GSM/GPRS transmitter, in *Proceedings of IEEE Symposium on VLSI Circuits*, Kyoto, Japan, June 2009, pp. 190–191.

5. C. Durdodt, M. Friedrich, C. Grewing et al., A low-IF RX two-point ΣΔ-modulation TX CMOS single-chip Bluetooth solution, *IEEE Trans. Microw. Theor. Tech.*, 49(9), 1531–1537, 2001.

6. B. Neurauter, G. Marzinger, A. Schwarz, and R. Vuketich, GSM900/DCS1800 fractional-N modulator with two-point modulation, in *Proceedings of IEEE MTT-S International Microwave Symposium Digest*, Vol. 1, Seattle, WA, June 2002, pp. 425–428.

7. S.T. Lee, S.-J. Fang, D.J. Allstot, A. Bellaouar, A.R. Fridi, and P.A. Fontaine, A quad-band GSM-GPRS transmitter with digital auto-calibration, *IEEE J. Solid-State Circ.*, 39(12), 2200–2214, 2004.

8. S. Lee, J. Lee, H. Park, K.-Y. Lee, and S. Nam, Self-calibrated two-point Delta-Sigma modulation technique for RF transmitters, *IEEE Trans. Microw. Theor. Tech.*, 58(7), 1748–1757, 2010.

9. R.B. Staszewski, J.L. Wallberg, S. Rezeq et al., All-digital PLL and transmitter for mobile phones, *IEEE J. Solid-State Circ.*, 40(12), 2469–2482, 2005.

10. T. Tokairin, M. Okada, M. Kitsunezuka, T. Maeda, and M. Fukaishi, A 2.1-to-2.8-GHz low-phase-noise all-digital frequency synthesizer with a time-windowed time-to-digital converter, *IEEE J. Solid-State Circ.*, 45(12), 2582–2590, 2010.

11. M. Youssef, A. Zolfaghari, H. Darabi, and A.A. Abidi, A low-power wideband polar transmitter for 3G applications, in *IEEE International Solid-State Circuits Conference (ISSCC) Digest of Technical Papers*, San Francisco, CA, February 2011, pp. 378–380.

12. D. Tasca, M. Zanuso, G. Marzin, S. Levantino, C. Samori, and A.L. Lacaita, A 2.9-to-4.0 GHz fractional-N digital PLL with bang-bang phase detector and 560fs$_{rms}$ integrated jitter at 4.5 mW power, *IEEE J. Solid-State Circ.*, 46(12), 2745–2758, 2011.

13. G. Marzin, S. Levantino, C. Samori, and A. Lacaita, A 20 Mb/s phase modulator based on a 3.6 GHz digital PLL with −36 dB EVM at 5 mW power, in *IEEE International Solid-State Circuits Conference (ISSCC) Digest of Technical Papers*, San Francisco, CA, February 2012, pp. 342–344.

14. J. Chen, L. Rong, F. Jonsson, Y. Geng, and L. Zheng, The design of all-digital polar transmitter based on ADPLL and phase synchronized ΔΣ modulator, *IEEE J. Solid-State Circ.*, 47(5), 1154–1164, 2012.

15. C.M. Hsu, M.Z. Straayer, and M.H. Perrott, A low-noise wide-BW 3.6-GHz digital ΔΣ fractional-N frequency synthesizer with a noise-shaping time-to-digital converter and quantization noise cancellation, *IEEE J. Solid-State Circ.*, 43(12), 2776–2786, 2008.

16. M. Lee, M.E. Heidari, and A.A. Abidi, A low-noise wideband digital phase-locked loop based on a coarse-fine time-to-digital converter with subpicosecond resolution, *IEEE J. Solid-State Circ.*, 44(10), 2808–2816, 2009.

17. E. Temporiti, C. Weltin-Wu, D. Baldi, M. Cusmai, and F. Svelto, A 3.5 GHz wideband ADPLL with fractional spur suppression through TDC dithering and feedforward compensation, *IEEE J. Solid-State Circ.*, 45(12), 2723–2736, 2010.

18. S.-K. Lee, Y.-H. Seo, H.-J. Park, and J.-Y. Sim, A 1 GHz ADPLL with a 1.25 ps minimum-resolution sub-exponent TDC in 0.18 μm CMOS, *IEEE J. Solid-State Circ.*, 45(12), 2874–2881, 2010.

19. M. Zanuso, S. Levantino, C. Samori, and A.L. Lacaita, A wideband 3.6 GHz digital ΔΣ fractional-N PLL with phase interpolation divider and digital spur cancellation, *IEEE J. Solid-State Circ.*, 46(3), 627–638, 2011.

20. F. Opteynde, A 40 nm CMOS all-digital fractional-N synthesizer without requiring calibration, in *IEEE International Solid-State Circuits Conference (ISSCC) Digest of Technical Papers*, San Francisco, CA, February 2012, pp. 346–347.

21. M.H. Perrott, Y. Huang, R.T. Baird et al., A 2.5-Gb/s multi-rate 0.25-μm CMOS clock and data recovery circuit utilizing a hybrid analog/digital loop filter and all-digital referenceless frequency acquisition, *IEEE J. Solid-State Circ.*, 41(12), 2930–2944, 2006.

22. P.-Y. Wang, J.-H. Zhan, H.-H. Chang, and H.-M. Chang, A digital intensive fractional-N PLL and all-digital self-calibration schemes, *IEEE J. Solid-State Circ.*, 44, 2182–2192, 2009.

23. R. He, C. Liu, X. Yu et al., A low-cost, leakage-insensitive semi-digital PLL with linear phase detection and FIR-embedded digital frequency acquisition, in *Proceedings of IEEE Asian Solid-State Circuits Conference (A-SSCC)*, Beijing, China, November 2010, pp. 197–200.

24. W. Yin, R. Inti, and P.K. Hanumolu, A 1.6 mW 1.6ps-rms-Jitter 2.5 GHz digital PLL with 0.7-to-3.5 GHz frequency range in 90 nm CMOS, in *Proceedings of IEEE Custom Integrated Circuits Conference (CICC)*, San Jose, CA, September 2010, pp. 1–4.

25. W. Yin, R. Inti, A. Elshazly, and P.K. Hanumolu, A TDC-less 7 mW 2.5 Gb/s digital CDR with linear loop dynamics and offset-free data recovery, in *IEEE International Solid-State Circuits Conference (ISSCC)*, San Francisco, CA, February 2011, pp. 440–441.

26. A. Sai, T. Yamaji, and T. Itakura, A 570fsrms integrated-jitter ring-VCO-based 1.21 GHz PLL with hybrid loop, in *IEEE International Solid-State Circuits Conference (ISSCC) Digest of Technical Papers*, San Francisco, CA, February 2011, pp. 98–99.

27. Y. Sun, N. Xu, M. Wang, W. Rhee, T.-Y. Oh, and Z. Wang, A 1.75 mW 1.1 GHz semi-digital fractional-N PLL with TDC-less hybrid loop control, *IEEE Microw. Wireless Comp. Lett.*, 22(12), 654–656, 2012.

28. M.H. Perrott, M.D. Trott, and C.G. Sodini, A modeling approach for ΔΣ fractional-N frequency synthesizers allowing straightforward noise analysis, *IEEE J. Solid-State Circ.*, 37(8), 1028–1038, 2002.

29. R.B. Staszewski, K. Muhammad, and P.T. Balsara, All-digital TX frequency synthesizer and discrete-time receiver for Bluetooth radio in 130-nm CMOS, *IEEE J. Solid-State Circ.*, 39(12), 2278–2291, 2004.

30. R.B. Staszewski, K. Waheed, F. Dulger et al., Spur-free multirate all-digital PLL for mobilephones in 65 nm CMOS, *IEEE J. Solid-State Circ.*, 46(12), 2904–2919, 2011.

18 Low-Noise PLL-Based Frequency Synthesizer

Youngwoo Jo, Dongmin Park,
Pyoungwon Park, and SeongHwan Cho

CONTENTS

18.1 INTRODUCTION

Frequency synthesizer is an essential building block in wireless transceivers. Without a stable frequency reference, reliable communication cannot be established. As with other circuits in wireless transceivers, specifications of the frequency synthesizer are becoming more stringent with the introduction of advanced communication standards of high data rate. For example, long term evolution (LTE) using orthogonal frequency division multiplexing (OFDM) with multilevel QAM requires much lower phase noise than what was required for a 3G standard. In addition to noise, power consumption is another important, if not the most important, factor that needs to be considered. As with any other analog circuits, there exists relationship between noise and power that must be carefully traded off to achieve a good frequency synthesizer. The goal of this chapter is to understand these relationships and investigate some of the low-noise techniques that can overcome existing trade-offs between power and noise. The remainder of the chapter is organized as follows. We first introduce the basics of phase-locked loop (PLL) by explaining the fundamentals of an integer-N PLL. Next, low-noise techniques are explored for integer-N PLLs and its components, which include subsampling PLL, injection-locked oscillator, and multiplying delay-locked loop (MDLL). Then, principles of fractional-N PLL based on delta-sigma modulator (DSM) will be studied, followed by several approaches that reduce the quantization noise of the DSM, which are noise canceling, noise shifting, and noise filtering. Finally, fractional-N synthesizers that incorporate low-noise integer-N PLL techniques will be discussed. The readers should note that the techniques investigated in this chapter can be applied for both analog and digital PLL-based frequency synthesizers. Experienced readers are recommended to skip Sections 18.2 and 18.4, which describe the basics of PLLs.

18.2   BASICS OF PLL

For simplicity, a PLL can be best described as the process of adjusting a clock at home or a watch on a wrist. As there is inevitable error in the clock, it must be calibrated every now and then by using a more accurate clock (usually from the Internet or TV, which receives the time information from a much more accurate clock, e.g., cesium clock in Boulder, CO). This is essentially what a PLL is; the clock being adjusted is the voltage-controlled oscillator (VCO) that produces the desired frequency and the more accurate clock is the reference oscillator that is usually a crystal oscillator. During the times when clock is not calibrated, noise will be accumulated in the clock, resulting in timing error. Obviously, the more often the clock is calibrated, the more accurate (or less noisy) it will be.

For a more rigorous understanding, consider the block diagram of a PLL shown in Figure 18.1. The VCO produces output signal whose frequency is proportional to the input control voltage and the divider generates a signal that is divided in phase and frequency. The phase detector produces phase difference between the divided clock and the reference clock, which is processed by the loop filter and fed to the VCO. Thus, a PLL is basically a negative feedback loop

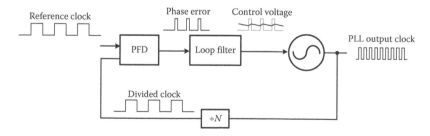

FIGURE 18.1   Basic block diagram of a PLL.

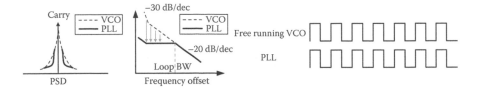

FIGURE 18.2 Typical phase noise of oscillator and PLL.

in which, when in lock, the frequency difference between the inputs to the PD becomes zero and f_{out} becomes N times f_{ref}.

As the main goal of this chapter is to reduce the noise of a PLL, let us first investigate the noise of individual components and look at how these noises are processed or filtered by the PLL before they show up at the output.

In a VCO, phase noise is generated due to the circuit's intrinsic thermal and $1/f$ noise. As VCO is an integrator from the input control voltage to output phase, thermal noise referred at the input is integrated to output phase noise. Thus, a white spectrum will be multiplied by $1/s$ and result in a phase noise plot shown in Figure 18.2. Another way to understand this mechanism is to think of an ordinary clock. Suppose you deliberately added some disturbance to the clock at some time instant so that the minute hand was pushed by 1 min behind. Obviously, even after the disturbance is gone, the clock will lag by 1 min forever. What has happened is that the impulse noise turned into a step error at the output, implying integration action that is inherent in an oscillator. Therefore, an oscillator will have phase noise that is integral of noise added at its input. The purpose of the PLL is to reduce or bound this noise by comparing the output clock with a less noisy clock as shown in Figure 18.2.

Divider usually does not add much noise. Its noise is much lower than that of the oscillator and can be neglected in most cases. (However, it can be important for far out-band noise.) One important thing about divider is that its resolution depends on the phase resolution of the oscillator. That is, if the divider operates based on the rising edges of the VCO, then the divider can only perform integer-valued division (thus the name integer-N frequency synthesizer). However, if it operates on falling edges as well, and assuming that the VCO output has 50% duty cycle, division values of $N + 0.5$ are possible, where N is an integer. If more phases are available, then resolution can be improved accordingly. (That is, if M phases are available within a VCO period, then divider resolution of $1/M$ is possible.) Such property will be important for fractional-N frequency synthesizers.

Phase frequency detector (PFD), charge pump (CP), and loop filters add thermal noise and $1/f$ noise to the PLL. These noises are multiplied by the division value N and also scaled by the PD gain. Thus, reducing N as well as increasing the PD gain helps reduce the noise at the output of the PLL.

The transfer functions from each of the noise sources to the PLL output are shown in Figure 18.3 for a type II PLL. It can be seen that while noise before the VCO goes through a low-pass filter, noise added after the VCO, the phase noise, is high-pass filtered.

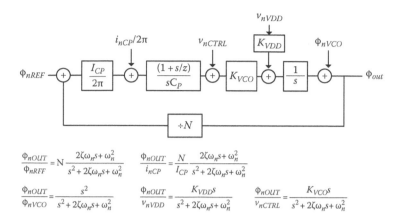

FIGURE 18.3 Noise transfer functions of a PLL.

18.3 LOW-NOISE INTEGER-N PLL TECHNIQUES

A typical way to lower the output noise of a PLL is to minimize the noise of the individual components and choose a loop bandwidth so that the filtered sum of the noises is minimized. In practice, the out-band noise of a PLL is dominated by the VCO and in-band by the PFD and CP noise. These noises are low-pass and high-pass filtered by the PLL, respectively, and show up at the output. For minimum noise, a rule of thumb is to choose a cutoff bandwidth as the frequency at which the VCO phase noise and PFD/CP noise cross, as shown in Figure 18.4. A larger bandwidth will increase the noise contribution from the PFD/CP while a lower bandwidth will increase that from the VCO.

Bandwidth optimization is a basic process that should be employed in all low-noise PLLs. In recent research, other techniques have been suggested to reduce the noise that overcomes power or bandwidth limitations. Some of the methods that are noteworthy are injection-locking, MDLL, and subsampling PLL.

18.3.1 INJECTION-LOCKED OSCILLATOR

When an oscillator is coupled to another oscillator in some way, its frequency can be synchronized to the other oscillator and in such case the oscillator is said to be injection-locked. Such phenomenon has been observed more than 300 years ago in pendulum clock by a Dutch scientist named Christiaan Huygens and more recently in electrical oscillators by Adler and others. An interesting property of injection-locking is that it can occur not only by the fundamental frequency of an oscillator but also in the subharmonics or superharmonics, and thus it can be employed in frequency multipliers and dividers. Another important property of injection-locked oscillator is that its phase noise can be reduced to that of the low-noise source as shown in Figure 18.5. In time domain, it can be considered that the clean source removes the accumulated jitter every time injection occurs. Thus, many researchers have exploited this property to achieve low-noise oscillators and PLLs. Unfortunately, there is a critical issue if injection-locked oscillator is to be used in a PLL, which is due to timing error of

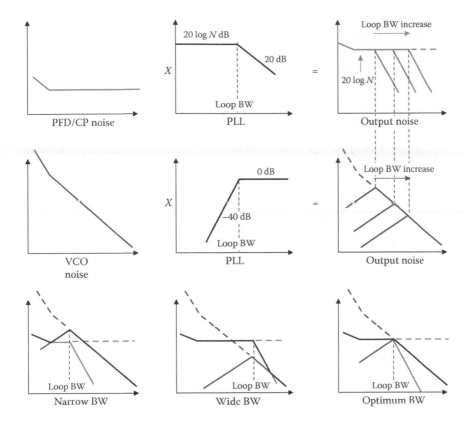

FIGURE 18.4 Optimum loop bandwidth for a PLL.

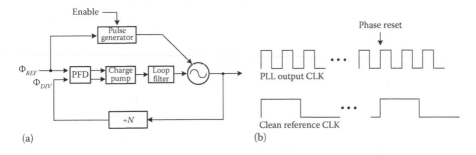

FIGURE 18.5 Injection-locked oscillator: (a) block diagram and (b) timing diagram.

the injection that leads to a large reference spur. In a PLL employing an injection-locked VCO, there are two conflicting paths that adjust the VCO phase, which are the feedback path of the PLL and the injection path. While the PLL tries to control the VCO phase so that its divided output is aligned with the reference, the reference injection path tries to align the VCO phase itself with the source. As there is propagation delay in both injection path and the divider, there will be conflict between these two paths and thus a large reference spur will be produced. To avoid timing errors, several techniques have been proposed, which basically add another feedback or calibration to adjust the timing [1]. However, it should be noted that even with the calibration, reference spur is no better than a PLL without injection-locking.

While such technique reduces the phase noise in integer-N PLLs, it cannot be applied to fractional-N PLLs.

18.3.2 MULTIPLYING DELAY-LOCKED LOOP

In an injection-locked oscillator, the rising edge of the output signal is *forced* to change its position by a low-noise reference. If one can *replace* the output signal by the reference instead of changing its position, injection strength will be maximized. Although replacement of an edge is not possible in an LC oscillator, it can be implemented in for a ring VCO shown in Figure 18.6. It can be seen that the ring VCO is actually a delay line in a feedback loop with a multiplexer so that the rising edge of the oscillator is periodically replaced by that of the reference clock [2]. Thus, jitter accumulation is removed in every reference clock. This architecture is called the multiplying DLL as the ring oscillator operates in a DLL mode. Nevertheless, this architecture also suffers from the same problem of timing mismatch between the reference injection and PLL.

18.3.3 SUBSAMPLING PLL

The two methods just described reduce the noise of the VCO that determines the outband noise of the PLL. Here, we look at a method that can reduce the in-band noise. As described in the previous section, the in-band noise coming from PFD and CP is inversely proportional to the PD gain. Hence, a straightforward method to improve the noise is to have higher gain in the PD. This can be accomplished by using large CP current, but that will result in increased power consumption. A power-efficient way to increase the PD gain is to use subsampling PD [3]. In a subsampling PLL shown in Figure 18.7, the output voltage of VCO is sampled by the reference frequency. As the slope of the VCO output is very high, even a slight jitter will cause large change in the output voltage. Thus, the gain of the PD is very high and consequently, in-band noise from PD and CP will be reduced. One problem of the subsampling PLL is that its locking range is limited. To solve this issue, a combined approach of using tristate PFD and subsampling PD has been proposed in [4], which increases the locking range and still achieves low in-band noise from the large gain of the subsampling PD.

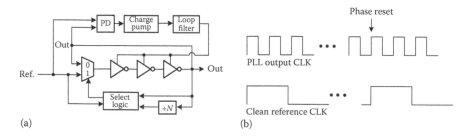

FIGURE 18.6 MDLL: (a) block diagram and (b) timing diagram.

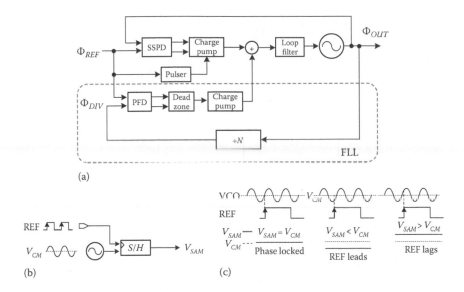

(a)

(b)

(c)

FIGURE 18.7 Subsampling PLL: (a) block diagram, (b) subsampling PD, and (c) timing diagram.

18.4 FRACTIONAL-*N* PLL

In an integer-*N* frequency synthesizer, frequency resolution is determined by the reference frequency. For example, if 10 kHz of resolution desired for output frequencies from 2.4 to 2.5 GHz, then a reference frequency of <10 kHz must be used, which leads to division value exceeding 240,000. Such requirement poses a couple of important problems. First is that a loop bandwidth is limited to $f_{ref}/10$, which result in very slow settling time and limited filtering of VCO phase noise. The second problem, which is more problematic, is the multiplication of noise. As seen in the previous section, in-band noise sources will be increased by a factor of 240,000, which translates to 108 dB. Such increase in noise is not acceptable, and in order to solve this issue, frequency resolution must somehow be decoupled from the reference frequency. This can be accomplished if fractional division is possible. While fractional divider can be implemented using phase interpolators, it suffers from mismatch and thus causes large fractional spur at the output. While there have been recent improvements in fractional dividers using digital-to-time converters, a more commonly used approach is using an oversampled DSM to control the division value of the multimodulus integer divider as shown in Figure 18.8. Basically, the DSM receives a fractional frequency control word between 0 and 1, and produces a quantized bit stream of 0 or 1 to change the division value by *N* or *N* + 1. The output bit stream is generated in such a way that its average is equal to the input word and the quantization noise is pushed to high frequency. As division modulus, together with the quantization noise, is averaged or low-pass filtered by the PLL, the desired frequency can be achieved with good spectral purity. Another way to understand this architecture is to consider the DSM as an oversampled analog-to-digital converter

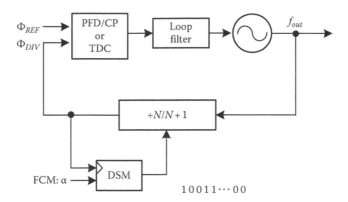

FIGURE 18.8 Block diagram of a DSM-based fractional-N PLL.

(ADC) since it quantizes the input signal. Consequently, the PLL can be viewed as an oversampled digital-to-analog converter (DAC) since it receives digital input and produces analog output either as the control voltage or as the output frequency.

It can be shown the phase noise at the output of the PLL due to quantization noise of the DSM can be represented as the following equation:

$$S_{\Phi out}(f) = S_{\Phi_{qn}}(f) \, | \, H(f) \, |^2 = \frac{Q(f) \, | \, H(f) \, |^2}{N_O^2 \cdot [2\sin(\pi(f/f_{REF}))]^2}$$

$$= \frac{| \, H(f) \, |^2}{N_O^2} \frac{\Delta^2}{12 f_{REF}} \left[2\sin\left(\frac{\pi f}{f_{REF}}\right) \right]^{2(m-1)} \qquad (18.1)$$

where
   $S_{\Phi_{qn}}$ is the PSD of the phase error due to quantization noise
   $Q(f)$ is the PSD of the output bit stream of the DSM
   $H(f)$ is the transfer function of the synthesizer from the reference phase input to
      the output

Note that when the quantization noise of the DSM is converted into phase in the PLL, it inherently goes through integration and normalization process. Thus, the quantization noise of an mth order DSM shows up at the output as if it were an $(m - 1)$th order noise shaping.

Another problem of DSM-based fractional-N frequency synthesizer is the fractional spur. In practice, the output bit stream of the DSM contains some periodic patterns depending on the input due to finite bit width of adders and accumulators in the DSM, the nonlinearity of the PLL as well as the fact that input is DC. While patterns of short period create spurs that can be filtered by the PLL, those with long period create spurs within the bandwidth of the PLL, severely deteriorating the output spectrum. In order to reduce these fractional spurs, several techniques exist such as dithering the LSB of the DSM using a pseudorandom bit sequence or resetting the register of the DSM. A more powerful method is to use quantization noise canceling or shifting that will be discussed in the next section.

18.5 LOW-NOISE FRACTIONAL-*N* PLL USING NOISE CANCELLATION

Although quantization noise of the DSM is pushed to high frequency and low-pass filtered by the PLL, it can be a problem if PLL has a wide-loop bandwidth. A large loop bandwidth is desirable when a frequency synthesizer is used as a high data rate phase or frequency modulator in a wireless transmitter or in a low-noise frequency synthesizer where VCO phase noise is dominant and in-band noise is low enough as shown in Figure 18.9. In such cases, bandwidth needs to be increased. While this can be done easily in an integer-*N* PLL, it cannot be done in a fractional-*N* PLL due to the quantization noise.

In order to suppress the quantization noise from the DSM, several approaches have been introduced in the past, which can be categorized as noise canceling, noise shifting, and noise filtering. Noise canceling exploits the fact that the quantization noise of the DSM is deterministic and can be subtracted while noise shifting simply exploits the property of oversampled converters that higher operating frequency will push the noise to higher frequencies. Noise filtering adds FIR filter to the quantization noise in a way that it does not disturb the loop stability or other core properties of the PLL. These techniques will be investigated in the following subsections.

18.5.1 NOISE CANCELLATION IN VOLTAGE DOMAIN

The basic concept of a quantization noise-canceling fractional-*N* PLL is shown in Figure 18.10. Since the output of DSM is a deterministic signal, quantization noise can be estimated precisely and canceled by the process shown in Figure 18.10a. In the noise-cancellation path, quantization noise is first calculated and integrated to achieve the corresponding phase error. Next, the phase error is multiplied by a gain and converted into charge by a DAC, which is a pulse-driven current DAC to match the charge from the CP. Note that the PFD and CP can be replaced by a TDC for digital PLLs, in which case DAC can be omitted and noise subtraction be done in digital domain. The timing diagram of the cancellation process is shown in Figure 18.10b,

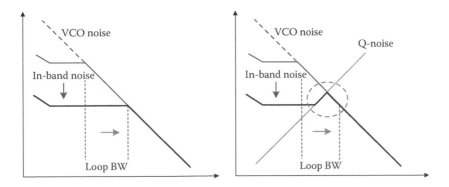

FIGURE 18.9 Effect of quantization noise on loop bandwidth.

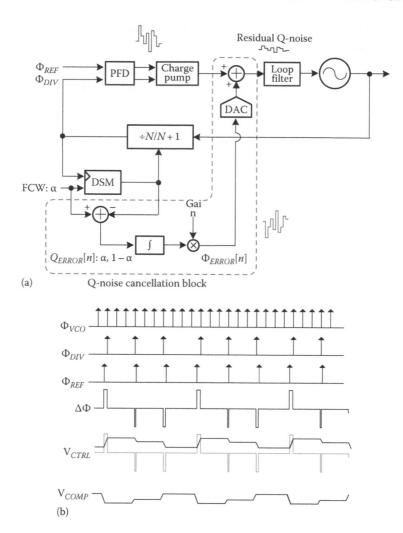

(a)

(b)

FIGURE 18.10 Quantization noise canceling fractional-N PLL: (a) block diagram and (b) timing diagram.

assuming that the DSM is a first order (i.e., accumulator). Despite the simple concept, several problems exist. First is that due to stringent noise requirements of the frequency synthesizer, the cancellation path must be very precise. As can be seen in the block diagram, the limitation comes from the performance of the DAC in its linearity and resolution. Note that while a DAC is absent in digital PLLs, same problem exists for the TDC. Second, the gain between the PLL and the cancellation path must be equal for perfect cancellation. As the PLL consists of analog blocks, the gain cannot be known a priori. Again, this is also a problem for digital PLL as well since TDC gain is unknown.

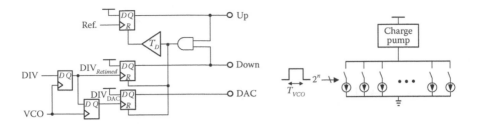

FIGURE 18.11 Circuit schematic of hybrid PFD and DAC.

In order to improve the performance of the DAC, a widely used technique is the dynamic element matching (DEM). By shuffling the unit cells of the DAC, mismatch can be averaged out and the linearity can be improved.

Although employing a high-performance DAC improves the quantization noise performance, its performance is still limited as there is inevitable mismatch between the output of the cancellation DAC and the output of the PFD/CP. To solve this issue, Meninger and Perrott [5] propose a cancellation method that shares the PFD/CP in the cancellation path to achieve inherent matching between cancellation signal and error waveform. It uses mismatch compensated hybrid PFD and DAC circuit to achieve self-aligned cancellation of quantization noise as shown in Figure 18.11, where quantization noise is subtracted using the same CP used in the PLL. The CP is driven by a pulse that has a constant pulse width equal to one VCO period but has different magnitude that corresponds to the quantization noise. As a result, 29 dB of quantization noise suppression is achieved for 1 MHz bandwidth without using any calibration.

If quantization noise cancellation technique is to be used in general, for example, to digital PLLs, there must be a systematic means to solve the gain mismatch between the cancellation path and the PLL. An adequate solution to this problem can be provided by using adaptive filter based on least mean square (LMS) algorithm as shown in Figure 18.12. It basically monitors the output of the canceled quantization noise by looking at the loop filter and by correlating it with the phase error, the gain of the DAC is adaptively adjusted so that the mean square error is minimized [6]. This technique can be useful for digital PLLs where TDC gain is unknown. One drawback of this approach is that a large settling time is necessary.

18.5.2 Noise Cancellation in Phase Domain: Fractional Divider

Quantization noise can also be canceled using digitally controlled delay line as shown in Figure 18.13. Instead of using DAC that removes the quantization noise in charge domain, the delay line does the same in phase domain and the delay line combined with the integer divider can be considered as a fractional divider [7]. In practice, similar problems of charge domain noise cancellation also exist. Precise control of delay line is difficult and thus nonlinearity, resolution, and gain matching need to be solved. Similarly, DEM- and LMS-based algorithms can be used to overcome these problems.

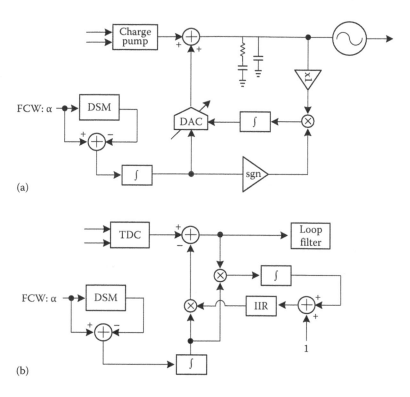

(a)

(b)

FIGURE 18.12   Concept of LMS algorithm applied to (a) analog PLL and (b) digital PLL.

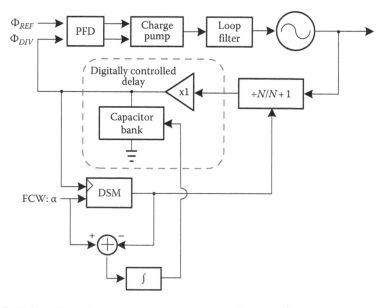

FIGURE 18.13   Block diagram of fractional-N PLL using fractional divider.

18.6 LOW-NOISE FRACTIONAL-*N* PLL USING HIGH-OSR DSM

Previous techniques of noise cancellation improve the noise performance of the PLL. Unfortunately, it requires precise analog circuitries for noise cancellation that require sophisticated analog circuitry and algorithms. In this section, we look at an alternative approach of reducing the quantization noise without using noise cancellation.

As can be seen in Equation 18.1, quantization noise is inversely related to the operating frequency of the DSM. Thus, if reference frequency is increased, then lower quantization noise can be achieved at the output of the PLL, as shown in Figure 18.14. With a third-order DSM, increasing the operating frequency results in noise reduction of 12 dB/octave. Furthermore, spur also moves to higher frequencies where it can be suppressed by the low-pass nature of the PLL.

18.6.1 FRACTIONAL-*N* FREQUENCY SYNTHESIZER USING QNS

One way to shift the quantization noise to higher frequency is shown in Figure 18.15. This architecture is fundamentally different from a conventional fractional-*N* synthesizer in that it does not use a multimodulus divider to achieve fractional division, but rather it directly injects modulated charge on to the loop filter to balance the residual error from the fixed integer divider [8]. Unlike the noise canceling architecture, the modulation path does not cancel the quantization noise, but rather it is responsible for the dithering operation and hence quantization step is not limited to the VCO period. While this approach removes the need for a fractional divider, amount of quantization noise shifting is limited by the frequency multiplier, which was implemented using a circulator for *M*-times multiplication.

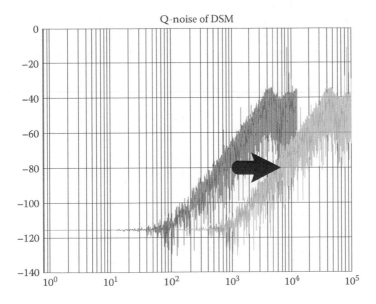

FIGURE 18.14 Effect of quantization noise shifting due to 10× reference frequency multiplication.

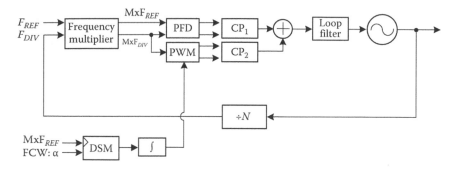

FIGURE 18.15   Block diagram of fractional-N PLL using reference frequency multiplier.

18.6.2 FRACTIONAL-N FREQUENCY SYNTHESIZER USING NESTED PLL

Another way to increase the operating frequency of the DSM is to split the feedback divider into two as shown in Figure 18.16. The first divider is a dual modulus divider while the second divider is a fixed divider. As can be seen, the DSM runs at N_2 times the reference frequency, and thus, its noise is shifted to higher frequencies. Unfortunately, such scheme will not work due to noise aliasing owing to the second divider. As shown in the timing diagram, the second divider performs moving sum and downsampling of the input edges. While moving sum provides a low-pass filter, downsampling folds high-frequency noise back to low frequency, which destroys quantization noise shaping provided by the DSM. In order to avoid such noise folding, an antialias filter is needed in the phase domain, which can be implemented using a PLL shown in Figure 18.17.

The block diagram of the nested PLL that employs another PLL in the feedback path is shown in Figure 18.18. As a PLL acts as a low-pass filter in phase domain, it suppresses the high-frequency noise of the DSM before it is aliased by the second divider.

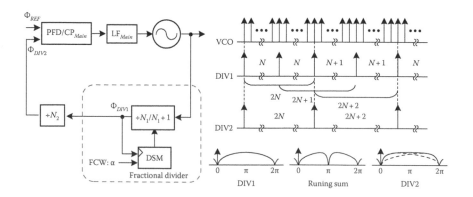

FIGURE 18.16   Nested PLL without antialias filter: (a) block diagram and (b) timing diagram and noise aliasing.

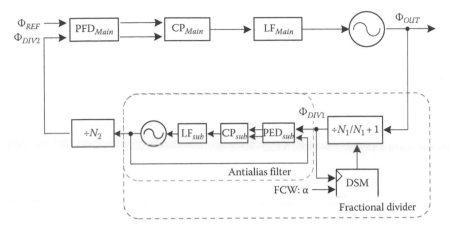

FIGURE 18.17   Implementation of antialias filter using nested PLL.

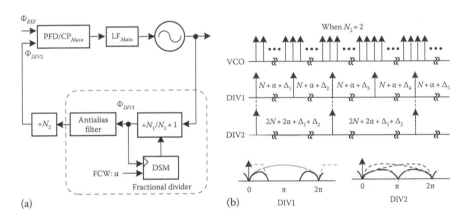

FIGURE 18.18   Nested PLL without antialias filter: (a) block diagram and (b) timing diagram and suppression noise aliasing.

The advantage of the fractional-N synthesizer using nested PLL is that quantization noise is reduced without any sophisticated analog circuits, for example, high-performance DACs, and also does not require any calibration. Moreover, fractional spurs are also pushed to higher frequencies and suppressed by the PLL. However, it does add some design complexity in terms of stability and trade-offs in noise and power consumption. In regard to stability, there are two feedback loops and thus stability of each PLL must be guaranteed and their effect on one another must be analyzed. Assuming that both PLLs meet the traditional stability requirements (i.e., enough phase margin with loop bandwidth $< f_{ref}/10$), it can be shown that the bandwidth of the sub-PLL must be larger than that of the main PLL by approximately two times. The fact that sub-PLL must have larger bandwidth than the main PLL is not surprising as in a conventional PLL the sub-PLL does not exist and thus its bandwidth can be considered very large.

When optimizing the parameters of the nested PLL such as the division value and loop bandwidth, noise and power trade-offs must be considered. In regard to choosing the division values, if N_2 is increased, quantization noise is pushed to higher frequencies, but at the cost of increased power consumption of the DSM and the sub-PLL as they operate at higher frequencies. For the loop bandwidth, there is a trade-off between aliased noise versus noise from the sub-PLL. In a nested PLL, there are additional noise sources compared with a conventional PLL, which are VCO and PFD/CP of the sub-PLL. While the noise of the PFD/CP is low-pass filtered twice by the main and the sub-PLL, VCO phase noise is high-pass filtered by the sub-PLL and then low-pass filtered by the main PLL as shown in Figure 18.19. If a ring VCO is used in the sub-PLL, then its phase noise will dominate the noise from the sub-PLL. Thus, a large loop bandwidth should be used to suppress this noise. However, a large loop bandwidth of the sub-PLL will increase the amount of noise folding. Thus, trade-off between these two noise sources must be carefully considered to achieve the minimum noise. In [9], division values have been set to 2 and 4/5 for a reference frequency of 32 MHz and output frequency of 2.4 GHz. Quantization noise suppression of 26 dB has been achieved.

Lastly, the main advantage of the nested PLL compared with the cascaded PLL, which we study next, is that the VCO phase noise of the sub-PLL can be reduced as a large loop bandwidth is possible due to the high reference frequency of $N_2 \cdot f_{ref}$. In cascaded PLLs, reference frequency is small and thus the amount of VCO phase noise filtering is limited.

18.6.3 FRACTIONAL-N FREQUENCY SYNTHESIZER USING CASCADED PLL

The operating frequency of the DSM can be increased by having a frequency multiplier before the fractional-N PLL. Another benefit of such architecture is that it helps to reduce the noise of PFD, CP, and the divider in the fractional-N PLL since its division value is reduced by the reference frequency multiplication factor. In detail, the rise/fall time of the PFD and divider is the same regardless of the reference frequency and hence the amount of white noise at every transition is the same, no matter what the reference frequency is. When the reference frequency is doubled, however, the white noise is generated twice as often and thus the absolute noise power at the PFD or divider output is increased by 3 dB. But the gain from the PFD and the divider to the output of the PLL is reduced by 6 dB due to the halved multiplication factor. Therefore, the white noise at the output of the PLL due to the PFD and divider is improved by 3 dB with the doubled reference frequency. Similarly, the white noise from the CP is also reduced by 3 dB as the reference frequency is doubled if the CP current is the same. A frequency multiplier can be implemented in many different ways. One of the most efficient ways to generate higher frequency with a given input signal is to use a frequency doubler if both rising and falling edges of the input signal are clean. While it can be simply implemented using an XOR gate and a delay, the duty cycle of the input signal must be 50% or else there will be large spur at the doubler output.

An alternative way of implementing a frequency multiplier is to use an integer-N PLL as shown in Figure 18.20. The techniques reviewed in the previous

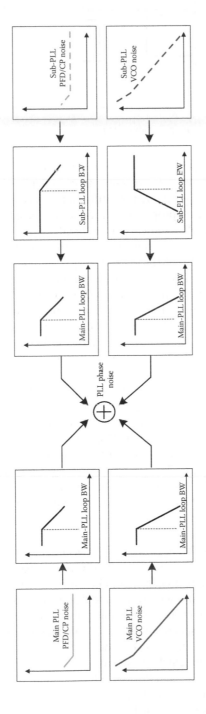

FIGURE 18.19 Noise transfer function of nested PLL.

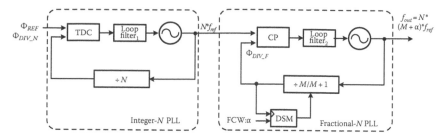

FIGURE 18.20 Block diagram of cascaded PLL.

sections for low-noise integer-N PLL can be used, such as subsampling, injection-locking, or MDLL. There are several issues to consider in choosing which technique to use for the low-noise frequency multiplier. If a ring oscillator is used for the integer-N PLL, then the appropriate technique will be either injection-locking or MDLL, as subsampling PLL reduces the in-band noise from the PD/CP and not the phase noise of the VCO that is dominant for a ring VCO-based PLL. Between, injection-locking and MDLL, MDLL is the right candidate since its injection strength is maximum and can easily be deployed in a ring oscillator. Note that if one is not concerned about the area, an LC-VCO can be used, which would result in lower noise and injection-locking and/or subsampling technique can be used to further improve the noise. However, there are some design issues that must be considered. First, if the output frequency of the integer-N PLL is sub-GHz, then implementing an LC-VCO is not very efficient in power or area as on-chip spiral inductors have low-Q and large area as frequency is reduced. While one can use an LC-VCO running at several GHz and use the divided output, it not only adds power consumption but injection-locking could occur by the LC-VCO of the fractional-N synthesizer, for example, through substrate coupling, as two LC-VCOs are physically close together. Even if injection-locking did not occur, undesired spurs can be created due to nonlinear mixing process near the vicinity of injection-locked regime. This phenomenon can occur not only between the fundamental outputs of the two LC-VCOs but among their subharmonics as well, and thus the frequency of the LC-VCOs must be chosen carefully. In this subsection, the MDLL technique investigated is more area efficient and does not suffer from such problems.

As mentioned earlier, an MDLL removes the accumulated jitter of a ring oscillator but suffers from timing mismatch between reference injection and the feedback loop that will incur a large reference spur. The timing errors in reference injection are described in Figure 18.21, where we assume that the PLL is locked to a frequency of $f_0 = 1/t_0$, and the delays incurred by feedback divider and the injection circuit are represented by t_{div} and t_{inj}, respectively. Assuming that there is no timing offset in the PFD, the divided output and the reference clock are aligned when the PLL is in lock. Thus, the VCO and reference clock are separated by t_{div} when the PLL is in lock. If there is to be zero timing error and thus no reference spur, the replaced edge of the VCO by the reference injection must be positioned t_0 from the previous edge, which implies that injection delay must be such that $t_d + t_i = t_0$. Unfortunately, meeting such

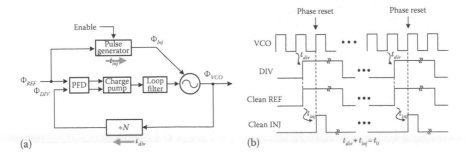

(a)

(b)

FIGURE 18.21 Injection-locked oscillator: (a) block diagram and (b) timing diagram and timing requirement.

requirement is very difficult due to inevitable process, temperature variation, and device mismatch, and thus timing will occur at every reference cycle. Moreover, as this timing error is due to direct injection, it is not filtered out by the loop filter and thus causes a large reference spur.

One way to mitigate this problem is to employ a replica VCO, as shown in Figure 18.22. The key idea is that by sharing the control voltage of the PLL, the frequencies of the two VCOs are the same but their phase can be different and independent. Hence, there will be no reference spur when the reference is injected to the replica.

Unfortunately, there will be mismatch between the VCOs, which will lead to frequency difference and hence reference spur will be created again. In [10], a dual-pulse ring oscillator (DPRO) is used, which results in mismatch-free VCOs by using single ring delay line but employing two pulses inside. In contrast to what the name implies, DPRO is not an oscillator in the sense that it cannot start to oscillate on its own, but rather it is a delay line closed in a stable feedback loop. An example of a DPRO with reference injection is shown in Figure 18.23, where there are four inverter stages and a mux. As there is an even number of inverter stages, the loop will not start to oscillate. However, if a start-up pulse whose width is smaller than the delay of four inverters is injected through the mux, and the mux is switched to

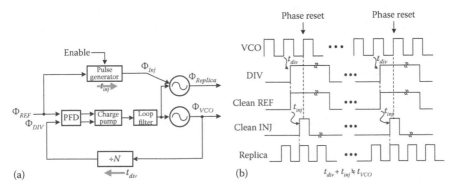

(a)

(b)

FIGURE 18.22 Injection-locked PLL using replica VCO: (a) block diagram and (b) timing diagram.

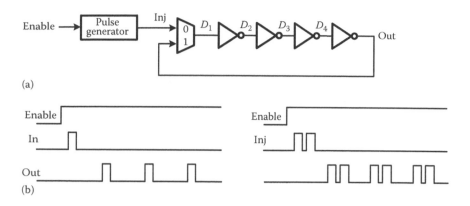

(a)

(b)

FIGURE 18.23   DPRO: (a) block diagram and (b) timing diagram.

form a loop afterward, then the injected pulse will run around the loop indefinitely. Therefore, the DPRO can be effectively seen as an oscillator. An interesting property of the DPRO is that more than one pulse can be running around the loop as long as the pulses are spaced enough apart so that they do not interact with each other. An example is shown in Figure 18.23, where two pulses are injected. A key observation that needs to be made is that these two pulses have the same frequency as they go through the same inverters, but their phase can be different. This is the desired trait of two perfect replica VCOs and thus a DPRO can be considered as such.

The block diagram of the integer-N PLL using DPRO is shown in Figure 18.24 together with its timing diagram. Initially, two pulses are generated and injected to the DPRO. The pulse in dotted line is used in the PLL for frequency locking while the other pulse in solid line is constantly replaced by the reference clock. Hence, the pulse used for PLL is noisy, like the VCO in conventional PLL, while the replaced pulse is much cleaner and thus used as a reference for the following fractional-N synthesizer. The divide-by-two that follows the DPRO translates these pulses into rising and falling edges to be used in the integer-N PLL and the fractional-N synthesizer.

As a result, the cascaded PLL provides state-of-the-art performance. The synthesizer generates 255 fs-rms jitter while dissipating 15 mW despite being implemented

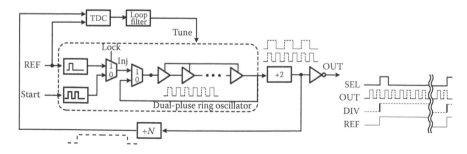

FIGURE 18.24   Block diagram of integer-N PLL using DPRO.

in 0.13 μm CMOS. The reference spur is below −80 dBc, which is about 25–30 dB lower than typical PLL-based synthesizers.

18.6.4 Techniques for Low-Power DSM

While frequency shifting technique provides a calibration-free and mismatch-free way of lowering the quantization noise, its drawback is that the burden now is shifted to the DSM. That is, if reference frequency of the PLL is increased by M, not only does the DSM run at M times the frequency, but its frequency resolution is decreased by M. To compensate for the loss in frequency resolution, higher resolution DSM is necessary by $\log_2 M$ bits. As the problem is digital in nature, use of advanced nanometer CMOS will eventually make it negligible in power and area. However, it is still worthwhile to investigate what can be done to reduce the area and power consumption of the DSM.

A higher resolution DSM requires large bit width in adders and accumulators, which dominate the power and area consumption. For example, a 16-bit input DSM requires adders of more than 20 bits for stability. To reduce the area and power consumption, a DSM architecture with reduced bit width adders can be used as shown in Figure 18.25b. The third-order 16b-input DSM consists of a two 4b-input first-order DSMs followed by an 8b-input third-order DSM. In the first-stage first-order DSM, the last 4 bits of the 16-bit input are converted to 1b carry output, which goes to the carry input of the following first-order DSM. The second-stage first-order DSM generates 1b carry output in the same manner. The output of the second stage is combined with the first 8 bits of the 16-bit input and the following third-order DSM generates the third-order noise-shaped 1b output. Note that it is easy to increase the resolution of the proposed DSM by cascading more 4b-input first-order DSMs, whose area and power consumption is small. As a result, this 16-bit input DSM occupies one-third the area and consumes one-sixth the power of the conventional DSM.

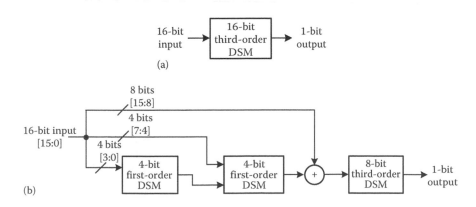

FIGURE 18.25 (a) Block diagram of conventional third-order DSM and (b) block diagram of power-efficient third-order DSM.

18.7 QUANTIZATION NOISE FILTERING IN FRACTIONAL-*N* PLLs

Another approach to reduce the quantization noise is to use filtering. While filtering seems like an easy, obvious way to reduce undesired noise at high frequencies, care must be taken as inserting poles in the PLL will hurt its stability. As quantization noise from the DSM is the main culprit, a desirable technique would be to filter out the quantization noise only without altering the overall loop dynamics of the PLL. This can be accomplished by the finite-impulse response (FIR) filter-embedded PLL architecture shown in Figure 18.26, where digital FIR filter is added between the DSM and the frequency divider. By summing the quantization errors in analog domain with multiple PFDs and charge pumps, the DC gain of the FIR filter can be maintained to unity and quantization noise can be suppressed. In [11], 15 dB of noise suppression is achieved despite having a large loop bandwidth of 2 MHz.

18.8 SUMMARY

Various techniques have been investigated for low-noise frequency synthesizers. For integer-*N* PLL, subsampling architecture increases the gain of the PD and thus can reduce the in-band noise, while MDLL and injection-locking reduces the accumulated jitter of VCO and thus can reduce the out-band noise. For fractional-*N* PLL, reducing the quantization noise and spur was the main problem, where techniques

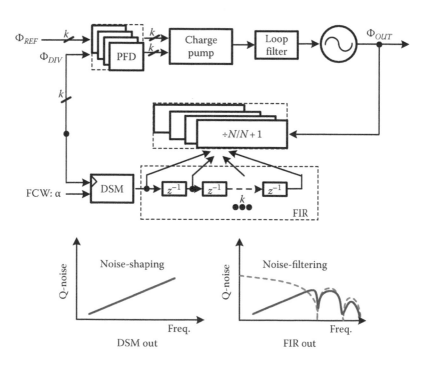

FIGURE 18.26 Block diagram of FIR-embedded PLL.

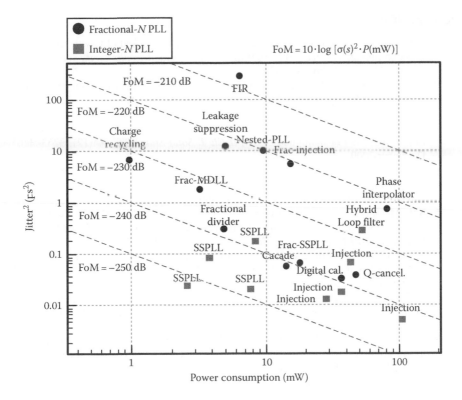

FIGURE 18.27 Figure of merit of PLLs.

such as noise canceling, noise shifting, and noise filtering have been explored. Note that it is difficult to apply low-noise integer-N PLL techniques to a fractional-N PLL because VCO output is not a subharmonic of the reference and thus there is inherent phase misalignment between the divided output of the VCO and the reference. Hence, injection-locking, MDLL, nor subsampling PD can be properly applied to a fractional-N PLL. However, some of the recent researches have shown ways to incorporate these techniques to fractional-N PLLs. For example in [12], a fractionally injection-locked PLL is introduced that exploits the multiphase output of a ring oscillator. In [13,14], fractional-N PLL using MDLL and subsampling is demonstrated, respectively, by employing a delay line to the reference clock so that phase misalignment can be removed. Lastly, a figure of merit (FoM) defined as FoM = 10·log [Jitter(s)^2·Power(mW)], is shown for recent state-of-the-art PLLs in Figure 18.27.

ACKNOWLEDGMENT

This research was supported by the Basic Science Research Program from the NRF of Korea funded by the Ministry of Education (NRF-2013R1A2A1A01014872).

REFERENCES

1. I.-T. Lee et al., A divider-less sub-harmonically injection-locked PLL with self-adjusted injection timing, *Dig. Tech. Papers, IEEE Int. Solid-State Circuits Conf.*, San Francisco, CA, pp. 414–415, February 2013.
2. R. Farjad-Rad et al., A low-power multiplying DLL for low-jitter multi-gigahertz clock generation in highly integrated digital chips, *IEEE J. Solid-State Circ.*, 37(12), 1804–1812, December 2002.
3. X. Gao et al., A low-noise sub-sampling PLL in which divider noise is eliminated and PD/CP noise is not multiplied by N^2, *IEEE J. Solid-State Circ.*, 44(12), 3253–3263, December 2009.
4. C.-W. Hsu et al., A 2.2 GHz PLL using a phase-frequency detector with an auxiliary sub-sampling phase detector for in-band noise suppression, *IEEE Custom Integrated Circuits Conf.*, pp. 1–4, San Jose, CA, September 2011.
5. S.E. Meninger and M.H. Perrott, A 1-MHz bandwidth 3.6 GHz 0.18-μm CMOS fractional-N synthesizer utilizing a hybrid PFD/DAC structure for reduced broadband phase noise, *IEEE J. Solid-State Circ.*, 41(4), 966–980, April 2006.
6. M. Gupta and B.-S. Song, A 1.8-GHz spur-cancelled fractional-N frequency synthesizer with LMS-based DAC gain calibration, *IEEE J. Solid-State Circ.*, 41(12), 2842–2851, December 2006.
7. D. Tasca et al., A 2.9–4.0-GHz fractional-N digital PLL with bang-bang phase detector and 560-fsrms integrated jitter at 4.5-mW power, *IEEE J. Solid-State Circ.*, 46(12), 2745–2758, December 2011.
8. W.-H. Chiu and T.-H. Lin, A 3.6 GHz 1 MHz-bandwidth ΔΣ fractional-N PLL with a quantization-noise shifting architecture in 0.18 μm CMOS, in *Symp. VLSI Circuits Dig.*, Kyoto, Japan, pp. 114–115, June 2011.
9. P. Park, D. Park, and S. Cho, A 2.4-GHz fractional-N frequency synthesizer with high-OSR ΔΣ modulator and nested PLL, *IEEE J. Solid-State Circ.*, 47(10), 2433–2443, October 2012.
10. D. Park and S. Cho, A 14.2 mW 2.55-to-3 GHz cascaded PLL with reference injection and 800 MHz delta-sigma modulator in 0.13 μm CMOS, *IEEE J. Solid-State Circ.*, 47(12), 2989–2998, December 2012.
11. X. Yu, Y. Sun, W. Rhee, and Z. Wang, An FIR-embedded noise filtering method for ΔΣ fractional-N PLL clock generators, *IEEE J. Solid-State Circ.*, 44(9), 2426–2436, September 2009.
12. P. Park, J. Park, H. Park, and S.H. Cho, An all-digital clock generator using a fractionally injection-locked oscillator in 65 nm CMOS, *Dig. Tech. Papers, IEEE Int. Solid-State Circuits Conf.*, San Francisco, CA, pp. 336–337, February 2012.
13. G. Marucci et al., A 1.7 GHz MDLL-based fractional-N frequency synthesizer with 1.4 ps RMS integrated jitter and 3 mW power using a 1b TDC, *Dig. Tech. Papers, IEEE Int. Solid-State Circuits Conf.*, San Francisco, CA, pp. 360–361, February 2014.
14. P.-C. Huang et al., A 2.3 GHz fractional-N dividerless phase-locked loop with −112 dBc/Hz in-band phase noise, *Dig. Tech. Papers, IEEE Int. Solid-State Circuits Conf.*, San Francisco, CA, pp. 362–363, February 2014.

19 Low-Phase Noise Quadrature Frequency Synthesizer for 60 GHz Radios

Wei Deng, Teerachot Siriburanon, Ahmed Musa, Kenichi Okada, and Akira Matsuzawa

CONTENTS

19.1 INTRODUCTION

The unlicensed bandwidth between 57 and 66 GHz is released for multi-Gb/s and broadband communications. Figure 19.1 shows the 60 GHz frequency band allocation in the United States, Canada, Japan, Korea, China, and Europe. As it is known, this is the widest portion of radiofrequency spectrum ever allocated in an exclusive way for wireless unlicensed application that allows multi-Gb/s wireless communications [1–5]. In addition, due to the high path loss at 60 GHz and the peak of resonance of

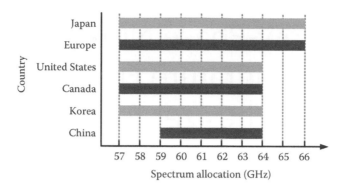

FIGURE 19.1   The 60 GHz band frequency allocation worldwide.

the oxygen molecule, several wireless personal area networks (WPANs) are allowed to operate closely without interfering. On the other hand, significant advancements in CMOS technology permit one to consider utilizing the 60 GHz band for high data rate commercial application including short-range, wireless HDTC transmission.

There have been already several standards for 60 GHz wireless communication, including IEEE 802.15.3c [6], IEEE 802.11ad [7], ECMA-387 [8], wirelessHD [9], and (Wireless Gigabit Alliance) WiGig [10] standards. Each is formed by a different community of potential users of technology. For example, an IEEE 802.11ad is one of the standards for implementing WLAN computer communication for Gb/s speed utilizing 60 GHz frequency band, and wirelessHD and ECMA are developed for streaming high-definition content between source and display devices. Most of the standards are focusing on very high speed communication (1–5 Gb/s) to enable wireless HDMI replacement.

Even though different standards are needed for different applications, an ability to operate across different standards is more preferable. These standards define four bands centered at 58.32/60.48/62.64/64.8 GHz, and ECMA-387 additionally defines channel-bonding bands centered at 59.4/61.56/63.72 GHz for higher data rates and more efficient constellations, as shown in Figure 19.2. In order to be compliant with these standards, millimeter-wave frequency synthesizers that can generate each of these carrier frequencies are required. Moreover, millimeter-wave frequency synthesizers with quadrature output for direct conversion architecture

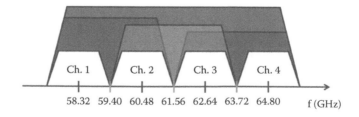

FIGURE 19.2   The 60 GHz band carrier frequency including channel bonding supporting capability.

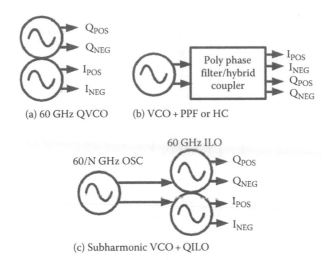

(a) 60 GHz QVCO   (b) VCO + PPF or HC

(c) Subharmonic VCO + QILO

FIGURE 19.3   Quadrature 60 GHz LO signal generation approaches.

and TDD-supporting capability for more efficient spectrum/lower cost become increasingly critical and challenging.

Recently, a few PLLs were designed for 60 GHz wireless communication system. The authors in [11] propose a 60 GHz PLL using a QVCO oscillating at the fundamental frequency. Later, a PLL with push–push VCO followed by a hybrid coupler [12] was published. In [13], a PLL with a subharmonic VCO that is used to inject a subharmonic quadrature injection-locked oscillator (ILO) is reported. Figure 19.3 depicts these simplified topologies for 60 GHz quadrature carrier frequency generation. As tabulated in Table 19.1, among these publications, PLL using subharmonic quadrature ILO is preferred due to the best phase noise performance at 60 GHz. However, due to inaccurate active/passive device modeling at millimeter-wave frequency band, it is difficult to guarantee the proper operation of the 60 GHz ILO, especially including process–voltage–temperature (PVT) variations. The typical lock range of quadrature injection-locked oscillator (QILO) is only about several hundred MHz. As a result, a small shift of free-running frequency caused by PVT variations can cause the QILO unable to lock to the input signal. Therefore, a calibration scheme is needed to tune the free-running frequency of QILO for robustness of its operation in various conditions.

TABLE 19.1

Comparison between Different Topologies for Generating 60 GHz Quadrature LO Signal for a Direct Conversion Transceiver

Architecture	Phase Noise	I/Q Error	Area
QVCO at 60 GHz	Poor	Good	Good
Push–push VCO at 30 GHz + PPF/HC	Moderate	Moderate	Moderate
Subharmonic injection	Good	Good (with QILO calibration)	Moderate

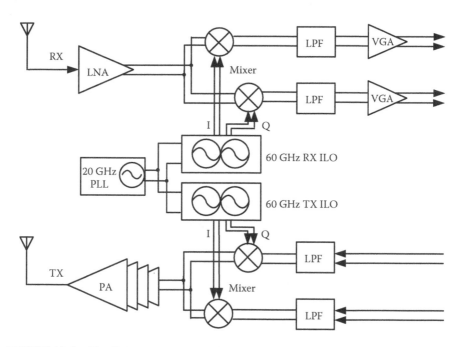

FIGURE 19.4   The 60 GHz direct conversion transceiver front end.

A conventional foreground calibration technique [11,14], which is operated only at startup, cannot track ILO frequency drift as temperature varies; thus, a background method that would automatically correct frequency drift by frequency feedback is needed inevitably. For TDD transceivers, one possible background calibration method is to calibrate the ILO for RX (TX) during TX (RX) time slots. Note that a typical TDD transceiver is composed of a transmitter and receiver that have separate ILOs for their own operation, as shown in Figure 19.4. In the TDD operation of an individual transceiver, once the transmitter is on, the receiver is off and vice versa. Therefore, once a transceiver is working in the transmitting mode, the free-running frequency of its ILO for the receiver part will be calibrated and vice versa.

As shown in Figure 19.5, if transceiver A (TRX A) is transferring data stream to transceiver B (TRX B), the data stream is divided into small packages. After the first package is transmitted to TRX B, an acknowledged package will be transmitted from TRX B back to TRX A to confirm that the first package was received successfully. TRX A will wait until acknowledged package is received completely before starting to transfer the next package. The most crucial part of timing that poses a time restriction for calibration occurs when an acknowledged packaged is being sent and received from TRX B and TRX A, respectively. The reason is because this length of time has shortest interframe space (SIFS) and shortest package length (PCK). As a result, the calibration time is restricted by two SIFS, each having a length of 3 μs and one PCK, with a length of 1.2 μs for the IEEE 802.11ad standard. The total time restriction for ILO calibration is, therefore, 7.2 μs. Conventional millimeter-wave calibration method, as shown in Figure 19.8, is infeasible for calibrating ILO since

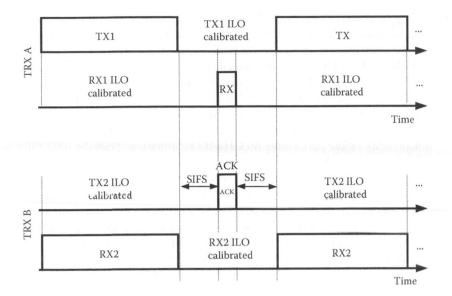

FIGURE 19.5   Calibration operation and timing of two transceivers.

two calibration procedures (step 1: calibrate ILFD and step 2: calibrate ILO) are required to carry out sequentially, which is time consuming.

In this chapter, a 60 GHz frequency synthesizer with background calibration is proposed for millimeter-wave TDD transceivers [15–17]. Only one calibration procedure is required, which significantly reduces calibration time cost. The proposed frequency synthesizer operates for a wide frequency range from 58.1 to 65.0 GHz, which supports all 60 GHz channels, including channel bonding defined by IEEE 802.15.3c, wirelessHD, IEEE 802.11ad, WiGig, and ECMA-387. This chapter is organized as follows: in Section 19.2, the proposed architecture is discussed. The following section describes the circuit implements in details. Section 19.4 demonstrates experimental results of proposed 60 GHz frequency synthesizer. Finally, conclusion is summarized in Section 19.5.

19.2   PROPOSED ARCHITECTURE

The basic approach adopted in this chapter uses a 20 GHz PLL to inject a 60 GHz ILO, which is operated as a frequency tripler. As mentioned previously, the correct operation of the 60 GHz ILO cannot be ensured due to inaccurate active/passive device modeling and PVT variations. Therefore, in order to guarantee that the 60 GHz ILO can acquire the injection locking to the 20 GHz PLL properly, it is necessary to automatically calibrate the free-running frequency of 60 GHz ILO to three times than the output frequency of the 20 GHz PLL.

During calibration, comparison between the derivative signal of 60 GHz ILO free-running frequency and 20 GHz PLL output frequency is involved. Figures 19.8 and 19.9 show a conceptual diagram of the 60 GHz frequency synthesizer with digital calibration using injection-locked frequency divider (ILFD), and a mixer for frequency reduction, respectively. Each one has its merits in terms of power dissipation, chip area, PVT tolerance, and calibration time. For TDD systems, transmitter (TX) and receiver (RX) operate in different time slots. Thus, the calibration of 60 GHz ILO for TX can be performed during RX time slots, and the calibration of 60 GHz ILO for RX can be performed during TX time slots. However, the time duration is limited to 7.2 μs only for each TX and RX slot according to IEEE 802.11ad, imposing a strict requirement for the calibration time of the 60 GHz ILO. In Figure 19.6, the ILFD for frequency reduction would be calibrated first to ensure its functionality before performing 60 GHz ILO calibration. Then the total calibration time is significantly increased, which demonstrates the infeasibility of calibration using ILFD for frequency down-conversion.

In order to reduce the calibration time, frequency down-conversion using mixer is employed as shown in Figure 19.7. The process of calibration is carried out when the 20 GHz PLL is locked. At the initial state or during TX time slots, the ILO for RX outputs its free-running frequency, and the output of ILO for RX is down-converted to a frequency around 20 GHz by a mixer, through doubling of the 20 GHz PLL injection signal. After two separate divider chains, the frequency coming from the output of mixer and the 20 GHz PLL are compared using digital calibration circuits. The output of digital calibration circuit directly controls the digital code for a digital-to-analog converter, giving rise to adjust the ILO free-running frequency. After several reference cycles, the 60 GHz ILO free-running frequency is closed to

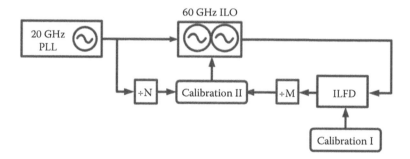

FIGURE 19.6 Conceptual diagram of 60 GHz frequency synthesizer with conventional calibration method.

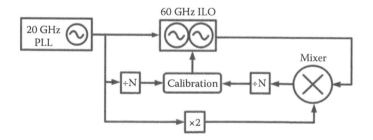

FIGURE 19.7 Conceptual diagram of 60 GHz frequency synthesizer with proposed calibration method.

three times the 20 GHz PLL output. Considering the calibration of ILO for RX (TX) is performed at TX (RX) time slots, the calibration method can run as a background calibration to monitor the 60 GHz ILO frequency over temperature drift.

19.3   CIRCUIT DESCRIPTION

The proposed 60 GHz frequency synthesizer mainly consists of a 20 GHz PLL, a 60 GHz quadrature ILO, a frequency down-converting circuit, and a digital calibration circuit. The block diagram of the proposed 60 GHz synthesizer is depicted in Figure 19.8.

19.3.1   20 GHz PLL

The 20 GHz PLL consists of a phase-frequency detector (PFD), a current-steering charge pump, a second-order on-chip low-pass filter, an LC-VCO with a tuning range of 17.9–21.7 GHz, followed by a divide-by-2 CML divider, a divide-by-3 E-TSPC divider, a digital divide-by-5 divider, and a digital multimodulus divider. The whole divide ratio can be controlled from 1620 to 1800 in steps of 30. This divide ratio combined with a 12 MHz frequency resulting from a halved off-chip 24 MHz reference enables synthesis of the required tones.

19.3.1.1   20G LC-VCO

The core VCO uses NMOS cross-couple architecture as shown in Figure 19.9. The inductor is designed using HFSS, and a model is constructed using measurement results from a standalone test structure. It is designed using a top metal to minimize via loss and has an inductance of about 240 pH and a quality factor of 14. The VCO is required to tune from 19 to 22 GHz to cover the required range, so the tuning part uses a 3-bit capacitor bank for discrete tuning and a varactor for continuous tuning. Switched varactors are used for discrete tuning since they have a higher quality factor than switched MIM capacitors at 20 GHz. Inverters are used to bias the gate of the varactors to avoid degradation in quality factor. The increase is due to the different gate biasing for both ON and OFF states, which would shift its capacitance curve in both states so that it is maximized. Other means to increase would result in a much more degradation of the quality factor and an increase in parasitic capacitance. In the same figure, M1 and M2 form a negative resistance to compensate for the loss in the tank circuit and to ensure a reliable startup. M3 and M4 make up the current source and provide high impedance at the common node to improve phase noise.

19.3.1.2   PFD/CP/Loop Filter

The PFD consists of two latches and a delay reset path. The delay reset path eliminates the dead zone of the PFD/CP, which reduces the resulting jitter. The charge pump utilizes a cascaded current source to improve current matching. Also, the reference current is supplied to allow for current adjustment. A second-order loop filter is adopted, followed by another filtering stage to further suppress ripples. Note that its pole is chosen to be much higher compared with the loop poles; this additional filtering does not impair loop stability. MOS capacitors with compensated

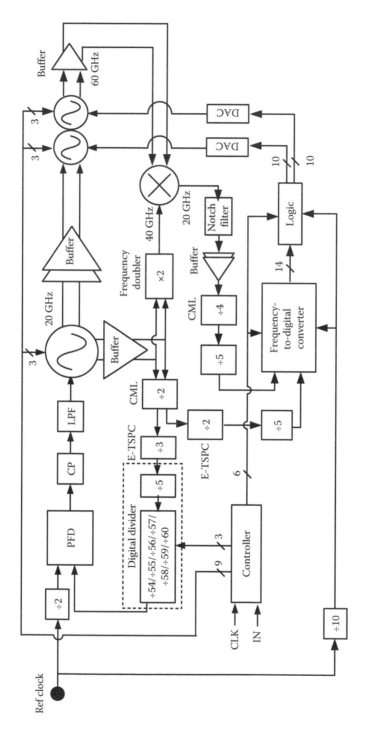

FIGURE 19.8   Block diagram of the proposed 60 GHz frequency synthesizer with background calibration.

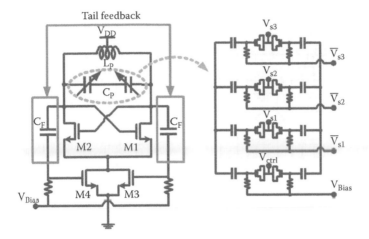

FIGURE 19.9   The 20 GHz LC-VCO with tail feedback.

method are adopted for the large capacitors in the loop filter since they benefit from higher capacitance density per area compared with MIM capacitors. Thus, the loop filter area is reduced significantly.

19.3.2   60 GHz ILO

The circuit schematic of 60 GHz ILO [2] is shown in Figure 19.10. A quadrature 60 GHz LO signal can be obtained due to a quadrature I/Q configuration. The free-running frequency can be adjusted by a switched capacitor bank for coarse tuning and a varactor that is controlled by the calibration circuit for fine tuning. The 20 GHz PLL signal is injected through tail transistor of I oscillator.

19.3.3   DOWN-CONVERTING SYSTEM

The down-converting system is composed of a frequency doubler, a mixer, and two notch filters. There are mainly two design considerations for the down-converting system. First, the occupied chip area should be minimized for low cost. Considering building blocks for down-converting system are operated in the millimeter-wave regime, on-chip impedance match and inductive peaking are necessary to improve the gain and output power of active circuits. In this work, miniaturized transmission line and custom-designed inductor help to reduce the whole chip area. Second, a sufficient large output power is critical to driving the following divider stage. In this section, detailed circuit implementations together with their simulation results are introduced.

19.3.3.1   Frequency Doubler

The frequency doubler operates as a push–push amplifier that combines two phases with 180° phase shift by combining output nodes. As a result, odd harmonics components are canceled out and second harmonic component survives. In this

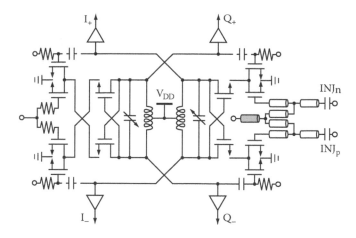

FIGURE 19.10   Circuit schematic of the quadrature ILO.

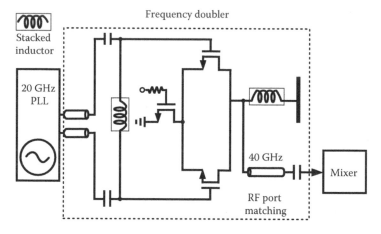

FIGURE 19.11   Circuit schematic of frequency doubler along with other building blocks.

implementation, a simple differential amplifier is used as shown in Figure 19.11, where output nodes are shorted. Even though the output signal is single ended, it is enough to carry frequency information for calibration. In order to transfer maximum power to load, the input impedance of this frequency doubler is matched to the output impedance of 20 GHz PLL buffers at 50 Ω. This is done by the use of transmission line and a custom-designed stacked inductor. The size of differential pair also affects the output amplitude. In this case, the width of transistors is chosen as 2 × 20 μm. Another stacked inductor is utilized at the output node to peak at high frequency. Meanwhile, the transmission line is utilized to match output node to RF port of the next-stage mixer. With 0 dBm input power from 20 GHz PLL, the simulated output power of the frequency multiplier is shown in Figure 19.12. The output power varies from about −9 to −7 dBm over the output frequency range from 38.88 to 43.2 GHz.

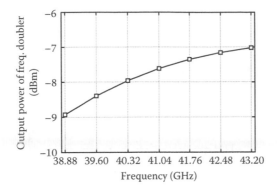

FIGURE 19.12   Simulated output power of the frequency doubler.

19.3.3.2   Mixer

A double-balanced mixer is widely used due to its excellent port-to-port isolations while maintaining reasonable conversion gain and noise performance. However, it requires differential RF inputs that would further complicate the topology of a frequency doubler or an additional balun might be needed. In this work, the mixer is designed using a single-balanced topology that mixes the single-ended RF output from the frequency doubler at frequency around 40 GHz, with differential LO output signal from ILO at frequency around 60 GHz. The detailed schematic of the single-balanced mixer with other building circuits is shown in Figure 19.13.

As mentioned earlier, a combination of stacked inductor at output node of the frequency doubler and transmission line is utilized to match the input impedance

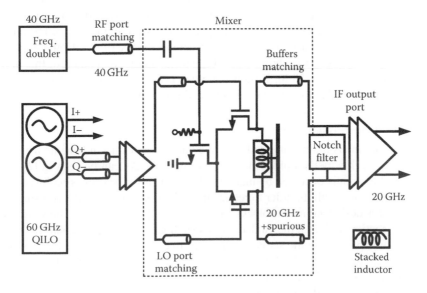

FIGURE 19.13   Circuit schematic of SSB mixer along with other building blocks.

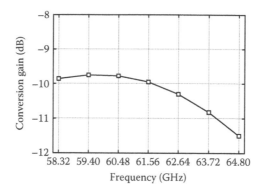

FIGURE 19.14   Simulated conversion gain of single-balanced mixer.

of RF port at the frequency of around 40 GHz. On the other hand, for matching the LO port, only transmission line is utilized since it is for higher frequency of 60 GHz. Another stacked inductor is used for the mixer core. The stacked inductor and transmission line are designed to match the impedance of notch filter and input impedance of buffers at the frequency around 20 GHz, which maintains 20 GHz signals and attenuates 60 GHz LO feed-through and other spurious signals. The size of differential pair is chosen as 2 × 10 μm, and the size of tail transistor is chosen as 2 × 26 μm. As shown in Figure 19.14, the simulated power conversion gain varies from −9.7 to −11.5 dB over the range of frequency from 58.32 to 64.8 GHz.

19.3.3.3   Notch Filter

Output signal of the mixer carries 20 GHz output signal and undesirable signals caused by strong LO input power. Thus, a passive LC band-rejection filter is implemented to attenuate the LO feed-through signal as shown in Figure 19.15. The inductor used in this filter is custom-designed with an inductance of 92 pH and a quality factor of 9. The capacitor is chosen as 75 fF. The resultant attenuation of this notch filter is shown in Figure 19.16. It shows an attenuation of about −20 dB at the frequency around 60 GHz to reduce the LO feed-through signal.

19.3.3.4   Transmission Line

Transmission line is mainly adopted for impedance match since it benefits from low loss. However, this method proves to be area intensive. As can be seen in [2], considerable chip area is occupied by matching blocks using a transmission line with a 10 μm signal line width. In order to reduce the chip area, a 6 μm width transmission is utilized for designing match blocks in this work [1]. Figure 19.17 shows the top and cross-sectional view of the guided microstrip line. It has a 6 μm signal line width and a 7 μm gap between the signal and ground line. The distance between the side grounds is 20 μm. Figure 19.18 shows the modeled and measured results including attenuation constant, phase constant, quality factor, and characteristic impedance of the transmission line adopted in this work.

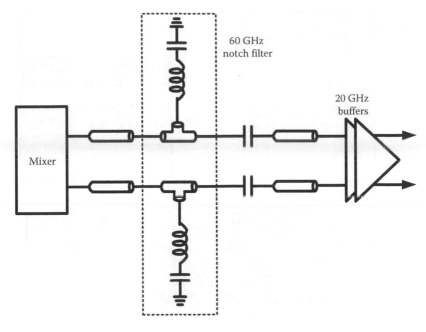

FIGURE 19.15   Schematic of notch filter along with other building blocks.

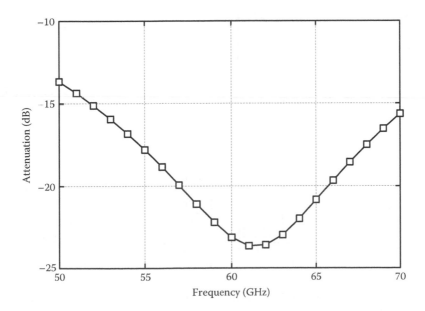

FIGURE 19.16   Simulated attenuation of the notch filter.

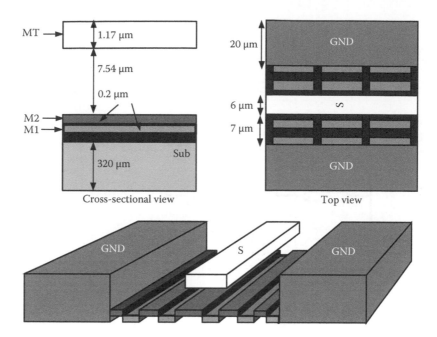

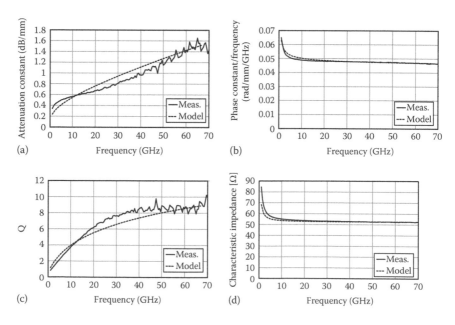

FIGURE 19.17   Structure of the transmission line.

FIGURE 19.18   Modeling results of (a) attenuation constant, (b) phase constant/frequency, (c) Q, and (d) characteristic impedance.

19.3.3.5   Summary

The full down-conversion system is shown in Figure 19.19. It is composed of three main blocks as discussed in detail in the previous chapters. The input impedance of the frequency doubler is designed to match output impedance of PLL buffers through the use of transmission line and a stacked inductor. The single-ended output of the frequency doubler feeds to RF port of mixer. The matching between two ports is also done with the use of a transmission line. The differential LO signals is input to one-stage buffers before input the LO ports of the mixer. The LO feed-through at the output of the mixer is filtered out by a notch filter. The output at frequency around 20 GHz is amplified by two-staged buffers. Then, it is divided to lower frequency by a divider chain and compared with the reference signal from 20 GHz PLL.

The amplitude of output signals from mixer carrying the 20 GHz output frequency is enhanced by two-stage buffers that are also designed using stacked inductors for gain peaking. From simulation, under a normal case of having 0 dBm input power from 20 GHz PLL and −10 dBm input from ILO, the output power over the range of frequency around 60 GHz is shown in Figure 19.20. The output power

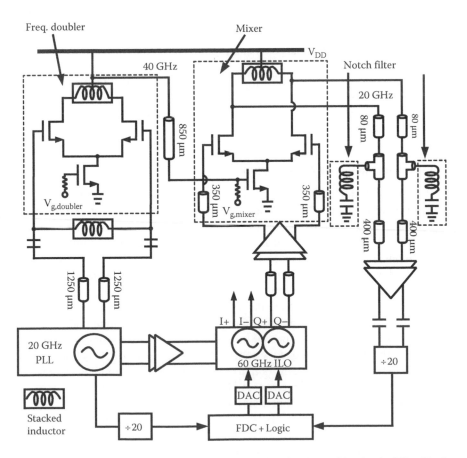

FIGURE 19.19   Full schematic of down-converting circuits along with other building blocks.

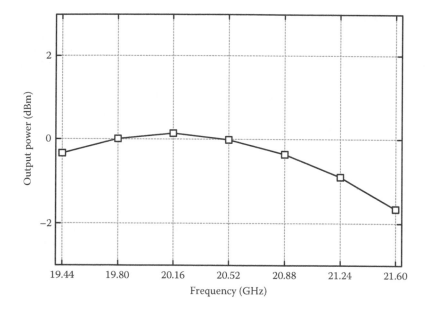

FIGURE 19.20   Simulated output power of the down-conversion system.

varies from −1.7 to 0.14 dBm in the range of frequency from 19.44 to 21.6 GHz. The strongest spurious frequency is lower than −21 dBm at frequency around 60 GHz. These output power levels are high enough to drive the next-stage dividers for digital calibration.

19.3.4 DIGITAL CALIBRATION CIRCUIT

The digital calibration circuit consists of a frequency-to-digital converter and digital logics. In order to increase the frequency resolution, digital calibration circuits are triggered at 1/10 of the reference clock. Derivative frequency signals coming from the 60 GHz ILO and 20 GHz PLL are measured using digital counters, respectively. Outputs of the digital counters, in the form of binary numbers, are compared in the following logic circuit during each reference cycle for digital circuits. The output of logic circuit directly controls the code for DAC, giving rise to adjust the ILO free-running frequency. If the derivative frequency coming from the 60 GHz ILO is greater than that coming from the 20 GHz PLL, the output code of digital logic circuit is decremented to speed down the 60 GHz ILO free-running frequency. Similarly, if the derivative frequency coming from the 60 GHz ILO is less than that coming from the 20 GHz PLL, the output code of digital logic circuit is incremented to speed up the 60 GHz ILO free-running frequency.

A possible calibration algorithm is outlined as follows. It is noted that TDD system allows foreground calibration. Thus, the entire calibration procedure can be divided into initial and intermittent calibration. Initially, ILO free-running frequencies are measured in the foreground by down-converting system and all-digital calibration circuits, and digital DAC codes representing the information of each frequency band

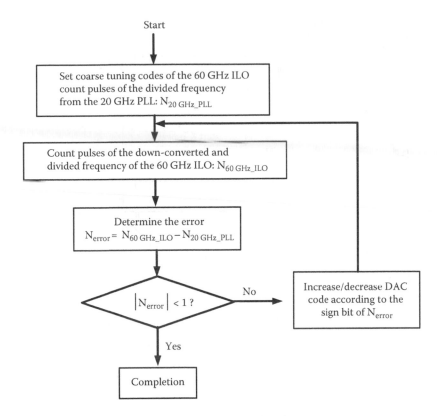

FIGURE 19.21   Flow chart of the ILO frequency calibration process.

(e.g., 58.32, 61.56 GHz, etc.) can be stored in memory. Thus, process variation can be calibrated during initial calibration. The flow chart of the ILO frequency calibration procedure is summarized in Figure 19.21. When millimeter-wave transceivers operate in TDD mode, the calibration system can monitor ILO free-running frequency drift mainly caused by temperature variation. In a short interval of time duration, the ILO free-running frequency shift is regarded as gradual and subtle changes. As a result, such a small frequency drift can be corrected easily within several calibration steps. The calibration algorithm mentioned earlier is not implemented on-chip but can be achieved easily by digital control circuits. An example of the calibration algorithm is shown in Figure 19.22.

19.4   EXPERIMENTAL RESULTS

The proposed 60 GHz frequency synthesizer is implemented in a standard 65 nm CMOS process. The microphotograph of the fabricated synthesizer is shown in Figure 19.23. The total chip area is 1.9 mm × 2 mm. PLL spectrum is measured with an Agilent E4448A PSA spectrum analyzer and a 50–75 GHz external mixer. The phase noise is evaluated using an Agilent E5052B SSA signal source analyzer and a 50–75 GHz external mixer.

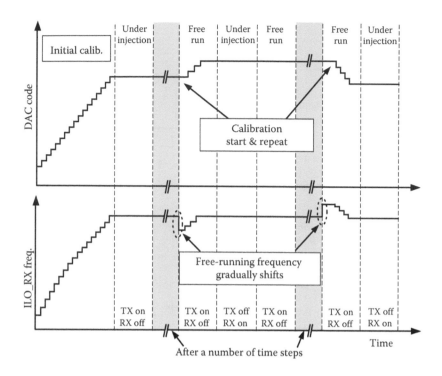

FIGURE 19.22   DAC code and output frequency of ILO for RX against time.

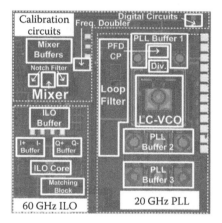

FIGURE 19.23   Chip microphotograph.

In order to assist in the evaluation of 60 GHz frequency synthesizer, a 60 GHz ILO TEG chip is also fabricated on the same die. As shown in Figure 19.24, the measured ILO free-running frequency range is 58.3–65.4 GHz and is covered with sufficient overlap between neighboring bands. The resulting frequency-tuning range can cover all 60 GHz bands. Even though the free-running frequency of ILO in the complete frequency synthesizer is slightly dropped due to undesired

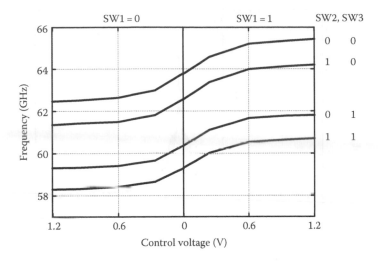

FIGURE 19.24   Measured free-running frequency range of ILO TEG.

parasitic capacitances, the measured tuning range of 58.1–65.0 GHz is still sufficient to cover all 60 GHz bands.

When the frequency of input reference clock is 24 MHz, the synthesizer can generate all 60 GHz channel bands (including channel bonding): 58.32, 59.4, 60.48,
61.56, 62.64, 63.72, and 64.8 GHz, which are defined by current 60 GHz wireless
standards. The frequency synthesizer consumes 72 mW from a 1.2 V power supply.
If the intermittent operation is disabled, the calibration circuits including down-
converting circuits and digital circuits consume 65 mW additionally.

The measured phase noise characteristic of the 60 GHz frequency synthesizer is
less than −114 dBc/Hz at 10 MHz offset across the entire frequency channel. The
measured reference spur level varies from −52 to −70 dBc. The output spectrum at
20.16 GHz is graphed in Figure 19.25 displaying the typical locked spectrum for the
20 GHz PLL. The graph demonstrates that the 12 MHz reference spur are kept 65 dB
below the carrier frequency, showing good matching between the charge pump currents. The output spectrum at 61.56 GHz is illustrated in Figure 19.26, indicating
the typical locked spectrum for the 60 GHz frequency synthesizer. As shown in
Figure 19.27, the typical measured phase noise is −95.7 dBc/Hz at 1 MHz offset,
from a carrier frequency of 61.56 GHz. Figure 19.28 shows the measured phase noise
across the entire frequency band.

Figure 19.29 shows the 60 GHz output spectra disabling and enabling the calibration. As can be seen, without carrying out calibration the ILO free-running
frequency is far from the third harmonic of the 20.52 GHz injected signal, causing failure of injection-locking operation. When enabling the calibration, the ILO
free-running frequency is corrected to vicinity of 61.56 GHz, thus guaranteeing the
injection-locking operation. The DAC signals were not brought off-chip and could
not be measured. It is noted that the output power indicated in Figure 19.27 does not
account for losses from cables and an external 50–75 GHz down-conversion mixer.
The calibrated output power is −8 dBm.

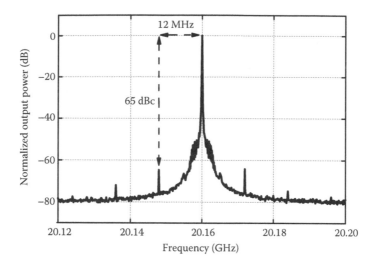

FIGURE 19.25   Measured spectrum at 20.16 GHz.

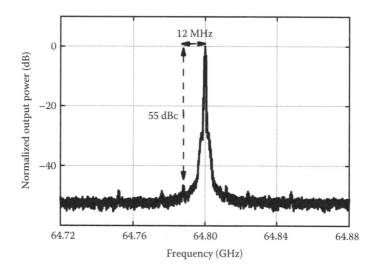

FIGURE 19.26   Measured spectrum at 64.80 GHz.

Table 19.2 compares the present 60 GHz frequency synthesizers/VCOs with the state-of-the-art publications [1,11–13,18–20]. This work could cover all 60 GHz channels and demonstrate very good phase noise performance with quadrature output phase (−96 dBc/Hz at 1 MHz and −117 dBc/Hz at 10 MHz at carrier frequency of 61.56 GHz). To the author's best knowledge, it is the first 60 GHz frequency synthesizer with frequency calibration for TDD transceivers.

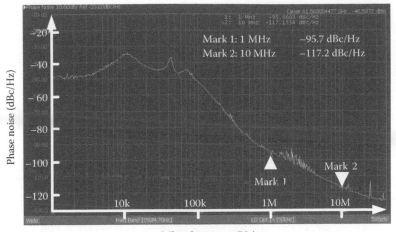

FIGURE 19.27   Measured phase noise at a carrier of 61.56 GHz.

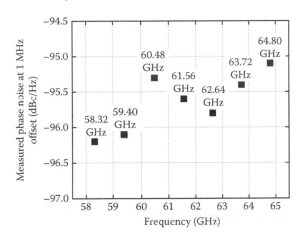

FIGURE 19.28   Measured phase noise performance across the entire frequency band.

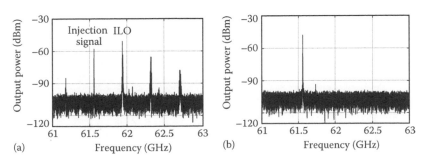

FIGURE 19.29   Measured 60 GHz output spectrum when (a) disabling the calibration and (b) enabling the calibration.

TABLE 19.2

Performance Comparison with Other State-of-the-Art 60GHz PLLs/VCO

Feature	[5] PLL	[12] PLL	[13] PLL	[2] PLL	[18] PLL	[19] VCO	[20] PLL	This Work PLL
	QVCO at 60 GHz	Push–Push VCO at 30 GHz + Coupler	Subharmonic Injection	Subharmonic Injection	Diff. VCO at 60 GHz	Mag. Coupled	VCO at 48 GHz	Subharmonic Injection + Frequency Calibration
CMOS Tech (nm)	45	90	65	65	90	65	65	65
f_{ref} (MHz)	100	203.2	36	36	60	65	54	24
Frequency (GHz)	57–66	59.4–64	58–63	58–64.8	61–63	56–60.35	42.1–53	58.1–65.0
Phase noise (dBc/MHz)	−75 at 1	−73 at 1, −112 at 10	−96 at 1, −113 at 10	−95 at 1	−80 at 1, −110 at 10	−95 at 1, −117 at 10	−95.6[a] at 1	−96 at 1, −117 at 10
Power (mW)	78	76	80	68	78	22	72	137 (Initial only) 72
Output type	Quad.	Quad.	Quad.	Quad.	Diff.	Quad.	Diff.	Quad.
Voltage/temperature tracking	No	—	No	No	No	—	No	Yes
Channel bonding supporting	No	No	No	No	No	—	No	Yes
60 GHz band coverage	All 4 bands	3 Bands	3 Bands	All 4 bands	1 Band	2 Bands	All 4 bands	All 4 bands

[a] Normalized to 62.64 GHz from 50.112 GHz measurement.

19.5   CONCLUSION

A 60 GHz frequency synthesizer with frequency calibration scheme that can support IEEE 802.15.3c, wirelessHD, IEEE 802.11ad, and ECMA-387 TX/RX front end is reported. A frequency calibration scheme is proposed to monitor the frequency drift caused by environmental variations. With careful design, the proposed synthesizer can be suited for millimeter-wave TDD transceivers.

REFERENCES

1. K. Okada, K. Kondou, M. Miyahara, M. Shinagawa, H. Asada, R. Minami, T. Yamaguchi et al., A full 4-channel 6.3 Gb/s 60 GHz direct-conversion transceiver with low-power analog and digital baseband circuitry, in: *IEEE International Solid-State Circuits Conference (ISSCC), Digest of Technical Papers*, San Francisco, CA, pp. 218–219, 2012.
2. K. Okada, K. Matsushita, K. Bunsen, R. Murakami, A. Musa, T. Sato, H. Asada, N. Takayama, N. Li, and S. Ito, A 60 GHz 16QAM/8PSK/QPSK/BPSK direct-conversion transceiver for IEEE 802.15.3c, in: *IEEE International Solid-State Circuits Conference (ISSCC), Digest of Technical Papers*, San Francisco, CA, pp. 160–161, 2011.
3. S. Emami, R.F. Wiser, E. Ali, M.G. Forbes, M.Q. Gordon, X. Guan, S. Lo, P.T. McElwee, J. Parker, and J.R. Tani, A 60 GHz CMOS phased-array transceiver pair for multi-Gb/s wireless communications, in: *IEEE International Solid-State Circuits Conference (ISSCC), Digest of Technical Papers*, San Francisco, CA, pp. 164–165, 2011.
4. T. Tsukizawa, N. Shirakata, T. Morita, K. Takana, J. Sato, Y. Morishita, M. Kanemaru et al., A fully integrated 60 GHz CMOS transceiver chipset based WiGig/IEEE802.11ad with build-in self calibration for mobile applications, in: *IEEE International Solid-State Circuits Conference (ISSCC), Digest of Technical Papers*, San Francisco, CA, pp. 230–231, 2013.
5. G.M.V. Vidojkovic, K. Khalaf, V. Szortyka, K. Vaesen, W.V. Thillo, B. Parvais, M. Libois et al., A low-power 57–66 GHz transceiver in 40 nm LP CMOS with −17 dB EVM at 7 gb/s, in: *IEEE International Solid-State Circuits Conference (ISSCC), Digest of Technical Papers*, San Francisco, CA, pp. 268–269, 2012.
6. 802.15.3c-2009. IEEE Std., October 2009. [Online]. Available: http://standards.ieee.org/getieee802/download/802.15.3c-2009.pdf.
7. IEEE802.11ad. IEEE Std. [Online]. Available: http://standards.ieee.org/develop/project/802.11ad.html.
8. ECMA. [Online]. Available: http://www.ecma-international.org/publications/files/ECMA-ST/ECMA-387.pdf.
9. WirelessHD. [Online]. Available: http://www.wirelesshd.org/pdfs/WirelessHD-Specification-Overview-v1.1May2010.pdf.
10. WiGig. [Online]. Available: http://wirelessgigabitalliance.org/specifications/.
11. K. Scheir, G. Vandersteen, Y. Rolain, and P. Wambacq, A 57–66 GHz quadrature PLL in 45 nm digital CMOS, in: *IEEE International Solid-State Circuits Conference (ISSCC), Digest of Technical Papers*, San Francisco, CA, pp. 494–495, 2009.
12. C. Marcu, D. Chowdhury, C. Thakkar, J.D. Park, L.K. Kong, M. Tabesh, Y. Wang, B. Afshar, A. Gupta, and A. Arbabian, A 90 nm CMOS low-power 60 GHz transceiver with integrated baseband circuitry, *IEEE Journal of Solid-State Circuits*, 44(12), 3434–3447, December 2009.
13. A. Musa, R. Murakami, T. Sato, W. Chaivipas, K. Okada, and A. Matsuzawa, A low phase noise quadrature injection locked frequency synthesizer for MM-wave applications, *IEEE Journal of Solid-State Circuits*, 46(11), 2635–2649, November 2011.

14. S. Pellerano, R. Mukhopadhyay, A. Ravi, J. Laskar, and Y. Palaskas, A 39.1–41.6 GHz ΔΣ fractional-N frequency synthesizer in 90 nm CMOS, in: *IEEE International Solid-State Circuits Conference (ISSCC), Digest of Technical Papers*, San Francisco, CA, pp. 484–485, 2008.

15. W. Deng, T. Siriburanon, A. Musa, K. Okada, and A. Matsuzawa, A 58.1–65.0 GHz frequency synthesizer with background calibration for millimeter-wave TDD, in: *IEEE European Solid-State Circuits Conference (ESSCIRC)*, Bordeaux, France, pp. 201–204, 2012.

16. T. Siriburanon, W. Deng, A. Musa, K. Okada, and A. Matsuzawa, Sub-harmonic injection-locking frequency synthesizer with frequency calibration scheme for use in 60 GHz TDD transceivers, in: *IEEE/ACM Asia South Pacific Design Automation Conference (ASP-DAC)*, Yokohama, Japan, 2013.

17. W. Deng, T. Siriburanon, A. Musa, K. Okada, and A. Matsuzawa, A sub-harmonic injection-locked quadrature frequency synthesizer with frequency calibration scheme for millimeter-wave TDD transceivers, *IEEE Journal of Solid-State Circuits*, 48(7), 1710–1720, July 2013.

18. H. Hoshino, R. Tachibana, T. Mitomo, N. Ono, Y. Yoshihara, and R. Fujimoto, A 60-GHz phase-locked loop with inductor-less prescaler in 90-nm CMOS, in: *IEEE European Solid-State Circuits Conference (ESSCIRC)*, Muenchen, Germany, pp. 472–475, 2007.

19. U. Decanis, A. Ghilioni, E. Monaco, A. Mazzanti, and F. Svelto, A mm-wave quadrature VCO based on magnetically coupled resonators, in: *IEEE International Solid-State Circuits Conference (ISSCC), Digest of Technical Papers*, San Francisco, CA, pp. 280–281, 2011.

20. D. Murphy, Q.J. Gu, Y.C. Wu, H.Y. Jian, Z. Xu, A. Tang, F. Wang, and M.C.F. Chang, A low phase noise, wideband and compact CMOS PLL for use in a heterodyne 802.15.3c transceiver, *IEEE Journal of Solid-State Circuits*, 46(7), 1606–1617, July 2011.

20 Digitally Controlled Oscillators for Wireless Applications

Chih-Ming Hung

CONTENTS

20.1 INTRODUCTION

In modern wireless communication systems, frequency selectivity plays a very important role to extract the desired radio signals out of the increasingly crowded spectrum. A typical transmitter and a receiver along with an interfering transmitter are shown in Figure 20.1. With the tightly spaced communication channels, tunable

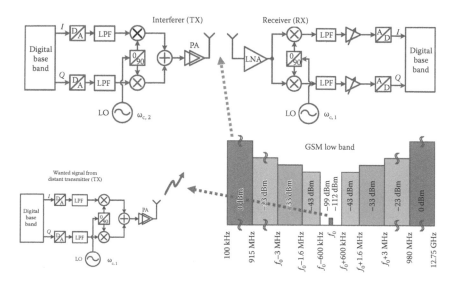

FIGURE 20.1 A modern communication system often encounters strong interference from nearby transmitters. The spectrum becomes very crowded due to increased network capacity. The interfering power level can be >100 dB higher than the wanted signal.

radiofrequency (RF) bandpass filters with sufficient out-of-channel and out-of-band attenuation are difficult to achieve. Therefore, a precise frequency translation of radio signals is needed to shift the desired RF signals down to a baseband frequency or an intermediate frequency (IF) to ease the suppression of interference.[1]

Frequency synthesizers have been widely used as the local oscillators in transceivers to perform the frequency translation as shown in Figure 20.1. Traditionally, for wireless applications, they have been mostly implemented based on a charge-pump phase-locked loop (CPPLL)[2] that suffers from nonidealities such as finite output impedance of the charge pump and voltage-dependent loop filter components resulting in nonlinearity in the loop dynamics and excessive spectral emission. In addition, because of the analog nature of the synthesizer circuits, when migrating to newer process technology nodes, significant engineering efforts are needed to redesign building blocks due to reduced voltage headroom, lower transistor early voltage, higher $1/f$ noise, etc.

With the advancement of digital CMOS processes, timing resolution has surpassed voltage resolution. It becomes possible to realize a PLL in a fully digital fashion to facilitate system-on-chip (SoC) integration with digital signal processing units. An all-digital PLL (ADPLL) for wireless communication systems was first published in 2003.[3] At the heart of the ADPLL lies the digitally controlled oscillator (DCO),[4] which is a counterpart of the voltage-controlled oscillator (VCO) in a conventional PLL. In this chapter, we will discuss design considerations for a DCO system from circuit design and frequency synthesizer system aspects using nanometer-scale CMOS technologies.

20.2 DCO VERSUS VCO

Both a DCO and a VCO serve the same purpose of generating oscillating RF signals. Figure 20.2 shows how they are exploited in PLLs, which help to improve output frequency stability and the phase noise within PLL loop bandwidth based on the spectral purity of the input reference source, F_{ref}. For most wireless applications, both types of oscillators need to achieve low phase noise across close-in and far-out frequency offsets away from the carrier frequency, sufficient frequency tuning range, low power consumption, and low sensitivity to supply and ground variations (defined as frequency pushing and quantified as $\Delta f/\Delta V_{supplyground}$). Similar design techniques can be applied to both DCOs and VCOs, and similar design trade-offs can be shared between them. For example, using a secondary LC tank[5] operating at the second harmonic of the main oscillation to reduce phase noise can be employed to both types of oscillators. Therefore, the same key parameters and figure of merit are often used for comparison in publications.

Although there are similarities in basic oscillator parameters, there is distinct difference in designing a DCO versus a VCO. Figure 20.3 shows the oscillator cores of these two types of oscillators. The topologies of LC tanks, cross-coupled transistors, and biasing circuitry are identical for a fair comparison. The DCO intentionally avoids any analog tuning voltage and is realized as an ASIC cell with only digital inputs and outputs. Unlike the VCO that relies on a developed tuning voltage (V_{ctl}) from a conventional PLL to control its oscillating frequency with a VCO gain, K_{VCO}, defined as $\Delta f/\Delta V_{vctl}$ usually in the order of hundreds of MHz/V, the DCO frequency is controlled by digital switching of capacitor units. Consider the $C-V$

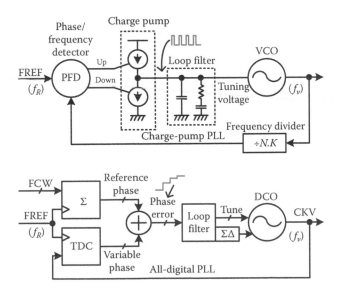

FIGURE 20.2 A DCO and a VCO deployed in an ADPLL and a CPPLL, respectively. DCO gain is typically designed to be lower than VCO gain, thus less susceptible to noise.

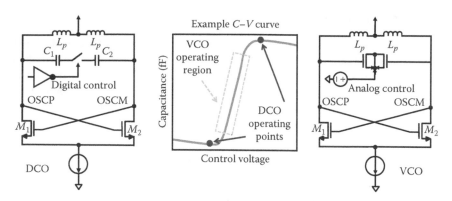

FIGURE 20.3 A schematic view for DCO and VCO cores. There is a distinct difference in the varactor design, requirements, and operating conditions.

curve shown in Figure 20.3; if the unit capacitors in the DCO always operate at one of the two flat portions where $\Delta C/\Delta V \approx 0$, the DCO should exhibit low susceptibility to frequency pushing and low vulnerability to power supply and ground noise. It can also be observed from Figure 20.2 that since there is no analog component in the ADPLL, there are less noise contributors to concern compared with the conventional PLL.

Modern CMOS processes allow creating extremely small yet well-controlled unit capacitors for DCO frequency tuning. The switchable capacitance, ΔC_{unit}, straight from an inversion-mode gate capacitor can be as low as 10's aF. Consider an implementation for a 2.4 GHz RF,[6] the corresponding frequency step size, denoted as K_{DCO}, is around 23 kHz, which is still too coarse for wireless applications. Therefore, techniques such as dithering are required to enhance the time-averaged frequency resolution. More details will be discussed in Section 20.3. Compared to a conventional PLL where dithering is applied to its feedback frequency divider shown in Figure 20.2, quantization noise introduced by dithering the DCO unit capacitor will not be attenuated by the PLL response. Hence, dithering in a DCO needs to be operated at a much higher speed than that in a conventional PLL. The dithering rate is a trade-off among ΔC_{unit}, quantization noise, and DCO frequency tuning range. More details will be discussed in later sections.

20.3   DCO SYSTEM DESIGN CONSIDERATIONS

In this section, DCO system design considerations including circuit topology, choice of oscillator core frequency, partition of capacitor banks, selection of device component, mismatch of unit capacitors, and reliability of circuit elements will be discussed. An example focusing on one cellular phone system that has stringent phase noise and frequency tuning range requirements to its LOs will be illustrated.

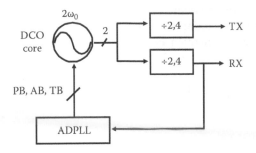

FIGURE 20.4 A DCO system is built as an ASIC cell with digital I/Os including the RF outputs.

20.3.1 DCO System Topology

Shown in Figure 20.4 is an example of a DCO system. It is built as an ASIC cell with only digital *I/Os* even for the RF outputs. The 10%–90% rise and fall time (T_{rise} and T_{fall}) of an inverter buffer in a low-power 28 nm CMOS process is <20 ps and is still <40 ps in an older 90 nm node. Therefore, modern CMOS processes are very capable of handling RF waveforms in the low GHz range. There are only two frequency division ratios in this example. It can be extended to have additional ratios with a proper frequency plan to cover more frequency bands such as those for 4G LTE standard. If quadrature outputs are required, at least one divisor of 2 is needed with 50% duty cycle input to the divide-by-2 circuit. Compared to a quadrature DCO or a differential DCO followed by a polyphase splitter, having one oscillator core operating at twice frequency cascaded with a frequency divider to generate quadrature phases is beneficial for low-power consumption, small silicon size, and low-phase noise especially under the constraint of low on-chip inductor quality factor (Q).

20.3.2 Oscillator Core Frequency

In a multimode multiband transceiver such as one for cellular phone systems, the needed frequency coverage is in excess of 120% (700–2700 MHz, not continuously) and will be extended to ~135% (700–3500 MHz) while the LO phase noise needs to be backward compatible with the older yet tougher 2G system requiring <–164 dBc/Hz at >20 MHz frequency offset from 914.8 MHz carrier. In light of the high-speed transistors from advanced CMOS processes, an economic solution is to increase the oscillator core frequency, and use frequency dividers to generate desired LO frequencies as illustrated in Figure 20.4. If the oscillator core can cover 3296–4340 MHz with margins on both sides, after division, both 2G and 3G standards can be well covered corresponding to ~90% (824–2170 MHz) frequency coverage. Additional techniques for extending frequency tuning range will be discussed in later sessions.

The benefit to phase noise by increasing the DCO core frequency can be understood with the well-known Leeson's model.[7] For simplicity, neglecting most parasitic elements except the parasitic series resistance, r_s, of the inductor, inductor quality factor $Q_L = \omega \cdot L/r_s$, where ω is the operating frequency and L is the series inductance. Leeson's phase noise model can be rewritten as

$$L(\Delta\omega) = 10 \cdot \log \frac{(1/2) \cdot \overline{i_{tot}^2} \cdot |H(\Delta\omega)|^2}{(1/2) \cdot V_{amp}^2} \propto \frac{F(\omega_{osc}/Q_{tank}\Delta\omega)^2}{I^2 R_p}, \qquad (20.1)$$

where

$H(\Delta\omega)$ and $\overline{i_{tot}^2}$ are noise transfer function and total noise power injected into the LC tank, respectively

I and V_{amp} are current consumption and voltage amplitude of the DCO, respectively

ω_{osc} and $\Delta\omega$ are resonant frequency of the DCO core and frequency offset from ω_{osc}, respectively

F is the amplifier noise factor

Q_{tank} and R_p are the tank quality factor and equivalent parallel tank resistance, respectively

Referring to low band frequencies as ω_{LB}, if ω_{osc} was increased from ω_{LB} to 4 ω_{LB}, the LC product for the tank needed to be decreased to 1/16 of its original value. Again, for simplicity, assuming that I, F, and L/r_s are constant, $R_p \approx Q_{tank}^2 \cdot r_s$ and $Q_{tank} \approx Q_L$, from Equation 20.1, increasing ω_{osc} from ω_{LB} to 4 ω_{LB} would increase Q_{tank} and R_p by four times, resulting in a 6 dB phase noise improvement. The larger R_p also implies that the large-signal transconductance, G_m, from the active transistors in the DCO core may be reduced to decrease the noise factor F, further improving the phase noise.

The aforementioned derivations assumed $Q_L \ll Q_C$, where Q_C is a quality factor of the LC tank capacitor. If ω_{osc} was increased significantly such that Q_L approaches or surpasses Q_C, the benefit of phase noise improvement will be reduced. Referring to Figure 20.5, L and C with parasitic series resistance can be

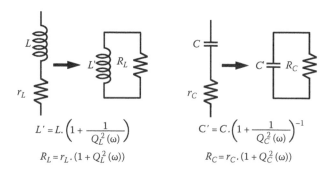

FIGURE 20.5 Transformation between series RL/RC and parallel RL/RC networks that is useful for LC resonator designs.

modeled as frequency-dependent parasitic shunt resistance. The total equivalent parallel tank resistance can be expressed as

$$R_p = \frac{R_L \cdot R_C}{R_L + R_C} \quad \text{and} \quad Q_{\text{tank}} = \omega C' R_p = \frac{R_p}{\omega L'} \qquad (20.2)$$

Observing R_p and Q_{tank} in Equation 20.2, one can assess how much to increase ω_{osc} based on Equation 20.1.

The inductor model used so far is rather ideal where Q_L increases linearly proportional to frequency. A more realistic yet simplified model is shown in Figure 20.6. With the parasitic elements, Q_L gradually saturates then drops at high frequencies due to reduced quality factor of the parasitic RC network, which is inversely proportional to frequency, as well as increased series inductance and resistance due to the parasitic parallel C_c and skin and proximity effects. Therefore, the design intuition from Equations 20.1 and 20.2 holds, but it is not quantitatively precise when operating near inductor peak-Q frequency.

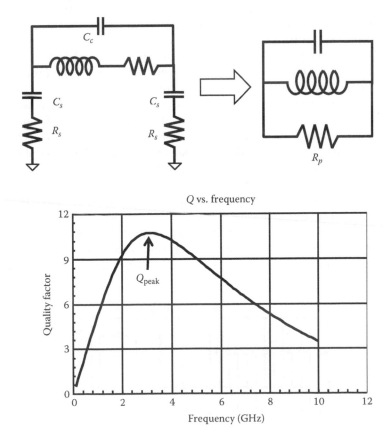

FIGURE 20.6 A simplified inductor model with parasitic elements. Intuitive understanding can be gained using the network transformation shown in Figure 20.5.

Another concern for increasing DCO core frequency is the trade-off with frequency tuning range. Since R_p does not linearly increase with frequency and Q_C is not $\gg Q_L$, the gm from cross-coupled transistors cannot be scaled down proportionally and the parasitic capacitance associated with tank varactor will be inevitably increased to maintain sufficient Q_C at a higher operating frequency. As a result, C_{tune}/C_{fix} ratio is reduced where C_{tune} is the tunable tank capacitance and C_{fix} is the fixed parasitic capacitance coming from all device elements connected to the LC tank. However, in advanced process nodes such as 40 and 28 nm standard digital CMOS, C_{fix} is becoming smaller because transistor length has been scaled down tremendously. Consequently, there is no major concern to increase operating frequency up to at least 12 GHz for typical commercial applications such as 5–6 GHz wireless local area network (WLAN) communication system.

20.3.3   DCO VARACTOR BANKS

The heart of an LC DCO is the digital-to-frequency conversion formed by capacitor banks that can be fully controlled by digital bits without any additional DAC. The principle is shown in Figure 20.7, where $-R$ is the negative resistance produced by an active circuitry, L_{tank} is the LC tank inductance, and $C_{n-1} - C_0$ is the capacitor array that can be individually switched between a high-capacitance and a low-capacitance state by the digital data bus $d_{n-1} - d_0$. The capacitor array can be arranged in a binary, a linear, a segmented or a combination of various encoding schemes. Large capacitor units give coarse frequency tuning steps, and small capacitor units provide a fine frequency tuning resolution. As mentioned in Section 20.2, the finest possible frequency step by switching the smallest unit capacitor is still too coarse. One solution is to dither the fine tuning bank LSB capacitor (Figure 20.8) or multiple of them by digital control bits in order to create a time-averaged unit capacitance smaller than that of a physical LSB capacitor.

In a VCO, the capacitor array is often segmented into a coarse tuning bank with discrete capacitance steps and a fine tuning varactor whose capacitance can be continuously varied. Since K_{VCO} for the fine tuning bank is desired to be as small as possible for low phase noise, the coarse tuning bank is required to cover essentially the full VCO frequency tuning range with its LSB frequency step smaller than the

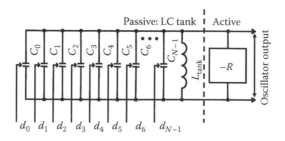

FIGURE 20.7   A conceptual diagram of a DCO. The varactors are individually added or subtracted from the LC tank.

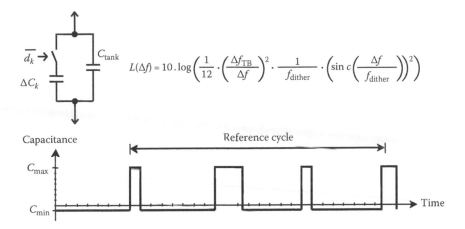

$$L(\Delta f) = 10 \cdot \log\left(\frac{1}{12} \cdot \left(\frac{\Delta f_{TB}}{\Delta f}\right)^2 \cdot \frac{1}{f_{dither}} \cdot \left(\operatorname{sin} c\left(\frac{\Delta f}{f_{dither}}\right)\right)^2\right)$$

FIGURE 20.8   One or few varactors are dithered to achieve a fine frequency step smaller than that from toggling a unit varactor.

total frequency tuning range of the fine tuning bank. Therefore, for a wide tuning range VCO, the coarse tuning bank needs to resemble a DAC with a large number of bits and a small differential nonlinearity (DNL), which is challenging. When DNL exceeds fine tuning bank frequency range, the VCO would suffer a discontinuity in its frequency range (Figure 20.9), resulting in a yield loss in production. For a DCO, there is a similarity in that K_{DCO} for the fine-step capacitor array needs to be small enough to limit the quantization noise[8] while large coarse- and fine-step capacitor arrays would introduce excessive parasitic capacitance from interconnect metal routed to all elements. To break the trade-off, the LC tank capacitor can be partitioned into three or more banks: for example, PVT Bank (PB) operating in the

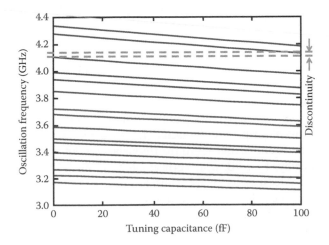

FIGURE 20.9   Frequency discontinuity commonly seen in a LC tank with segmented varactor banks when the coarse tuning bank is far greater than the fine-tuning bank.

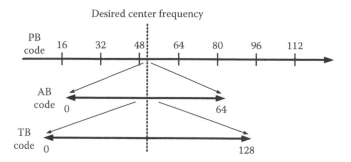

FIGURE 20.10 An example partition of DCO varactor banks with appropriate frequency overlaps to eliminate the frequency discontinuity illustrated in Figure 20.9.

beginning of synthesizer locking procedure to calibrate for process, voltage and temperature variations, Acquisition Bank (AB) operating after PVT calibration to reach a finer resolution smaller than a communication channel bandwidth, and Tracking Bank (TB) starting after AB operations for synthesizer tracking, and data modulation. An example is demonstrated in Figure 20.10.

There is no constraint for encoding of each bank, but there is a good practice to serve as a guideline. Starting with TB, its unit capacitor and frequency step is determined by the device component of choice that will be further discussed in a later section. In addition to covering the DNL of AB, it is responsible for PLL tracking considering supply disturbance, frequency pushing and frequency drift over temperature, inaccuracy in the acquisition process up to ±0.5 AB LSB, as well as phase or frequency modulation in a transmitter. Therefore, minimizing DNL of TB is critical. A natural choice would be a thermometer-coded array. In the case of polar modulation,[9] even for a relatively narrow-band system such as GSM with 200 kHz channel bandwidth, the phase modulation path can have about 1 MHz bandwidth. As mentioned earlier, there is a trade-off with parasitic capacitance that limits the total TB array size. Hence, total TB frequency range should be budgeted to be between ±4 and ±5 LSB of AB for a GSM DCO. For a wider bandwidth system such as WCDMA, the TB array size needs to be increased even when system-level techniques are applied[10] to reduce the phase modulation bandwidth.

PB needs to cover a wide frequency tuning range. Even the LC tank inductance could be programmed by software, for each inductance, PB often spans >1 GHz tuning range at oscillator core frequency. If the frequency resolution is limited by ~5 fF unit capacitor deployed for PB array corresponding to 5–10 MHz frequency step, PB needs to have at least 8 bits in size. Thus, a binary-coded capacitor array whose DNL can be controlled to be within ±1 LSB (i.e., monotonic) is a natural choice. The PB unit capacitor can be further decreased with trade-offs from choices of device components and sensitivity to parasitic capacitance from surrounding environment. More details will be discussed in the later section.

AB has a medium frequency step and is intended to buffer between PB and TB. Its total frequency tuning range and resolution are bounded by PB DNL and TB array

TABLE 20.1

A Summary of the DCO Varactor Banks

Notation	Varactor Banks	Coding	Δf at 3600 MHz
PB	PVT	8-bit binary	7.5 MHz
AB	Acquisition	64-bit unit	540 kHz (ΔC = 500 aF)
TB	Tracking	128-bit unit	40 kHz (ΔC = 50 aF)

size, respectively. The AB array is normally not large. Hence, a thermometer-coded capacitor bank is common. Table 20.1 summarizes such partition with an example.

20.3.4 DCO CORE

There are a variety of oscillator topologies such as Colpitts and a differential cross-coupled transistor pairs that can be used for the DCO core. With an example based on the latter that is the most commonly used topology for wireless applications, a simplified DCO core schematic is shown in Figure 20.11. L_p is the equivalent half-circuit inductor, C_{PB}, C_{AB}, and C_{TB} are the PB, AB, and TB capacitor arrays, respectively, M_1 and M_2 form the cross-coupled negative resistance pair, L_s and C_s form the high-impedance secondary LC tank, M_0 is used for controlling the DCO current consumption, and C_{GND} is to set an AC ground at drain node of M_0. Having a low supply voltage design in mind, when the oscillator voltage swing becomes large, M_1 and M_2 alternately go into linear region for some time in each oscillation period presenting low parallel resistance to the primary LC tank, thereby degrading the overall tank Q. As a result, the phase noise performance is deteriorated. The secondary LC tank is used to preserve primary LC tank Q as much as possible by having high impedance in series with the gm pair[5] at the source node. The resulting oscillator voltage waveform is less clipped even though M_1 and M_2 still periodically operate in linear region. Instead of tying C_s to the real ground net as originally depicted in [5], since C_{GND} is large, locally connecting C_s in parallel with L_s has a benefit that there are less parasitic L and C from interconnect to offset the target resonance. M_0 can be biased either in linear or saturation region. Operating M_0 in linear region with a large shunt capacitor, C_{GND}, connected in parallel is advantageous for reducing thermal noise contribution from M_0. The supply rejection and frequency pushing will be degraded. However, combining a low-drop-out regulator (LDO) that is usually present for DCO supply, frequency pushing of <20 MHz/V measured at LDO output and from 2 GHz carrier frequency is typical and sufficient.

In Figure 20.11, the gm pair is formed using NMOS transistors. In some process nodes such as one of the commercially available 28 nm processes, PMOS may be a better choice because its driving strength and transistor noise are better than NMOS. For applications that are sensitive to current consumption, a fully complementary CMOS topology is often preferred. However, if phase noise performance is demanding, the secondary LC tank may need to be duplicated so that high impedance is

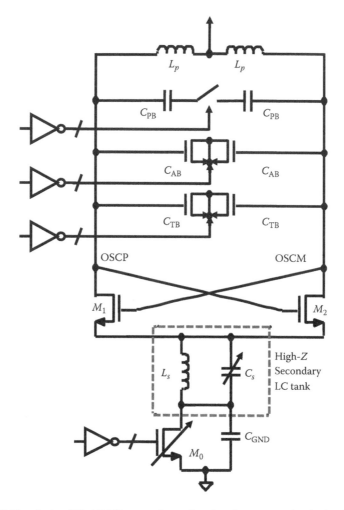

FIGURE 20.11 A simplified DCO core schematic using the varactor banks in Figure 20.10 and a modified LC network for the high-Z tank.

seen from source nodes of both NMOS and PMOS to reduce loading to the tank Q. Multiple secondary LC tanks certainly have a penalty for silicon area.

20.3.4.1 Design Considerations for Capacitor Arrays

In standard CMOS processes, there are a few types of passive capacitors that can be categorized into two groups: fixed-value capacitors and varactors. The former requires switches to have high- and low-capacitance states while the latter can be continuously tuned by a control voltage. Fixed-value capacitors include metal-insulator-metal (MIM) capacitors where a dedicated thin layer of insulator is used as the dielectric between metal plates, and metal-oxide-metal (MoM) capacitors where metal fingers, interlayer dielectric (ILD) and intermetal dielectric (IMD) of the back-end metal system are used to form capacitors. The scaling of CMOS processes has increased MoM

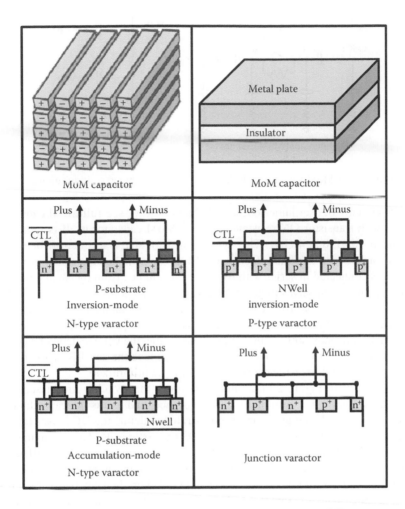

FIGURE 20.12   Candidate capacitor components available in standard CMOS processes to form DCO varactors.

capacitance/area density significantly to ~5 fF/μm² in 28 nm using 5 lower-level metal layers. As a result, most designs in modern CMOS nodes adopt MoM capacitors to avoid a mask adder for MIM capacitors. Varactors include accumulation (e.g., Npoly over NWell) and inversion (transistor structure) types of gate capacitors, as well as junction capacitors. For DCO operations, junction varactors are not suitable since its K_{DCO} in reverse bias is small, and there are not flat regions in its C–V curve (relatively high K_{VCO} across all bias ranges). The basic structures of these capacitors are shown in Figure 20.12. Gate capacitors are shown in differential configuration.

20.3.4.2   MoM Varactors

To use MoM capacitors as tunable varactors, switches are needed as illustrated in Figure 20.13 for both differential and single-ended configurations. When the

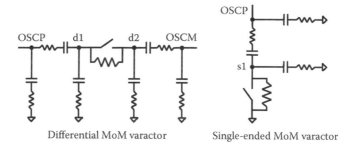

Differential MoM varactor   Single-ended MoM varactor

FIGURE 20.13 A MoM varactor in differential and single-ended configurations.

switches are closed, oscillator nodes, OSCP and OSCM, see full MoM capacitance in series with transistor ON resistance. Q of a MoM capacitor itself is typically in the range of 80–120 at 4 GHz. The switch transistor size is a trade-off between its parasitic capacitance and the effective Q. When the switches are open, each MoM capacitor is connected in series with parasitic capacitance from the switches and the MoM capacitors themselves to ground. Substrate contacts need to be placed nearby to increase Q of these parasitic capacitors. For DCO operations, it is desired for each bit of the capacitor array to have a maximized C_{max}/C_{min} ratio (K_{DCO}) in order to increase the frequency tuning range. In the same time, the voltage dependence to the digital control lines (K_{VCO}) should be lowered to reduce the sensitivity to any noise coupling. That is, all parasitic capacitance should be minimized.

The structure in Figure 20.13 cannot guarantee the switches are biased appropriately at all time for minimum ON resistance and parasitic capacitance. In Figure 20.14, several configurations are illustrated to define the DC biases for nodes d1, d2, and s1. Fundamentally, when an NMOS transistor is used as the main switch, in ON state, its drain (and source for differential type) node should be pulled down to a low-voltage potential. At this moment, parasitic capacitance is less a concern since it is effectively shorted by the main switch. When the switched varactor is in low-capacitance state, the internal nodes should be set up to increase reverse bias of the source/drain junction diodes, thereby reducing their parasitic capacitance. As the total parasitic capacitance of the internal nodes is significantly less than the MoM capacitance, d1, d2, and s1 can experience near the full voltage swing of OSCP/OSCM whose peak-to-peak voltage can be as large as two times VDD for low-phase noise resulting in transistor reliability concerns. There will be more discussions in a later section. Solutions include reducing the pull-up voltage or increasing the control voltage at its logic low state.

The pull-down transistors, PD_1 and PD_2, only need to provide near-ground voltage potential. Minimum L and W are often used to also reduce their parasitic capacitance. For pull-up paths, the resistance needs to be sufficiently high relative to magnitude of the MoM capacitor reactance, preferably >10×. The pull-up resistance can be implemented with MOS transistors (Figure 20.14a) or resistors with switches (Figure 20.14b through d). A PMOS transistor in Figure 20.14a with minimum width and $W/L < 0.1$ operating in triode region can provide ~70 kΩ resistance in 40 nm. However, the total drain capacitance, C_{dd}, is still significant compared to the LSB

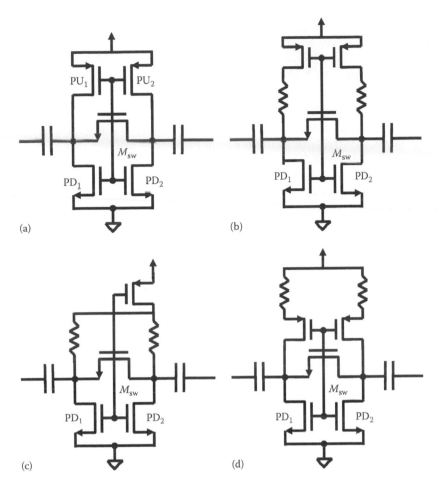

FIGURE 20.14 Multiple configurations that can be used to setup DC biases for the internal MoM varactor nodes.

unit capacitance of MoM capacitor array degrading the oscillator frequency tuning range. In Figure 20.14b, resistors are used to provide the DC bias. The PMOS are switches that can be implemented with minimum-sized transistors and can be further combined as only one switch (Figure 20.14c). In cost-sensitive designs where silicide-blocked polysilicon resistors are not available, MOSFET, NWell or silicided polysilicon resistors with significant parasitic capacitance, process variation, and temperature and voltage coefficients may be the only choices. These resistors need to be oversized to account for PVT corners further contributing parasitic capacitance. The resistors and PMOS switches may be swapped (Figure 20.14d) to reduce the effective parasitic capacitance, although the benefit is limited. The pull-up paths in Figure 20.14a, b, and d are applicable to a single-ended configuration.

The small K_{DCO} requirements for AB and TB prevent MoM varactors from to be used because the relatively large bias-dependent parasitic capacitance would overwhelm the MoM varactors. In addition, as the MoM unit capacitor is scaled down, it becomes very sensitive to the adjacent layout environment such that the exact capacitance is very difficult to be controlled. Therefore, MoM varactor is not preferred to be used for DCO fine-tune bits.

20.3.4.3 Accumulation-Mode MOS Varactor

An accumulation-mode MOS (AMOS) varactor can be implemented using an NPoly-NWell MOSCAP structure[11] as illustrated in Figure 20.12. When gate bias voltage is increased, an accumulation layer is formed underneath the gate area. When gate bias voltage is decreased, charges beneath the gate area are pushed out forming a depletion region. Under high-frequency operation, inversion layer cannot be formed because the process of generation and recombination is too slow. As an example, a measured small-signal C_{max}/C_{min} ratio of >3 with $Q > 120$ at 4 GHz can be achieved using $W/L = 2.5/0.25$ μm/μm. The quality factor can be approximated by the following equation:

$$Q_{AMOS} = \frac{1}{\omega r_s C} = \frac{1}{(\omega C_{ox}/12)(R_{sheet,NWell} \cdot L^2 + R_{sheet,Poly} \cdot W^2)}, \tag{20.3}$$

where
    C_{ox} is the capacitance density of the gate oxide region
    R_{sheet} is the sheet resistance

Q_{AMOS} can be improved by scaling down L and W. However, due to increased parasitic capacitance from interconnect, C_{max}/C_{min} ratio would suffer. Over process corners, unlike a MoM capacitor whose variation is typically >±15% and can be up to ±45% if considering foundry porting, the intrinsic corner variation of C_{ox} is typically within ±2.5% since gate oxide thickness is one of the best controlled parameters in CMOS processes. There is parasitic capacitance surrounding an AMOS structure coming from gate fringing capacitance and metal interconnect. The total variation over process corners can be over ±10%. When operating AMOS varactors under a large voltage swing, a section of the small-signal C–V curve will be averaged (Figure 20.15) generating an effective capacitance. Sweeping DC offset of the applied signal results in an effective large-signal C–V curve under a given voltage swing. From Figure 20.15, it can be observed that when the voltage swing, A, increases from 0 to 1 V, the C_{max}/C_{min} ratio decreases within a given gate bias range. Unless the bias range can be increased, such effect is not favorable since it conflicts with the requirement for low-phase noise performance where a large voltage swing is desired.

At the cathode terminal of the AMOS varactor, there is a large area of NWell-to-Psubstrate (NWPS) diode contributing significant parasitic capacitance to the varactor. In modern process such as 28 nm, at 0 V bias, the parasitic capacitance from diode perimeter (sidewall) is in the order of 0.5 fF/μm and is significantly larger than that from diode area at bottom interfacing PWell (~0.35 fF/μm²). That is, for a small varactor size of, for example, 1/0.25 μm/μm, its parasitic capacitance from NWPS diode at 0 V bias can easily exceed 50% of the gate capacitance.

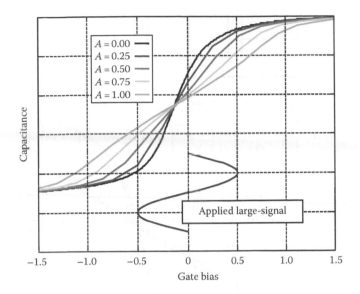

FIGURE 20.15 Modulating a small-signal CV curve of an accumulation-mode varactor with a voltage waveform results in a large-signal CV curve whose slope is dependent of the voltage amplitude.

From [11], if a MOS varactor was designed properly, the interconnect resistance at both top and bottom plates should dominate the total parasitic resistance especially when resistance per contact is as high as >100 Ω in 28 nm and beyond. Because a DCO requires small varactor sizes and has a tight tuning range constraint, using more contacts and metal to reduce interconnect resistance is not favorable since that will significantly increase parasitic capacitance. The parasitic resistance can be partially reduced by employing a layout technique for the differential configurations as shown in Figure 20.12. During normal operation, to the first order, the ac current flows between OSCP and OSCM through interconnect of gate terminal, silicide of gate, channel region, and silicide on diffusions but not going through any contact and back-end metal of diffusions (CTL node), thereby improving the quality factor.

20.3.4.4 Inversion-Mode MOS Varactor

A MOSFET transistor with its source and drain shorted together forms an inversion-mode MOS (IMOS) varactor as illustrated in Figure 20.12. The operation is exactly the same as a standard NMOS transistor that when gate bias voltage increases, an inversion layer is created under the gate region. Compared to an AMOS varactor, there are several distinct different characteristics:

1. *C–V curve*: It can be observed that the *C–V* curve of an IMOS varactor (Figure 20.16) is shifted from that of an AMOS varactor. In modern CMOS process nodes, since the core oxide is thin, transistors with a thick oxide (≥ 1.8 V) are typically available for digital *I/O* buffers. The difference in threshold voltages (V_t) between core and *I/O* transistors also offers

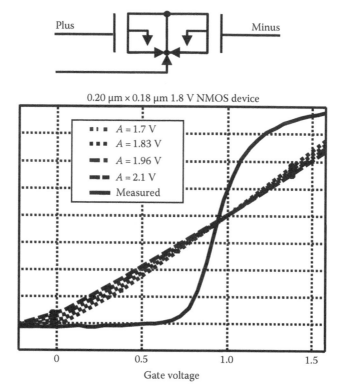

FIGURE 20.16 Large-signal CV curves of an inversion-mode varactor. Utilizing the non-zero body bias, the curves are centered around half V_{DD}, which is ideal for switched-varactor operations and is less sensitive to PVT variation.

a flexibility to DCO designs. Since the control voltage is applied to the source/drain node, without deep NWell, V_t of an NMOS transistor will be increased due to the body effect when the control voltage is above 0 V. The C–V curves in Figure 20.16 are derived from a 1.8 V NMOS. Under large-signal operations, C_{max} and C_{min} that are set by digital control voltages at 0 and 1.4 V are centered around half of the supply voltage (1.4 V), which is ideal for the switched operations in a DCO. It can also be observed that V_t variation over process corners has a minimal impact to $\Delta C/LSB$. That is, this type of varactor is relatively insensitive to process variation. When using PMOS transistors as IMOS varactors, the polarity is simply reversed. If the NWell is further shorted to the source/drain diffusion, the body effect can also be eliminated.

2. *Channel resistance*: An inversion layer is relatively thin and has a higher resistance compared to that from an accumulation layer in an AMOS varactor. The former is generally three to five times larger than the latter, resulting in a lower Q for IMOS varactors. In a DCO, most of the LC tank capacitance is from PB. Therefore, lower Q for AB and TB is practically tolerable.

3. *Parasitic capacitance at diffusion node*: At source/drain diffusion terminal, the parasitic capacitance of an IMOS varactor comes from source/drain to body (NWell or Psubstrate) junction diodes if the body terminal is not tied to the source/drain diffusion. Since the diffusion area is small in advanced process nodes, such parasitic capacitance is usually insignificant. However, if body terminal is shorted to the source/drain diffusion, the parasitic capacitance contributed by NWPS diode is the same as an AMOS varactor.

20.3.4.5 Switchable Inductor

Wide frequency tuning range of an LC oscillator has always been a great challenge. For example, the cellular frequency range has been extended from between 850 and 2170 MHz to between 700 and 2700 MHz, and soon may be between 600 and 3650 MHz with more than 40 bands. Using only capacitor arrays to tune oscillator frequencies would require a very large C/L ratio that may become impractical while consuming large amount of current with poor phase noise. Inductor tuning by shorting a fraction of the inductor[12] is one technique to reduce the C/L ratio. A penalty is that the switch for shorting a segment of the inductor has to be large for low ON resistance, resulting in an excessive parasitic capacitance lessening the benefit of having a switched inductor.

Magnetic tuning[13,14] is an alternative way to change the effective inductance without introducing significant parasitic capacitance and phase noise degradation. An example DCO utilizing a magnetically tuned inductor and having a negligible size impact is illustrated in Figure 20.17. L_p is half-circuit inductance of the primary differential spiral inductor with its center tap connected to the supply. The tuning inductor L_t is a one-turn ring around L_p. An NMOS transistor is connected in series to open or short the ring. With a careful design, the impedance of the parasitic capacitance between L_p and L_t is sufficiently high at the oscillation frequency. Therefore, when the switch is open, L_p would perform as if L_t is physically not present. The switch size can also be made large so that when its ON resistance is transformed to the primary side, the impact to the overall Q is minimized. The center tap of the ring is connected through a resistor and an inverter to VDD/VSS in order to properly set the bias conditions for the

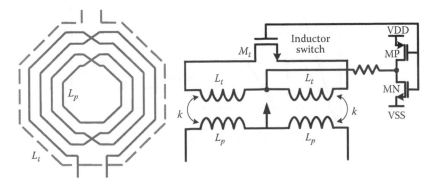

FIGURE 20.17   A magnetically tuned inductor to increase the DCO frequency tuning range with a minimum size impact.

NMOS switch, similar to the pull-up and pull-down paths for MoM varactors. Since the loop is passive, that is, no active current, closing the switch M_t creates a current flow in L_t opposite to that in L_p. That means the mutual inductance is always negative. As a first-order estimation, assuming the transformer windings are lossless and M_t is an ideal switch, the effective inductance L'_p with the switch closed is

$$L'_p = L_p(1 - k^2), \tag{20.4}$$

where k is the coupling factor between L_p and L_t, which is primarily determined by the physical dimensions and location of L_t relative to L_p. L_t can have multiple turns and can be placed inside, outside, or overlapping with L_p. Multiple tuning inductors can be designed to have finer tuning steps. For a planar spiral inductor, the outer turns are responsible for a large portion of the inductance. Placing L_t near the outer turns will have a higher effective k value. In this example implementation with one-turn L_t, L_p can be tuned between 1 and 0.8 nH.

Since parasitic resistance of the primary inductor remains the same when L_p is reduced to L'_p, inductor Q is degraded. The Q is further reduced considering the series resistance of M_t. In a current-limited oscillator design, this means a phase noise degradation of $30 * \log(Q'/Q)$ would be expected (Equation 20.1), where Q' is the effective inductor Q when the switch M_t is closed. However, for the oscillator application, a properly designed inductor should have a Q curve peaking slightly higher than the operating frequency range. As a result, sensitivity of Q' to the inductance switching is reduced because near peak-Q frequency, inductor quality factor is significantly dependent of Q of the inductor parasitic capacitance. As shown in Figure 20.18, the operating frequency is centered between the two peak-Q

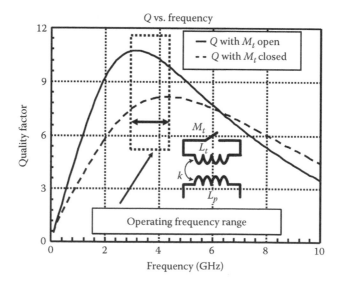

FIGURE 20.18 Electrical characteristic and operating conditions of the magnetically tuned inductor.

frequencies when toggling M_t. The peak Q only drops from 10.8 to 8 when M_t is closed. This technique for extending oscillator frequency tuning range is nearly process independent because the inductors have very large geometries relative to the lithography precision in deep-submicron processes.

20.3.4.6 An Example DCO

From previous sections, there are plenty of options available when designing a DCO. An example[8] with specific performance targets can better demonstrate the design considerations. Aiming at a Global System for Mobile (GSM), the target parameters and their associated constraints are listed in Table 20.2.

Starting with the supply voltage constraint, the majority of the tank varactors should be designed with MoM capacitors in order to reduce the required range of tuning voltage. Supply frequency pushing is also simultaneously minimized such that the DCO is less sensitive to its surroundings in a SoC. The output frequency range is wide between 824 and 1990 MHz (~82.9%). From Figure 20.4, when frequency dividers with different division ratios are considered, the required frequency coverage for the DCO core is reduced to between 3296 and 3980 MHz, which is ~18.8% referring to the center of the frequency range (3638 MHz). Taking process variation into account, in particular, MoM capacitors have ±25% variation across process fabs, the frequency tuning range needs to be about 25% as shown in Figure 20.19 where single-ended inductance L_p is switched between 0.8 and 1 nH, C_{max}/C_{min} ratio of MoM varactor is 3.5, process variation of metal interconnect is 12%.

Since varactor Q is typically higher than inductor Q, a higher tank inductance is desired to increase the effective tank parallel resistance. Considering Equations 20.1 and 20.2, as well as the specification that HB phase noise requirement is significantly relaxed, it is beneficial to use the inductor switching technique described in Section 20.3.4.5 to improve LB phase noise. Including all parasitic capacitance in the LC tank, the effective C_{max}/C_{min} is reduced to 1.7 (3/1.75 pF/pF single-ended). If the inductance can be changed by around 20%, it can be calculated that differential L_{max} can be as large as 2 nH. For HB, the phase noise and current consumption would be degraded by ~2 dB and 9 mA, respectively, due to a 25% lower inductor Q.

TABLE 20.2
Design Targets and Trade-Offs of a DCO for a GSM System

System Parameters	Requirements	Design Parameters
Frequency range	824–1990 MHz	$L, C, L_{max}/L_{min}, C_{max}/C_{min}$
Channel bandwidth	200 kHz	AB step size
Peak frequency deviation	67.7 kHz	TB step size
Phase noise	−164 dBc/Hz at 20 MHz offset from 914 MHz carrier	$Q, L/C$, gm sizing, supply current
	−141 dBc/Hz at 3 MHz offset from 1980 MHz carrier	
Supply voltage	1.4 V	Device component selection
DCO size	Minimum	Number of inductor turns

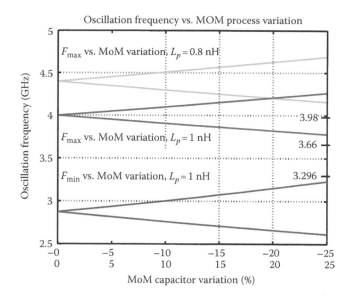

FIGURE 20.19   A large frequency tuning range is required to cover ±25% MoM capacitor process variation. A magnetically tuned inductor helps to reduce the stress on varactor design.

The primary inductor is made of a three-turn inductor to reduce its size as well as the keep-out region. The smaller inductor diameter helps to reduce magnetic flux pick up from potential noise aggressors, although its Q is somewhat lower than a single-turn inductor.

The frequency step for AB capacitor array should be similar to channel bandwidth, and the total TB frequency coverage is from a combination of AB step size and peak frequency deviation of phase modulation path as described in Section 20.3.3. With an internal supply regulator, the DCO supply voltage is programmable between 1.3 and 1.5 V. That means the ideal large-signal $C-V$ curve of AB and TB should be centered around 0.7 V if the digital control voltage is switched between 0 and 1.4 V. Further considering the small frequency steps of AB and TB, 1.8 V NMOS transistors are chosen to form IMOS varactors. The W/L of TB varactor is determined to be 0.2/0.12 μm/μm, which is not the minimum dimension offered by the process technology in order to reduce the DNL and variability. The measured TB DNL is about 1%. The technique of dynamic element matching (DEM) commonly used for DAC designs[15] can also be applied to further reduce the overall DNL. Although there is no flat region in the large-signal $C-V$ curves of AB and TB, their capacitance is only a small fraction compared to the overall tank capacitance. Therefore, the phase noise impact due to higher K_{VCO} from the digital control lines is still low. From Figure 20.16, it can be observed that when V_t varies, K_{DCO} is relatively constant as the large-signal $C-V$ curves are near straight lines. Therefore, K_{DCO} is fairly insensitive to process variation. The overall parameters for the capacitor arrays are shown in Table 20.1.

Since the intention of the secondary LC tank is to provide high impedance to reduce loading from the gm pair to the main LC tank, its Q does not need to be very high as long as its impedance in series with the linear-region gm-pair transistors is relatively high compared to the main LC tank. In this particular example, $Q \sim 8$ between 7 and 8 GHz is sufficient. Consequently, the bandwidth of the secondary LC tank is wide enough such that frequency tuning is not required. If considering multimode multiband operations, tuning the secondary tank would become necessary to achieve optimal performance. With the total parasitic capacitance of ~500 fF at source node of the gm pair, 0.9 nH is needed for L_s. The equivalent parallel resistance is >350 Ω, which is about 2.5 times that from the main LC tank.

20.3.5 Reliability

For a stringent phase noise requirement, both signal power level and signal-to-noise ratio (SNR) need to be very high so that the absolute noise power from the far-out spectral skirt is still above the broadband phase noise floor. This imposes concerns on gate oxide integrity (GOI), channel hot carrier (CHC), and metal electromigration (EM) in an advanced digital CMOS process since the gate oxide for core transistors is thin with a low-voltage rating, the channel length is short, and the metal thickness is small. Although increasing the DCO core frequency helps to relax the required SNR and internal voltage/current swings at oscillation nodes, there are still concerns on marginality.

20.3.5.1 Gate Oxide Integrity

Referring to Figure 20.11, assuming the oscillator nodes, OSCP and OSCM, have 1 V amplitude, in order to be better than −156 dBc/Hz at 20 MHz offset from 4 GHz carrier, the noise level of DCO core must be $< 16 \, \text{nV}/\sqrt{\text{Hz}}$ for transistors operating under a sinusoidal waveform, which is challenging. With 1.4 V supply, the maximum absolute voltage on drain and gate nodes is 2.4 V on 1.3 V rated 65 nm transistors. Applying a statistical model on the drain and gate voltage waveforms shown in Figure 20.20, the effective V_{GS} overstress voltage, denoted as V_{GOI}, can be calculated using the following equation[16]:

$$V_{GOI} = \frac{1}{\beta} \cdot \ln \left(\frac{1}{T} \cdot \int_0^T e^{\beta V_{GS}(t)} dt \right), \tag{20.5}$$

where
β is a constant
T is the oscillation period

The calculated V_{GOI} is 1.9 V, which means equivalently, V_{GS} stays at 1.9 V for 25% of oscillation period T, 1.4 V for another 25% of T, and 0 V for the rest of T.

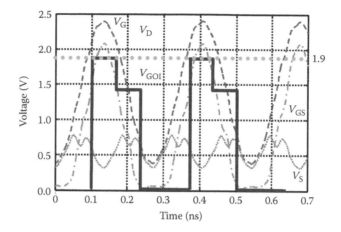

FIGURE 20.20 Reliability stress on the DCO cross-coupled pair. The equivalent oxide stress can be calculated to be ~1.9 V with 25% duty cycle on a 65 nm core oxide transistor, which is high.

Applying burn-in with multiple acceleration factors and monitoring parametric shifts of the DCO such as phase noise, an empirical model can be fitted to predict the DCO lift time. For cellular phone systems, combining the duty-cycled operations, the phone life time can be >5 years.

20.3.5.2 Channel Hot Carrier

The transistor degradation due to CHC includes decreased channel current, increased threshold voltage, and decreased transconductance over time.[17] The transistors of concern are again the cross-coupled gm pair, M_1 and M_2, in Figure 20.11. CHC life time is dependent of several factors such as V_{gs}, V_{ds}, and channel length. In general, the worst-case operating condition is when $V_{gs} \sim= V_{ds}$. In the DCO waveform shown in Figure 20.20, such condition occurs when the waveforms have the largest absolute value of slope. That is, the duration for such condition is very short. For the extreme voltage conditions such as $V_d = 2.4$ V and $V_{gs} = 0$ V, there is no CHC concern. However, there is a nonconductive stress (NCS) concern.[18] The channel length for M_1 and M_2 are not the minimum allowed dimension of this process in order to further reduce the CHC concern.

20.3.5.3 Metal EM

The interconnect metal in recent digital CMOS processes is mostly made of copper (Cu), which has high conductivity. However, the Cu layers are very thin in order to reduce the parasitic capacitance between adjacent metal lines for high-density and high-speed logic operations. Consequently, the EM capability for each lower-level Cu layer is limited. The EM degradation leads to an increase of metal sheet resistance over time impacting circuit performance.

EM concerns can generally be resolved by increasing metal width and/or metal stack. However, on-chip spiral inductors are one exception that the metal cannot be arbitrarily changed. As shown in Figure 20.18, the DCO operates around the inductor peak-Q frequencies. There is only a limited flexibility to increase the inductor metal width or number of metal layers. When stacking an aluminum (Al) layer on top of Cu layer to increase inductor Q, significant amount of current will be distributed in the Al layer. Compared to Cu, Al can only handle <15% current density for the same sheet resistance and the same width. Therefore, the more current flowing in the Al metal layer, the worse the EM reliability is. For the same metal width, if the total thickness for Cu and Al is 1 μm each, stacking Al on top of Cu would result in 38% current flowing in Al layer. Characterizing from test structures, the effective EM capability can handle >100 mA total AC current, which is sufficient for a typical DCO design.

20.4 INTEGRATION WITH AN ALL-DIGITAL PHASE-LOCKED LOOP

When integrating a DCO into an ADPLL, there are nonidealities that need to be taken into account. Three impairments will be discussed in the following sections.

20.4.1 PHASE NOISE DUE TO ΣΔ DITHERING

As explained in Section 20.3, dithering on the TB varactors is required to achieve a fine frequency resolution. By nature of the PLL, the varactors will be dithered at an update rate the same as the input reference frequency f_R. Assuming the digital input is a series of impulses and TB capacitance changes instantly following the impulses, the PLL is effectively making a white dithering on TB varactors at f_R rate. However, in reality, both the digital input and the varactor capacitance switching are not impulses but square waves corresponding to a zero-order hold operation, the white noise assumption is not strictly correct and the quantization noise needs to be multiplied by a sinc function. The quantization noise can be expressed as

$$L(\Delta f) = 10 \cdot \log\left(\frac{1}{12} \cdot \left(\frac{\Delta f_{TB}}{\Delta f}\right)^2 \cdot \frac{1}{f_R} \cdot \left(\sin c\left(\frac{\Delta f}{f_R}\right)\right)^2\right), \tag{20.6}$$

where
   Δf is the frequency offset
   Δf_{TB} is the frequency step size from TB varactors

If f_R is 26 MHz and Δf_{TB} is 20 kHz, at $\Delta f = 400$ kHz, the quantization noise can be calculated to be −111 dBc/Hz, which is far insufficient for applications such as cellular phone communication systems. Using a higher rate deriving from DCO frequency to replace f_R and applying ΣΔ dithering to reduce Δf_{TB}, the quantization noise can be reduced to below −160 dBc/Hz if f_R becomes 225 MHz and 8-bit dithering was applied to reduce Δf_{TB}.

   Another nonideality is the finite rise and fall time (T_{rise} and T_{fall}) of the digital inputs to the dithered TB bits. To investigate its impact, a simulation approach was taken. A Verilog-A model was built for the MASH ΣΔ block to cosimulate with the transistor-level DCO. A simplified DCO circuit operating at 3.6 GHz with 450 MHz dithering clock and >100 times larger TB varactors were used to exacerbate the effects for illustration purpose. The results are shown in Figure 20.21. It can be clearly seen that when T_{rise} and T_{fall} are increased from 100 to 200 ps, the close-in

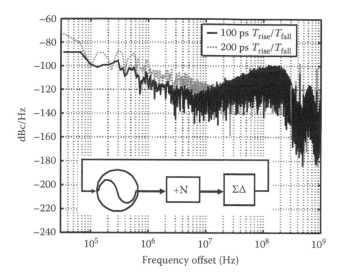

FIGURE 20.21 DCO quantization noise due to circuit nonideality is negligible when rise and fall time of the digital dithering is <100 ps, which is natural in modern CMOS processes.

phase noise has a trend of increasing 10 dB. Below 100 ps, no significant impairment can be observed, thus not shown in the figure. This is in line with modern CMOS process capability that typical 10%–90% T_{rise} and T_{fall} are <20 ps for 40 nm node and beyond.

20.4.2 PHASE NOISE DUE TO DCO TUNING WORD TOGGLING

In a conventional PLL, when changing the control voltage of a VCO in order to adjust its phase or frequency, it is quite a disturbing event to the oscillation and momentarily causes phase noise degradation until the loop settles again. Such effect is exacerbated in a DCO due to the sample-mode operation where the oscillating frequency is commanded to change in discrete-time events. The worst moment to change the DCO frequency is at the instances when the LC tank energy is fully stored in the capacitor because any capacitance change and the associated charge redistribution is visible to the LC tank. That is, the disturbance to the PLL is the highest when the oscillation waveform is at its maximum (Figure 20.22). This is opposite to the case when any perturbation to a free-running oscillator is not desired.[19] An intuitive solution is to control the timing of the digital inputs to the varactors so that the perturbations are minimized. The retiming ensures that the digital inputs are at optimal time near the oscillator zero-crossings. The necessity of such retiming is application driven.

20.4.3 INL OF K_{DCO}

From Equation 20.6, when the frequency step size is reduced for TB varactor, namingly lower K_{DCO}, the quantization noise can be lowered. An instinctive method is to

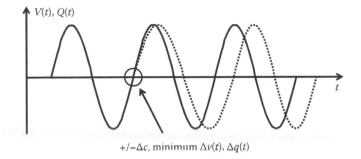

FIGURE 20.22 Disturbance to DCO operations due to the varactor switching.

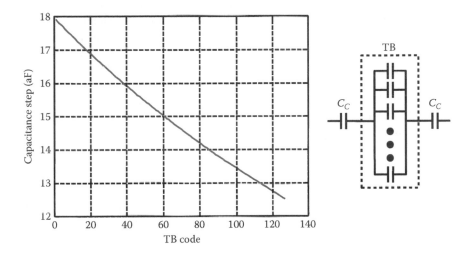

FIGURE 20.23 K_{DCO} nonlinearity versus tuning code.

add AC coupling capacitors, C_C, as shown in Figure 20.23. The size of C_C has to be limited in that if it was too large, K_{DCO} will not be reduced. The effective capacitance step with C_C can be expressed as

$$\Delta c = \frac{C_{TB} \cdot C_C}{C_{TB} + C_C} - \frac{(C_{TB} - \Delta C_{TB}) \cdot C_C}{(C_{TB} - \Delta C_{TB}) + C_C}, \tag{20.7}$$

where
ΔC_{TB} is the capacitance change when toggling one TB bit
C_{TB} is the total TB capacitance for a given input code including all parasitic capacitance from the TB varactor array

If ΔC_{TB} is 50 aF, C_{max}/C_{min} ratio for each TB bit is 3 and C_C has the same capacitance as the total TB array, the Δc for 128 TB bits can be calculated and plotted in

Figure 20.23. The effective Δc is indeed significantly smaller than 50 aF but the INL is between 30% and 45% of the new effective TB LSB size at extreme codes. For frequency division multiple access (FDMA) systems where there is limited or no time to recalibrate K_{DCO}, the INL will cause phase modulation to fail even for a lower-order modulation such as Gaussian Minimum Shift Keying (GMSK). One way to recover the K_{DCO} error is to have predistortion of the nonlinear curve in Figure 20.23, provided that the curve is well deterministic over voltage and temperature variations assuming uncertainty from process corners has been calibrated during factory or power-on self-tests.

20.5   SUMMARY

In this chapter, we reviewed the commonality and difference between a LC DCO and a LC VCO. Most of the techniques developed for VCOs can be applied to DCOs. However, the difference between them leads to specific optimizations for DCOs, in particular, varactor banks. When integrated a DCO into an ADPLL, the trade-offs impact system performance such as quantization noise due to $\Sigma\Delta$ dithering, modulation errors due to DNL and INL, and close-in and far-out phase noise due to wide bandwidth modulations. The nonideality from imperfect square waveforms controlling DCOs was also examined that for most applications it should have negligible effects to overall ADPLL performance.

REFERENCES

1. B. Razavi, *RF Microelectronics*, 2nd ed., Chapter 4, Prentice Hall Communications Engineering and Emerging Technologies Series, Upper Saddle River, NJ, October 2011.
2. F. Gardner, *Phaselock Techniques*, 3rd ed., Chapter 12, Wiley-Interscience, River Street, NJ, July 2005.
3. R.B. Staszewski, C.-M. Hung, K. Maggio, J. Wallberg, D. Leipold, and P.T. Balsara, All-digital phase-domain TX frequency synthesizer for bluetooth radios in 0.13 μm CMOS, *IEEE ISSCC*, Piscataway, NJ, pp. 272–274, February 2004.
4. R.B. Staszewski, D. Leipold, C.-M. Hung, and P.T. Balsara, A first digitally-controlled oscillator in a deep-submicron CMOS process for multi-GHz wireless applications, *IEEE RFIC Symposium*, Piscataway, NJ, pp. 81–84, June 2003.
5. E. Hegazi, H. Sjöland, and A.A. Abidi, A filtering technique to lower LC oscillator phase noise, *IEEE Journal of Solid-State Circuits*, 36(12), 1921–1930, December 2001.
6. R.B. Staszewski, C.-M. Hung, D. Leipold, and P.T. Balsara, A first multigigahertz digitally-controlled oscillator for wireless applications, *IEEE Transactions on Microwave Theory and Techniques*, 51(11), 2154–2164, November 2003.
7. D.B. Leeson, A simple model of feedback oscillator noise spectrum, *IEEE Proceedings*, Piscataway, NJ, pp. 329–330, February 1966.
8. C.-M. Hung, R.B. Staszewski, N. Barton, M.-C. Lee, and D. Leipold, A digitally-controlled SAW-less oscillator system for cellular handsets, *IEEE Journal of Solid-State Circuits*, 41(5), 1160–1170, May 2006.
9. R.B. Staszewski, J. Wallberg, S. Rezeq et al., All-digital PLL and GSM/edge transmitter in 90 nm CMOS, *IEEE ISSCC*, Piscataway, NJ, pp. 316–318, February 2005.

10. J. Zhuang, K. Waheed, and R.B. Staszewski, A technique to reduce phase/frequency modulation bandwidth in a polar RF transmitter, *IEEE Transactions on Circuits and Systems I*, 57(8), 2196–2207, August 2010.
11. C.-M. Hung, Y.-C. Ho, I.-C. Wu, and K.K. O, High-Q capacitors implemented in a CMOS process for low-power wireless applications, *IEEE Transactions on Microwave Theory and Techniques*, 46(5), 505–511, May 1998.
12. S.-M. Yim and K.K. O, Demonstration of a switched resonator concept in a dual-band monolithic CMOS LC-tuned VCO, *IEEE CICC*, Piscataway, NJ, pp. 205–208, 2001.
13. C.-M. Hung and N. Barton, A low-phase-noise wide-tuning-range digitally-controlled LC oscillator using a switchable inductor, *Electronics Letters*, 45(17), 890–892, August 2009.
14. J. Yin and H.C. Luong, A 57.5-to-90.1 GHz magnetically-tuned multi-mode CMOS VCO, *IEEE CICC*, Piscataway, NJ, pp. 1–4, 2012.
15. K. Chan, J. Zhu, and I. Galton, Dynamic element matching to prevent nonlinear distortion from pulse-shape mismatches in high-resolution DACs, *IEEE Journal of Solid-State Circuits*, 43(9), 2067–2078, September 2008.
16. B. Hunter, Gate oxide reliability: The statistical dependence of oxide failure rates on V_{dd} and t_{ox} variations, Texas Instruments Inc., Dallas, TX, Technical Paper, September 1998.
17. R. Khamankar, H. Bu, C. Bowen et al., An enhanced 90 nm high performance technology with strong performance improvements from stress and mobility increase through simple process changes, *IEEE VLSI Technology Symposium*, Piscataway, NJ, pp. 162–163, June 2004.
18. V. Reddy, N. Barton, S. Martin et al., Impact of transistor reliability on RF oscillator phase noise degradation, *IEEE IEDM*, Piscataway, NJ, pp. 1–4, December 2009.
19. A. Hajimiri and T.H. Lee, A general theory of phase noise in electrical oscillators, *IEEE Journal of Solid-State Circuits*, 33(2), 179–194, February 1998.

Index